Carlo Cercignani

The Boltzmann Equation and Its Applications

With 42 Illustrations

Springer-Verlag
New York Berlin Heidelberg
London Paris Tokyo

Carlo Cercignani
Department of Mathematics
Politecnico di Milano
20133 Milano (I)
Italy

AMS Classifications: 76P05, 82A40

Library of Congress Cataloging-in-Publication Data
Cercignani, Carlo.
The Boltzmann equation and its applications / Carlo Cercignani.
p. cm. — (Applied mathematical sciences; v.67)
Includes index.
1. Transport theory. 2. Rarefied gas dynamics. I. Title.
II. Series: Applied mathematical sciences (Springer-Verlag New York Inc.)
QA1.A647 vol. 67
[QC175.2]
510 s—dc19
[530.1′38] 87-26654

Portions of this work previously appeared as *Theory and Application of the Boltzmann Equation.*

Printed and bound by Arcata Graphics/Halliday, West Hanover, Massachusetts.
Printed in the United States of America.

9 8 7 6 5 4 3 2 1

ISBN 0-387-96637-4 Springer-Verlag New York Berlin Heidelberg
ISBN 3-540-96637-4 Springer-Verlag Berlin Heidelberg New York

To Silvana

PREFACE

Statistical mechanics may be naturally divided into two branches, one dealing with equilibrium systems, the other with nonequilibrium systems. The equilibrium properties of macroscopic systems are defined in principle by suitable averages in well-defined Gibbs's ensembles. This provides a framework for both qualitative understanding and quantitative approximations to equilibrium behaviour. Nonequilibrium phenomena are much less understood at the present time. A notable exception is offered by the case of dilute gases. Here a basic equation was established by Ludwig Boltzmann in 1872.

The Boltzmann equation still forms the basis for the kinetic theory of gases and has proved fruitful not only for a study of the classical gases Boltzmann had in mind but also, properly generalized, for studying electron transport in solids and plasmas, neutron transport in nuclear reactors, phonon transport in superfluids, and radiative transfer in planetary and stellar atmospheres. Research in both the new fields and the old one has undergone a considerable advance in the last thirty years.

In the last ten years, a new wave of interest has surrounded the Boltzmann equation, stemming from its unique role in the theory of time-dependent phenomena in large systems. In fact, the Boltzmann equation appears as a prototype of a reduced description taking into account only partial information about the underlying microscopic state (fully described by the coordinate and momenta of all the molecules), but nevertheless undergoing an autonomous time evolution. Thus the problem of the rigorous derivation of the Boltzmann equation from the microscopic description has attracted a certain amount of interest among physicists and mathematicians. This, in turn, has revived the interest in the theory of existence and uniqueness of the solutions of the Boltzmann equation, since this problem has proved to be intimately tied with the previous one.

This justifies the appearance of the present book (which tries to present a unified approach to the problems arising in the different fields mentioned above) by exploiting the similarities whenever they exist and underlining the differences when necessary. The main line of exposition, however, is tied to the classical equation established by Boltzmann, and hence the detailed descriptions of some applications almost exclusively refer to monatomic neutral gases. Appropriate references are given, however, to papers dealing with similar problems arising in other fields, with particular concern for neutron transport, gas mixtures, and polyatomic gases.

The material dealing with the basic properties and applications known before 1975 was covered in a previous book by the author*; thus, it was natural to incorporate in the present book the material of the first seven chapters of that book. This material is updated in an extensive Appendix, covering the developments from 1975 to 1987.

But the main feature of the book is Chapter VIII, completely rewritten and covering the important studies resulting from new mathematical approaches to the old problem of existence and uniqueness, and the new ones tied up with the question of validity. There is still no satisfactory proof for existence of solutions under general reasonable initial conditions. The material presented here indicates, however, that a great deal of progress has been achieved in recent times. There is no doubt that the better understanding resulting from this progress will be essential to the further penetration to be expected in the next few years.

It is hoped that the book will be useful as a textbook for an advanced course in kinetic theory and as a reference for mathematicians, physicists, and engineers interested in the kinetic theory of gases and its applications.

Milano, Italy
September 1987

CARLO CERCIGNANI

**Theory and Application of the Boltzmann Equation*, Scottish Academic Press, Edinburgh, 1975.

CONTENTS

VIII. THEOREMS ON THE SOLUTIONS OF THE BOLTZMANN EQUATION

THE BOLTZMANN EQUATION AND ITS APPLICATIONS

I BASIC PRINCIPLES OF THE KINETIC THEORY OF GASES

1. Introduction

According to the molecular theory of matter, a macroscopic volume of gas (say, 1 cm^3) is a system of a very large number (say, 10^{20}) of molecules moving in a rather irregular way. In principle, we may assume, ignoring quantum effects, that the molecules are particles (mass points or other systems with a small number of degrees of freedom) obeying the laws of classical mechanics. We may also assume that the laws of interaction between the molecules are perfectly known so that, in principle, the evolution of the system is computable, provided suitable initial data are given. If the molecules are, for example, mass points, the equations of motion are:

$$\dot{\boldsymbol{\xi}}_i = \mathbf{X}_i \qquad\qquad (1.1a)$$
$$\dot{\mathbf{x}}_i = \boldsymbol{\xi}_i$$

or

$$\ddot{\mathbf{x}}_i = \mathbf{X}_i \qquad\qquad (1.1b)$$

where $\mathbf{x}_i$ is the position vector of the i-th particle ($i = 1, \ldots, N$) and $\boldsymbol{\xi}_i$ its velocity vector; both $\mathbf{x}_i$ and $\boldsymbol{\xi}_i$ are functions of the time variable t and the dots denote, as usual, differentiation with respect to t. Here $\mathbf{X}_i$ is the force acting upon the i-th particle divided by the mass of the particle. Such a force will in general be the sum of the resultant of external forces (e.g., gravity or, if the observer is not inertial, apparent forces such as centrifugal or Coriolis forces) and the forces describing the action of the other particles of the system on the i-th particle. As we said before, the expression of such forces must be given as a part of the description of the mechanical system.

In order to compute the time evolution of the system, one would have to solve the $6N$ first-order differential equations, Eq. (1.1a) in the $6N$ unknowns constituting the components of the $2N$ vectors $(\mathbf{x}_i, \boldsymbol{\xi}_i)$ ($i = 1, \ldots, N$). A prerequisite for this is the knowledge of the $6N$ initial conditions:

$$\mathbf{x}_i(0) = \mathbf{x}_i^0; \qquad \dot{\mathbf{x}}_i(0) = \boldsymbol{\xi}_i(0) = \boldsymbol{\xi}_i^0 \qquad\qquad (1.2)$$

where the components of $\mathbf{x}_i^0$ and $\boldsymbol{\xi}_i^0$ are $6N$ given constants which describe the initial state of the system.

However, solving the above initial value problem for a number of particles of a realistic order of magnitude (say, $N \simeq 10^{20}$) is an impossible and useless task, for the following reasons:

(1) We have to know the initial data $\mathbf{x}_i^0$ and $\boldsymbol{\xi}_i^0$; that is, the positions and velocities of all the molecules at $t = 0$, and obtaining these data appears difficult, even in principle. In fact, it would involve the simultaneous measurement of the positions and velocities of all the molecules at $t = 0$.

(2) The information on the initial data, if available in spite of the above remark, is enormous and the duration of a human life would be insufficient to utter a faint fraction of these data (assume that one may give the six data for each particle in one second and observe that there are less than 10^8 seconds in a year, that is less than 10^{10} seconds in the lifetime of a human being).

(3) Even if we could obtain the data and feed them into a computer, it seems impossible to imagine a computer capable of solving so many equations. (Think of the number of computer cards required to supply the initial data!).

(4) No matter how accurately we measure or give the initial data, the latter cannot be infinitely accurate; for example, we shall never consider more than 100 decimal figures in our computations, thus introducing truncation errors of order 10^{-100}.

As a consequence, one should consider the evolution, not of a system, but of an ensemble of identical systems whose initial data differ from each other by quantities of the order of the accepted errors. It is possible to see (Section 5) that, for a system of about 10^{20} molecules, truncation errors of the order of 10^{-100} in the computations would make it impossible to compute the motion of these molecules for more than one millionth of a second.

(5) Even if we could work with infinitely many figures (!), we should include all the particles of the Universe in our computations. In fact according to Borel (see Section 5) the displacement of 1 gram of matter by 1 cm on a not too distant star (say, Sirius) would produce a change of force larger than 10^{-100} times a typical force acting on the molecule and we would then again be back to the difficulty mentioned under 4), unless we want to include all the particles of the Universe (!) in our computations.

(6) Even if we overcome the above difficulties and compute the subsequent evolution of the considered system, this detailed information would be useless because knowledge of where the single molecules are and what their velocities are is information which, in this form, does not tell us what we really want to know; for example, the pressure exerted on a wall by a gas at a given density and temperature.

The conclusion is that the only significant and useful results are those about the behavior of many systems in the form of statistics, that is information about probable distributions. This information can be obtained by averaging over our ignorance (meaning the incapability of macroscopic bodies to detect certain microscopic details of another macroscopic body) or the errors introduced by neglecting the influence of other bodies in the Universe.

As a result, only averages can be computed and are what matter, provided they are related to such macroscopic quantities as pressure, temperature, stresses, heat flow, etc. This is the basic idea of statistical mechanics.

The first kind of averaging which appears in any treatment of mechanics based upon statistical ideas is, as suggested by the above considerations, over our ignorance of initial data. However, other averaging processes or limiting procedures are usually required, to take into account the interactions of particles in a statistical fashion. These interactions also include the interaction of the molecules of a fluid with the solid boundaries, which bound the region where the fluid flows and are also formed of molecules.

When we deal with statistical mechanics, therefore, we talk about probabilities instead of certainties: that is, in our description, a given particle will not have a definite position and velocity, but only different probabilities of having different positions and velocities. In particular, this is true for the kinetic theory of gases, that is, the statistical mechanics of gas molecules, and the theories of transport of particles (neutrons, electrons, photons, etc.). Under suitable assumptions, the information required to compute averages for these systems can be reduced to the solution of an equation, the so-called Boltzmann equation. In the case of neutrons the equation is frequently called the transport equation, while the name of transfer equation is in use for the case of photons (radiative transfer).

The main aim of this book is to provide an introduction to the mathematical techniques and concepts related to the Boltzmann equation and, in particular, to the boundary value problems which arise in connection with such an equation.

2. Probability

As mentioned before, probability concepts are of basic importance in the kinetic theory of gases and, more generally, in statistical mechanics. As is well known, the main problem in the applications of probability theory is that of assigning probabilities to elementary events. In some cases, this assignment is trivial, as illustrated by the familiar experience of throwing (unloaded) dice or tossing coins and asking for the probability of getting one of the numbers from one to six or of getting heads and tails.

The probability of getting a certain result is a number between zero and one, which, roughly speaking, can be experimentally interpreted as the relative frequency of that result in a long series of trials (in the case of tossing a coin, $P(H) = P(T) = \frac{1}{2}$, where $P(H)$ and $P(T)$ denote the probabilities of getting heads and tails respectively). If the events are mutually exclusive, then the sum of the probabilities of all possible events must be one, since one of the events will certainly happen (either heads or tails, in the above example).

It is to be stressed, however, that the variables which appear in statistical mechanics usually range through a continuous set of values, instead of being restricted to a discrete set (such as the set of two elements, heads and tails, which describe the result of tossing a coin). Therefore, strictly speaking, the probability of obtaining any given value of the continuum of possible values will be, in general, zero; on the other hand, the "sum" of the probabilities must be one. There is nothing strange (or, at least, new) in this, since it is the exact parallel of the statement that a geometrical point has no length, while a segment, which is a set of points, has a nonzero length. Therefore we have to talk about the probability of obtaining a result which lies in an infinitesimal interval (or, more generally, set) instead of one having a fixed value: this probability will also, in general, be an infinitesimal quantity of the same order as the length of interval, or measure of the set. Thus in the case of n continuous variables $z_1, z_2, \ldots, z_n$, that is, a vector variable $\mathbf{z} = (z_1, z_2, \ldots, z_n)$, we have to introduce a probability density $P(\mathbf{z})$ such that $P(\mathbf{z})\, d^n\mathbf{z}$ is the probability that $\mathbf{z}$ lies between $\mathbf{z}$ and $\mathbf{z} + d\mathbf{z}$, with $d^n\mathbf{z}$ denoting the volume of an infinitesimal cell, also denoted by $dz_1\, dz_2 \ldots dz_n$. In this case the property that the "sum" of probabilities is one becomes:

$$\int_Z P(\mathbf{z})\, d\mathbf{z} = 1 \tag{2.1}$$

where Z is the region of the n-dimensional space in which $\mathbf{z}$ varies (possibly the whole n-dimensional space) and we omit the superscript n in the volume element, since no confusion arises.

What is the use of a probability density? The answer is simple: a probability density is needed to compute averages: if we know the probability density $P(\mathbf{z})$ we can compute the average value of any given function $\varphi(\mathbf{z})$ of the vector $\mathbf{z}$. As a matter of fact, we can define averages as follows

$$\langle \varphi(\mathbf{z}) \rangle = \overline{\varphi(\mathbf{z})} = \int_Z P(\mathbf{z})\varphi(\mathbf{z})\, d\mathbf{z} \tag{2.2}$$

where brackets or a bar is conventional notation for averaging. In other words, in order to compute the average value of a function $\varphi(\mathbf{z})$ we integrate it over all values of $\mathbf{z}$, weighting each $d\mathbf{z}$ with the probability density for a value in $d\mathbf{z}$ to be realized. It is clear that this definition is in agreement with our intuition about averages.

Speaking about probability densities is very useful, but at first sight would seem to present a serious inconvenience. It can turn out to be convenient to consider, in some instances, the highly idealized case, in which some variable is known with complete certainty. Then, if $\mathbf{z}$ definitely has the value $\mathbf{z}_0$, the probability density for any $\mathbf{z} \neq \mathbf{z}_0$ will obviously be zero. On the other hand,

Eq. (2.1) has to be satisfied. No ordinary function can satisfy these two requirements. If we want to include certainty as a particular case of probability, we have to enlarge our concept of function.

The required generalization is achieved by means of the so called "generalized functions" or "distributions". Generalized functions can be defined in many ways, for example, as ideal limits of sequences of sufficiently regular functions in the same way as real numbers are ideal limits of sequences of rational numbers. We can say, therefore, that a generalized function $g(\mathbf{z})$ "is" a sequence, $\{g_m(\mathbf{z})\}$ ($m = 1, 2, 3, \ldots$), of ordinary functions, in the same sense as a real number α "is" the sequence, for example, of the rationals $\{\alpha_m\}$ obtained by truncating after m figures the decimal representation of α. In the same way as we never deal with an irrational number in a computation, but only with the rational approximations to it, we never deal with the "values assumed by a generalized function", but only with the sequence of functions approximating it. However, in the same way as we speak of sum, product, etc. of real numbers and find it very useful to operate on them, so we can introduce operations on generalized functions and find it useful to use them.

The most important concept is that of the "scalar product" of a generalized function $g(\mathbf{z})$ defined by a sequence $\{g_m(\mathbf{z})\}$ with a sufficiently smooth ordinary function $\varphi(\mathbf{z})$ (test function):

$$\langle g, \varphi \rangle = \lim_{m \to \infty} \int_Z g_m(\mathbf{z})\varphi(\mathbf{z})\, d\mathbf{z} \tag{2.3}$$

for any test function $\varphi(\mathbf{z})$ such that the indicated limit exists.

It is usually assumed that the set of such test functions is dense in the set of continuous functions (i.e. any continuous function can be approximated as closely as we wish by a linear combination of test functions). The existence of the limit in the right hand side of Eq. (2.3) is the condition for $\{g_m(\mathbf{z})\}$ to be legitimately called a generalized function or distribution (with respect to the chosen class of test functions).

One can obviously define addition for generalized functions; if $g(\mathbf{z})$ and $h(\mathbf{z})$ are generalized functions defined by the sequences $\{g_m(\mathbf{z})\}$ and $\{h_m(\mathbf{z})\}$, then their sum $g(\mathbf{z}) + h(\mathbf{z})$ is defined by the sequence $\{g_m(\mathbf{z}) + h_m(\mathbf{z})\}$. In general, however, one cannot define the product of two generalized functions $g(\mathbf{z})$ and $h(\mathbf{z})$, but only the product of a generalized function $g(\mathbf{z})$ and an ordinary smooth function $\psi(\mathbf{z})$; such a product is the generalized function associated with the sequence $\{\psi(\mathbf{z})g_m(\mathbf{z})\}$, a test function $\varphi(\mathbf{z})$ for $\psi(\mathbf{z})g(\mathbf{z})$ being such that $\psi(\mathbf{z})\varphi(\mathbf{z})$ is a test function for $g(\mathbf{z})$.

Finally, one can define the integral of a generalized function $g(\mathbf{z})$ over the region Z, within which $\mathbf{z}$ varies, provided the function $u(\mathbf{z})$ which is identically equal to unity in Z is a test function for $g(\mathbf{z})$.

In such a case we set:

$$\int_Z g(\mathbf{z})\,d\mathbf{z} = \langle g, u\rangle = \lim_{m\to\infty}\int_Z g_m(\mathbf{z})\,d\mathbf{z} \tag{2.4}$$

If $\varphi(\mathbf{z})$ is a test function for $g(\mathbf{z})$, one can consider the generalized function $g(\mathbf{z})\varphi(\mathbf{z})$, which admits $u(\mathbf{z}) \equiv 1$ as a test function; according to Eq. (2.4) the integral of $g(\mathbf{z})\varphi(\mathbf{z})$ is well defined and given by:

$$\int_Z g(\mathbf{z})\varphi(\mathbf{z})\,d\mathbf{z} = \lim_{m\to\infty}\int g_m(\mathbf{z})\varphi(\mathbf{z})\,d\mathbf{z} = \langle g, \varphi\rangle \tag{2.5}$$

Hence the integral of the product $g(\mathbf{z})\varphi(\mathbf{z})$ over Z equals the "scalar product" $\langle g, \varphi\rangle$ defined by Eq. (2.3).

The simplest example of a generalized function is given by the so-called Dirac delta function, which is illustrated by the probability density corresponding to the abovementioned case when the n-dimensional vector $\mathbf{z}$ has the value $\mathbf{z}_0$ with certainty.

We can define the delta function $\delta(\mathbf{z} - \mathbf{z}_0)$ by means of the sequence:

$$\delta_m(\mathbf{z} - \mathbf{z}_0) = \frac{m^n}{\tau_n} H\left(\frac{1}{m} - |\mathbf{z} - \mathbf{z}_0|\right) \tag{2.6}$$

where $H(x)$ denotes the Heaviside step function, equal to one for $x > 0$ and zero for $x < 0$, while τ_n is the volume of the unit sphere in n dimensions, related to the area of the same sphere, ω_n, by $\tau_n = \omega_n/n$ with ω_n given by (see Appendix to this Chapter):

$$\omega_n = \begin{cases} \dfrac{2\pi^{n/2}}{\left(\dfrac{n}{2} - 1\right)!} & (n \text{ even}) \\[2ex] \dfrac{2\pi^{(n-1)/2}}{\left(\dfrac{n}{2} - 1\right)\left(\dfrac{n}{2} - 2\right)\cdots\dfrac{1}{2}} & (n \text{ odd}) \end{cases} \tag{2.7}$$

In particular for $n = 2$, $\omega_2 = 2\pi$, $\tau_2 = \pi$, while for $n = 3$, $\omega_3 = 4\pi$, $\tau_3 = 4\pi/3$. For $n = 1$ Eq. (2.7) is meaningless, but, clearly, $\tau_1 = 2$ (the unit sphere in one dimension reduces to a segment from -1 to 1). Hence, in the one dimensional case, Eq. (2.6) becomes:

$$\delta_m(z - z_0) = \frac{m}{2} H\left(\frac{1}{m} - |z - z_0|\right), \tag{2.8}$$

and the sequence is illustrated in Fig. 1 in the case $z_0 = 0$. $\delta_m(z)$ is zero for z outside the interval $(-1/m, 1/m)$ and equal to $m/2$ inside; the integral of

$\delta_m(z)$ from $-\infty$ to $+\infty$ is the area of the rectangle under the profile; that is, $(2/m)(m/2) = 1$, which is unity for any m. It is clear that the limit of the sequence $\{\delta_m(z)\}$ cannot be an ordinary function: as a matter of fact, the indicated limit gives 0 for $z \neq z_0$ and $+\infty$ for $z = z_0$ and can hardly be thought of as defining a function.

Analogously, in n dimensions, $\delta_m(\mathbf{z} - \mathbf{z}_0)$ is zero outside the sphere with center at $\mathbf{z}_0$ and radius $1/m$ and equal to m^n/τ_n, that is, the inverse of the volume of such a sphere, inside it. The integral of $\delta_m(\mathbf{z} - \mathbf{z}_0)$ over the

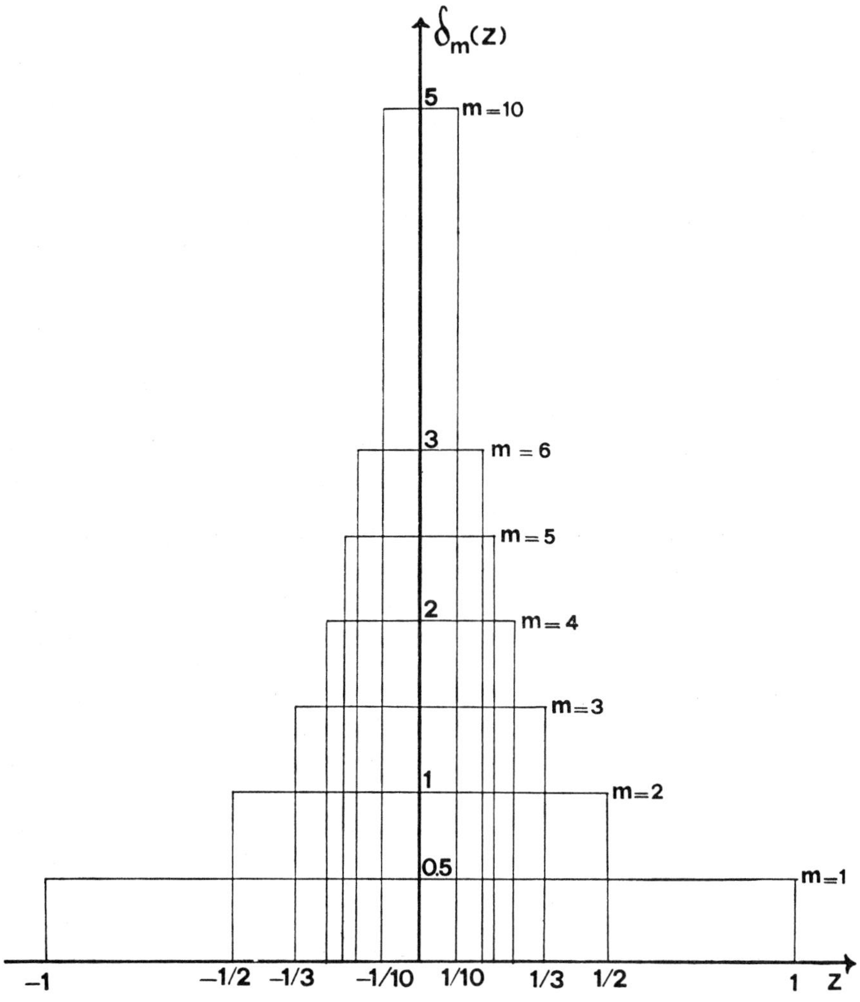

Fig. 1. Sketch of a sequence of functions approximating the delta function.

n-dimensional space is, accordingly, unity:

$$\int \delta_m(\mathbf{z} - \mathbf{z}_0)\, d\mathbf{z} = 1. \tag{2.9}$$

More generally, if we have a function continuous in a neighborhood of $\mathbf{z}_0$ we have:

$$\int \delta(\mathbf{z} - \mathbf{z}_0)\varphi(\mathbf{z})\, d\mathbf{z} = \lim_{m\to\infty} \int \delta_m(\mathbf{z} - \mathbf{z}_0)\varphi(\mathbf{z})\, d\mathbf{z} = \varphi(\mathbf{z}_0) \tag{2.10}$$

as the appropriate average of $\varphi(\mathbf{z})$ when the probability is concentrated at a single point, $\mathbf{z} = \mathbf{z}_0$. Incidentally, Eq. (2.10) shows that all the continuous functions are test functions for the sequence $\{\delta_m(\mathbf{z} - \mathbf{z}_0)\}$.

In order to prove Eq. (2.10), we observe that, by the continuity of $\varphi(\mathbf{z})$, we can find, for any given $\varepsilon > 0$, an m such that in the closed sphere $|\mathbf{z} - \mathbf{z}_0| \leqslant 1/m$, the following inequality is uniformly valid

$$-\varepsilon < \varphi(\mathbf{z}) - \varphi(\mathbf{z}_0) < \varepsilon, \tag{2.11}$$

and integrating over n-dimensional space after multiplying by $\delta_m(\mathbf{z} - \mathbf{z}_0) \geqslant 0$ gives:

$$-\varepsilon < \int \delta_m(\mathbf{z} - \mathbf{z}_0)\varphi(\mathbf{z})\, d\mathbf{z} - \varphi(\mathbf{z}_0) < \varepsilon \tag{2.12}$$

where repeated use has been made of Eq. (2.9). Eq. (2.10) follows from Eq. (2.12) *via* the definition of limit.

Before returning to general considerations on probability densities we remark that, although a sequence $\{g_m\}$ defines a generalized function, the latter does not define the former uniquely; that is, different sequences can have the same generalized function as an ideal limit, in the same way as different sequences of rational numbers can have the same real number as their ideal limit. Thus, the delta function could also be defined by the sequence $\{\delta_m\}$ given by

$$\delta_m(\mathbf{z} - \mathbf{z}_0) = (m\pi^{-\frac{1}{2}})^n \exp[-m^2(\mathbf{z} - \mathbf{z}_0)^2] \tag{2.13}$$

Another sequence converging to the delta function in the one-dimensional case is

$$\delta_m(z - z_0) = m/\{\pi[1 + (z - z_0)^2 m^2]\} \tag{2.14}$$

The only difference between these sequences and the one considered above arises in connection with the test functions which must satisfy additional conditions besides being continuous at $\mathbf{z} = \mathbf{z}_0$.

The treatises on generalized functions are now numerous [1–3] although they frequently make use of different approaches to this concept; the above sketch should suffice for understanding the very limited use made of generalized functions in this book.

We return now to the discussion about the use of probability densities, to introduce a suitable way of measuring the deviation from an average value, defined by Eq. (2.2). The local deviation from the average is $\varphi(\mathbf{z}) - \overline{\varphi(\mathbf{z})}$, but we want to define an average deviation; the average of the local deviation is not useful because

$$\int [\varphi(\mathbf{z}) - \overline{\varphi(\mathbf{z})}]P(\mathbf{z})\, d\mathbf{z} = \int \varphi(\mathbf{z})P(\mathbf{z})\, d\mathbf{z} - \overline{\varphi(\mathbf{z})} \int P(\mathbf{z})\, d\mathbf{z}$$

$$= \overline{\varphi(\mathbf{z})} - \overline{\varphi(\mathbf{z})} = 0 \tag{2.15}$$

where Eqs. (2.2) and (2.1) have been used.

An average measure of the departures from the average value of $\varphi(\mathbf{z})$ can be obtained by evaluating the so called mean-square deviation, whose square is given by:

$$\overline{[\varphi(\mathbf{z}) - \overline{\varphi(\mathbf{z})}]^2} = \int [\varphi(\mathbf{z}) - \overline{\varphi(\mathbf{z})}]^2 P(\mathbf{z})\, d\mathbf{z} \tag{2.16}$$

Another possible way of measuring the deviation of $\mathbf{z}$ from $\mathbf{z}_0$, once the probability density is given, is by computing the average of $\overline{|\varphi(\mathbf{z}) - \overline{\varphi(\mathbf{z})}|}$ instead of $\overline{[\varphi(\mathbf{z}) - \varphi(\mathbf{z})]^2}$. In general, the result will not be the square root of the previous one, but the square of the average will not be larger than the average of the square.

If $P(\mathbf{z}) = \delta(\mathbf{z} - \mathbf{z}_0)$ then $\overline{\varphi(\mathbf{z})} = \varphi(\mathbf{z}_0)$ because of Eqs. (2.2), (2.5), (2.10) and by applying the same equations with $[\varphi(\mathbf{z}) - \overline{\varphi(\mathbf{z})}]^2$ and $|\overline{\varphi(\mathbf{z})} - \varphi(\mathbf{z})|$ in place of $\varphi(\mathbf{z})$, we find:

$$\begin{aligned} \overline{[\varphi(\mathbf{z}) - \overline{\varphi(\mathbf{z})}]^2} &= [\varphi(\mathbf{z}_0) - \varphi(\mathbf{z}_0)]^2 = 0 \\ \overline{|\varphi(\mathbf{z}) - \overline{\varphi(\mathbf{z})}|} &= |\varphi(\mathbf{z}_0) - \varphi(\mathbf{z}_0)| = 0. \end{aligned} \tag{2.17}$$

that is, in the case of a probability density equal to the delta function, the average deviation, no matter how defined, turns out to be zero, as appropriate for a probability density meant to represent certainty.

3. Phase space and Liouville's theorem

In order to discuss the behavior of a system of N mass points satisfying Eqs. (1.1a), it is very convenient to introduce the so called phase space; that is, a $6N$-dimensional space where the Cartesian coordinates are the $3N$ components of the N position vectors $\mathbf{x}_i$ and the $3N$ components of the N velocities $\boldsymbol{\xi}_i$.

In this space, the state of a system at a given time t, if known with absolute accuracy, is represented by a point whose coordinates are the $6N$ values of the components of the position vectors and velocities of the N particles. (Frequently, the momenta of the particles are used in place of their velocities, but the difference will not matter for our purposes.) Let us introduce the $6N$-dimensional vector $\mathbf{z}$ which gives the position of the representative point in phase space; clearly, the components of $\mathbf{z}$ are respectively given by the $3N$ components of the N three-dimensional vectors $\mathbf{x}_i$ and the $3N$ components of the N three-dimensional vectors $\boldsymbol{\xi}_i$. The evolution equation for $\mathbf{z}$ is from Eqs. (1a)

$$\dot{\mathbf{z}} = \frac{d\mathbf{z}}{dt} = \mathbf{Z}, \tag{3.1}$$

where $\mathbf{Z}$ is a $6N$-dimensional vector, whose components are respectively given by the $3N$ components of the N three-dimensional vectors $\boldsymbol{\xi}_i$ and the $3N$ components of the N three-dimensional vectors $\mathbf{X}_i$. Given the initial state, that is a point $\mathbf{z}_0$ in phase space, Eq. (3.1) determines $\mathbf{z}$ at subsequent times (provided the conditions for existence and uniqueness of the solution are satisfied).

If the initial data are not known with absolute accuracy, we must introduce a probability density $P_0(\mathbf{z})$ which gives us the distribution of probability for the initial data and we can try to set up the problem of computing the probability density at subsequent times, $P(\mathbf{z}, t)$. In order to achieve this, we must find an evolution equation for $P(\mathbf{z}, t)$; this can easily be done, as we shall see, provided the forces are known; that is, if the only uncertainty is in the initial data.

An intuitive way of deriving the equation satisfied by $P(\mathbf{z}, t)$ is the following. We replace the representative point by a continuous distribution with density proportional to the probability density; in such a way, the system of mass points is replaced by a sort of fluid with density proportional to P and velocity $\dot{\mathbf{z}} = \mathbf{Z}$. Hence conservation of mass will give:

$$\frac{\partial P}{\partial t} + \operatorname{div}(P\mathbf{Z}) = 0 \tag{3.2}$$

where, as usual, for any vector $\mathbf{u}$ of the phase space, we write

$$\operatorname{div} \mathbf{u} = \sum_{i=1}^{6N} \frac{\partial u_i}{\partial z_i} \equiv \frac{\partial}{\partial \mathbf{z}} \cdot \mathbf{u} \tag{3.3}$$

Eq. (3.2) is the Liouville equation; note that the components of $\mathbf{z}$ are independent variables.

But:

$$\operatorname{div}(P\mathbf{Z}) = \mathbf{Z} \operatorname{grad} P + P \operatorname{div} \mathbf{Z} \tag{3.4}$$

where, as usual, $\operatorname{grad} P \equiv \partial P/\partial \mathbf{z}$ is the vector with components $\partial P/\partial z_i$. Hence P satisfies the equation:

$$\frac{\partial P}{\partial t} + \mathbf{Z} \cdot \operatorname{grad} P + P \operatorname{div} \mathbf{Z} = 0 \tag{3.5}$$

Usually, div $\mathbf{Z} = 0$. In fact, since $\mathbf{x}_i$ and $\boldsymbol{\xi}_i$ are independent variables:

$$\operatorname{div} \mathbf{Z} = \sum_{i=1}^{N} \left(\frac{\partial}{\partial \mathbf{x}_i} \cdot \boldsymbol{\xi}_i + \frac{\partial}{\partial \boldsymbol{\xi}_i} \cdot \mathbf{X}_i \right) = \sum_{i=1}^{N} \frac{\partial}{\partial \boldsymbol{\xi}_i} \cdot \mathbf{X}_i \tag{3.6}$$

If the force per unit mass is velocity-independent, then also $(\partial/\partial\boldsymbol{\xi}_i) \cdot \mathbf{X}_i = 0$, and div $\mathbf{Z} = 0$ as stated. Note, however, that for some velocity dependent forces $(\partial/\partial\boldsymbol{\xi}_i) \cdot \mathbf{X}_i = 0$; the most notable case being that of the Lorentz force acting on a charged particle in a magnetic field. We shall always consider forces such that div $\mathbf{Z} = 0$ (typically, velocity-independent forces). Hence we write the Liouville equation in the following form:

$$\frac{\partial P}{\partial t} + \mathbf{Z} \cdot \frac{\partial P}{\partial \mathbf{z}} = 0 \tag{3.7}$$

Eq. (3.7) can be of course rewritten in terms of the variables $\mathbf{x}_i$, $\boldsymbol{\xi}_i$:

$$\frac{\partial P}{\partial t} + \sum_{i=1}^{N} \boldsymbol{\xi}_i \cdot \frac{\partial P}{\partial \mathbf{x}_i} + \sum_{i=1}^{N} \mathbf{X}_i \cdot \frac{\partial P}{\partial \boldsymbol{\xi}_i} = 0 \tag{3.8}$$

where $\partial P/\partial \mathbf{x}_i$ are the gradients in the three-dimensional space of the positions of the i-th particle and $\partial P/\partial \boldsymbol{\xi}_i$ are the gradients in the three-dimensional space of velocities of the i-th particle.

In order to give a more accurate derivation of the Liouville equation, we observe that the evolution equation, Eq. (3.1) defines, at each instant, a mapping of the phase space into itself; in this mapping, to each point $\mathbf{z}_0$ there corresponds the point $\mathbf{z} = \mathbf{z}(\mathbf{z}_0, t)$ reached at time t by a point which was at $\mathbf{z}_0$ at $t = 0$. The mapping is one-to-one, if the equations have, as we shall assume, a unique solution, corresponding to given initial data, for both $t > 0$ and $t < 0$. (Existence and uniqueness for $t < 0$ follow from the corresponding properties for $t > 0$ if the equations are time reversible, as in the case of velocity-independent forces.)

The probability for the point representing the system to be found in a region R of the phase space at time t is

$$\operatorname{Prob}(\mathbf{z} \in R) = \int_R P(\mathbf{z}, t)\, d\mathbf{z} \tag{3.9}$$

where $\mathbf{z} \in R$ is the usual notation for "$\mathbf{z}$ belonging to R". The abovementioned probability will be equal to the probability that the representative point was, at $t = 0$, in the region R_0 consisting of the points $\mathbf{z}_0$ which are the inverse images of the points $\mathbf{z} \in R$ in the mapping considered above. In fact a point can be in R at time t if, and only if, it was in R_0 at $t = 0$. Hence:

$$\int_R P(\mathbf{z}, t)\, d\mathbf{z} = \int_{R_0} P_0(\mathbf{z}_0)\, d\mathbf{z}_0, \tag{3.10}$$

where the set of values of $\mathbf{z} \in R$ coincides with the set of points $\mathbf{z} = \mathbf{z}(\mathbf{z}_0, t)$ with $\mathbf{z}_0 \in R_0$. We can exploit this fact to change the integration variables from the components of $\mathbf{z}$ to those of $\mathbf{z}_0$ in the first integral. We obtain

$$\int_R P(\mathbf{z}, t)\, d\mathbf{z} = \int_{R_0} P(\mathbf{z}(\mathbf{z}_0, t), t) J(\mathbf{z}/\mathbf{z}_0)\, d\mathbf{z}_0 \tag{3.11}$$

where $J(\mathbf{z}/\mathbf{z}_0)$ is the Jacobian determinant of the old variables with respect to the new ones (or the absolute value of such a determinant, if the latter turns out to be negative). Comparison of Eqs. (3.11) and (3.10) gives, due to the arbitrariness of R_0,

$$P(\mathbf{z}(\mathbf{z}_0, t), t) J(\mathbf{z}/\mathbf{z}_0) = P_0(\mathbf{z}_0) \tag{3.12}$$

Since the right hand side is time-independent, the total derivative of the left hand side with respect to time must vanish identically:

$$J\left(\frac{\partial P}{\partial t} + \frac{\partial \mathbf{z}}{\partial t} \cdot \frac{\partial P}{\partial \mathbf{z}}\right) + P\frac{\partial J}{\partial t} = 0 \tag{3.13}$$

where the arguments of P are $\mathbf{z}, t$ and those of J are $\mathbf{z}_0, t$. Since $J \neq 0$, as will be shown below, and because of Eq. (3.1) (note that $\partial\mathbf{z}/\partial t$ in Eq. (3.13) is the derivative of $\mathbf{z}$ with respect to time for constant initial data; i.e. what was denoted by $\dot{\mathbf{z}}$ in Eq. (3.1)), we obtain

$$\frac{\partial P}{\partial t} + \mathbf{Z} \cdot \frac{\partial P}{\partial \mathbf{z}} + \frac{P}{J}\frac{\partial J}{\partial t} = 0 \tag{3.14}$$

We now compute $\partial J/\partial t$. Let J_{rs} be the cofactor of $\partial z_r/\partial z_s{}^0$ in the Jacobian determinant; then by the rule for differentiating determinants:

$$\begin{aligned}\frac{\partial J}{\partial t} &= \sum_{r,s=1}^{6N} \frac{\partial}{\partial t}\left(\frac{\partial z_r}{\partial z_s{}^0}\right) J_{rs} = \sum_{r,s=1}^{6N} \frac{\partial}{\partial z_s{}^0}\left(\frac{\partial z_r}{\partial t}\right) J_{rs} \\ &= \sum_{r,s=1}^{6N} \frac{\partial Z_r}{\partial z_s{}^0} J_{rs} = J \sum_{r,s=1}^{6N} \frac{\partial Z_r}{\partial z_s{}^0}\frac{\partial z_s{}^0}{\partial z_r} \\ &= J \sum_{r=1}^{6N} \frac{\partial Z_r}{\partial z_r} = J \operatorname{div} \mathbf{Z}\end{aligned} \tag{3.15}$$

where we have replaced $\partial z_r/\partial t$ by Z_r according to Eq. (3.1), and used the fact that the cofactor J_{rs} is given by $J_{rs} = J\, \partial z_s{}^0/\partial z_r$, as is well known and easily follows from Laplace's theorem on determinants and the chain rule for differentiation.

Inserting Eq. (3.15) into Eq. (3.13) gives Eq. (3.5) again; that is, the Liouville equation, Eq. (3.2). In particular, if div $\mathbf{Z} = 0$, Eq. (3.7) follows.

We note, incidentally, that Eq. (3.15) shows that $J =$ const. for div $\mathbf{Z} = 0$ and, since $J = 1$ for $t = 0$, it follows that $J = 1$ for any t. As a consequence, the volume of a region in phase space is invariant with respect to the mapping induced by the motion of the system provided div $\mathbf{Z} = 0$ (Liouville's theorem). In fact, we have writing Eq. (3.11) for the case $P = 1$

$$\int_R d\mathbf{z} = \int_{R_0} J(\mathbf{z}/\mathbf{z}_0)\, d\mathbf{z}_0 = \int_{R_0} d\mathbf{z}_0 \tag{3.16}$$

The Liouville equation, Eq. (3.7) or Eq. (3.8), simply states that P takes at time t, at the point $\mathbf{z}$ the value which it took at time $t = 0$ at the phase space point $\mathbf{z}_0$ which is carried into $\mathbf{z}$ by the motion described by Eqs. (3.1) or (1.1a), provided div $\mathbf{Z} = 0$. In fact the left hand side of Eq. (3.7) is the time derivative of P in a reference frame convected with velocity $\dot{\mathbf{z}} = \mathbf{Z}$ in phase space. Accordingly, if the probability is constant in a region R_0 it will be the same constant in the region R_t into which R_0 is mapped by the flow induced in phase space by the motion of the system; in particular, if the probability density is zero everywhere except at $\mathbf{z} = \mathbf{z}_0$ for $t = 0$ it will be zero everywhere except at $\mathbf{z} = \mathbf{z}(\mathbf{z}_0, t)$ at subsequent times. This fact can be expressed by saying that the solution corresponding to an initial datum which is a delta function (centered at $\mathbf{z} = \mathbf{z}_0$), is a delta function (centered at $\mathbf{z} = \mathbf{z}(\mathbf{z}_0, t)$) for $t \geqslant 0$.

Accordingly, the Liouville equation is an alternative formulation of the equations of motion, which contains information not only on a given motion but also on the motions close to the latter, in a sense shortly to be explained. If the initial data are given with absolute certainty, P is a delta function at $t = 0$ and the solution of the Liouville equation will give a delta function for subsequent times: the point $\mathbf{z} = \mathbf{z}(\mathbf{z}_0, t)$ at which the delta function is centered gives the solution of the equations of motion (we note that in order to apply the Liouville equation to this highly idealized case, recourse must be had to the notion of the derivative of a generalized function which, for the sake of brevity, was not considered in Section 2; see, however, a previous book of the author [4] and the quoted books on generalized functions [1–3]). If, as is more realistic, just the probability density of the initial data is given, the Liouville equation determines not only the most probable motion but also the distribution of deviations from the latter.

4. Hard spheres and rigid walls. Mean free path

In the previous sections, we considered the case of mass points which continuously interact with each other according to the equations of motion (1.1). It is frequently convenient to consider limiting cases in which the system has only discrete interactions with finite impulses (hard collisions); in such a case the forces are not describable by means of ordinary functions and the Liouville equation must be handled in a different way. The limiting case of hard collision is useful because it gives a more intuitive idea of the evolution of the system and is a good approximation to the strong repulsive forces which actual molecules mutually exert when they are close to each other. These considerations lead to the concept of a gas of hard spheres; that is, a system of many "billiard balls" which do not interact at a distance and collide according to the laws of elastic impact. The diameter σ of the spheres is equivalent to the range of the force through which actual molecules interact; as a matter of fact a gas of rigid spheres can also be pictured as a system of mass points which do not interact when their mutual distance is larger than σ but interact with a formally infinite repulsive central force when this distance becomes exactly equal to σ so that a closer approach is impossible.

Another example of instantaneous interaction is considered when a molecule is assumed to be elastically reflected by a solid wall. This model is more unrealistic than the hard spheres model, because a solid wall has macroscopic dimensions and certainly shows a very detailed structure at a microscopic level. As we shall see in detail in Chapter III, this structure will prevent an elastic collision on a regular geometric surface representing the wall in a macroscopic description.

In spite of this, it is useful to consider the case of perfectly elastic reflections on a rigid wall for illustrative purposes.

If $\mathbf{n}$ is the normal to the wall (assumed at rest), the effect of a collision will be to change the sign of the normal component while leaving the tangential component unmodified. Thus, if $\boldsymbol{\xi}'$ denotes the velocity of a molecule before the collision and $\boldsymbol{\xi}$ the velocity after the impact, $\boldsymbol{\xi}'$ and $\boldsymbol{\xi}$ will be related as follows

$$\boldsymbol{\xi} = \boldsymbol{\xi}' - 2\mathbf{n}(\mathbf{n} \cdot \boldsymbol{\xi}') \tag{4.1}$$

Eq. (4.1) simply expresses the fact that the molecules are specularly reflected by the wall (see Fig. 2). If the wall is not at rest, but moves with velocity $\mathbf{u}_0$ with respect to the reference frame chosen to describe the motion, then Eq. (4.1) will apply to the velocities relative to the wall; that is, $\boldsymbol{\xi}$ and $\boldsymbol{\xi}'$ must be replaced by $\boldsymbol{\xi} - \mathbf{u}_0$ and $\boldsymbol{\xi}' - \mathbf{u}_0$. Hence Eq. (4.1) will be replaced by:

$$\boldsymbol{\xi} = \boldsymbol{\xi}' - 2\mathbf{n}[\mathbf{n} \cdot (\boldsymbol{\xi} - \mathbf{u}_0)] \tag{4.2}$$

In the case of a collision between two identical rigid spheres, the equations which relate the velocities after impact, $\boldsymbol{\xi}_1$ and $\boldsymbol{\xi}_2$ to those before impact, $\boldsymbol{\xi}_1'$ and $\boldsymbol{\xi}_2'$, are:

$$\begin{aligned} \boldsymbol{\xi}_1 &= \boldsymbol{\xi}_1' - \mathbf{n}[\mathbf{n} \cdot (\boldsymbol{\xi}_1' - \boldsymbol{\xi}_2')] \\ \boldsymbol{\xi}_2 &= \boldsymbol{\xi}_2' - \mathbf{n}[\mathbf{n} \cdot (\boldsymbol{\xi}_2' - \boldsymbol{\xi}_1')] \end{aligned} \tag{4.3}$$

where $\mathbf{n}$ is the unit vector directed along the line joining the centers of the two spheres (orientation does not matter). Eqs. (4.3) can be derived by the following considerations. Conservation of momentum and energy must be satisfied:

$$\begin{aligned} \boldsymbol{\xi}_1 + \boldsymbol{\xi}_2 &= \boldsymbol{\xi}_1' + \boldsymbol{\xi}_2' \\ \xi_1^2 + \xi_2^2 &= \xi_1'^2 + \xi_2'^2 \end{aligned} \tag{4.4}$$

Introduce a unit vector $\mathbf{n}$ directed along $\boldsymbol{\xi}_1 - \boldsymbol{\xi}_1'$; this direction bisects the directions of $\boldsymbol{\xi}_1$ and $-\boldsymbol{\xi}_1'$ (therefore, $\mathbf{n}$ is the unit vector directed along the line joining the centers at the moment of impact, since the impulse is assumed to be directed along such a line, as appropriate for a central interaction). Because of our definition of $\mathbf{n}$ we have

$$\boldsymbol{\xi}_1 = \boldsymbol{\xi}_1' - \mathbf{n}C \tag{4.5}$$

where C is a scalar to be determined. The first of Eqs. (4.4) gives:

$$\boldsymbol{\xi}_2 = \boldsymbol{\xi}_2' + \mathbf{n}C \tag{4.6}$$

Inserting Eqs. (4.5) and (4.6) into Eq. (4.4) leads us to the following result

$$\xi_1'^2 - 2\mathbf{n}\cdot\boldsymbol{\xi}_1'C + C^2 + \xi_2'^2 + 2\mathbf{n}\cdot\boldsymbol{\xi}_2'C + C^2 = \xi_1'^2 + \xi_2'^2 \tag{4.7}$$

Cancelling equal terms from each side, we have

$$-\mathbf{n}\cdot(\boldsymbol{\xi}_1' - \boldsymbol{\xi}_2')C + C^2 = 0 \tag{4.8}$$

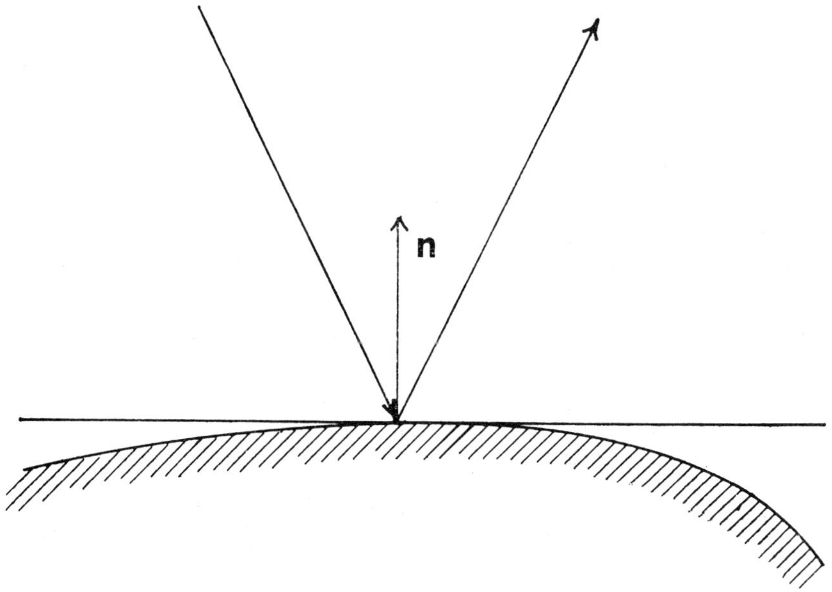

Fig. 2. Specular reflection of a mass point.

Hence, dismissing the case $C = 0$ which corresponds to a trivial solution of the conservation equations (absence of interaction), we have:

$$C = \mathbf{n} \cdot (\boldsymbol{\xi}_1' - \boldsymbol{\xi}_2') \tag{4.9}$$

Inserting this value into Eqs. (4.5) and (4.6), Eqs. (4.3) follow.

We want to show, now, that the Liouville theorem (conservation of volume in phase space) remains valid for the instantaneous interactions considered in this section. We observe that both Eq. (4.1) and Eqs. (4.3) are reversible: that is, one can solve for the primed variables to find

$$\boldsymbol{\xi}' = \boldsymbol{\xi} - 2\mathbf{n}(\mathbf{n} \cdot \boldsymbol{\xi}) \tag{4.10}$$

and

$$\begin{aligned} \boldsymbol{\xi}_1' &= \boldsymbol{\xi}_1 - \mathbf{n}[\mathbf{n} \cdot (\boldsymbol{\xi}_1 - \boldsymbol{\xi}_2)] \\ \boldsymbol{\xi}_2' &= \boldsymbol{\xi}_2 - \mathbf{n}[\mathbf{n} \cdot (\boldsymbol{\xi}_2 - \boldsymbol{\xi}_1)] \end{aligned} \tag{4.11}$$

respectively. These are the same equations as Eqs. (4.1) and Eq. (4.3), with the primed and unprimed variables interchanged. To derive Eq. (4.10), it is sufficient to compute $\mathbf{n} \cdot \boldsymbol{\xi}'$ first (and this is easily done by scalar multiplication of Eq. (4.1) by $\mathbf{n}$) and then to employ the result $\mathbf{n} \cdot \boldsymbol{\xi}' = -\mathbf{n} \cdot \boldsymbol{\xi}$ in Eq. (4.1). Analogously, to obtain Eqs. (4.11), we first compute the value of $\mathbf{n} \cdot (\boldsymbol{\xi}_1' - \boldsymbol{\xi}_2')$ in terms of the unprimed variables; to this end, we deduce by subtraction the relation between relative velocities

$$\boldsymbol{\xi}_1 - \boldsymbol{\xi}_2 = \boldsymbol{\xi}_1' - \boldsymbol{\xi}_2' - 2\mathbf{n}[\mathbf{n} \cdot (\boldsymbol{\xi}_1' - \boldsymbol{\xi}_2')] \tag{4.12}$$

Scalar multiplications by $\mathbf{n}$ yields $\mathbf{n} \cdot (\boldsymbol{\xi}_1' - \boldsymbol{\xi}_2') = -\mathbf{n} \cdot (\boldsymbol{\xi}_1 - \boldsymbol{\xi}_2)$ and insertion of this result into Eqs. (4.3) produces the desired relations, Eqs. (4.11).

We observe now that in both cases (interaction with the wall and impact between two rigid spheres), the components of the velocities involved undergo a linear transformation described by some matrix A (3×3 or 6×6), whose elements depend on $\mathbf{n}$. In both cases comparison of the direct and inverse transformations (Eqs. (4.1) and (4.10); Eqs. (4.3) and (4.11)) shows that the inverse matrix A^{-1} equals A, i.e., A^2 is the identity matrix. Hence the square of the determinant of A (which is simply the Jacobian J_1 of the linear transformation) is unity, so that $J_1 = \pm 1$.

It is not difficult to see that actually $J_1 = -1$. In the first case, in fact, we can use as variables the normal component which is reversed in the impact, and the components along two axes in the tangential plane, which are left unchanged, when the result then follows.

In the second case, we can first transform to the variables $\bar{\boldsymbol{\xi}}' = \frac{1}{2}(\boldsymbol{\xi}_1' + \boldsymbol{\xi}_2')$, $\bar{\boldsymbol{\xi}} = \frac{1}{2}(\boldsymbol{\xi}_1 + \boldsymbol{\xi}_2)$ (the velocity of the center of mass before and after the impact) and $\mathbf{V}' = \boldsymbol{\xi}_1' - \boldsymbol{\xi}_2'$, $\mathbf{V} = \boldsymbol{\xi}_1 - \boldsymbol{\xi}_2$ (the relative velocity before and after the impact). The new variables undergo the following transformation in the impact:

$$\bar{\boldsymbol{\xi}} = \bar{\boldsymbol{\xi}}'$$
$$\mathbf{V} = \mathbf{V}' - 2\mathbf{n}(\mathbf{n} \cdot \mathbf{V}') \tag{4.13}$$

the first of these equations being the first of Eqs. (4.4), and the second, Eq. (4.12). Since $\bar{\boldsymbol{\xi}}'$ does not change, the Jacobian of the transformation (4.13) is simply given by the Jacobian of the transformation from $\mathbf{V}'$ to $\mathbf{V}$ which is -1 (since the matrix is the same as in the previous case of interaction with a solid wall). Hence:

$$J(\bar{\boldsymbol{\xi}}, \mathbf{V} \mid \bar{\boldsymbol{\xi}}', \mathbf{V}') = -1 \tag{4.14}$$

But if J_0 denotes the Jacobian of the transformations $(\boldsymbol{\xi}_1', \boldsymbol{\xi}_2') \mapsto (\bar{\boldsymbol{\xi}}', \mathbf{V}')$ and $(\boldsymbol{\xi}_1, \boldsymbol{\xi}_2) \mapsto (\bar{\boldsymbol{\xi}}, \mathbf{V})$ (they obviously have the same Jacobian determinant), we have for the Jacobian J_1 of the transformation (4.3):

$$\begin{aligned} J_1 = J(\boldsymbol{\xi}_1, \boldsymbol{\xi}_2 \mid \boldsymbol{\xi}_1', \boldsymbol{\xi}_2') &= J(\boldsymbol{\xi}_1, \boldsymbol{\xi}_2 \mid \bar{\boldsymbol{\xi}}, \mathbf{V}) J(\bar{\boldsymbol{\xi}}, \mathbf{V} \mid \bar{\boldsymbol{\xi}}', \mathbf{V}') J(\bar{\boldsymbol{\xi}}', \mathbf{V}' \mid \boldsymbol{\xi}_1', \boldsymbol{\xi}_2') \\ &= J_0 \cdot (-1) \cdot (1/J_0) = -1 \end{aligned}$$

The Jacobian J of the phase space transformation corresponding to a hard impact is the product of the two Jacobians J_1 and J_2 corresponding to the two transformations of space and velocity variables, respectively. We have seen that $J_1 = -1$ and we shall see that $J_2 = -1$; hence $J = 1$ and invariance of the space volume is proved, on account of Eq. (3.16).

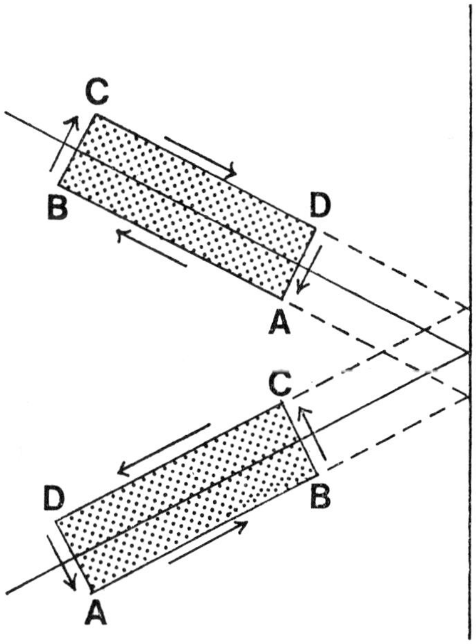

Fig. 3. Specular reflection of a volume element at a flat wall.

In order to prove that $J_2 = -1$ it is sufficient to consider the case of reflection of a point from a fixed wall, because the collision between two hard spheres reduces to the latter case by means of the transformation $\bar{\mathbf{x}} = \frac{1}{2}(\mathbf{x}_1 + \mathbf{x}_2)$, $\mathbf{r} = \mathbf{x}_1 - \mathbf{x}_2$, $\bar{\mathbf{x}}' = \frac{1}{2}(\mathbf{x}_1' + \mathbf{x}_2')$, $\mathbf{r}' = \mathbf{x}_1' - \mathbf{x}_2'$ by noting that $\bar{\mathbf{x}}' \mapsto \bar{\mathbf{x}}$ is a rigid motion and $\mathbf{r}' \mapsto \mathbf{r}$ is equivalent to a reflection from a fixed wall.

We first consider the case of the reflection of mass points from a plane wall; in this case the mapping induced by the motion does not deform an infinitesimal space region but changes its orientation (see Fig. 3; it is sufficient to consider the plane case, because changes take place in the plane of $\mathbf{n}$ and $\boldsymbol{\xi}'$). Hence $J_2 = -1$ in this case.

The case of curved surfaces is more complicated, because deformation takes place for finite regions (see Fig. 4). An infinitesimal region, however, is transformed (in the infinitesimally short period within which the collision takes place) into a region of the same volume and opposite orientation. To see this, we observe that we can consider the transformation to take place in a plane (the plane of motion). Since the reflection laws depend only on the

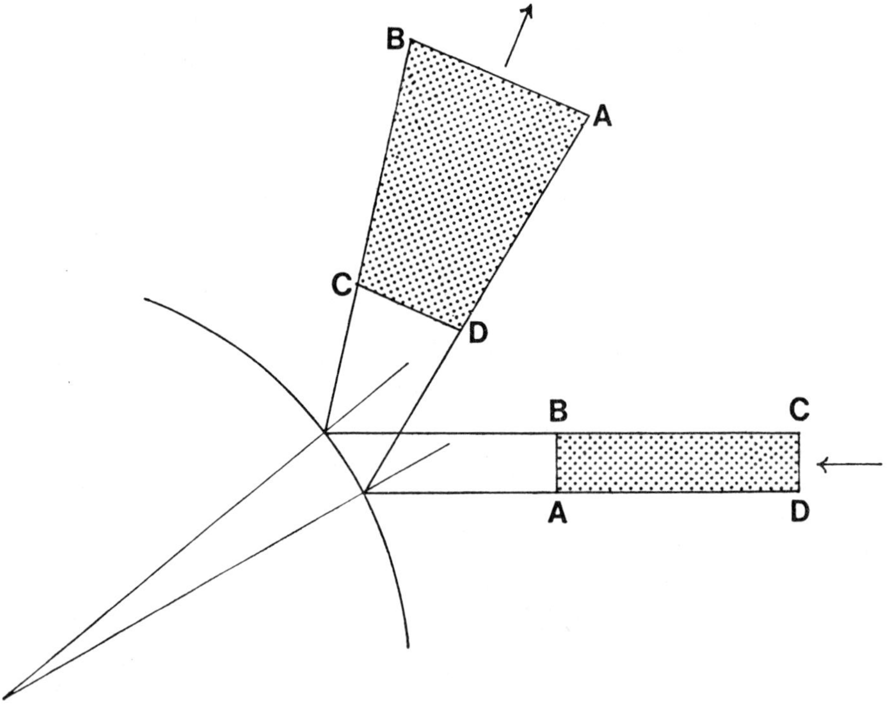

Fig. 4. Specular reflection of a volume element at a curved wall.

orientation of the tangent to the line upon which reflection takes place, the Jacobian will contain first order derivatives of the tangent unit vector; that is, at most, second derivatives of the transformed coordinates with respect to the initial ones. Hence the boundary may be replaced by its osculating circle; in this case the Jacobian, the ratio of the volume of an infinitesimal region after the impact to the volume of the corresponding region before the impact, could be, *a priori*, any finite nondimensional function of the radius of the circle and the incidence angle. But it cannot depend on the radius because one cannot form nondimensional functions containing a single length; hence the Jacobian must be the same for any curvature; that is, it must be equal to the value, $J_1 = -1$ for a reflection on a flat wall (corresponding to the limiting case of an infinitely large radius).

In connection with hard spheres it is convenient to introduce the notion of a free path. It is the distance travelled by a sphere S_1 between two consecutive collisions. This distance, of course, depends upon the number n of spheres per unit volume, the velocity of the chosen sphere S_1 and the velocity of the sphere S_2 with which S_1 will have the next impact. Accordingly, only the notion of mean free path will be meaningful.

A simple estimate of the value of the mean free path l of a hard sphere is obtained by assuming the other spheres at rest and surrounding each of them by a sphere of radius equal to the diameter σ of the particles, while the travelling sphere S_1 is represented by a point (Fig. 5). Then, if S_1 travels a distance l on average between two impacts, this means that there is only one molecule, namely S_1, in a cylinder of base $\pi\sigma^2$ and height l or:

$$n\pi\sigma^2 l \simeq 1 \tag{4.16}$$

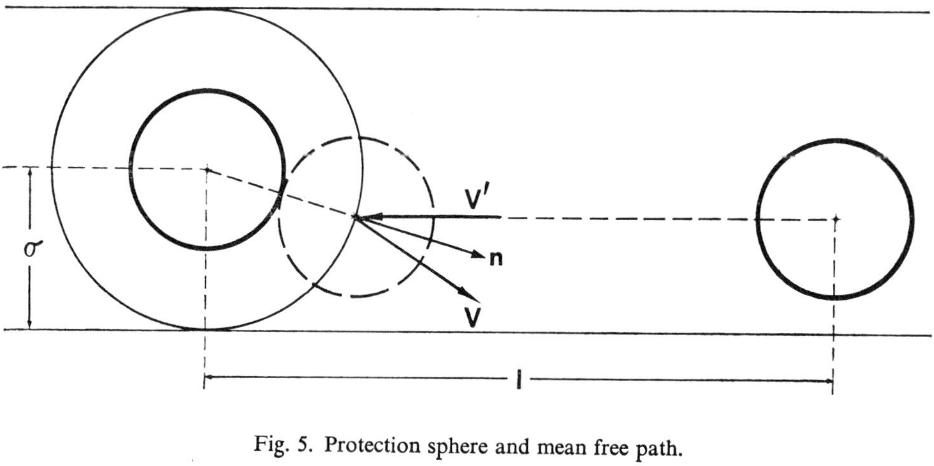

Fig. 5. Protection sphere and mean free path.

Hence the mean free path l is approximately given by:

$$l \simeq \frac{1}{n\pi\sigma^2} \tag{4.17}$$

5. Scattering of a volume element in phase space

Let us consider a mass point moving along an axis between two rigid walls. Let x denote the abscissa of the point and ξ its velocity, $\pm l$ the abscissa of the walls. We represent the evolution of the system in its own phase plane (see Fig. 6). Let us assume that the initial state is not exactly known, but that it is known that the point corresponding to the initial state is within the rectangle

$$\begin{gathered} 0 \leqslant x \leqslant \Delta x \\ \xi_0 \leqslant \xi \leqslant \xi_0 + \Delta\xi \end{gathered} \tag{5.1}$$

That is, the initial data for x and ξ are affected by errors smaller than Δx and $\Delta\xi$, respectively. We want to study the deformation of this rectangle as a consequence of the motion or, alternatively, the evolution of the uncertainty on the state of the system. A point located at (x, ξ) at $t = 0$ will be transformed, as a consequence of the motion, into the point $(x + \xi t, \xi)$ if we disregard the walls for a moment (thus the motion takes place at constant velocity). The Jacobian matrix of the considered transformation is

$$\mathsf{A}_t = \begin{Vmatrix} 1 & t \\ 0 & 1 \end{Vmatrix}$$

and the Jacobian determinant is obviously unity, as was to be expected from the Liouville theorem. After a time t has elapsed, the position of the mass point is known with an error $\Delta x + t\,\Delta\xi$ and its velocity with the initial error $\Delta\xi$. Fig. 6 illustrates the situation: the initial rectangle is deformed, because points with larger ξ move faster, and become a parallelogram with the same base and height and hence the same area as the initial rectangle. After a time $t = \tau$ we obtain the oblique distribution shown in Fig. 6, which becomes more oblique with increasing time. Since we have neglected the effect of the walls, the parallelogram will be outside the segment $(-l, l)$ (see the unshaded area corresponding to $t = 3\tau$). In order to pass to the case of the motion with reflection, it is sufficient to consider the motion in the absence of the latter, to cut the strip $\xi_0 < \xi < \xi_0 + \Delta\xi$ into portions of length $2l$ and then to place the portions alternately on the portion of the strip between $-l$ and l and its specular reflection with respect to the x-axis. In fact a reflection sends

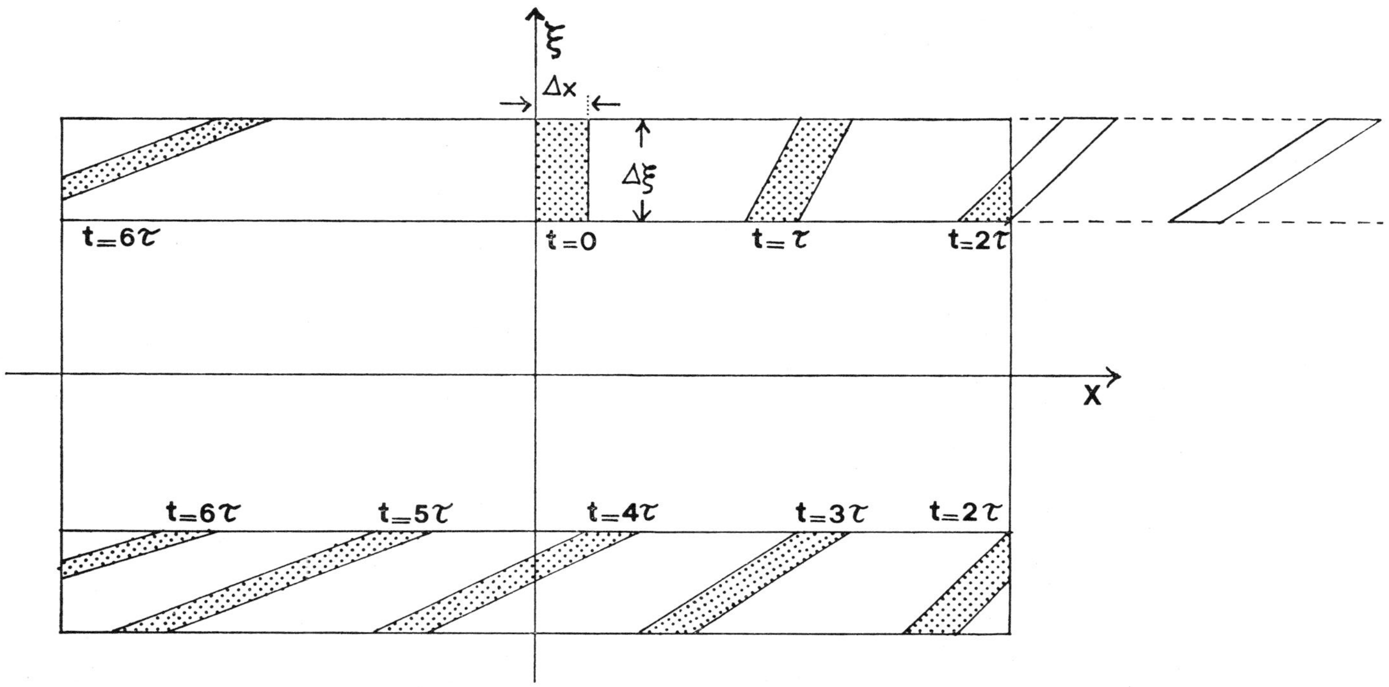

Fig. 6. Time evolution, in phase space, of a mass point moving on a segment. The initial velocity uncertainty $\Delta\xi$ makes the volume element more and more oblique as time goes on.

the points (l, ξ) into $(l, -\xi)$ and the point $(-l, -\xi)$ into $(-l, \xi)$. By this procedure we obtain the oblique distributions corresponding to $t = 2\tau, 3\tau, 4\tau, 5\tau, 6\tau$ (shown in Fig. 6), etc.

During the motion, the region becomes thinner and thinner and finally breaks up in layers. In fact at some time points with velocity $\xi_0 + \Delta\xi$ undergo the $(n + 1)$-th reflection while points with velocity ξ_0 have not yet undergone the n-th reflection (see Fig. 7).

More generally, at some time, points with velocity $\xi_0 + \Delta\xi$ will have undergone several reflections more than the points with velocity ξ_0, and we shall obtain the situation sketched in Fig. 8. The region has become subdivided into a large number of "strings" which tend to spread throughout the two rectangles, $\xi_0 < \xi < \xi_0 + \Delta\xi$, $-\xi_0 - \Delta\xi < \xi < -\xi_0$ $(-l < x < l)$.

The case of a mass point moving in more than one dimension is more complicated because a geometrical representation of phase space is not available. In the case of motion in two dimensions, the phase space is four-dimensional; however, by neglecting the speed variable, whose error remains constant in time, we can represent the evolution in phase space by a three-dimensional picture. It is sufficient to plot the trajectories by recording the coordinates x and y of a mass point as abscissa and ordinate of a cartesian frame while recording the angle α between the velocity vector and the x-axis as the third coordinate.

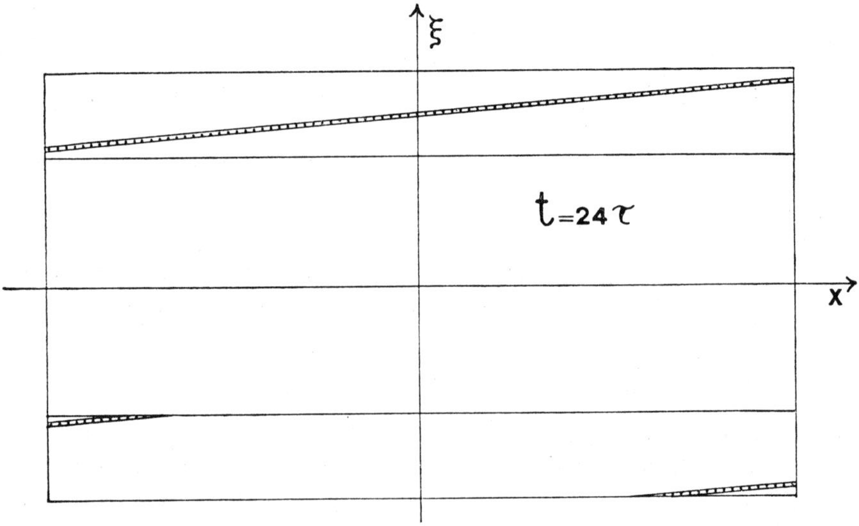

Fig. 7. A later stage in the time evolution, whose starting sequence was depicted in Fig. 6. Points with velocity $\xi_0 + \Delta\xi$ undergo the $(n + 1)$-th reflection while the points with velocity ξ_0 have not undergone the n-th reflection.

The equations giving the movement in the absence of walls are:

$$
\begin{aligned}
x &= x_0 + tR\cos\alpha \\
y &= y_0 + tR\sin\alpha \qquad (5.3)\\
\alpha &= \alpha_0
\end{aligned}
$$

where x_0, y_0, α_0 are the initial values of x, y, α and R is a constant related to the speed.

If the initial data are between x_0, y_0, α_0 and $x_0 + \Delta x$, $y_0 + \Delta y$, $\alpha_0 + \Delta\alpha$, the representative point, at time t, will be in a parallelepiped (if $\Delta\alpha \ll 1$) with two sides (of length Δx, Δy) parallel to the (x, y) plane, but very oblique because the bases move at different velocities. Again, after a sufficiently long time, the parallelepiped will be a very long thin ribbon and, if we introduce the effect of reflection against the walls of a rectangular box, the ribbon will be cut into a large number of strings. An exactly similar situation arises for motion in a three-dimensional box.

In the case of reflection from a curved surface, additional effects of scattering arise, particularly if the surface is convex and has a small radius of curvature. The latter case is particularly important, because it corresponds to the collision between two hard spheres; a given sphere, as noticed at the end of Section 4, may be replaced by a point colliding with a spherical

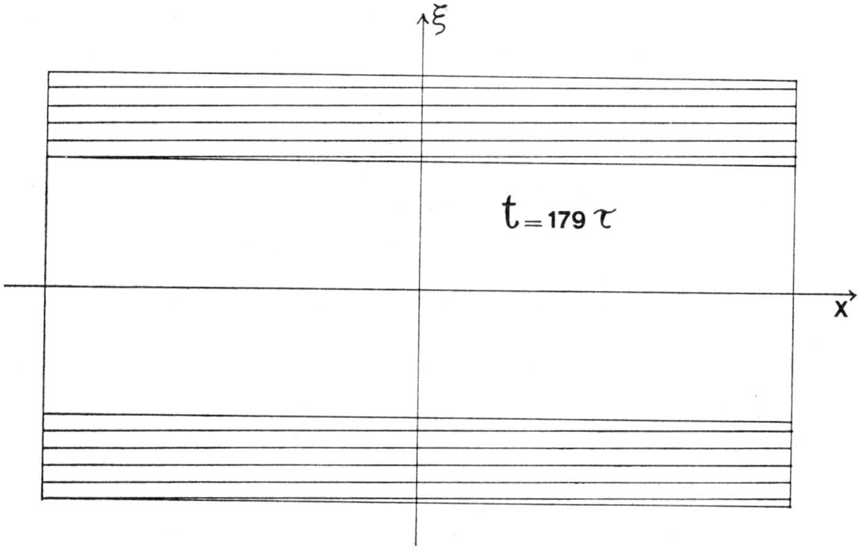

Fig. 8. Further evolution of the situation for the system considered in Figs. 6 and 7. The initial rectangle has become subdivided into a large number of "strings" which tend to be dense in the accessible part of phase space.

surface whose radius σ is equal to the diameter of the hard spheres (assumed to be identical). During the free path following a given reflection, the volume element (in physical space, not in phase space) becomes larger by a factor l^2/σ^2 (where l is the mean free path length, introduced at the end of Section 4) as a result of the divergence of two initially parallel straight lines impinging upon two different points and, hence, at different incidence angles (see Fig. 4 where the effect is proportional to l/σ because the motion is assumed to take place in a perfectly determined plane). As a consequence, m collisions produce a magnification of the volume of the region where the particle can be found, of the order of $(l/\sigma)^{2m}$. This volume is of course the projection from the 6-dimensional phase space of the particle onto physical space; hence the region occupied in phase space will be very "long" ($(l/\sigma)^m$ times the initial "length") and consequently very thin $(\sigma/l)^m$ times the initial "thickness"). If the motion takes place between solid walls, the region representing our system in phase space must become folded several times. If $l/\sigma > 10$ (which is a safe estimate for the ratio between the diameter and the mean free path of the molecules of a not-too-dense gas) the reduction factor in the thickness of the strings forming the phase extension of our system will be of the order of 10^{-1000} after one thousand collisions; that is, after one millionth of a second for hard spheres representing the molecules of a gas at standard pressure and temperature. The number of strings into which the phase extension of the possible positions of a given molecule is subdivided, is, as a consequence, of the order of 10^{1000}.

E. Borel, to whom the above considerations are due [5, 6], added a striking remark. Until now we have assumed that there was just the uncertainty affecting the initial data, but the differential equations of motion have been taken to be exactly known and solvable. This means not only that we know exactly the laws of interaction between any two particles, but also that we include in the differential equations of motion the forces exerted on the particles of our system by any other mass of the Universe, capable of modifying the motion in a significant way. To appreciate the meaning of the latter sentence, we note that displacing a mass of 1 gram by 1 centimeter on a star (say, Sirius) produces a change in the gravitational field on the earth, which is larger than 10^{-100} times the typical forces experienced in everyday life. It is therefore impossible, unless we want to consider the entire Universe as our system, to avoid fluctuations of the order of $10^{-100}\%$. But the foliated structure acquired by a region of phase space, made up of points which were very close to each other initially, is too fine to be preserved; after one millionth of a second the 10^{1000} sheets, whose theoretical thickness is about 10^{-1000} times the initial thickness, overlap and the image of an initially small region practically fills the whole accessible part of phase space and not only a region, to which an abstract computation would assign a volume equal to the initial one. In this way, the conservation of density in phase space disappears and we obtain a distribution rather less dense but much more extended than the initial one.

The same remark applies to the errors in computations; if we work with 10^{100} exact figures, errors of the order of 10^{-100} are introduced and the overlapping takes place.

Averaging over the initial data is not sufficient; we must average over the details of interaction in order to investigate the collective properties of our molecules which are stable with respect to small perturbations. This is the task of statistical mechanics. It should be noted, however, that our considerations apply to systems of rigid spheres and can probably be extended to systems of mass points interacting with strong repulsive forces of short range, but that they require careful consideration before they can be extended to systems of points interacting with attractive forces, which tend to introduce stability and, as a consequence, a more orderly behavior. Physically, this means that we are on safe ground as far as gases are concerned, while the treatment of liquids and solids requires completely different considerations.

6. Time averages, ergodic hypothesis and equilibrium states

If we accept the hypothesis of a gas composed of a great number of molecules (mass points or hard spheres), any physical quantity referring to the gas must be determined, at a given time instant, by the set of values of the coordinates and velocities of the molecules. The problem of computing these values leads to the methodological problems discussed in Sections 1 and 5.

In order to avoid, to a certain extent, these difficulties, we observe that the measurement of a physical quantity is not performed instantaneously, but requires a certain interval of time, which, no matter how small it may appear to us, would, as a rule, be very large with respect to the typical time scales involved in the evolution of the system. The latter will be subjected, during the time interval required for the measurement, to substantial changes resulting in wild variations of the quantity which we want to measure. Thus we will have to compare experimental data not with separate values of functions of the coordinates and velocities, but with the averages of such functions over relatively large intervals of time. This means that, rather than the probability density P introduced in Section 3, one should introduce its time average over a time interval τ:

$$\bar{P}(\mathbf{z}, t) = \frac{1}{\tau}\int_t^{t+\tau} P(\mathbf{z}, t_1)\, dt_1 \tag{6.1}$$

If τ is chosen to be constant for all times and phase space points, we have, by integrating the Liouville equation, Eq. (3.7), with respect to time and observing that differentiation will respect to $\mathbf{z}$ and averaging over τ are interchangeable operations:

$$\frac{1}{\tau}\,[P(\mathbf{z}, t + \tau) - P(\mathbf{z}, t)] + \mathbf{Z} \cdot \frac{\partial \bar{P}}{\partial \mathbf{z}} = 0 \tag{6.2}$$

provided $\mathbf{Z}$ is time-independent (i.e. forces are time-independent). But Eq. (6.1) gives

$$\frac{\partial \bar{P}}{\partial t} = \frac{P(\mathbf{z}, t + \tau) - P(\mathbf{z}, t)}{\tau} \tag{6.3}$$

and Eq. (6.2) shows that $\bar{P}(\mathbf{z}, t)$ satisfies the Liouville equation exactly as $P(\mathbf{z}, t)$. The only difference is that the initial datum will be

$$\bar{P}(\mathbf{z}, 0) = \frac{1}{\tau}\int_0^\tau P(\mathbf{z}, t_1)\, dt_1 \tag{6.4}$$

which is, in general, a smoother function than $P(\mathbf{z}, 0)$ (which is already smooth because it embodies the uncertainty of the initial coordinates and velocities). Hence time averaging reduces to a further averaging of the initial data. If τ is sufficiently large, the new averaging will tend to spread the probability throughout the region accessible to the system, as is clear from the examples of Section 5. In particular, if the system is in a macroscopic equilibrium; that is, any macroscopic measurement yields results independent of the time interval chosen in which to perform it, we can take τ as large as we wish, so that we may let $\tau \to \infty$.

In such a case $\partial\bar{P}/\partial t \to 0$ according to Eq. (6.3) provided $P(\mathbf{z}, 0)$ is bounded almost everywhere; in fact, the values taken by $P(\mathbf{z}, t + \tau)$ and $P(\mathbf{z}, t)$ are the same as those taken by $P(\mathbf{z}, 0)$, and hence bounded, due to the fact that the motion carries the initial values of $P(\mathbf{z}, 0)$ into different points, without changing these values, as we saw in Section 3.

Accordingly, in equilibrium and provided the limit $\tau \to \infty$ is taken in Eq. (6.1), $\bar{P}$ will be independent of time and will satisfy the steady Liouville equation:

$$\mathbf{Z} \cdot \frac{\partial \bar{P}}{\partial \mathbf{z}} = 0 \tag{6.5}$$

or

$$\sum_{i=1}^{N} \boldsymbol{\xi}_i \cdot \frac{\partial \bar{P}}{\partial \mathbf{x}_i} + \sum_{i=1}^{N} \mathbf{X}_i \cdot \frac{\partial \bar{P}}{\partial \boldsymbol{\xi}_i} = 0 \tag{6.6}$$

If we were able to determine the general solution $\bar{P} = \bar{P}(\mathbf{z}) = \bar{P}(\mathbf{x}_i, \boldsymbol{\xi}_i)$ of the steady Liouville equation, the problem of computing the equilibrium distribution would seem to be solved. It is not difficult to give a recipe for finding this general solution, but it is practically impossible to use it. To begin with, we observe that a function $\bar{P}(\mathbf{z})$ satisfies Eq. (6.4) if and only if it is a constant of motion for our system; that is, a function of the coordinates and velocities whose time derivative identically vanishes when $\mathbf{z}$ varies according to the equations of motion, Eq. (3.1).

In fact, if $\mathbf{z} = \mathbf{z}(t)$ satisfies Eq. (3.1), then

$$\frac{d\bar{P}}{dt} = \frac{\partial \bar{P}}{\partial \mathbf{z}} \cdot \frac{d\mathbf{z}}{dt} = \mathbf{Z} \cdot \frac{\partial \bar{P}}{\partial \mathbf{z}} \tag{6.7}$$

and $d\bar{P}/dt = 0$ implies Eq. (6.5), and *vice versa.*

If there are k functionally independent constants of motion, $I_1, I_2, I_3, \ldots, I_k$, then the general solution of Eq. (6.5) or (6.6) will be

$$\bar{P} = f(I_1, I_2, \ldots, I_k) \tag{6.8}$$

There is, however, a difficulty which is hard to overcome. Namely, that a system of N points without constraints involving $3N$ degrees of freedom has $6N - 1$ constants of motion. (In fact the $6N$ scalar relations corresponding to the vector relation $\mathbf{z} = \mathbf{z}(\mathbf{z}_0, t)$ can be solved to yield $\mathbf{z}_0 = \mathbf{z}_0(\mathbf{z}, t)$ and one of the components then used to eliminate t.) The computation of these $6N - 1$ functions of $\mathbf{z}$ is as hopeless a task as the computation of the time evolution of the system. Assume, however, that we know these functions and that for a given system in equilibrium the value taken by I_j is $I_j{}^0$; then we would have:

$$\bar{P} = A \prod_{j=1}^{6N-1} \delta(I_j - I_j^{(0)}) \tag{6.9}$$

where A is a normalizing factor to be chosen in such a way that $\int \bar{P}\, dz = 1$. Eq. (6.9) gives a probability density which is very close to a "certainty" density and could be used if only we could measure the $6N - 1$ values experimentally, a task involving the same difficulties already discussed in connection with the determination of initial data.

There is, however, a circumstance which alleviates all the difficulties. One of the constants of motion of the system is the total energy, which is a well known and physically meaningful quantity. For systems of points repelling each other with short range forces, all the other constants of motion turn out to be different from the total energy, in the sense that they are very peculiar functions, whose level surfaces in phase space pass as close as we wish to any point of a given surface of constant energy (linear and angular momentum can be ruled out as constants of motion by enclosing the system in a rigid box at rest, or taken into account on the same basis as energy).

To give an example, consider a mass point in a rectangular box with specularly reflecting walls (Fig. 9). The motion is given by Eq. (5.3) provided we understand that we introduce the effect of reflection from the walls. This can be done by the following formal device: we have from Eqs. (5.3), if

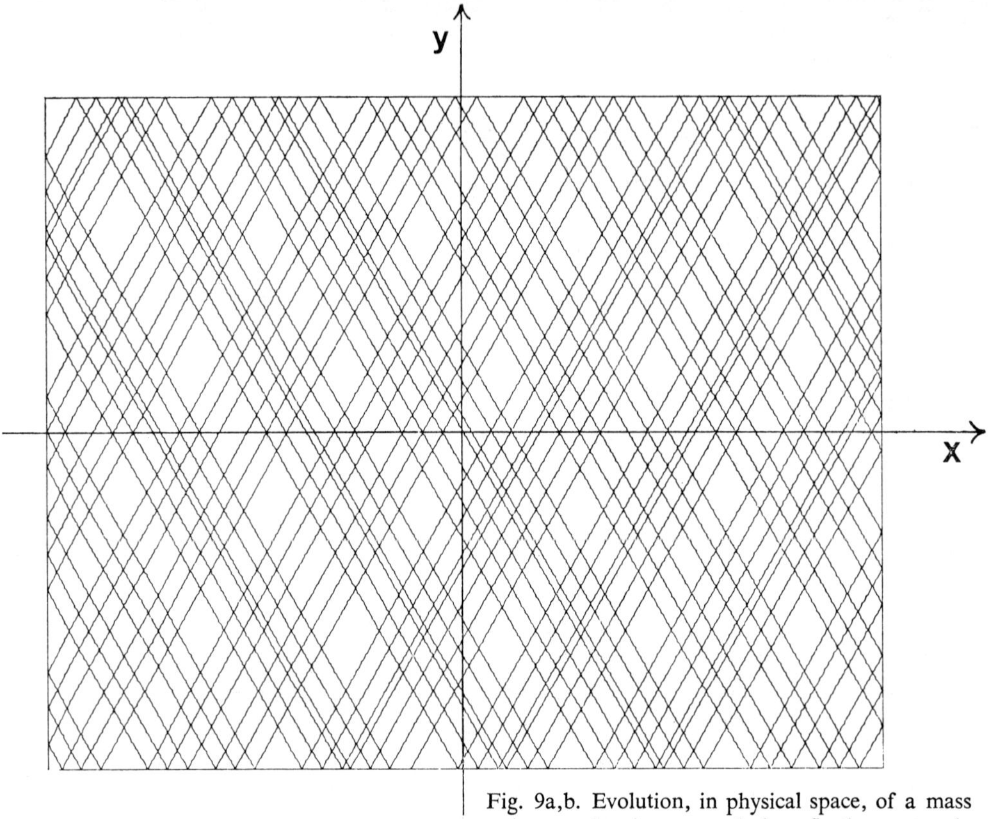

Fig. 9a,b. Evolution, in physical space, of a mass point moving in a specularly reflecting rectangle box. The two figures refer to the physical x,y-plane and the auxiliary u,v-plane.

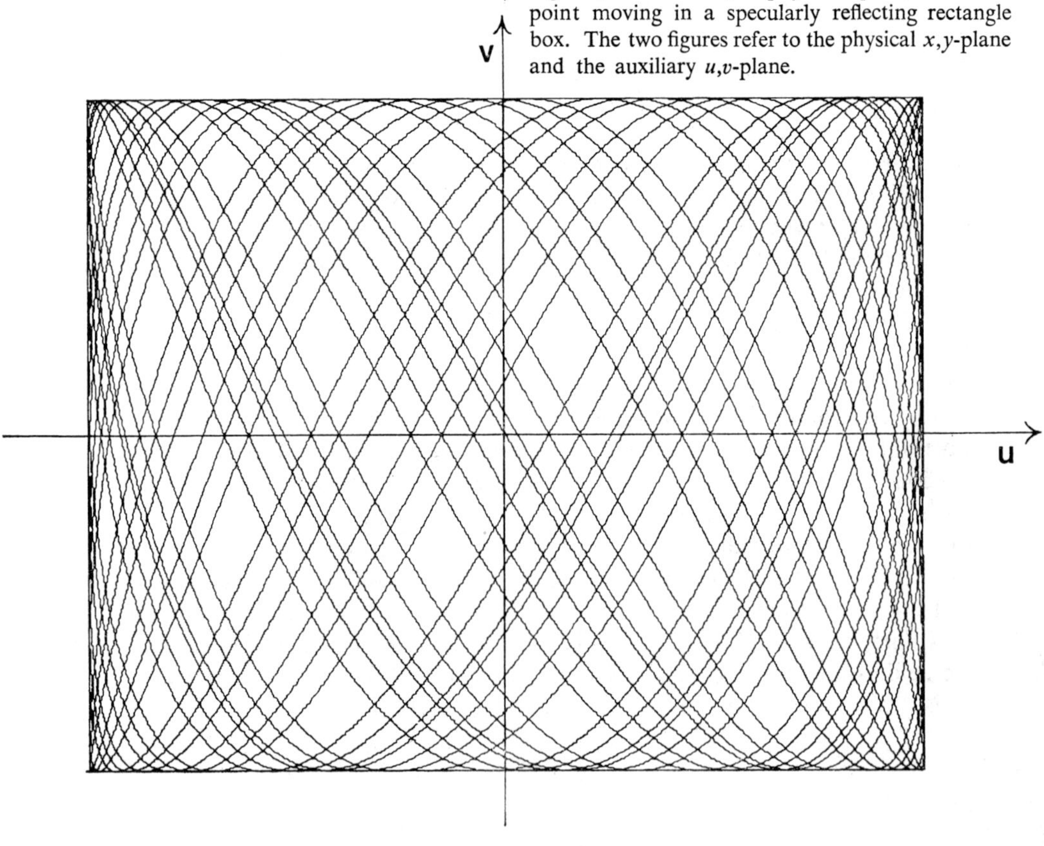

a, b are the sides of the rectangle:

$$\begin{aligned} u &\equiv \sin\left(\frac{\pi x}{2a}\right) = \sin\left(\frac{\pi}{2a}(x_0 + Rt\cos\alpha_0)\right) \\ v &\equiv \sin\left(\frac{\pi y}{2b}\right) = \sin\left(\frac{\pi}{2b}(y_0 + Rt\sin\alpha_0)\right) \end{aligned} \tag{6.10}$$

where $-a \leqslant x \leqslant a$, $-b \leqslant y \leqslant b$ and hence x and y may be replaced by the new coordinates u and v. In such a way the effect of reflection will be automatically taken into account, because when $x_0 + Rt\cos\alpha_0 > a$, the value of u will correspond to the point with abscissa $2a - (x_0 + Rt\cos\alpha_0)$ as required by the fact that the mass point has been reflected at $x = a$.

In this example, the constant of motion, besides the ones which have been already taken into account (i.e., total energy and $\alpha = \alpha_0$) is obtained by eliminating t between Eqs. (6.10). The projection of a level surface onto the physical plane is none other than the trajectory corresponding to the given initial data; it is clear from Fig. 9 that the trajectory will pass rather close to any point of the rectangle, unless the ratio $(b\cos\alpha_0)/(a\sin\alpha_0)$ is a rational number; if this ratio is an irrational number, the trajectory passes as close as we wish to any point of the rectangle. This is also clear in the u, v plane, where the trajectories are curved rather than piecewise straight; in this plane, Eqs. (6.10) represent the trajectories of a motion obtained by combining two orthogonal harmonic motions of frequencies $R/(4a)$ and $R/(4b)$, and hence the well-known Lissajous figures arise, which, as is well known, are dense in the square $-1 < u < 1$, $-1 < v < 1$, if the ratio between the frequencies is an irrational number.

It is clear that constants of motion of the same type as the one corresponding to the trajectory considered above do not preclude access to any region of phase space. If, in addition, they do not privilege any region, that is, if the corresponding trajectories pass the same number of times (in a very long time interval) through each region having a given extension, they do not restrict any physically measurable quantity and hence are to be neglected in the computation of the equilibrium probability density $\bar{P}(\mathbf{z})$.

In particular, if all the constants of motion except the total energy are such that they do not privilege any region of phase space (*ergodic hypothesis*), then the equilibrium distribution will depend only on the total energy E, and will take the form:

$$\bar{P}(\mathbf{z}) = A\,\delta(E - E_0) \tag{6.11}$$

if E_0 is the (experimentally measurable) value of the total energy.

The ergodic hypothesis which leads to Eq. (6.11) has recently been proved to be correct (and hence transformed into an ergodic *theorem*) by Sinai [7–8] for a system of rigid spheres in a specularly reflecting box at rest.

His proof is complicated and lengthy; accordingly, it is beyond the scope of the present book. We shall only attempt to make the result plausible by means of intuitive arguments. First, it is rather obvious that no region of physical space is privileged, if we exclude very special sets of initial data of zero measure (this is already true for our simple example of one point in a rectangle box); hence $\bar{P}$ will not depend upon any of the space coordinates $\mathbf{x}_i$, but only upon the N velocities $\boldsymbol{\xi}_i$. In each collision, a function of the velocities of the two particles involved remains constant, if, and only if, it depends only on the total momentum $(m_1\boldsymbol{\xi}_1' + m_2\boldsymbol{\xi}_2' = m_1\boldsymbol{\xi}_1 + m_2\boldsymbol{\xi}_2)$ and the total kinetic energy $(\frac{1}{2}m_1\xi_1'^2 + \frac{1}{2}m_2\xi_2'^2 = \frac{1}{2}m_1\xi_1^2 + \frac{1}{2}m_2\xi_2^2)$ of the pair of molecules. Since this is true for any pair i, j $(i, j = 1, \ldots, N)$ the functions which are conserved and depend on the initial data in a stable fashion must be functions of the total momentum $\sum_{j=1}^N m_j\boldsymbol{\xi}_j$ and the total kinetic energy $\frac{1}{2}\sum_{j=1}^N m_j\xi_j^2$ (as a matter of fact, all the constants of motion must not change no matter what particles happen to collide, and a collision between two given particles at a given instant is highly dependent upon the initial data and hence out of control if the latter data are known only with finite accuracy). Collisions with the walls, however, do not conserve total momentum and, therefore, $\bar{P}$ must depend upon the total kinetic energy alone. Hence, Eq. (6.11) follows, if we assume that E_0 can be measured with complete accuracy.

This kind of argument can be extended to molecules strongly repelling each other with short range forces, in such a way that the potential energy associated with these forces is negligible compared with kinetic energy, except in regions of negligible volume (which are occupied for a very short time). A rigorous proof, at present, is missing, but it seems reasonable to expect that Sinai's theorem can be extended to this case. In the case of a combination of both repulsive and attractive forces, the validity of the ergodic hypothesis does not seem obvious and it is not correct, in general, for systems with a finite number of degrees of freedom (whether the limit $N \to \infty$ restores ergodicity is an open problem). We shall consider only the case of mechanical systems describing the behavior of sufficiently rarefied gases and we can take the validity of the ergodic hypothesis for granted.

Accordingly, we shall take the probability density given by Eq. (6.11) to describe a gas in equilibrium. In order to emphasize the dependence upon the number of particles, we shall write P_N in place of $\bar{P}$ (the bar can be dispensed with, since no confusion arises) and A_N in place of A; hence, thanks to the assumption of negligible potential energy except in volumes of negligible extensions, we can write:

$$P_N = A_N\,\delta\left(\sum_{j=1}^{N} \mu_j\xi_j^2 - 2Ne\right) \tag{6.12}$$

where, if $\bar{m} = (\sum_{j=1}^N m_j)/N$ is the average mass, $\mu_j = m_j/\bar{m}$ and we have introduced the energy per unit mass

$$e = \frac{E_0}{\sum_{j=1}^{N} m_j} = \frac{E_0}{N\bar{m}}$$

which is a quantity endowed with a macroscopic significance and hence can be assumed to be known exactly. As was noted before, A_N is a constant to be computed in such a way as to satisfy the condition:

$$A_N \int \delta\left(\sum_{j=1}^{N} \mu_j \xi_j^2 - 2Ne\right) d\boldsymbol{\xi}_1 \cdots d\boldsymbol{\xi}_N \int d\mathbf{x}_1 \cdots d\mathbf{x}_N = 1 \tag{6.13}$$

Now, the integral over the space variables $\mathbf{x}_1, \ldots, \mathbf{x}_n$ gives a factor V for each of them, if V is the volume of the enclosing box, while the integral over velocities is better evaluated if we introduce $\mathbf{y}_j = \sqrt{\mu_j}\, \boldsymbol{\xi}_j$ in Eq. (6.13) and write:

$$A_N V^N \prod_{j=1}^{N} \mu_j^{-\frac{3}{2}} \int \delta\left(\prod_{j=1}^{N} \mathbf{y}_j^2 - 2Ne\right) d\mathbf{y}_1 \cdots d\mathbf{y}_N = 1 \tag{6.14}$$

If we pass to polar coordinates in the $3N$-dimensional space where the components of the $\mathbf{y}_j$ are Cartesian coordinates, we have:

$$\sum_{j=1}^{N} \mathbf{y}_j^2 = \rho^2 \qquad d\mathbf{y}_1 \cdots d\mathbf{y}_n = \rho^{3N-1}\, d\rho\, d^{3N}\boldsymbol{\Omega} \tag{6.15}$$

where ρ is the radial coordinate and $d^{3N}\boldsymbol{\Omega}$ the surface element of the unit sphere in $3N$ dimensions. Consequently, Eq. (6.14) becomes:

$$A_N V^N \prod_{j=1}^{N} \mu_j^{-\frac{3}{2}} \int \delta(\rho^2 - 2Ne)\rho^{3N-1}\, d\rho \int d^{3N}\boldsymbol{\Omega} = 1 \tag{6.16}$$

Denoting by ω_{3N} the surface of unit sphere in $3N$ dimensions (see Eq. (2.6) and Appendix) and using Eq. (2.9) in one dimension (with $z = \rho^2$, $dz = 2\rho\, d\rho$), we have:

$$A_N \frac{V^N}{2}\left(\prod_{j=1}^{N} \mu_j^{-\frac{3}{2}}\right)(2Ne)^{(3N-2)/2}\omega_{3N} = 1 \tag{6.17}$$

Therefore

$$P_N = \frac{2\prod_{j=1}^{N} \mu_j^{\frac{3}{2}}}{\omega_{3N}(2Ne)^{(3N-2)/2}V^N}\, \delta\left(\sum_{j=1}^{N} \mu_j \xi_j^2 - 2Ne\right) \tag{6.18}$$

This formula solves the problem of computing the probability density in phase space for a gas in statistical equilibrium. However, it does not look very illuminating in this form; we have to manipulate it in order to extract useful information. Suppose that we select a molecule at random and ask for the probability density $P_N^{(1)}$ that it has velocity between $\boldsymbol{\xi}$ and $\boldsymbol{\xi} + d\boldsymbol{\xi}$ and position between $\mathbf{x}$ and $\mathbf{x} + d\mathbf{x}$ without considering the other molecules; this means that we have to "sum" over all possible positions and velocities of all the molecules except one. We can assume that the latter is specified by $\mathbf{x}_1$, $\boldsymbol{\xi}_1$ since this can be always obtained by a proper labelling of the molecules. Then we must integrate over all the values of the coordinates and velocities of the particles labelled with 2, 3, . . . , N:

$$\begin{aligned} P_N^{(1)} &= \int P_N \, d\boldsymbol{\xi}_2 \cdots d\boldsymbol{\xi}_N \int d\mathbf{x}_2 \cdots d\mathbf{x}_N \\ &= A_N V^{N-1} \int \delta\left(\sum_{j=1}^{N} \mu_j \xi_j^2 - 2Ne\right) d\boldsymbol{\xi}_1 \cdots d\boldsymbol{\xi}_N \\ &= A_N V^{N-1} \int \delta\left(\sum_{j=2}^{N} \mu_j \xi_j^2 - (2Ne - \mu_1 \xi_1^2)\right) d\boldsymbol{\xi}_2 \cdots d\boldsymbol{\xi}_N \end{aligned} \tag{6.19}$$

The integral to be computed is exactly the same as before, except that it is over a $(3N - 3)$-dimensional space, rather than over a $3N$-dimensional one, and $2Ne$ is replaced by $2Ne - \mu_1 \xi_1^2$.

Hence we obtain in place of Eq. (6.17)

$$\begin{aligned} P_N^{(1)} &= A_N \frac{V^{N-1}}{2} \left(\prod_{j=2}^{N} \mu_j^{-\frac{3}{2}}\right) (2Ne - \mu_1 \xi_1^2)^{(3N-5)/2} \omega_{3N-3} \\ &\qquad\qquad (\text{for } \xi_1^2 < 2Ne/\mu_1) \\ P_N^{(2)} &= 0 \qquad (\text{for } \xi_1^2 > 2Ne/\mu_1) \end{aligned} \tag{6.20}$$

where the second relation is due to the fact that, if $\xi_1^2 > 2Ne/\mu_1$, the argument of the delta function in Eq. (6.19) is never zero and hence the integral vanishes.

The result expressed by Eq. (6.20) is that, although a random molecule is likely to be at any point of the region occupied by the gas ($P_N^{(1)}$ does not depend upon $\mathbf{x}_1$), the distribution of velocities is by no means uniform. If we open a little hole at any point of the boundary and let a very few molecules (few with respect to the total number) escape, and we measure their velocities and plot the distribution of the molecules according to their velocities, then, in the limit of an infinitely repeated sampling, we find that the probability density for the molecule velocity to lie in the interval ($\boldsymbol{\xi}_1$, $\boldsymbol{\xi}_1 + d\boldsymbol{\xi}_1$) does depend upon $\boldsymbol{\xi}_1$ and is given by Eq. (6.20).

Actually one has to be a little more careful, since if one extracts molecules from the gas, Eq. (6.20) cannot be applied, since it refers to a gas of exactly N molecules. If the sample is a small one, however, this should give no trouble; that is, we expect that if N is very large, Eq. (5.9) is not essentially altered if N is changed by a small fraction of its value. Mathematically, this means that the limit for $N \to \infty$ must exist; physically, that one can use the limit as the significant distribution function. (The latter statement is justified by the large number of molecules which are usually considered, the former by a direct computation of the limiting form of Eq. (6.20)). If we insert the expression of A_N, as given by Eq. (6.17), into Eq. (6.19), we obtain:

$$P_N^{(1)} = V^{-1}\mu_1^{\frac{3}{2}}\left(1 - \frac{\mu_1\xi_1^{\ 2}}{2Ne}\right)^{(3N-5)/2} \frac{1}{(2e)^{\frac{3}{2}}} \frac{\omega_{3N-3}}{N^{\frac{3}{2}}\omega_{3N}} \tag{6.21}$$

Now:

$$\begin{aligned}\lim_{N\to\infty}\left(1 - \frac{\xi_1^{\ 2}}{2Ne}\right)^{(3N-5)/2} &= \left\{\lim_{N\to\infty}\left[1 - \frac{(\mu_1\xi_1^{\ 2}/2e)}{N}\right]^N\right\}^{\frac{3}{2}} \cdot \lim_{N\to\infty}\left(1 - \frac{\mu_1\xi_1^{\ 2}}{2Ne}\right)^{-\frac{5}{2}} \\ &= \left[\exp\left(-\frac{\mu_1\xi_1^{\ 2}}{2e}\right)\right]^{\frac{3}{2}} = \exp\left(-\frac{3\mu_1\xi_1^{\ 2}}{4e}\right)\end{aligned} \tag{6.22}$$

and, as shown in the Appendix:

$$\lim_{N\to\infty}\frac{\omega_{3N-3}}{N^{\frac{3}{2}}\omega_{3N}} = \left(\frac{3}{2\pi}\right)^{\frac{3}{2}} \tag{6.23}$$

Hence, if we put

$$P^{(1)} = \lim_{N\to\infty} P_N^{(1)}, \tag{6.24}$$

we obtain:

$$P^{(1)} = \left(\frac{4\pi e}{3\mu_1}\right)^{-\frac{3}{2}} V^{-1}\exp\left(-\frac{3\mu_1\xi_1^{\ 2}}{4e}\right) \tag{6.25}$$

If we had considered another molecule with mass $m_2 = \mu_2\bar{m}$ we would have found a similar distribution except for the fact that μ_2 would have appeared in place of μ_1. The parameter e, the energy per unit mass, is a parameter which appears in all the velocity distributions, no matter what is the mass of the chosen molecule; this means that e is a quantity which takes the same value for each of several gases which are in contact with each other and are in thermal equilibrium. In other words, e must be a function of the temperature of the mixture and may depend on the average mass $\bar{m}$ but not on the masses of the single molecules. A connection with the macroscopic form of the state equation for a perfect gas, to be discussed later (Chapter II, Section 8) will show that

$$e = \tfrac{3}{2}RT \tag{6.26}$$

where R is the Boltzmann constant for the mixture of gases, and it is related to the average molecular mass $\bar{m}$ and Boltzmann's universal constant k ($k = 1.38 \times 10^{-23}\, J/°K$) by:

$$R = k/\bar{m} \tag{6.27}$$

(The fact that k is a universal constant follows from the circumstance that if we bring two gases with equal temperature T into contact, T remains the same, while the total energy $N\bar{m}e$ is the sum of the two previous ones, say $N_1\bar{m}_1e_1$ and $N_2\bar{m}_2e_2$ and $N\bar{m} = N_1\bar{m}_1 + N_2\bar{m}_2$, $N = N_1 + N_2$.)

If we insert Eq. (6.26) into Eq. (6.25), the latter reduces to Maxwell's formula for the velocity distribution, to be discussed in detail in the third chapter.

In addition to the *one particle distribution function* $P_N^{(1)}$ (or its limit $P^{(1)}$) one has occasion to consider the *two particle distribution function* $P_N^{(2)}$ defined as the probability density that two randomly chosen molecules have velocities between $\boldsymbol{\xi}_1$ and $\boldsymbol{\xi}_1 + d\boldsymbol{\xi}_1$ and $\boldsymbol{\xi}_2$ and $\boldsymbol{\xi}_2 + d\boldsymbol{\xi}_2$ respectively, and positions between $\mathbf{x}_1$ and $\mathbf{x}_1 + d\mathbf{x}_1$ and $\mathbf{x}_2$ and $\mathbf{x}_2 + d\mathbf{x}_2$ respectively, without regard to the $N - 2$ remaining molecules. This means that we have to "sum" P_N over all the possible positions and velocities of all the molecules except two. Then

$$\begin{aligned}
P_N^{(2)}(\mathbf{x}_1, \mathbf{x}_2; \boldsymbol{\xi}_1, \boldsymbol{\xi}_2) &= \int P_N \, d\boldsymbol{\xi}_3 \cdots d\boldsymbol{\xi}_N \, d\mathbf{x}_3 \cdots d\mathbf{x}_N \\
&= V^{N-2} A_N \prod_{j=3}^{N} \mu_j^{-\frac{3}{2}} \omega_{3N-6} \tfrac{1}{2} [2Ne - (\mu_1 \xi_1^2 + \mu_2 \xi_2^2)]^{(3N-8)/2} \\
&= \frac{\omega_{3N-6}}{N^3 \omega_{3N}} (\mu_1 \mu_2)^{\frac{3}{2}} V^{-2} (2e)^{-3} \left(1 - \frac{\mu_1 \xi_1^2 + \mu_2 \xi_2^2}{2Ne}\right)^{(3N-8)/2} \\
&\qquad\qquad (\text{for } \xi_1^2 + \xi_2^2 < 2Ne)
\end{aligned}$$

$$P_N^{(2)}(\mathbf{x}_1, \mathbf{x}_2; \xi_1, \boldsymbol{\xi}_2) = 0 \qquad (\text{for } \xi_1^2 + \xi_2^2 > 2Ne) \tag{6.28}$$

where calculations similar to the previous ones have been made. Now if we let $N \to \infty$ and use the limits (the first is analogous to Eq. (6.22), the second one can be found in the Appendix)

$$\lim_{N\to\infty} \left(1 - \frac{x}{N}\right)^{(3N-8)/2} = \exp(-\tfrac{3}{2}x) \tag{6.29}$$

$$\lim_{N\to\infty} \frac{\omega_{3N-6}}{N^3 \omega_{3N}} = \left(\frac{3}{2\pi}\right)^3 \tag{6.30}$$

we find

$$\begin{aligned}
P^{(2)} = \lim_{N\to\infty} P_N^{(2)} &= \left(\frac{4\pi e}{3\mu_1\mu_2}\right)^{-3} V^{-2} \exp\left[-\frac{3}{4e}(\mu_1\xi_1^2 + \mu_2\xi_2^2)\right] \\
&= P^{(1)}(\mathbf{x}_1, \boldsymbol{\xi}_1) P^{(2)}(\mathbf{x}_2, \boldsymbol{\xi}_2)
\end{aligned}$$

We have found, therefore, the remarkable result that in a gas containing an extremely large number of molecules, in an equilibrium state, two randomly selected molecules do not show any correlation; that is, the probability density of finding the first molecule in a given place with a given velocity (i.e. in a state which we may call state one) and simultaneously the second one at another given place with another given velocity (state two) is simply the product of the probability density of finding the first molecule in state one and the second molecule in any state, times the probability density of finding the second molecule in state two and the first one in any state. This means that the fact of finding one molecule in a given state has no influence upon the probability of finding the second one in another given state, and is expressed by saying that the states of the molecules (in the equilibrium case of a gas comprising molecules which are rigid spheres, with negligible total volume, or mass points, with negligible potential energy except in regions of negligible extension) are statistically uncorrelated. An elementary example of random events, which are also statistically uncorrelated, is given by the results of throwing a pair of dice; the fact that we obtain a certain number from one die does not influence the number which is shown by the other die, and the probability of getting a given combination (say, six from the first die and one from the second one) is $\frac{1}{36}$, i.e. the product of getting a given number from the first and any number from the second (which is $\frac{1}{6}$) times the product of getting a given number from the second and any number from the first (which is also $\frac{1}{6}$) (note that in this example we make a distinction between the two dice—as if, for example, one were blue and the other red—and consequently we consider a "one and six" to be an event different from a "six and one").

These results refer to the distribution functions of a dynamical system which can be assumed to represent a monatomic perfect gas in equilibrium.

By "monatomic" we mean that our molecules have no internal degrees of freedom. The state of each molecule is completely described by three space coordinates and three velocity components (as is the case for mass points and smooth hard spheres). By "perfect" (or "ideal", though the latter term is frequently used in the sense of "non-viscous") we mean that the potential energy of intermolecular forces is negligible unless we consider particles which are closer than some distance σ (the "molecular diameter"), which is usually negligibly small with respect to any other length of interest and such that the total volume corresponding to the regions of significant potential energy, of order $N\sigma^3$, is negligible compared with the volume V of the enclosing box. These two facts (negligible potential energy outside some regions and negligible extensions of these regions) do not mean negligible interactions; rather they mean that (at our level of description) the only important effects of the strong repulsive force which arises when two molecules are at a distance of the order of σ are the deviation of the molecules from their rectilinear paths and the change of their speeds (which are not negligible effects at all). These facts are reflected in the circumstance that Eq. (6.12) can be valid only in the regions where potential energy is negligible, while we integrated over all the volume V in Eqs. (6.13), (6.19), (6.28). This means that the results are correct in the limiting case when $\sigma \to 0$, $N \to \infty$ in such a way that $N\sigma^3/V \to 0$ (perfect gas limit).

Appendix

In this appendix we compute the surface and volume of a sphere in n-dimensional space as well as some limits connected with these quantities.

Let

$$z_1^2 + z_2^2 + \cdots + z_n^2 = R^2 \tag{A.1}$$

be the equation of a spherical surface in a n-dimensional space. The sphere has radius R and its center at the origin. The volume of this sphere is the volume of the region inside the surface defined by Eq. (A.1):

$$\begin{aligned}\tau_n(R) &= \iint\cdots\int_{z_1^2+z_2^2+\cdots+z_n^2\leq R} dz_1\, dz_2 \cdots dz_n \\ &= R^n \iint\cdots\int_{y_1^2+\cdots+y_n^2\leq 1} dy_1\, dy_2 \cdots dy_n\end{aligned} \tag{A.2}$$

where the second expression is obtained by means of the change of variables $z_k = Ry_k$ $(k = 1, \ldots, n)$. Hence, as was to be expected:

$$\tau_n(R) = \tau_n R^n \tag{A.3}$$

where $\tau_n \equiv \tau_n(1)$ is the volume of the unit sphere. The area of the surface of the sphere is best defined, by specializing a general definition due to Minkowski, as the limit of the ratio of the difference between the volumes of two neighboring spheres to their separation distance δ, when $\delta \to 0$. Hence:

$$\omega_n(R) = \lim_{\delta\to 0} \frac{\tau_n(R+\delta) - \tau_n(R)}{\delta} = \frac{d\tau_n}{dR} = n\tau_n R^{n-1} \tag{A.4}$$

In particular, if ω_n denotes the surface of the unit sphere:

$$\omega_n = n\tau_n \tag{A.5}$$

a result mentioned in Section 2.

In order to compute τ_n, and hence ω_n, by means of Eq. (A.5), we introduce polar coordinates in the (y_{n-1}, y_n) plane:

$$\begin{aligned} y_{n-1} &= \rho \cos \varphi \\ y_n &= \rho \sin \varphi \end{aligned} \tag{A.6}$$

Hence ($dy_{n-1}\,dy_n = \rho\,d\rho\,d\varphi$):

$$\tau_n = \underset{y_1^2+y_2^2+\cdots+y_{n-2}^2+\rho^2\le 1}{\iint\cdots\iiint} dy_1\,dy_2\cdots dy_{n-2}\rho\,d\rho\,d\varphi$$

$$= \underset{y_1^2+y_2^2+\cdots+y_{n-2}^2+\rho^2\le 1}{\iint\cdots\int} dy_1\,dy_2\cdots dy_{n-2}\int_0^{\sqrt{1-(y_1^2+\cdots+y_{n-2}^2)}}\rho\,d\rho\int_0^{2\pi}d\varphi$$

$$= \pi \underset{y_1^2+y_2^2+\cdots+y_{n-2}^2\le 1}{\iint\cdots\int} [1-(y_1^2+y_2^2+\cdots+y_{n-2}^2)]\,dy_1\,dy_2\cdots dy_{n-2} \tag{A.7}$$

If we introduce spherical coordinates in the $(n-2)$-dimensional space $(y_1, \ldots, y_{n-2})$ we have ($\rho^2 = y_1^2 + \cdots + y_{n-2}^2$, $d^{n-2}\boldsymbol{\Omega}$ = surface element of the unit sphere, $dy_1\cdots dy_{n-2} = \rho^{n-3}\,d\rho\,d^{n-2}\boldsymbol{\Omega}$):

$$\tau_n = \pi\int_0^1(1-\rho^2)\rho^{n-3}\,d\rho\int d^{n-2}\boldsymbol{\Omega}$$

$$= \pi\left(\frac{1}{n-2}-\frac{1}{n}\right)\omega_{n-2} = \frac{2\pi}{n(n-2)}\,\omega_{n-2} \tag{A.8}$$

and because of Eq. (A.5)

$$\omega_n = \frac{2\pi}{n-2}\,\omega_{n-2} \tag{A.9}$$

which is a simple recursion formula for ω_n. From Eq. (A.9) we obtain, by recursion for even n:

$$\omega_n = \frac{\pi}{\frac{n}{2}-1}\,\omega_{n-2} = \frac{\pi}{\frac{n}{2}-1}\,\frac{\pi}{\frac{n-2}{2}-1}\cdots\frac{\pi}{\frac{4}{2}-1}\,\omega_2$$
$$= \frac{\pi^{(n/2)-1}}{\left(\frac{n}{2}-1\right)!}\,2\pi = \frac{2\pi^{n/2}}{\left(\frac{n}{2}-1\right)!} \tag{A.10}$$

and for odd n

$$\omega_n = \frac{\pi}{\frac{n}{2}-1}\,\frac{\pi}{\frac{n-2}{2}-1}\cdots\frac{\pi}{\frac{5}{2}-1}\,\omega_3$$
$$= \frac{\pi^{(n-3)/3}}{\left(\frac{n}{2}-1\right)\left(\frac{n}{2}-2\right)\cdots\left(\frac{3}{2}\right)}\,4\pi = \frac{2\pi^{(n-1)/2}}{\left(\frac{n}{2}-1\right)\left(\frac{n}{2}-2\right)\cdots\frac{3}{2}\frac{1}{2}} \tag{A.11}$$

Eq. (2.6) of the main text follows from these results.

We now want to compute the following limit

$$\alpha = \lim_{n\to\infty} \frac{\omega_n\sqrt{n}}{\omega_{n-1}} \tag{A.12}$$

We shall assume that this limit exists, as could be easily shown by a more detailed consideration. If this is the case, then:

$$\lim_{n\to\infty} \frac{\omega_n(n-2)}{\omega_{n-2}} = \lim_{n\to\infty} \frac{\omega_n\sqrt{n}}{\omega_{n-1}} \lim_{n\to\infty} \frac{\omega_{n-1}\sqrt{n-1}}{\omega_{n-2}} \lim_{n\to\infty} \frac{n-2}{\sqrt{n(n-1)}}$$

$$= \left(\lim_{n\to\infty} \frac{\omega_n\sqrt{n}}{\omega_{n-1}}\right)^2 = \alpha^2 \tag{A.13}$$

Eq. (A.9) shows that the left hand side is trivially equal to 2π; hence

$$\alpha = \sqrt{2\pi} \tag{A.14}$$

As a simple consequence, we compute the following limit, which occurred in Section 6:

$$\lim_{N\to\infty} \frac{\omega_{3N-3}}{N^{\frac{3}{2}}\omega_{3N}}$$

$$= \lim_{N\to\infty}\left[\frac{\omega_{3N-3}}{\sqrt{3N-2}\,\omega_{3N-2}} \frac{\omega_{3N-2}}{\sqrt{3N-1}\,\omega_{3N-1}} \frac{\omega_{3N-1}}{\sqrt{3N}\,\omega_{3N}} \sqrt{\frac{3(3N-1)(3N-2)}{N^2}}\right]$$

$$= \frac{1}{\alpha}\frac{1}{\alpha}\frac{1}{\alpha} 3^{\frac{3}{2}} = \left(\frac{3}{2\pi}\right)^{\frac{3}{2}} \tag{A.15}$$

where Eq. (A.12) has been used three times, with $n = 3N - 2$, $n = 3N - 1$ and $n = 3N$.

Analogously:

$$\lim_{N\to\infty} \frac{\omega_{3N-6}}{N^3\omega_{3N}} = \lim_{N\to\infty} \frac{\omega_{3N-6}}{N^3 \dfrac{2\pi}{3N-2} \dfrac{2\pi}{3N-4} \dfrac{2\pi}{3N-6}\, \omega_{3N-6}}$$

$$= \frac{1}{(2\pi)^3} \lim_{N\to\infty} \frac{(3N-2)(3N-4)(3N-6)}{N^3}$$

$$= \left(\frac{3}{2\pi}\right)^3$$

where Eq. (A.9) has been used three times with $n = 3N$, $3N - 2$ and $3N - 4$.

A more compact treatment of the above points can be obtained by means of the Gamma function, whose use we have avoided.

References

[1] I. M. Gelfand and G. E. Shilov, "Generalized Functions", vol. I, Academic Press, New York (1964).

[2] J. M. Lighthill, "Fourier Analysis and Generalized Functions", Cambridge University Press, Cambridge (1958).

[3] L. Schwartz, "Théorie des Distributions", Hermann, Paris, Part I (1957), Part II (1959).

[4] C. Cercignani, "Mathematical Methods in Kinetic Theory", Plenum Press (1969).

[5] E. Borel, "Introduction géométrique à quelques théories physiques", Gauthier-Villars, Paris (1914).

[6] E. Borel, "Mecanique statistique classique", Gauthier-Villars, Paris (1925).

[7] Ya. Sinai, "Uspekhi Matematicheskikh Nauk" **25,** 141 (1970), Translated in "Russian Mathematical Surveys", **25,** 137 (1970).

[8] Ya. Sinai, "Westnik Moskovskogo Gosudarstvennogo Universitata", Math. Series N. 5 (1962).

II THE BOLTZMANN EQUATION

1. The problem of nonequilibrium states

In the previous chapter we saw that the problem of describing the state of thermal equilibrium of a monatomic perfect gas can be nicely solved; in particular, we found a very simple formula for the one-particle distribution $P^{(1)}$ in the form of a Maxwellian. This result has a large variety of applications in the statistical description of matter in the gaseous state.

It is clear, however, that the state of thermal equilibrium is a very special one, since in practice we frequently have to deal with nonequilibrium gases which can be, for example, surrounding a body in flight in the atmosphere or flowing in a pipe. The conditions of such gases are very different from those appropriate to gases enclosed in a box and kept at uniform temperature and pressure. Can we handle these more general problems of nonequilibrium gases by the kind of argument used in Chapter I, Section 6? The answer can be "no" or "yes", depending on the precise meaning attached to the question. It is certainly hopeless, for example, to try to evaluate the distribution function $P^{(1)}$ for a gas in a general nonequilibrium situation by means of simple arguments. However, we can do something: we can derive an equation satisfied by this quantity $P^{(1)}$ under the same assumptions as were used in Chapter I, Sect. 6, except, of course, for the condition of statistical equilibrium. This equation is called the *Boltzmann equation*. Then we can hope to attack the Boltzmann equation on purely mathematical grounds by means of more or less standard procedures of either an approximate or exact nature.

We must say from the beginning that the Boltzmann equation turns out to be particularly difficult to solve even for very simple nonequilibrium situations. We shall, however, have much more to say about this later, since the procedures for solving the Boltzmann equation will be treated in some detail in the subsequent chapters.

What is the motivation for setting up and solving the Boltzmann equation? We can distinguish two main kinds of application. The first one is concerned with deducing the macroscopic behaviour of gases from the microscopic model, when the mean free path (defined at the end of Section 4 of the previous chapter) is much smaller than other typical lengths of the problem. These applications are therefore a particular instance of the basic problem of statistical mechanics, which is to bridge the gap between the atomic structure of matter and its continuum-like behaviour at a macroscopic level. Typical results of these researches are the explanation of the macroscopic behaviour of gases, and computations of viscosity and heat conduction coefficients from postulated laws of interaction between pairs of molecules [1–3]. Besides their intrinsic importance, these researches are of interest because they constitute a model of what one should be able to do for other states of aggregation of matter (liquids, solids, many phase systems). The second kind of application is not connected with deducing a macroscopic theory in the ordinary sense, but, rather, with the study of the behaviour of gases, when the mean free path is no longer negligibly small compared to a characteristic dimension of the flow geometry. It is clear that in such conditions one cannot expect a "macroscopic behaviour" easily described in terms of such quantities as density, mass velocity, temperature, pressure, etc., although all these concepts retain a meaning and the final results are in terms of measurable and practical quantities, such as the drag on an object moving in a rarefied atmosphere. In these conditions, the use of the one-particle distribution function proves useful, and the Boltzmann equation becomes of paramount importance as an equation capable of encompassing the whole range of rarefactions and consequent behaviour from the fluid-like regime of a moderately dense gas to the free-molecular regime, where the molecular encounters are practically negligible.

Our first aim will be that of deriving the Boltzmann equation for a system of rigid spheres, when the system is not in statistical equilibrium. This means that we return to the beginning of Section 6 of the previous chapter, where we proved that the time average $\bar{P}$ of the probability density in phase space, P, satisfies the Liouville equation. The time interval τ will not be fixed, but should typically be of the order of magnitude of the time required by a molecule to move a distance of the order of its own diameter (or range of interaction, for mass points interacting at distance). This time average does not introduce any essentially new feature; accordingly, we shall write P in place of $\bar{P}$ in the following.

At a first sight, it seems impossible to describe the approach to equilibrium, because a volume in phase space will be conserved even after a time average; as a consequence, a uniform distribution on the energy surface cannot be reached for $t \to \infty$. We avoided this problem in Section 6 by taking the time interval, over which we averaged, to be infinity; thus equilibrium was essentially characterized by those uniformities, which are not valid at a given instant or during a short time interval, but reveal themselves in the average behavior during an extremely long time interval. If we want to describe the approach to equilibrium, or, more generally, states of non-equilibrium, it is not possible to let $\tau \to \infty$ (on the contrary, τ must be kept very short); hence, we must introduce some new features in our description. A first important remark is that a gas is made of a great number of *identical* molecules (or, in the case of a mixture, of a small number of sets of molecules, each set consisting of a great number of identical molecules) and macroscopic experiments are not capable of distinguishing between identical molecules (pressure on a wall, for instance, does not depend on the labels which we attach to the molecules hitting that wall). Hence the macroscopically measurable averages will be of the type:

$$\bar{\varphi} = \int \varphi(\mathbf{x}_1, \mathbf{x}_2, \ldots, \mathbf{x}_N; \boldsymbol{\xi}_1, \boldsymbol{\xi}_2, \ldots, \boldsymbol{\xi}_N) \times \\ \times P(\mathbf{x}_1, \mathbf{x}_2, \ldots, \mathbf{x}_N; \boldsymbol{\xi}_1, \boldsymbol{\xi}_2, \ldots, \boldsymbol{\xi}_N; t) \prod_{i=1}^{N} d\mathbf{x}_i \, d\boldsymbol{\xi}_i \quad (1.1)$$

where $\varphi(\mathbf{x}_1, \mathbf{x}_2, \ldots, \mathbf{x}_N; \boldsymbol{\xi}_1, \boldsymbol{\xi}_2, \ldots, \boldsymbol{\xi}_N)$ can be assumed to remain unchanged when two molecules are exchanged at will (e.g. $\mathbf{x}_1 \to \mathbf{x}_2$, $\mathbf{x}_2 \to \mathbf{x}_1$, $\boldsymbol{\xi}_1 \to \boldsymbol{\xi}_2$, $\boldsymbol{\xi}_2 \to \boldsymbol{\xi}_1$) at least in the case of a simple gas (the case of a mixture requires obvious modifications).

Then the integral appearing in Eq. (1.1) can be written as follows:

$$\bar{\varphi} = \int \varphi(\mathbf{x}_1, \mathbf{x}_2, \ldots, \mathbf{x}_N; \boldsymbol{\xi}_1, \boldsymbol{\xi}_2, \ldots, \boldsymbol{\xi}_N) \times \\ \times S[P(\mathbf{x}_1, \mathbf{x}_2, \ldots, \mathbf{x}_N; \boldsymbol{\xi}_1, \boldsymbol{\xi}_2, \ldots, \boldsymbol{\xi}_N; t] \prod_{i=1}^{N} d\mathbf{x}_i \, d\boldsymbol{\xi}_i \quad (1.2)$$

where $S[P]$ denotes the function obtained from P by complete symmetrization with respect to the N molecules. That is, the sum of $N!$ P's with arguments corresponding to the $N!$ permutations of the N molecules, divided by $N!$.

This means that for the purposes of computing measurable quantities, we may replace P by $S[P]$. It is clear that it would be very convenient to work with $S[P]$ from the beginning, if possible. It is very simple to modify our description in such a way as to make it possible to replace P by $S[P]$. In fact, it is sufficient to observe that $S[P]$ satisfies the Liouville equation, because the latter is symmetrical with respect to identical molecules (identity includes, of course, identical interaction laws between the molecules). Hence the only change required in order to use $S[P]$ in place of P will be in the initial data again, the initial datum for $S[P]$ being of course the symmetrized expression $S[P_0]$ of the initial datum P_0 for P.

According to this discussion, we shall use $S[P]$ in the following, but we shall denote it by P_N in order to emphasize the number of molecules. The only difference with respect to the previously used probability density is that P_N must always be assumed to be symmetrical in the arguments referring to identical molecules. This condition implies a further scattering of the volume elements in phase space, but still does not allow us to understand the possibility of an approach to equilibrium.

We can, however, take advantage of two facts: first, N can be allowed to tend to infinity, as was done in the case of equilibrium, and, second, it is sufficient to compute s-particle distribution functions:

$$P_N^{(s)} = \int P_N \prod_{i=s+1}^{N} d\mathbf{x}_i \, d\boldsymbol{\xi}_i \tag{1.3}$$

with $s \ll N$ (usually $s = 1, 2$, as in Section 6 of Chapter I). This second circumstance is related to the fact that the interesting averages $\bar{\varphi}$ refer to functions which are symmetrical sums of terms, each of which contains the coordinates and velocities of one or two molecules. When $N \to \infty$ the volume in phase space loses its meaning (note that the volume of a cube of side a, a^n, tends to 0, 1, ∞ when $n \to \infty$ according to $a < 1$, $a = 1$, $a > 1$; something which contradicts the possibility of an arbitrary choice of the unit length!). In particular, the "conservation of volume" in an infinite-dimensional phase space does not impede a filling of subspaces of arbitrary (finite) dimension.

Accordingly, the limit $N \to \infty$ gives us the possibility of describing the approach to equilibrium. However, if we consider a finite volume V and let the number of molecules $N \to \infty$, the proper volume ($\simeq N\sigma^3$) of the molecules will tend to infinity, which is absurd ($N\sigma^3 < V$), unless we let $\sigma \to 0$. Hence, statistical mechanics essentially refers to the limit $N \to \infty$, $\sigma \to 0$. Still, a variety of possibilities is open. We may consider, for instance, the case of a gas for which in the limit $N\sigma^3 \to b > 0$, or the case of a gas for which $N\sigma^3 \to 0$; since $N\sigma^3$ is, apart from trivial numerical factors, the proper volume of molecules, $N\sigma^3 \to 0$ means that the molecules occupy a negligible fraction of the available volume. In this case we say that the gas is perfect (or ideal).

The case of a perfect gas, in turn, offers two possibilities: either $N\sigma^2$ remains finite or $N\sigma^2 \to 0$. In the first case the mean free path $l(\simeq V/(\pi N\sigma^2)$ according to Eq. (I.4.17)) is finite; that is to say the collisions are significant (Boltzmann gas); in the second case the mean free path is infinite, which corresponds to the collisions being negligible (Knudsen gas). The most interesting case is, of course, that of a finite mean free path (the case of a Knudsen gas can be obtained as a limit of the Boltzmann gas by letting $l \to \infty$). In order to give an idea of the actual orders of magnitude, we note that, in standard conditions, for $V \simeq 1$ cm^3, $N \simeq 10^{20}$, while $\sigma \simeq 10^{-8}$ cm: hence $N\sigma^3 = 10^{-4}$ cm^3, $N\sigma^2 = 10^4$ cm^2 (for lower densities, both $N\sigma^3$ and $N\sigma^2$ decrease and when $l \simeq 1$ cm $N\sigma^2 \simeq 1$ cm^2, $N\sigma^3 \simeq 10^{-8}$ cm^3 in a volume $V \simeq 1$ cm^3).

For consistency, the mass m of a molecule is to be assumed to tend to zero while the total mass remains finite

$$m \to 0 \qquad Nm \to M < \infty \tag{1.4}$$

In this limit, the Liouville equation is formally equivalent to a system of infinitely many equations where the unknowns are the s-particle distribution functions ($1 \leqslant s \leqslant \infty$ since $N \to \infty$) as we shall see later. This system, in turn, will be shown to possess a particular solution for which the s-particle distribution function is factored into the product of s factors, each equal to the one particle distribution function and which satisfy the Boltzmann equation.

2. Equations for the many particle distribution functions for a gas of rigid spheres

In this section we derive the equations satisfied by the s-particle distribution functions ($s = 1, \ldots, N$) for a gas of N rigid spheres contained within a region R of volume V. We have for the probability density $P_N = P_N(\mathbf{x}_i, \boldsymbol{\xi}_i, t)$:

$$P_N = 0 \qquad (|\mathbf{x}_i - \mathbf{x}_j| < \sigma, i \neq j) \tag{2.1}$$

since the spheres cannot penetrate into each other; accordingly the P_N, in general, will be discontinuous at the points of phase space where $|\mathbf{x}_i - \mathbf{x}_j| = \sigma$, the limit from one side being zero and from the other side ($|\mathbf{x}_i - \mathbf{x}_j| > \sigma$), generally speaking, different from zero. Accordingly, when we consider P_N for some $|\mathbf{x}_j - \mathbf{x}_i| = \sigma$, we shall always understand the limiting value from "outside"; that is, from the region $|\mathbf{x}_j - \mathbf{x}_i| > \sigma$. In the latter region the state of molecules corresponds to inertial motion; hence the Liouville equation, Eq. (I.3.8) reduces to:

$$\frac{\partial P_N}{\partial t} + \sum_{i=1}^{N} \boldsymbol{\xi}_i \cdot \frac{\partial P_N}{\partial \mathbf{x}_i} = 0 \qquad (|\mathbf{x}_i - \mathbf{x}_j| > \sigma, i \neq j) \tag{2.2}$$

where according to the discussion in Section 1, P_N is symmetric with respect to an exchange of the N molecules.

Let us integrate Eq. (2.2) over its domain of validity, with respect to the coordinates and velocities of $N - s$ particles; without loss of generality, we shall integrate with respect to the particles numbered from $s + 1$ to N. If we introduce the s-particle distribution function P_N defined by Eq. (1.3), Eq. (2.2) gives:

$$\frac{\partial P_N^{(s)}}{\partial t} + \sum_{i=1}^{s} \int \boldsymbol{\xi}_i \cdot \frac{\partial P_N}{\partial \mathbf{x}_i} \prod_{l=s+1}^{N} d\mathbf{x}_l \, d\boldsymbol{\xi}_l + \sum_{j=s+1}^{N} \int \boldsymbol{\xi}_j \cdot \frac{\partial P_N}{\partial \mathbf{x}_j} \prod_{l=s+1}^{N} d\mathbf{x}_l \, d\boldsymbol{\xi}_l = 0 \tag{2.3}$$

where integration extends to the whole space with respect to $\boldsymbol{\xi}_l$ ($l = s + 1, \ldots, N$) and to the region R deprived of the sets $|\mathbf{x}_i - \mathbf{x}_l| < \sigma$ ($i = 1, \ldots, N$, $i \neq l$) with respect to $\mathbf{x}_l$ ($l = s + 1, \ldots, N$). Terms with $1 \leqslant i \leqslant s$ have been separated from those with $s + 1 \leqslant i \leqslant N$ for later convenience.

A typical term in the first sum in Eq. (2.3) contains the integral of a derivative with respect to a variable, $\mathbf{x}_i$, over which one does not integrate; it is not possible, however, to exchange the orders of integration and differentiation because the domain has boundaries ($|\mathbf{x}_i - \mathbf{x}_l| = \sigma$) depending upon $\mathbf{x}_i$. To obtain the correct result, a boundary term has to be added:

$$\int \boldsymbol{\xi}_i \cdot \frac{\partial P_N}{\partial \mathbf{x}_i} \prod_{l=s+1}^{N} d\mathbf{x}_l \, d\boldsymbol{\xi}_l$$

$$= \boldsymbol{\xi}_i \cdot \frac{\partial}{\partial \mathbf{x}_i} \int P_N \prod_{l=s+1}^{N} d\mathbf{x}_l \, d\boldsymbol{\xi}_l - \sum_{j=s+1}^{N} \int P_N \boldsymbol{\xi}_i \cdot \mathbf{n}_{ij} \, d\sigma_{ij} \, d\boldsymbol{\xi}_j \prod_{\substack{l=s+1 \\ l \neq j}}^{N} d\mathbf{x}_l \, d\boldsymbol{\xi}_l$$

$$= \boldsymbol{\xi}_i \cdot \frac{\partial P_N^{(s)}}{\partial \mathbf{x}_i} - \sum_{j=1}^{N} \int P_N^{(s+1)} \boldsymbol{\xi}_j \cdot \mathbf{n}_{ij} \, d\sigma_{ij} \, d\boldsymbol{\xi}_j \tag{2.4}$$

where $\mathbf{n}_{ij}$ is the outer normal to the sphere $|\mathbf{x}_i - \mathbf{x}_j| = \sigma$ (with its center at $\mathbf{x}_j$), $d\sigma_{ij}$ the surface element on the same sphere and $P_N^{(s+1)}$ is the $(s + 1)$-particle distribution function with arguments $(\mathbf{x}_k, \boldsymbol{\xi}_k)$ $(k = 1, 2, \ldots, s, j)$.

A typical term in the second sum in Eq. (2.3) can be immediately integrated by means of the Gauss theorem, since it involves the integration of a derivative taken with respect to one of the variables of integration. We find:

$$\int \boldsymbol{\xi}_j \cdot \frac{\partial P_N}{\partial \mathbf{x}_j} \prod_{l=s+1}^{N} d\mathbf{x}_l \, d\boldsymbol{\xi}_l$$

$$= \sum_{i=1}^{s} \int P_N^{(s+1)} \boldsymbol{\xi}_j \cdot \mathbf{n}_{ij} \, d\sigma_{ij} \, d\boldsymbol{\xi}_j + \sum_{\substack{k=s+1 \\ i \neq j}}^{N} \int P_N^{(s+2)} \boldsymbol{\xi}_j \cdot \mathbf{n}_{kj} \, d\sigma_{ij} \, d\boldsymbol{\xi}_j \, d\mathbf{x}_k \, d\boldsymbol{\xi}_k +$$

$$+ \int P_N^{(s+1)} \boldsymbol{\xi}_j \cdot \mathbf{n}_j \, dA_j \, d\boldsymbol{\xi}_j \tag{2.5}$$

where dA_j is the surface element of the boundary of the region R in the three-dimensional subspace described by $\mathbf{x}_j$, and $\mathbf{n}_j$ is the unit vector normal to such a surface element and pointing into the gas. The last term in Eq. (2.5) is the contribution from the solid boundary of R; if the molecules are specularly reflected there, then the term under consideration is obviously zero because $\boldsymbol{\xi}_j \cdot \mathbf{n}_{ij}$ changes its sign under specular reflection. The boundary term, however, is zero under more general assumptions; it is sufficient to assume that the effect of an interaction of a rigid sphere with the wall is independent of the evolution of the state of the other spheres and that no particles are captured by the solid walls. We shall discuss this point at the end of this section and neglect the boundary term for the moment.

Inserting Eqs. (2.4) and (2.5) into Eq. (2.3), we find:

$$\frac{\partial P_N^{(s)}}{\partial t} + \sum_{i=1}^{s} \boldsymbol{\xi}_i \cdot \frac{\partial P_N^{(s)}}{\partial \mathbf{x}_i} = \sum_{i=1}^{s} \sum_{j=s+1}^{N} \int P_N^{(s+1)} \mathbf{V}_{ij} \cdot \mathbf{n}_{ij} \, d\sigma_{ij} \, d\boldsymbol{\xi}_j -$$

$$- \frac{1}{2} \sum_{\substack{k,j=s+1 \\ i \neq j}}^{N} \int P_N^{(s+2)} \mathbf{V}_{kj} \cdot \mathbf{n}_{kj} \, d\sigma_{kj} \, d\boldsymbol{\xi}_j \, d\mathbf{x}_k \, d\boldsymbol{\xi}_k \quad (2.6)$$

where $\mathbf{V}_{ij} = \boldsymbol{\xi}_i - \boldsymbol{\xi}_j$ is the relative velocity of the i-th particle with respect to the j-th and we have taken into account that $\boldsymbol{\xi}_i \cdot \mathbf{n}_{ij}$ can be replaced by $(\frac{1}{2})\mathbf{V}_{ij} \cdot \mathbf{n}_{ij}$ in the second sum because of the antisymmetry of $\mathbf{n}_{ij}$ with respect to its own indexes. $\mathbf{x}_j$ and $\boldsymbol{\xi}_j$ are integration variables in Eq. (2.6); hence the sums over j are made up of identical terms, as well as the sum over k in the second integral. We shall write $\mathbf{x}_*$, $\boldsymbol{\xi}_*$ in place of $\mathbf{x}_j$, $\boldsymbol{\xi}_j$ in order to emphasize that the index j is a dummy, and $\mathbf{x}_0$, $\boldsymbol{\xi}_0$ in place of $\mathbf{x}_k$, $\boldsymbol{\xi}_k$, while we shall simply write $\mathbf{V}_i$, $\mathbf{n}_i$, $d\sigma_i$ for $\mathbf{V}_{ij}$, $\mathbf{n}_{ij}$, $d\sigma_{ij}$ and $\mathbf{V}_0$, $\mathbf{n}_0$, $d\sigma_0$ for $\mathbf{V}_{kj}$, $\mathbf{n}_{kj}$, $d\sigma_{kj}$. Accordingly we obtain:

$$\frac{\partial P_N^{(s)}}{\partial t} + \sum_{i=1}^{s} \boldsymbol{\xi}_i \cdot \frac{\partial P_N^{(s)}}{\partial \mathbf{x}_i} = (N-s) \sum_{i=1}^{s} \int P_N^{(s+1)} \mathbf{V}_i \cdot \mathbf{n}_i \, d\sigma_i \, d\boldsymbol{\xi}_* -$$

$$- \frac{(N-s)(N-s-1)}{2} \int P_N^{(s+2)} \mathbf{V}_0 \cdot \mathbf{n}_0 \, d\sigma_0 \, d\boldsymbol{\xi}_* \, d\boldsymbol{\xi}_0 \, d\mathbf{x}_0 \quad (2.7)$$

where the arguments of $P_N^{(s+1)}$ are $(\mathbf{x}_1, \boldsymbol{\xi}_1, \ldots, \mathbf{x}_s, \boldsymbol{\xi}_s, \mathbf{x}_*, \boldsymbol{\xi}_*, t)$ and those of $P_N^{(s+2)}$ are $(\mathbf{x}_1, \boldsymbol{\xi}_1, \ldots, \mathbf{x}_s, \boldsymbol{\xi}_s, \mathbf{x}_*, \boldsymbol{\xi}_*, \mathbf{x}_0, \boldsymbol{\xi}_0, t)$.

We observe now that multiple collisions (i.e. simultaneous contacts of more than two spheres) contribute nothing to the above integrals (at least, if $P_N^{(s+1)}$, $P_N^{(s+2)}$ are ordinary integrable functions). In fact, the integrals with respect to $d\sigma_i$, $d\sigma_0$ are extended over the surface

$$|\mathbf{x}_* - \mathbf{x}_i| = \sigma \qquad (i = 0, 1, \ldots, s; s \leqslant N) \quad (2.8)$$

with center at $\mathbf{x}_i$, but not over the whole surface, because we must cut out those parts which are inside the other similar surfaces

$$|\mathbf{x}_* - \mathbf{x}_j| = \sigma \qquad (j = 0, 1, \ldots, N; j \neq i) \quad (2.9)$$

(This is due to the fact that $P_N = 0$ inside such surfaces). Multiple collisions correspond to the boundary of the integration region, and hence to a one-dimensional subset; accordingly, their contribution to the integrals is zero, unless singularities occur, which we have excluded by using a smoothed P_N.

It is important, now, to separate each of the integrals appearing in Eq. (2.7) into the corresponding integrals extended to the subsets $\mathbf{V}_i \cdot \mathbf{n}_i < 0$ and $\mathbf{V}_i \cdot \mathbf{n}_i > 0$ $(i = 0, 1, \ldots, s)$, respectively. The first set corresponds to molecules entering a collision, while the second set corresponds to molecules which have just collided (remember that $P_N^{(s+1)}$ is the limiting value from outside the sphere $|\mathbf{x}_* - \mathbf{x}_i| = \sigma$ with its center at $\mathbf{x}_i$, so that the value of $P_N^{(s+1)}$ for $|\mathbf{x}_i - \mathbf{x}_*| = \sigma$ and $\mathbf{V}_i \cdot \mathbf{n}_j > 0$ corresponds to the limit from the state just after collision, since $\mathbf{n}_i = (\mathbf{x}_i - \mathbf{x}_*)/\sigma$). Accordingly, Eq. (2.7) becomes

$$\frac{\partial P_N^{(s)}}{\partial t} + \sum_{i=1}^{s} \boldsymbol{\xi}_i \cdot \frac{\partial P_N^{(s)}}{\partial \mathbf{x}_i}$$
$$= (N-s)\sum_{i=1}^{s}\left[\int^{(+)} P_N^{(s+1)} |\mathbf{V}_i \cdot \mathbf{n}_i|\, d\sigma_i\, d\boldsymbol{\xi}_* - \int^{(-)} P_N^{(s+1)} |\mathbf{V}_i \cdot \mathbf{n}_i|\, d\sigma_i\, d\boldsymbol{\xi}_*\right] -$$
$$- \frac{(N-s)(N-s-1)}{2}\left[\int^{(+)} P_N^{(s+2)} |\mathbf{V}_0 \cdot \mathbf{n}_0|\, d\sigma_0\, d\boldsymbol{\xi}_*\, d\boldsymbol{\xi}_0\, d\mathbf{x}_0 -\right.$$
$$\left. - \int^{(-)} P_N^{(s+2)} |\mathbf{V}_0 \cdot \mathbf{n}_0|\, d\sigma_0\, d\boldsymbol{\xi}_*\, d\boldsymbol{\xi}_0\, d\mathbf{x}_0\right] \qquad (s = 1, \ldots, N) \tag{2.10}$$

where the $(+)$ and $(-)$ superscripts in the integrals correspond to $\mathbf{V}_i \cdot \mathbf{n}_i \gtrless 0$ $(i = 0, 1, \ldots, s)$ respectively.

The equations used so far are incomplete, because we have not used the laws of elastic impact, Eqs. (I.4.3) or (I.4.11).

According to these laws, the velocities after collision, $\boldsymbol{\xi}_i$ and $\boldsymbol{\xi}_*$, are related to the ones before collision, $\boldsymbol{\xi}_i'$ and $\boldsymbol{\xi}_*'$, by

$$\boldsymbol{\xi}_i = \boldsymbol{\xi}_i' - \mathbf{n}_i(\mathbf{n}_i \cdot \mathbf{V}_i')$$
$$\boldsymbol{\xi}_* = \boldsymbol{\xi}_*' + \mathbf{n}_i(\mathbf{n}_i \cdot \mathbf{V}_i') \qquad (\mathbf{V}_i' = \boldsymbol{\xi}_i' - \boldsymbol{\xi}_*') \tag{2.11}$$

or by

$$\boldsymbol{\xi}_i' = \boldsymbol{\xi}_i - \mathbf{n}_i(\mathbf{n}_i \cdot \mathbf{V}_i)$$
$$\boldsymbol{\xi}_*' = \boldsymbol{\xi}_* + \mathbf{n}_i(\mathbf{n}_i \cdot V_i) \qquad (\mathbf{V}_i = \boldsymbol{\xi}_i - \boldsymbol{\xi}_*) \tag{2.12}$$

where $\mathbf{n}_i = (\mathbf{x}_i - \mathbf{x}_*)/\sigma$, the unit vector directed along the line joining the centers of the two spheres, coincides with the outer normal to the sphere $|\mathbf{x}_i - \mathbf{x}_*| = \sigma$ with center at $\mathbf{x}_*$ and variable $\mathbf{x}_i$ or the inner normal to the sphere $|\mathbf{x}_* - \mathbf{x}_i| = 0$ with center at $\mathbf{x}_i$ and variable $\mathbf{x}_*$. $\mathbf{V}_i'$ and $\mathbf{V}_i$ are, of course, the relative velocities before and after collision, and are related to each other by Eq. (I.4.12):

$$\mathbf{V}_i = \mathbf{V}_i' - 2\mathbf{n}_i(\mathbf{n}_i \cdot \mathbf{V}_i') \tag{2.13}$$

Accordingly, any molecule entering a collision with velocity $\boldsymbol{\xi}_i'$ at $\mathbf{x}_i$ is at the same time (or a vanishingly short time later) in an after-collision state with velocity $\boldsymbol{\xi}_i$ related to $\boldsymbol{\xi}_i'$ and $\mathbf{n}_i = (\mathbf{x}_i - \mathbf{x}_*)/\sigma$ by Eq. (2 .11); accordingly:

$$P_N^{(s+1)}(\mathbf{x}_1, \boldsymbol{\xi}_1, \ldots, \mathbf{x}_i, \boldsymbol{\xi}_i, \ldots, \mathbf{x}_s, \boldsymbol{\xi}_s, \mathbf{x}_*, \boldsymbol{\xi}_*, t)$$
$$= P_N^{(s+1)}(\mathbf{x}_1, \boldsymbol{\xi}_1, \ldots, \mathbf{x}_i, \boldsymbol{\xi}_i - \mathbf{n}_i(\mathbf{n}_i \cdot \mathbf{V}_i), \ldots, \mathbf{x}_s, \boldsymbol{\xi}_s, \mathbf{x}_*, \boldsymbol{\xi}_* + \mathbf{n}_i(\mathbf{n}_i \cdot \mathbf{V}_i), t)$$
$$(i = 0, 1, \ldots, s; 1 \leqslant s \leqslant N - 1) \quad (2.14)$$

We first examine the term involving $P_N^{(s+2)}$ in Eq. (2.11) and claim that it is zero; that is that:

$$\int^{(+)} P_N^{(s+2)} |\mathbf{V}_0 \cdot \mathbf{n}_0| \, d\sigma_0 \, d\boldsymbol{\xi}_* \, d\boldsymbol{\xi}_0 \, d\mathbf{x}_0 = \int^{(-)} P_N^{(s+2)} |\mathbf{V}_0 \cdot \mathbf{n}_0| \, d\sigma_0 \, d\boldsymbol{\xi}_* \, d\boldsymbol{\xi}_0 \, d\mathbf{x}_0 \quad (2.15)$$

In fact, changing the variables from $\boldsymbol{\xi}_0$ and $\boldsymbol{\xi}_*$ to $\boldsymbol{\xi}_0'$ and $\boldsymbol{\xi}_*'$ given by Eqs. (2.12) with ($i = 0$) and taking Eq. (2.14) (with $i = 0$ and s replaced by $s + 1$) into account, the *left* hand side of Eq. (2.15) becomes:

$$\text{l.h.s.} = \int^{(-)} P_N^{(s+2)'} |\mathbf{V}_0' \cdot \mathbf{n}_0| \, d\sigma_0 \, d\boldsymbol{\xi}_*' \, d\boldsymbol{\xi}_0' \, d\mathbf{x}_0 \quad (2.16)$$

where the arguments of $P_N^{(s+2)'}$ are the same as in $P_N^{(s+2)}$ with $\boldsymbol{\xi}_*$ and $\boldsymbol{\xi}_0$ replaced by $\boldsymbol{\xi}_*'$ and $\boldsymbol{\xi}_0'$, and we have taken into account that the absolute value of the Jacobian determinant of the transformation from $\boldsymbol{\xi}_*$, $\boldsymbol{\xi}_0$ to $\boldsymbol{\xi}_*'$, $\boldsymbol{\xi}_0'$ is 1 (see Section 4 of Chapter I). The integral extends to the hemisphere $\mathbf{V}_0' \cdot \mathbf{n}_0 < 0$, because Eq. (2.13) (with $i = 0$) implies

$$\mathbf{V}_0 \cdot \mathbf{n}_0 = -\mathbf{V}_0' \cdot \mathbf{n}_0 \quad (2.17)$$

We can now drop the primes in Eq. (2.16), since $\boldsymbol{\xi}_0'$ and $\boldsymbol{\xi}_*'$ are integration variables, when we find that:

$$\text{l.h.s.} = \int^{(-)} P_N^{(s+2)} |\mathbf{V}_0 \cdot \mathbf{n}_0| \, d\sigma_0 \, d\mathbf{x}_0 \, d\boldsymbol{\xi}_0 \, d\boldsymbol{\xi}_* = \text{r.h.s.} \quad (2.18)$$

and Eq. (2.15) is proved.

A similar argument cannot be used in connection with the integrals involving $P_N^{(s+1)}$. We can use, however, Eq. (2.14) to express both integrals in terms of the distribution function holding before the collision, or, alternatively, in terms of the one prevailing after the collision. In fact, Eq. (2.14) is perfectly reversible and can be used for expressing either the distribution function of molecules which have just collided in terms of the one for molecules entering a collision, or the latter in terms of the former. It is obvious that the first way is to be used if we want to predict the future from the past and not *vice versa;* it is equally clear, however, that, at this point, we commit ourselves to a definite time arrow; that is to say we introduce a difference between past and future, as we shall see in more detail later.

According to this discussion, we use Eq. (2.14) to obtain:

$$\int^{(+)} P_N^{(s+1)} |\mathbf{V}_i \cdot \mathbf{n}_i| \, d\sigma_i \, d\boldsymbol{\xi}_* = \int^{(+)} P_N^{(s+1)\prime} |\mathbf{V}_i \cdot \mathbf{n}_i| \, d\sigma_i \, d\boldsymbol{\xi}_* \tag{2.19}$$

where $P_N^{(s+1)\prime}$ means the value taken by $P_N^{(s+1)}$ when the arguments $\boldsymbol{\xi}_i$ and $\boldsymbol{\xi}_*$ are replaced by $\boldsymbol{\xi}_i'$ and $\boldsymbol{\xi}_*'$ given by Eqs. (2.12). If we insert Eqs. (2.19) and (2.15) into Eq. (2.10), we obtain

$$\frac{\partial P_N^{(s)}}{\partial t} + \sum_{i=1}^{s} \boldsymbol{\xi}_i \cdot \frac{\partial P_N^{(s)}}{\partial \mathbf{x}_i}$$
$$= (N - s)\sigma^2 \sum_{i=1}^{s} \left[\int^{(+)} P_N^{(s+1)\prime} |\mathbf{V}_i \cdot \mathbf{n}_i| \, d\mathbf{n}_i \, d\boldsymbol{\xi}_* - \int^{(-)} P_N^{(s+1)} |\mathbf{V}_i \cdot \mathbf{n}_i| \, d\mathbf{n}_i \, d\boldsymbol{\xi}_* \right]$$
$$(s = 1, \ldots, N) \tag{2.20}$$

where we have replaced $d\sigma_i$ by its expression $\sigma^2 \, d\mathbf{n}_i$ in terms of the radius σ of the sphere given by Eq. (2.8) and the element of solid angle $d\mathbf{n}_i$. We may even abolish the index i in $\mathbf{n}_i$ provided the argument $\mathbf{x}_*$ in the i-th integral is replaced by

$$\mathbf{x}_* = \mathbf{x}_i - \mathbf{n}\sigma \tag{2.21}$$

Finally, we may transform the two integrals extended to $\mathbf{V}_i \cdot \mathbf{n} < 0$ and $\mathbf{V}_i \cdot \mathbf{n} > 0$ into a single integral by changing, for example, $\mathbf{n}$ into $-\mathbf{n}$ in the second integral. Thus we have

$$\frac{\partial P_N^{(s)}}{\partial t} + \sum_{i=1}^{s} \boldsymbol{\xi}_i \cdot \frac{\partial P_N^{(s)}}{\partial \mathbf{x}_i} = (N - s)\sigma^2 \sum_{i=1}^{s} \int [P_N^{(s+1)\prime} - P_N^{(s+1)}] \, |\mathbf{V}_i \cdot \mathbf{n}| \, d\mathbf{n} \, d\boldsymbol{\xi}_*$$
$$(s = 1, \ldots, N) \tag{2.22}$$

where integration is extended to the hemisphere $\mathbf{V}_i \cdot \mathbf{n} > 0$ and the arguments in $P_N^{(s+1)\prime}$ are the same as those in $P_N^{(s+1)}$ except for $\boldsymbol{\xi}_i$, $\boldsymbol{\xi}_*$ which are replaced by $\boldsymbol{\xi}_i'$, $\boldsymbol{\xi}_*'$ given by Eqs. (2.12) and $\mathbf{x}_i - \mathbf{n}\sigma$ which is replaced by $\mathbf{x}_i + \mathbf{n}\sigma$.

We stress the fact that Eq. (2.22) was derived only under the assumptions of the symmetrical dependence of $P_N^{(s)}$ upon the molecules and a sufficient regularity of $P_N^{(s)}$ (the latter is required in order to neglect the contribution of a *line* to a surface integral, i.e. to neglect the effect of triple collisions). In addition, we neglected the surface integral in Eq. (2.5).

To justify the fact that we disregarded the boundary integral, we must discuss the boundary conditions satisfied by P_N when $\mathbf{x}_i$ belongs to the boundary ∂R of the region R in which the gas is enclosed. This is a topic which will be considered in detail in the next chapter. Here we assume that a molecule hitting the solid boundary ∂R at some point $\mathbf{x}$ with some velocity $\boldsymbol{\xi}'$ re-emerges at practically the same point with some other velocity $\boldsymbol{\xi}$, the duration of the molecule-wall interaction being negligible.

The nature of the interaction determines a probability density $R(\boldsymbol{\xi}' \to \boldsymbol{\xi}; \mathbf{x}, t)$ of a transition from a velocity $\boldsymbol{\xi}'$ to a velocity $\boldsymbol{\xi}$ at point $\mathbf{x}$ and time t; we shall assume that this probability is independent of the state of the other molecules and that no particles are captured by the solid walls.

The probability that the j-th molecule emerges from the surface element dA_j about $\mathbf{x}_j$ during the time interval dt with velocity between $\boldsymbol{\xi}_j$ and $\boldsymbol{\xi}_j + d\boldsymbol{\xi}_j$ when the l-th molecule $(l \neq j)$ is in the volume element $d\mathbf{x}_l$ with velocity between $\boldsymbol{\xi}_l$ and $\boldsymbol{\xi}_l + d\boldsymbol{\xi}_l$ is

$$d^*\mathscr{P} = P_N \, |\boldsymbol{\xi}_j \cdot \mathbf{n}_j| \, dt \, dA_j \, d\boldsymbol{\xi}_j \prod_{\substack{l=1 \\ l \neq j}}^{N} d\mathbf{x}_l \, d\boldsymbol{\xi}_l \tag{2.23}$$

where $\mathbf{n}_i$ is the normal unit vector pointing into the gas at $\mathbf{x}_j$. In fact, this is the probability that we find the molecule in the cylinder filled by points leaving dA_j during the time interval dt (see Fig. 10) with velocity between $\boldsymbol{\xi}_j$

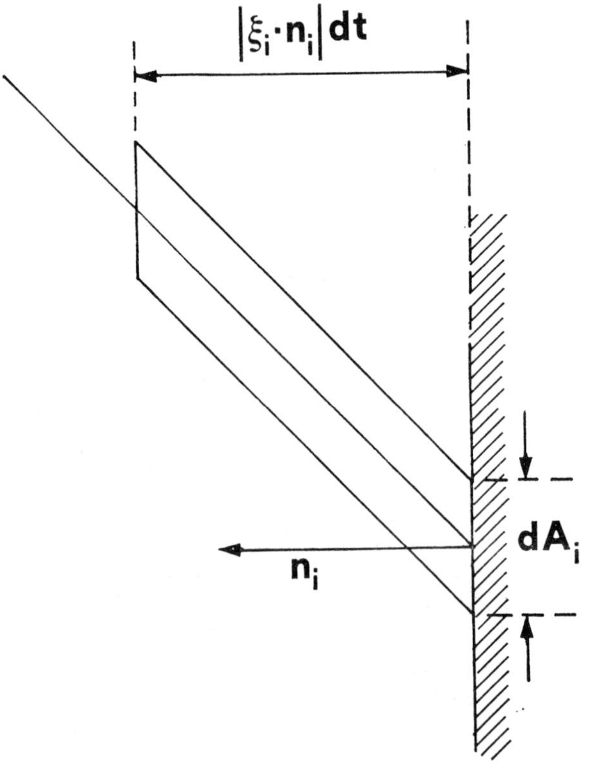

Fig. 10. Sketch for the computation of the number of molecules of velocity $\boldsymbol{\xi}_i$ hitting a surface element dA_i of normal unit vector $\mathbf{n}_i$ during the time interval dt.

and $\boldsymbol{\xi}_j + d\boldsymbol{\xi}_j$ when the l-th molecule ($l \neq j$) is in the volume element $d\mathbf{x}_l$ with velocity between $\boldsymbol{\xi}_l$ and $\boldsymbol{\xi}_l + d\boldsymbol{\xi}_l$, and the two probabilities are obviously the same thing.

Analogously, the probability that the same molecule impinges upon the same surface element with velocity between $\boldsymbol{\xi}_j'$ and $\boldsymbol{\xi}_j' + d\boldsymbol{\xi}_j'$ during dt when the l-th molecule is in $d\mathbf{x}_l$ with velocity between $\boldsymbol{\xi}_l$ and $\boldsymbol{\xi}_l + d\boldsymbol{\xi}_l$ ($l \neq j$)

$$d^*\mathscr{P} = P_N' \, |\boldsymbol{\xi}_j' \cdot \mathbf{n}_j| \, dt \, dA_j \, d\boldsymbol{\xi}_j' \prod_{\substack{l=1 \\ l \neq j}}^{N} d\boldsymbol{\xi}_l \, d\mathbf{x}_l \tag{2.24}$$

where the arguments of P_N' are the same as those of P_N except for the fact that $\boldsymbol{\xi}_j'$ replaces $\boldsymbol{\xi}_j$. If we multiply $d^*\mathscr{P}'$ by the probability of a scattering from the wall from velocity $\boldsymbol{\xi}_j'$ to a velocity between $\boldsymbol{\xi}_j$ and $\boldsymbol{\xi}_j + d\boldsymbol{\xi}_j$, $R(\boldsymbol{\xi}_j' \to \boldsymbol{\xi}_j; \mathbf{x}_j, t)\, d\boldsymbol{\xi}_j$ and "sum", that is integrate, over all the possible values of $\boldsymbol{\xi}_j'$, we must obtain $d^*\mathscr{P}$:

$$d^*\mathscr{P} = d\boldsymbol{\xi}_j \int_{\boldsymbol{\xi}_j' \cdot \mathbf{n}_j < 0} R(\boldsymbol{\xi}_j' \to \boldsymbol{\xi}_j; \mathbf{x}_j, t) \, d^*\mathscr{P}' \qquad (\boldsymbol{\xi}_j \cdot \mathbf{n}_j > 0) \tag{2.25}$$

or, using Eqs. (2.23) and (2.24) and cancelling the common factor $d\boldsymbol{\xi}_j(\prod d\boldsymbol{\xi}_l \, d\mathbf{x}_l) \, dA_j \, dt$:

$$\begin{aligned} &|\boldsymbol{\xi}_j \cdot \mathbf{n}_j| \, P_N(\mathbf{x}_i, \boldsymbol{\xi}_1, \ldots, \mathbf{x}_j, \boldsymbol{\xi}_j, \ldots, \mathbf{x}_N, \boldsymbol{\xi}_N, t) \\ &\quad = \int_{\boldsymbol{\xi}_j' \cdot \mathbf{n}_j < 0} R(\boldsymbol{\xi}_j' \to \boldsymbol{\xi}_j; \mathbf{x}_j, t) P_N(\mathbf{x}_1, \boldsymbol{\xi}_1, \ldots, \mathbf{x}_j, \boldsymbol{\xi}_j, \ldots, \mathbf{x}_N, \boldsymbol{\xi}_N, t) \times \\ &\qquad \times |\boldsymbol{\xi}_j' \cdot \mathbf{n}_j| \, d\boldsymbol{\xi}_j' \qquad (\mathbf{x}_j \in \partial R; \boldsymbol{\xi}_j \cdot \mathbf{n}_j > 0) \end{aligned} \tag{2.26}$$

This is the boundary condition satisfied by P_N at the solid boundaries, under the assumptions of instantaneous interaction. In particular, if the wall specularly reflects the molecules, we have:

$$R(\boldsymbol{\xi}' \to \boldsymbol{\xi}; \mathbf{x}, t) = \delta(\boldsymbol{\xi}' - \boldsymbol{\xi} + 2\mathbf{n}(\mathbf{n} \cdot \boldsymbol{\xi})) \tag{2.27}$$

In general, R must satisfy some restrictions, which will be discussed in the next chapter. The only restriction to be presently considered is the one related to the assumption that no molecules are captured by the walls; this means that any molecule impinging upon the wall eventually re-emerges with some velocity $\boldsymbol{\xi}$, and consequently the "sum" of the elementary probabilities $R(\boldsymbol{\xi}' \to \boldsymbol{\xi}) \, d\boldsymbol{\xi}$ over all possible values of $\boldsymbol{\xi}$ must be unity:

$$\int_{\boldsymbol{\xi} \cdot \mathbf{n} > 0} R(\boldsymbol{\xi}' \to \boldsymbol{\xi}; \mathbf{x}, t) \, d\boldsymbol{\xi} = 1 \qquad (\mathbf{x} \in \partial R, \boldsymbol{\xi}' \cdot \mathbf{n} < 0) \tag{2.28}$$

Hence Eq. (2.26) gives, after integration over $\boldsymbol{\xi}_j$ and using Eq. (2.28):

$$\int_{\boldsymbol{\xi}_j\cdot\mathbf{n}_j>0} |\boldsymbol{\xi}_j\cdot\mathbf{n}_j|\, P_N\, d\boldsymbol{\xi}_j = \int_{\boldsymbol{\xi}_j'\cdot\mathbf{n}_j<0} P_N'\, |\boldsymbol{\xi}_j'\cdot\mathbf{n}_j|\, d\boldsymbol{\xi}_j' = \int_{\boldsymbol{\xi}_j\cdot\mathbf{n}_j<0} P_N\, |\boldsymbol{\xi}_j\cdot\mathbf{n}_j|\, d\boldsymbol{\xi}_j \qquad (\mathbf{x}_j\in\partial R) \quad (2.29)$$

where the third expression comes from replacing $\boldsymbol{\xi}_j'$ by $\boldsymbol{\xi}_j$. Eq. (2.29) can be rewritten as follows:

$$\int \boldsymbol{\xi}_j\cdot\mathbf{n}_j P_N\, d\boldsymbol{\xi}_j = 0 \qquad (\mathbf{x}_j\in\partial R) \qquad (2.30)$$

where $\boldsymbol{\xi}_j$ unrestrictedly varies throughout both the half-spaces $\boldsymbol{\xi}_j\cdot\mathbf{n}_j \gtrless 0$. Finally, by integrating Eq. (2.30) with respect to dA_j and the coordinates and velocities of $N-s-1$ molecules (other than the j-th one), we obtain:

$$\int \boldsymbol{\xi}_j\cdot\mathbf{n}_j P_N^{(s+1)}\, d\boldsymbol{\xi}_j\, dA_j = 0 \qquad (2.31)$$

which proves that the last term in Eq. (2.5) is indeed zero, under our assumptions.

3. The Boltzmann equation for rigid spheres

The previous section was devoted to deriving the equations satisfied by $P_N^{(s)}$ $(s = 1, \ldots, N)$, Eq. (2.22). In particular, for $s = 1$, we have:

$$\frac{\partial P_N^{(1)}}{\partial t} + \boldsymbol{\xi}_1\cdot\frac{\partial P_N^{(1)}}{\partial \mathbf{x}_1} = (N-1)\sigma^2 \int [P_N^{(2)\prime} - P_N^{(2)}]\, |\mathbf{V}_1\cdot\mathbf{n}|\, d\mathbf{n}\, d\boldsymbol{\xi}_* \qquad (3.1)$$

This equation shows that the time evolution of the one-particle distribution function, $P_N^{(1)}$, depends upon the two-particle distribution function, $P_N^{(2)}$. In order to have a closed form equation for $P_N^{(1)}$, it is necessary to express $P_N^{(2)}$ in terms of $P_N^{(1)}$; a simple intuitive way of doing this is to assume the absence of correlation and to write:

$$P_N^{(2)}(\mathbf{x}_1, \boldsymbol{\xi}_1, \mathbf{x}_*, \boldsymbol{\xi}_*, t) = P_N^{(1)}(\mathbf{x}_1, \boldsymbol{\xi}_1, t) P_N^{(1)}(\mathbf{x}_*, \boldsymbol{\xi}_*, t) \qquad (3.2)$$

This relation was obtained in the case of thermal equilibrium for $N \to \infty$. If we accept it even in the case of non-equilibrium and insert Eq. (3.2) into Eq. (3.1), an equation involving $P_N^{(1)}$ alone is found. This is essentially the "*stosszahlansatz*" used by Boltzmann [4–7, 1] to derive the equation for $P_N^{(1)}$, which is accordingly called the Boltzmann equation.

We have no right, however, to postulate Eq. (3.2) because $P_N^{(2)}$ is determined by another equation (Eq. (2.22) with $s=2$) involving $P_N^{(3)}$, and the latter in turn, by another equation, Eq. (2.22) with $s=3$, involving $P_N^{(4)}$, etc. The least requirement is, therefore, to show that Eq. (3.2) is not at variance with the equation regulating the time evolution of $P_N^{(s)}(s \geqslant 2)$. Now, we cannot prove this statement, at least if we take it literally. In fact, as pointed out in connection with Eq. (I.6.30), Eq. (3.2) means that the states of the two molecules considered are statistically uncorrelated. Now, this makes sense for any two randomly chosen molecules of the gas, since they do not interact when they are far apart, and therefore behave independently. In particular, this seems true for two molecules which are going to collide, because they are just two random molecules whose paths happen to cross; but the same statistical independence is far from being true for two molecules which have just collided. We note, however, that Eq. (3.1) involves $P_N^{(2)}$ for molecules that are entering a collision, because we used Eq. (2.14) to eliminate the values of $P_N^{(s+1)}$ corresponding to after-collision states. This remark is important, but problems still arise because Eq. (2.22) for $P_N^{(s)}$ is valid provided $\mathbf{x}_i$ $(i = 1, \ldots, s)$ is outside the sets $|\mathbf{x}_i - \mathbf{x}_j| \leqslant \sigma$ $(j = 1, \ldots, s; j \neq i)$; the volume of these sets grows linearly with s, being proportional to $s\sigma^3$. These sets are, however, negligible in the limit $\sigma \to 0$, $N \to \infty$ for a fixed s (or even if we let s grow with N $(s \leqslant N)$, provided $N\sigma^3 \to 0$ as is the case for a perfect gas). We conclude that Boltzmann's *ansatz*, Eq. (3.2), is not true in a literal sense, but could become true for $N \to \infty$, $\sigma \to 0$ provided we specify that Eq. (3.2) is valid almost everywhere, that is to say it ceases to be valid only in exceptional sets of zero measure (among which is to be included the set of after-collision states).

Accordingly, we must prove that Eq. (3.2) (for $N \to \infty$, $\sigma \to 0$) is not at variance with the equations governing the time evolution of $P_N^{(s)}$ $(s \geqslant 2)$. We shall prove more, namely that the factorization property:

$$P^{(s)} = \prod_{i=1}^{s} P^{(1)}(\mathbf{x}_i, \boldsymbol{\xi}_i, t) \tag{3.3}$$

where

$$P^{(s)} = \lim_{N \to \infty} P_N^{(s)} \tag{3.4}$$

is not at variance with Eqs. (2.2) provided $\sigma \to 0$ in such a way that $N\sigma^2$ is bounded (hence $N\sigma^3 \to 0$). In order to prove this, we shall assume that the limit shown in Eq. (3.4) exists for any finite s and that the resulting function $P^{(s)}$ is sufficiently smooth.

Then, if we fix s and let $N \to \infty$, $\sigma \to 0$ in Eq. (2.22), in such a way that $N\sigma^2$ is bounded, we obtain:

$$\frac{\partial P^{(s)}}{\partial t} + \sum_{i=1}^{s} \boldsymbol{\xi}_i \cdot \frac{\partial P^{(s)}}{\partial \mathbf{x}_i} = (N\sigma^2) \sum_{i=1}^{s} \int [P^{(s+1)\prime} - P^{(s+1)}] \, |\mathbf{V}_i \cdot \mathbf{n}| \, d\mathbf{n} \, d\boldsymbol{\xi}_* \qquad (s = 1, 2, 3, \ldots) \tag{3.5}$$

where the arguments in $P^{(s+1)\prime}$ and $P^{(s+1)}$ are the same as above, except for the fact that $\mathbf{x}_* = \mathbf{x}_*{}' = \mathbf{x}_i$, in agreement with Eq. (2.21) for $\sigma \to 0$. Eqs.

(3.5) give a complete description of the time evolution of a Boltzmann gas, provided the initial value problem is well posed for this system of equations.

A particular solution of Eqs. (3.5) can be found in the form given by Eq. (3.3) provided the one-particle distribution function satisfies

$$\frac{\partial P}{\partial t} + \boldsymbol{\xi} \cdot \frac{\partial P}{\partial \mathbf{x}} = (N\sigma^2) \int (P\, P_*{}' - PP_*)\, |\mathbf{V} \cdot \mathbf{n}|\, d\mathbf{n}\, d\boldsymbol{\xi}_* \tag{3.6}$$

where we have written $\boldsymbol{\xi}$ and $\mathbf{x}$ in place of $\boldsymbol{\xi}_1$ and $\mathbf{x}_1$, P in place of $P^{(1)}$, while P_*, P', $P_*{}'$ denote that the argument $\boldsymbol{\xi}$ appearing in P is to be replaced by $\boldsymbol{\xi}_*$, $\boldsymbol{\xi}'$, $\boldsymbol{\xi}_*{}'$, respectively. The above statement is easily verified by substituting Eq. (3.3) into Eq. (3.5), provided Leibnitz's rule for differentiating a product is used when evaluating the time derivative of $P^{(s)}$.

Hence, if the system of Eqs. (3.5) admits a unique solution for a given initial datum, we conclude that the solution corresponding to a datum satisfying the "chaos assumption":

$$P^{(s)} = \prod_{i=1}^{s} P^{(1)}(\mathbf{x}_i, \boldsymbol{\xi}_i, 0) \qquad (t = 0) \tag{3.7}$$

will remain factored for all subsequent times and the one-particle distribution function $P = P^{(1)}$ will satisfy the Boltzmann equation. Therefore the factorization assumption, Eq. (3.3), is not inconsistent with the dynamics of rigid spheres in the limit $N \to \infty$, $\sigma \to 0$ ($N\sigma^2$ bounded) and leads to the Boltzmann equation.

A problem remains open: why should we assume that Eq. (3.7) is valid at $t = 0$? The answer is not completely clear, but one may argue according to one of the following justifications:

(1) It is conjectured that, in the limit $\sigma \to 0$, $N \to \infty$, one loses the possibility of describing a transient during which an arbitrarily given initial function relaxes to a factored distribution. This conjecture does not seem very reasonable and, in particular, seems to imply that the initial value problem for the system of Eqs. (3.5) is not well posed unless the initial datum satisfies Eq. (3.7).

(2) The initial datum is not specially prepared but comes from a previous evolution during which the gas also satisfies Eqs. (3.5). It is sufficient, therefore, that Eq. (3.3) was valid at some time instant in the remote past for it to be valid subsequently (in particular, if the gas was removed from an equilibrium state by an interaction with a solid boundary).

(3) Even if we consider all the possible initial data, most of them will be factored [8] in the following sense. If $P^{(1)}$ is given, the average value of the negative of the logarithm of $P^{(s)}$ (which is a measure of the likelihood of a

distribution, as shown in the Appendix) is a maximum for $P^{(s)} = \prod_{i=1}^{s} P^{(1)}(\mathbf{x}_i, \boldsymbol{\xi}_i)$. That is

$$\overline{\log P^{(s)}} = \int P^{(s)} \log P^{(s)} \prod_{i=1}^{s} d\mathbf{x}_i \, d\boldsymbol{\xi}_i \tag{3.8}$$

is a maximum for $P^{(s)} = \prod_{i=1}^{s} P^{(1)}(\mathbf{x}_i, \boldsymbol{\xi}_i)$. This is verified easily because $w(z) = z \log z - z \geqslant -1$ ($w'(z) = \log z$ is zero if and only if $z = 1$, $w''(z) = 1/z > 0$ for $z > 0$ and $w(1) = -1$) and hence (set $z = x/y$)

$$x \log x - x \geqslant x \log y - y \tag{3.9}$$

or any $x, y \geqslant 0$, equality being valid if and only if $x = y$; if we insert $P^{(s)}$ in place of x and $\prod_{i=1}^{s} P^{(1)}(\mathbf{x}_i, \boldsymbol{\xi}_i)$ in place of y and integrate both sides we find

$$\int P^{(s)} \log P^{(s)} \prod_{i=1}^{s} d\mathbf{x}_i \, d\boldsymbol{\xi}_i \geq s \int P^{(1)}(\mathbf{x}, \boldsymbol{\xi}) \log P^{(1)}(\mathbf{x}, \boldsymbol{\xi}) \, d\mathbf{x} \, d\boldsymbol{\xi} \tag{3.10}$$

where equality holds if and only if $P^{(s)} = \prod_{i=1}^{s} P^{(1)}$ and use has been made of the fact that $\int P^{(s)} \prod_{i=1}^{s} d\mathbf{x}_i \, d\boldsymbol{\xi}_i = \int P^{(1)} \, d\mathbf{x} \, d\boldsymbol{\xi} = 1$ and of the symmetry of $P^{(s)}$ with respect to the interchange of molecules. Eq. (3.10) proves that the expression given in Eq. (3.8) is a minimum for $P^{(s)} = \prod_{i=1}^{s} P^{(1)} (\mathbf{x}_i, \boldsymbol{\xi}_i)$.

If one of these justifications, or a combination of them, is valid, then Eq. (3.6), the Boltzmann equation, fully describes the time evolution of a Boltzmann gas. It is to be noted that this result depends heavily upon the fact that we used Eq. (2.4) to express the distribution function of molecules which have just collided in terms of that of molecules entering a collision, and not *vice versa*. If we had made the opposite choice, we would have found exactly the same equation as Eq. (3.6) except for the fact that the right hand side would have a minus sign in front of it! This result seems paradoxical, and is such, if we maintain that Eq. (3.6) describes the evolution of the system for any set of initial data. In fact, we assumed a smoothness of the distribution functions in the limit $N \to \infty$, $\sigma \to 0$; this smoothness, if assumed at $t = 0$, will remain true if and only if the system does not evolve, in average, towards a more inhomogeneous distribution, but, rather, towards a more homogeneous one. This means that we are describing those processes which lead from an "unprobable" distribution to a more probable one (compatibly with the given boundary conditions) and not *vice versa*. We remarked that the factorization property, Eq. (3.3), fails in some sets of negligibly small measure, among which we expect to find the after collision states; this expectation is correct if we are describing an evolution process from an "irregular" to a "regular" distribution function (notice that a "regular" distribution function means a disordered homogeneous state and hence a "probable" distribution function). If the evolution, on the contrary, goes from a disordered state to an ordered one, this means that there must be a strong correlation between two molecules entering a collision, and we must reverse our argument and obtain the "anti-Boltzmann equation". That is, Eq. (3.6) modified by a minus sign in front of the right hand side! This circumstance is related to the fact that the scattering of the volume element in phase space (Chapter I, Section 5) is a perfectly reversible property, and so, while points close to each other are scattered away, some points which were far apart are brought together during the evolution of the system. If the latter is in the most disordered microscopic state this mixing process does not change the macroscopic state (described by symmetric averages), because the chain of events leading to an ordered state is extremely unprobable, although dynamically possible. If the microscopic state, however, has some degree of order, the scattering of the volume element in phase space will tend to mix it up and produce a disordered state. This tendency is not a strictly dynamical property, but only a consequence of the fact that the number of disordered states having the same macroscopic averages is overwhelmingly larger than the number of ordered states. An idea of the meaning of this statement is gained by considering the experiment of shaking a vessel containing a large number of black and white marbles: if the two kinds of marbles are initially well separated, they will get finely mixed as a consequence of shaking, but it will be practically impossible to obtain the ordered state, in which the two kinds are separated, by shaking a vessel containing a highly homogeneous mixture. The latter process, however, is not dynamically impossible, because it is the former process in a reversed order and the equations of dynamics are time reversible (if we project a film of the first process backwards, we obtain the second one; if we look at the details of it, we cannot discover any contradiction of the laws of dynamics). When we derived the Boltzmann equation, we managed to screen these unprobable processes off; the set of the after-collision states has been excluded from the equations and all the other states have been assumed to be sufficiently disordered, that is, to be described by smooth distribution functions (a collision creates a certain degree of correlation, and hence of order). The Boltzmann equation is not capable of describing the reversed process, because we renounced the description of after-collision states which become the before-collision states in the reversed process.

Before concluding this section, let us look more carefully at the equation which we have obtained, Eq. (3.6). We note that it is a nonlinear integro-partial differential functional equation, where the specification "functional" refers to the fact that the unknown function P appears in the integral term not only with the arguments $\boldsymbol{\xi}$ (the current velocity variable) and $\boldsymbol{\xi}_*$ (the integration variable) but also with the arguments $\boldsymbol{\xi}'$ and $\boldsymbol{\xi}_*'$. The latter variables are related to $\boldsymbol{\xi}$ and $\boldsymbol{\xi}_*$ by the condition of being transformed into $\boldsymbol{\xi}$ and $\boldsymbol{\xi}_*$ by the effect of a collision, according to Eqs. (2.12):

$$\begin{aligned} \boldsymbol{\xi}' &= \boldsymbol{\xi} - \mathbf{n}(\mathbf{n}\cdot\mathbf{V}) \\ \boldsymbol{\xi}_*' &= \boldsymbol{\xi}_* + \mathbf{n}(\mathbf{n}\cdot\mathbf{V}) \end{aligned} \qquad (\mathbf{V} = \boldsymbol{\xi} - \boldsymbol{\xi}_*) \tag{3.11}$$

The integral on the right hand side of Eq. (3.6), which is called the collision term, is extended to all the values of $\boldsymbol{\xi}_*$ and to the hemisphere $|\mathbf{n}| = 1$, $\mathbf{V} \cdot \mathbf{n} > 0$. We observe that it could be equivalently extended to the whole unit sphere provided that the result was divided by 2, because changing a into $-\mathbf{n}$ does not alter the integrand.

Frequently, when dealing with the Boltzmann equation, one introduces a different unknown f which is related to P by:

$$f = NmP = MP \tag{3.12}$$

where N is the number of molecules, m is the mass of a molecule and M the total mass. The meaning of f is an (expected) mass density in the phase space of a single particle, that is to say the (expected) "mass per unit volume" in the six-dimensional space described by $(\mathbf{x}, \boldsymbol{\xi})$. We note that because of the normalization condition

$$\int P \, d\mathbf{x} \, d\boldsymbol{\xi} = 1 \tag{3.13}$$

we have

$$\int f \, d\mathbf{x} \, d\boldsymbol{\xi} = M \tag{3.14}$$

It is clear that in terms of f we have:

$$\frac{\partial f}{\partial t} + \boldsymbol{\xi} \cdot \frac{\partial f}{\partial \mathbf{x}} = \frac{\sigma^2}{m} \int (f' f_*' - f f_*) \, |\mathbf{V} \cdot \mathbf{n}| \, d\mathbf{n} \, d\boldsymbol{\xi}_* \tag{3.15}$$

where $f_* = f(\boldsymbol{\xi}_*)$, $f_*' = f(\boldsymbol{\xi}_*')$, $f' = f(\boldsymbol{\xi}')$. This is the form of the Boltzmann equation for a gas of rigid spheres which will be used in what follows.

The above considerations could be repeated if an external force per unit mass, $\mathbf{X}$, acts on the molecules, the only influence of this force being that one should add a term $\mathbf{X} \cdot \partial f / \partial \boldsymbol{\xi}$ to the left-hand side of Eq. (3.15). Since we shall usually consider cases when the external action on the gas is exerted through solid boundaries (surface forces), we shall not usually write the above-mentioned term describing body forces; it should be kept in mind, however, that such simplification implies neglecting, *inter alia*, gravity.

4. Generalizations

In the previous section, following a paper of the author [9] it was shown that, under certain assumptions, the Boltzmann equation follows from the Liouville equation for a gas of identical rigid spheres in the Boltzmann limit, defined by $N \to \infty$, $\sigma \to 0$, $N\sigma^2$ finite. Three possible generalizations suggest themselves: (1) molecules interacting with an at-distance force, (2) systems composed of several species of molecules such as a mixture of gases, (3) polyatomic gases, (4) dense gases ($N \to \infty$, $\sigma \to 0$, $N\sigma^3$ finite).

At first sight, the case of molecules interacting with an at-distance force seems to yield equations completely different from the Boltzmann equation. In fact, the Liouville equation, Eq. (I.3.8), can be written as follows:

$$\frac{\partial P_N}{\partial t} + \sum_{i=1}^{N} \boldsymbol{\xi}_i \cdot \frac{\partial P_N}{\partial \mathbf{x}} + \sum_{i,j=1}^{N} \mathbf{X}_{ij} \cdot \frac{\partial P_N}{\partial \boldsymbol{\xi}_i} = 0 \tag{4.1}$$

Here we assume that the force per unit mass acting on the i-th molecule, $\mathbf{X}_i$, is the resultant of $N-1$ two-body forces $\mathbf{X}_{ij}$ ($\mathbf{X}_{ii} = 0$) due to the interaction with remaining molecules and such that $\mathbf{X}_{ij} = \mathbf{X}(\mathbf{x}_i, \mathbf{x}_j)$ depends on the coordinates $\mathbf{x}_i$ and $\mathbf{x}_j$ alone.

If we integrate Eq. (4.1) with respect to the coordinates and velocities of $N - s$ molecules and use Eq. (1.3) defining the s-particle distribution function $P_N^{(s)}$, we obtain

$$\frac{\partial P_N^{(s)}}{\partial t} + \sum_{i=1}^{s} \boldsymbol{\xi}_i \cdot \frac{\partial P_N^{(s)}}{\partial \mathbf{x}_i} + \sum_{i=1}^{s}\sum_{j=1}^{s} \mathbf{X}_{ij} \cdot \frac{\partial P_N^{(s)}}{\partial \boldsymbol{\xi}_j} + \\ + (N-s)\sum_{i=1}^{s} \frac{\partial}{\partial \boldsymbol{\xi}_i} \cdot \int P_N^{(s+1)} \mathbf{X}_i \, d\mathbf{x}_* \, d\boldsymbol{\xi}_* = 0 \tag{4.2}$$

where $\mathbf{X}_i = \mathbf{X}(\mathbf{x}_i, \mathbf{x}_*)$. In order to obtain Eq. (4.2), it is sufficient to observe that the terms of the last sum in Eq. (4.1) with $i \geqslant s+1$ integrate to zero because they can be transformed into a surface integral at infinity in the velocity space of the i-th molecule (we assume that $P \to 0$ when $\boldsymbol{\xi}_i \to \infty$), while the terms with $j \geqslant s+1$ give identical contributions: finally, terms involving space derivatives with respect to $\mathbf{x}_i$ ($i \geqslant s+1$) are transformed into surface integrals extended to the physical boundary of the system, which are assumed to be zero by the same kind of arguments used in Section 2.

Eqs. (4.2) constitute the so-called BBGKY-hierarchy (from the names of Bogoliubov [10], Born and Green [11], Kirkwood [12, 13], Yvon [14]). It is not obvious how to handle these equations in the Boltzmann limit. There is another limit, however, in which Eq. (4.2) lends itself to the derivation of a simple result. If each of the forces $\mathbf{X}_{ij}$ is uniformly small, of order ε, in such a way that we may let $N \to \infty$ and $\varepsilon \to 0$ and keep $N\varepsilon$ finite (i.e. the order of magnitude of the total force is finite), then we obtain from Eq. (4.2):

$$\frac{\partial P^{(s)}}{\partial t} + \sum_{i=1}^{s} \boldsymbol{\xi}_i \cdot \frac{\partial P^{(s)}}{\partial \mathbf{x}_i} + N \sum_{i=1}^{s} \frac{\partial}{\partial \boldsymbol{\xi}_i} \cdot \int P^{(s+1)} \mathbf{X}_i \, d\mathbf{x}_* \, d\boldsymbol{\xi}_* = 0 \tag{4.3}$$

where $P^{(s)} = \lim_{N\to\infty} P_N^{(s)}$, as above. This system of equations however, possesses a particular solution having the factorization property expressed by Eq. (3.3), as is verified by direct substitution; the one-particle probability density $P = P^{(1)}$ satisfies:

$$\frac{\partial P}{\partial t} + \boldsymbol{\xi} \cdot \frac{\partial P}{\partial \mathbf{x}} + \bar{\mathbf{X}} \cdot \frac{\partial P}{\partial \boldsymbol{\xi}} = 0 \tag{4.4}$$

Here

$$\bar{\mathbf{X}}(\mathbf{x}) = N\int P(\mathbf{x}_*, \boldsymbol{\xi}_*, t)\mathbf{X}(\mathbf{x}, \mathbf{x}_*)\, d\mathbf{x}_*\, d\boldsymbol{\xi}_*$$
$$= \int n(\mathbf{x}_*, t)\mathbf{X}(\mathbf{x}, \mathbf{x}_*)\, dx_* \tag{4.5}$$

where

$$n(\mathbf{x}_*, t) = N\int P(\mathbf{x}_*, \boldsymbol{\xi}_*, t)\, d\boldsymbol{\xi}_* \tag{4.6}$$

is the number density in physical space; that is, the number of molecules per unit volume, in a neighbourhood of $\mathbf{x}_*$ at time t.

Eq. (4.4) is a remarkable equation, called the Vlasov equation. It is completely different from the Boltzmann equation and is useful to describe the short time behavior of a system of weakly interacting mass points; this is the case of a rarefied gas whose particles interact with relatively weak, long range forces, such as the electrons of an ionized gas (Coulomb force) or the stars of a stellar system (gravitational force). In an ordinary gas, however, the intermolecular force is rather strong when the molecules are close to each other; hence the model of hard collisions, though extremely crude, is closer to a significant description of the state of affairs than the model of a continuously distributed, weak force.

In the kinetic theory of gases, it is customary to consider some molecular models which take the molecular interaction into account in a more or less accurate fashion. One of these is the hard sphere model which was discussed in detail before; the other models are based on mass points interacting with central, hence conservative, forces and differ from each other only in the form of the expression for the potential U of these forces. The simplest assumption is $U(\rho) = k\rho^{1-n}$ where ρ is the distance between two interacting molecules, and the force $\mathbf{X} = -\text{grad}\, U$ is assumed to be repulsive ($k > 0$). Considerable use, especially in the computation of transport coefficients, has been made of the Lennard-Jones model, which includes both a repulsive and an attractive part (see Fig. 11):

$$U = \frac{k}{\rho^{n-1}} - \frac{k'}{\rho^{n'-1}} \qquad (n > n') \tag{4.7}$$

with the typical choice $n = 13$, $n' = 7$. Other models replace the first of these terms by an exponential in ρ or by a rigid sphere potential of the form 0 for $\rho > \sigma$, ∞ for $\rho < \sigma$. The force corresponding to a potential of the Lennard-Jones type, Eq. (4.7), is well approximated by a power law potential for short distances ($\rho \leqslant (k/k')^{1/(n-n')}$) and may be replaced by a cutoff force with potential:

$$U(\rho) = \begin{cases} k\rho^{1-n} & \rho \leqslant \sigma \\ k\sigma^{1-n} & \rho \geqslant \sigma \end{cases} \tag{4.8}$$

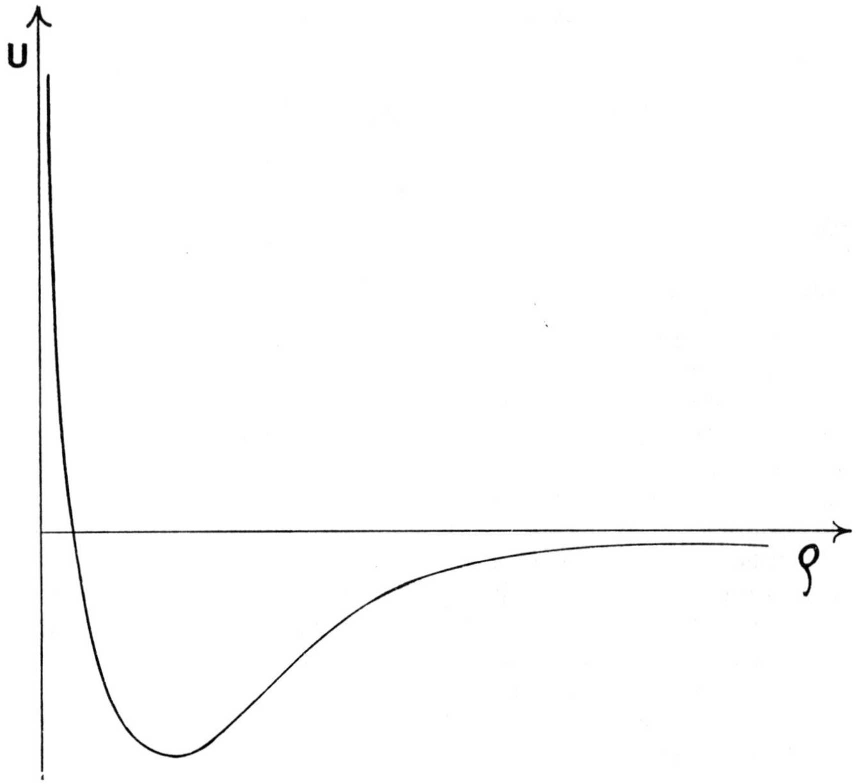

Fig. 11a,b. Intermolecular potential and force.

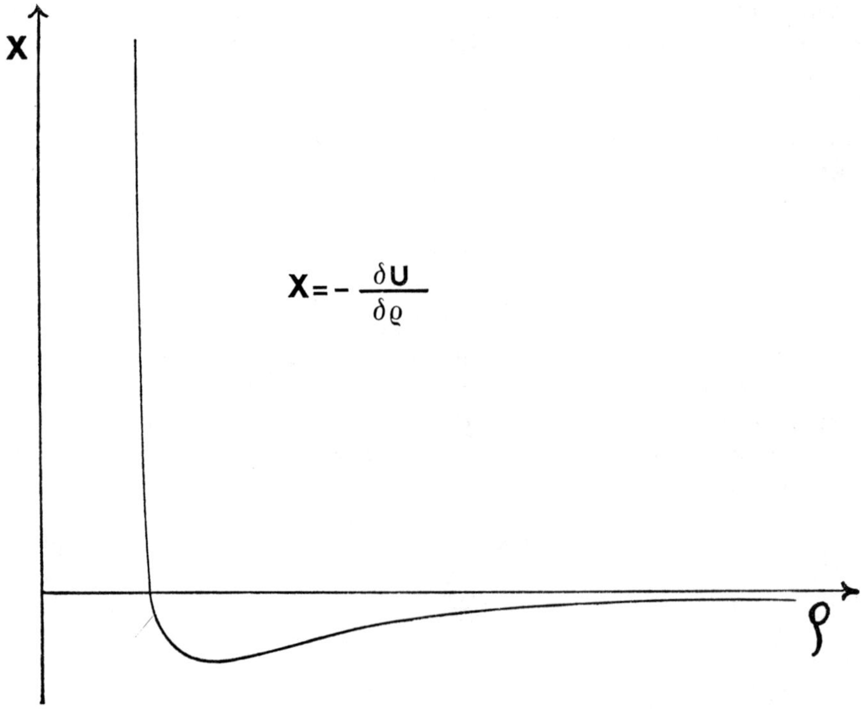

If we adopt such a cutoff potential, it is possible to derive the Boltzmann equation under the assumption $\sigma \to 0$, $N \to \infty$, $N\sigma^2$ finite, provided $U(\sigma)$ is small and of the order of the molecular mass $m \simeq \sigma^2 Ml/V$ (V volume, l mean free path). The latter circumstance is reasonably well satisfied by monatomic gases, since $U(\sigma)/(mR)$ (where R is the gas constant) is of the order of magnitude of a typical temperature (ranging from 10°K to 230°K).

In order to prove what we asserted, we introduce the truncated distribution functions:

$$^{\sigma}P_N^{(s)} = \int_{D_\sigma} P_N \prod_{l=s+1}^{N} d\mathbf{x}_l \, d\boldsymbol{\xi}_l \tag{4.9}$$

where the domain of integration excludes those regions where P_N would be zero by definition if the molecules were rigid spheres of radius $\sigma/2$. Particular cases of such truncated distribution functions were considered by H. Grad [15].

We can then repeat the derivation given in Sections 2 and 3 except for two facts:

(a) Multiple collisions are not a set of zero measure in the set of all collisions, because now collisions are replaced by finite duration interactions. If we let $N\sigma^3 \to 0$, however, as is correct for a Boltzmann gas, the measure tends to zero, because the probability of a triple collision is small like $N\sigma^3/V$ (V volume). Hence, for a Boltzmann gas, it is safe to neglect multiple collisions, and treat each collision as a two-body problem, even if we are not dealing with a gas of rigid spheres.

(b) A molecule leaves the protection sphere ($\rho = \sigma$) of another molecule (see Fig. 12) at a point different from the one at which the molecule entered the same sphere. The law of scattering can be written in the form given by Eq. (2.14), provided $\mathbf{n}_i$ is directed along the apse line of the orbit of the "bullet" molecule with respect to the "target" molecule (see Fig. 12; the apse line is the line through the target molecule and the point of closest approach) and $\mathbf{x}_i$, $\mathbf{x}_*$, t on the right hand side are replaced by some $\mathbf{x}_i'$, $\mathbf{x}_*'$, t' differing from $\mathbf{x}_i$, $\mathbf{x}_*$, t by terms small of the order of σ. The latter correction disappears when $\sigma \to 0$. There is an additional point, however, which will be presently considered. Let $\mathbf{n}^{(\mathrm{in})}$ and $\mathbf{n}^{(\mathrm{out})}$ be the normal unit vectors at the points where the "bullet" molecules reach and leave the protection sphere; then

$$\sigma^2 \, |\mathbf{V}' \cdot \mathbf{n}^{(\mathrm{in})}| \, d\mathbf{n}^{(\mathrm{in})} = \sigma^2 \, |\mathbf{V} \cdot \mathbf{n}^{(\mathrm{out})}| \, d\mathbf{n}^{(\mathrm{out})} \tag{4.10}$$

because the trajectories are fully symmetrical with respect to the apse line and $V' = V$. But, in general,

$$\frac{|\mathbf{V}' \cdot \mathbf{n}^{(\mathrm{in})}| \, d\mathbf{n}^{(\mathrm{in})}}{|\mathbf{V} \cdot \mathbf{n}| \, d\mathbf{n}} = \frac{|\mathbf{V} \cdot \mathbf{n}^{(\mathrm{out})}| \, d\mathbf{n}^{(\mathrm{out})}}{|\mathbf{V} \cdot \mathbf{n}| \, d\mathbf{n}} \tag{4.11}$$

will not be unity. Accordingly, if we want to use **n** throughout, we must compute this ratio. We note that an elementary geometrical argument gives

$$\sigma^2 \, |\mathbf{V} \cdot \mathbf{n}^{(\text{out})}| \, d\mathbf{n}^{(\text{out})} = V r \, dr \, d\varepsilon \tag{4.12}$$

where V is the relative speed and r, ε the polar coordinates in a plane orthogonal to **V** so that $r\,dr\,d\varepsilon$ is the surface element into which the surface element $\sigma^2\,d\mathbf{n}^{(\text{out})}$ of the protection sphere is projected (see Fig. 13). When $\sigma\mathbf{n}$ traverses the protection sphere, the point (r, ε) traverses the corresponding disk twice, but the image of a point is only once in the positive hemisphere ($\mathbf{V} \cdot \mathbf{n} > 0$). The quantity r is none other than the impact parameter; that is, the distance of closest approach of the two particles, had they continued their motion without interacting. The problem is to compute r as a function of V

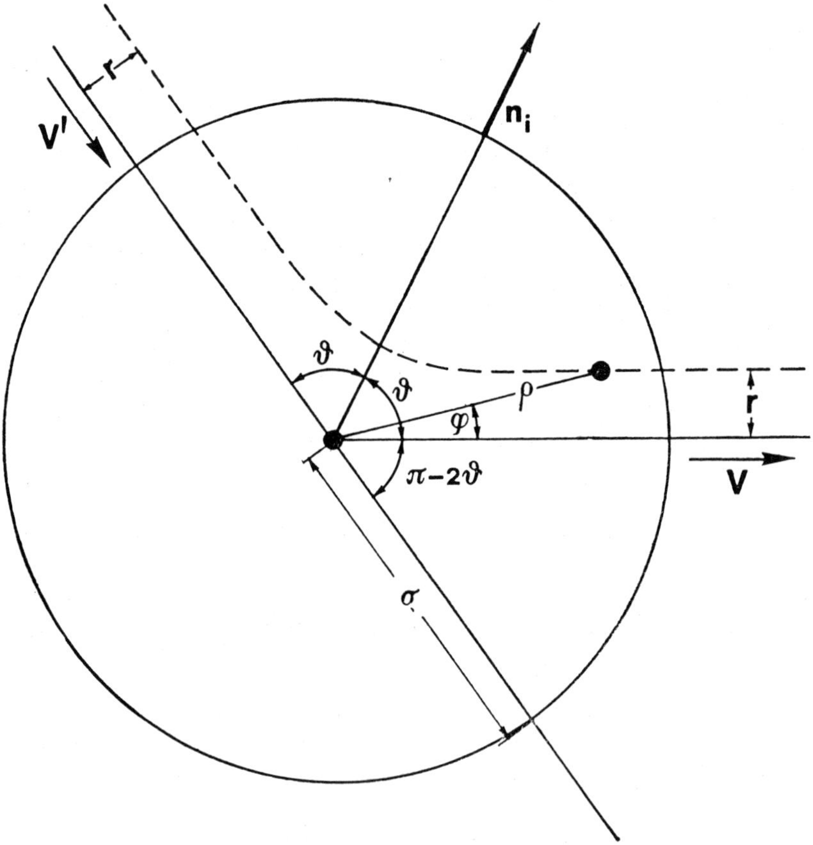

Fig. 12. Nomenclature for a two-body interaction.

and θ, where θ is the angle between $\mathbf{n}$ and $\mathbf{V}$; in fact $d\mathbf{n} = \sin\theta\, d\theta\, d\varepsilon$, and consequently:

$$\sigma^2 |\mathbf{V} \cdot \mathbf{n}^{(\text{out})}|\, d\mathbf{n}^{(\text{out})} = Vr \frac{\partial r}{\partial \theta}\, d\theta\, d\varepsilon = B(\theta, V)\, d\theta\, d\varepsilon = Vs(\theta, V)\, d\mathbf{n} \quad (4.13)$$

where

$$B(\theta, V) = Vr \frac{\partial r}{\partial \theta} \quad (4.14)$$

$$s(\theta, V) = \frac{1}{\sin\theta}\, r \frac{\partial r}{\partial \theta} \quad (4.15)$$

and $s(\theta, V)$ is called the differential scattering cross section, since it has the dimensions of an area; for rigid spheres $r = \sigma \sin\theta$, $B(\theta, V) = V\sigma^2 \sin\theta \cos\theta$, $s(\theta, V) = \sigma^2 \cos\theta$. Details of the computation of $B(\theta, V)$

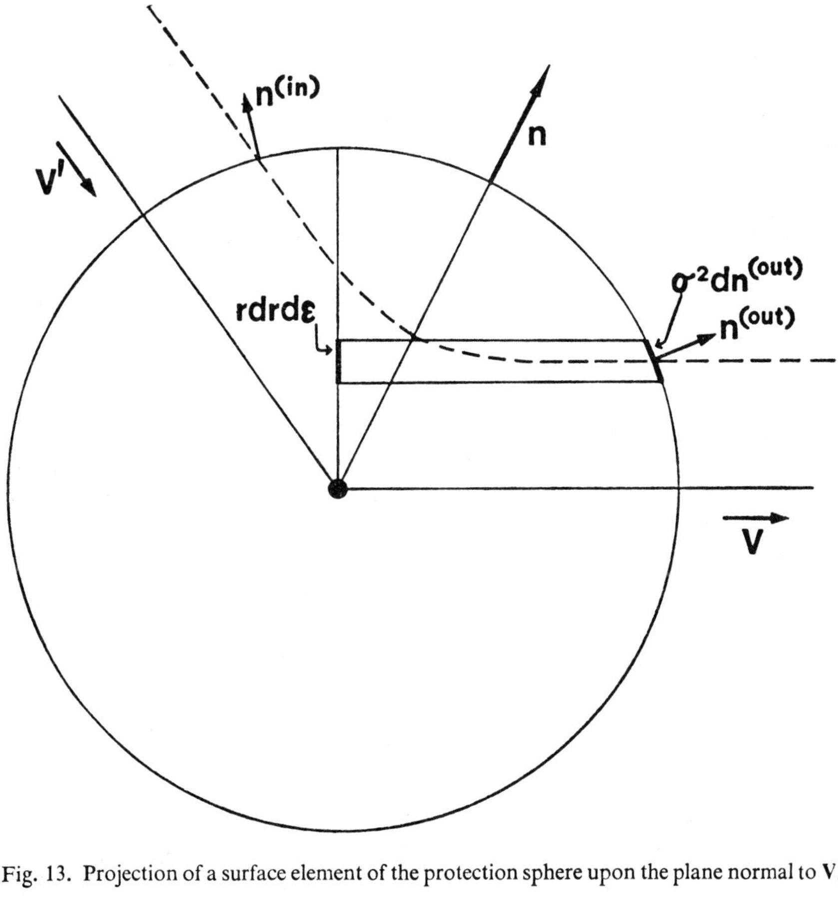

Fig. 13. Projection of a surface element of the protection sphere upon the plane normal to V.

or, equivalently, $s(\theta, V)$ for a given potential $U(\rho)$ will be given in the next section.

If we take into account these remarks, the Boltzmann equation follows in the limit $N \to \infty$, $\sigma \to 0$, $N\sigma^2$ finite, exactly as for the case of rigid spheres, provided $|\mathbf{V} \cdot \mathbf{n}^{(\text{out})}|\, d\mathbf{n}^{(\text{out})} = B(\theta, V)\, d\theta\, d\varepsilon$ replaces $|\mathbf{V} \cdot \mathbf{n}|\, d\mathbf{n}$. Accordingly, the Boltzmann equation for mass points interacting with a central force can be written as follows:

$$\frac{\partial f}{\partial t} + \boldsymbol{\xi} \cdot \frac{\partial f}{\partial \mathbf{x}} = \iiint (f'f_*' - ff_*) B(\theta, V)\, d\theta\, d\varepsilon\, d\boldsymbol{\xi}_* \tag{4.16}$$

There is a further point to be discussed concerning the fact that we have used a cutoff potential, but we defer the discussion until Section 9.

The next point to be discussed is the way of treating a mixture of different gases. If the molecules are rigid spheres, then the only difference can be that the molecules have different radii and masses, but for points interacting with at-distance forces we can have differences in the laws of interaction and in the values of the parameters appearing in these laws. In the statistical treatment, a first difference arises in connection with the N-body distribution function P_N, which can be symmetrized with respect to the molecules of each species, but not with respect to all the molecules of the mixture; hence a difference arises in the s-particle distribution functions which will differ according to the N-s molecules with respect to which we integrate. In particular, if there are n species, there will be n different one-particle distribution functions, and $n(n + 1)/2$ two-particle distribution functions. The notation becomes complicated but there is no new idea, except for the obvious fact that we must derive n equations for the n one-particle distribution functions f_j ($j = 1, \ldots, n$). The result is

$$\frac{\partial f_j}{\partial t} + \boldsymbol{\xi} \cdot \frac{\partial f_j}{\partial \mathbf{x}} = \sum_{i=1}^{n} \frac{1}{m_i} \iiint (f_j' f_{i*}' - f_j f_{i*}) B_{ij}(\theta, V)\, d\theta\, d\varepsilon\, d\boldsymbol{\xi}_* \qquad (j = 1, \ldots, n) \tag{4.17}$$

where m_i is the mass of the molecules of the i-th species, $B_{ij}(\theta, V)$ is defined by Eq. (4.14) provided $r = r(\theta, V)$ is computed from the law of interaction between the i-th and j-th species, and the arguments $\boldsymbol{\xi}'$, $\boldsymbol{\xi}_*'$ in f_j' and f_{i*}' in the i-th term are computed in terms of $\boldsymbol{\xi}$, $\boldsymbol{\xi}_*$, θ, ε from the laws of conservation of momentum and energy (see next section).

The description of the behavior of polyatomic gases may also be reduced to the case of a mixture of gases, with a suitable modification. We observe that Eq. (4.17) can be written as follows:

$$\frac{\partial f_j}{\partial t} + \boldsymbol{\xi} \cdot \frac{\partial f_j}{\partial \mathbf{x}} = \sum_{i=1}^{n} \frac{1}{m_i} \iiint [f_j(\boldsymbol{\xi}') f_i(\boldsymbol{\xi}_*') - f_j(\boldsymbol{\xi}) f_i(\boldsymbol{\xi}_*)] \times \times W_{ij}(\boldsymbol{\xi}, \boldsymbol{\xi}_* \mid \boldsymbol{\xi}', \boldsymbol{\xi}_*')\, d\boldsymbol{\xi}_*\, d\boldsymbol{\xi}'\, d\boldsymbol{\xi}_*' \tag{4.18}$$

where $\boldsymbol{\xi}'$, $\boldsymbol{\xi}_*'$, $\boldsymbol{\xi}$, $\boldsymbol{\xi}_*$ are now independent of each other (i.e. they are not related by the conservation laws) and:

$$W_{ij}(\boldsymbol{\xi}, \boldsymbol{\xi}_* \mid \boldsymbol{\xi}', \boldsymbol{\xi}_*') = S_{ij}(\theta, V)\delta(m_j\boldsymbol{\xi} + m_i\boldsymbol{\xi}_* - m_j\boldsymbol{\xi}' - m_i\boldsymbol{\xi}_*') \times \times \delta(m_j\xi^2 + m_i{\xi_*}^2 - m_j\xi'^2 - m_i{\xi_*}'^2) \quad (4.19)$$

where

$$\theta = \cos^{-1}\left[\frac{(\boldsymbol{\xi}' - \boldsymbol{\xi}) \cdot (\boldsymbol{\xi} - \boldsymbol{\xi}_*)}{|\boldsymbol{\xi}' - \boldsymbol{\xi}|\,|\boldsymbol{\xi} - \boldsymbol{\xi}_*|}\right]$$
$$S_{ij}(\theta, V) = \frac{B_{ij}(\theta, V)}{2V\cos\theta\sin\theta}(m_i + m_j)^2 m_i m_j \quad (4.20)$$

In particular, for rigid spheres of equal diameter σ and mass m, $S_{ij} = 2\sigma^3 m^4$. $W_{ij}(\boldsymbol{\xi}, \boldsymbol{\xi}_* \mid \boldsymbol{\xi}', \boldsymbol{\xi}_*')$ is the probability density of a collision which carries the molecules i and j from velocities $\boldsymbol{\xi}'$, $\boldsymbol{\xi}_*'$ into velocities $\boldsymbol{\xi}$, $\boldsymbol{\xi}_*$. Conservation of momentum and energy is now taken care of by the delta functions appearing in Eq. (4.19). The fact that Eq. (4.18) reduces to Eq. (4.17), when Eq. (4.19) is used, is easily verified by effecting the (trivial) integration with respect to $\boldsymbol{\xi}_*'$, and then first changing the integration variable from $\boldsymbol{\xi}'$ to $\boldsymbol{\lambda} = m_j(\boldsymbol{\xi} - \boldsymbol{\xi}')(d^3\boldsymbol{\xi}' = m_j^{-3}\,d^3\boldsymbol{\lambda})$ and then to the corresponding polar coordinates λ, θ, φ, with the polar axis along $\mathbf{V}$

$$(d^3\boldsymbol{\lambda} = \lambda^2\,d\lambda\sin\theta\,d\theta\,d\varphi,\ m_j\xi^2 + m_i{\xi_*}^2 - m_j\xi'^2 - m_i{\xi_*}'^2 = 2\lambda V\cos\theta - (m_i + m_j)\lambda^2/(m_i m_j)).$$

Integration with respect to λ yields the desired result, because letting

$$t = \frac{m_i + m_j}{m_i m_j}\left(\lambda - \frac{m_i m_j}{m_i + m_j}V\cos\theta\right)^2$$

gives

$$\int \delta\left(\frac{m_i m_j}{m_i + m_j}\lambda^2 - 2\lambda V\cos\theta\right)\lambda^2\,d\lambda$$
$$= \frac{1}{2}\int \delta\left(t - \frac{m_i m_j}{m_i + m_j}V^2\cos^2\theta\right) \times$$
$$\times \left(\sqrt{\frac{m_i m_j}{m_i + m_j}t} + \frac{m_i m_j}{m_i + m_j}V\cos\theta\right)^2 \frac{dt}{\sqrt{t}}\sqrt{\frac{m_i m_j}{m_i + m_j}}$$
$$= \frac{2(m_i m_j)^2}{(m_i + m_j)^2}V\cos\theta \quad (4.21)$$

where Eq. (I.2.10) has been used.

With a slight modification, Eq. (4.18) can be extended to a mixture in which a collision can transform the two colliding molecules of species k, l into two molecules of different species i, j (a very particular kind of reaction). In such a case the relations for velocities before and after the encounter are different from the ones used so far but we may still write a set of Boltzmann equations for the n species:

$$\frac{\partial f_j}{\partial t} + \boldsymbol{\xi} \cdot \frac{\partial f_j}{\partial \mathbf{x}} = \sum_{l,k,i=1}^{m} \frac{1}{m_i} \int (f_i' f_k' - f_j f_i) W_{ij}^{kl}(\boldsymbol{\xi}, \boldsymbol{\xi}_* \mid \boldsymbol{\xi}', \boldsymbol{\xi}_*') \, d\boldsymbol{\xi}_* \, d\boldsymbol{\xi}' \, d\boldsymbol{\xi}_*' \tag{4.22}$$

where W_{ij}^{kl} gives the probability density that a transition from velocities $\boldsymbol{\xi}'$, $\boldsymbol{\xi}_*'$ to velocities $\boldsymbol{\xi}$, $\boldsymbol{\xi}_*$ takes place in a collision which transforms two molecules of species l, k respectively, into two molecules of species j, i respectively. In the case that such reactive collisions do not occur, $W_{ij}^{kl} = W_{ij}\, d_{ik}\, \delta_{jl}$ and W_{ij} reduces to the previously given expression (δ_{ik} is the Kronecker symbol, $\delta_{ik} = 1$ if $i = k$, $\delta_{ik} = 0$ if $i \neq k$).

A suitable picture of a molecule of a polyatomic gas is the following [16, 2]. The molecule is a mechanical system, which differs from a mass point by having a succession of internal states, which can be identified by a label j assuming integral values (a continuous set of internal states could also be considered, however). In the simplest cases (the only ones to be considered here) these states differ from each other because the molecule has, besides kinetic energy, an internal energy taking different values E_j in each of the different states. A collision between two molecules, besides changing the velocities, can also change the internal states of the molecules and, consequently, the internal energy enters in the energy balance. From the viewpoint of writing evolution equations for the statistical behavior of the system, it is convenient to think of a single polyatomic gas as a mixture of different monatomic gases. Each of these gases is formed by the molecules corresponding to a given internal energy, and a collision changing the internal state of at least one molecule is considered as a reactive collision of the kind considered above, $W_{ij}^{kl}(\boldsymbol{\xi}, \boldsymbol{\xi}_* \mid \boldsymbol{\xi}', \boldsymbol{\xi}_*')$ being the probability density of a collision transforming two molecules with internal states k, l respectively, and velocities $\boldsymbol{\xi}'$, $\boldsymbol{\xi}_*'$ respectively, into molecules with internal states i, j respectively, and velocities $\boldsymbol{\xi}$, $\boldsymbol{\xi}_*$, respectively.

This treatment can be further extended to include chemical reactions in which molecules are broken up and atoms are rearranged, or nuclear reactions in which nuclei are broken with the resulting emission and absorption of neutrons by nuclei. Analogous considerations apply to ionization phenomena and processes of emission and absorption of radiation. Additional terms are needed in the j-th equation, one for each reaction which involves the j-th species; the quantity generalizing W_{ij} to this case will give the probability density that a reaction takes place which either creates or destroys a particle of the j-th species.

We come now to the last point which is the generalization of the Boltzmann equation to dense gases. In this case $N \to \infty$, $\lim (N\sigma^3) > 0$; hence $N\sigma^2 \to \infty$ and Eq. (2.2) becomes singular in the limit. Accordingly we only neglect s with respect to N and consider s as ranging from 1 to ∞. The resulting equations still have the form shown in Eq. (3.5) but the arguments $\mathbf{x}_*$ and $\mathbf{x}_*'$ in the i-th term of the right hand side will keep the expressions:

$$\mathbf{x}_* = \mathbf{x}_i - \mathbf{n}\sigma \qquad \mathbf{x}_*' = \mathbf{x}_i + \mathbf{n}\sigma \tag{4.23}$$

rather than collapsing into $\mathbf{x}_i$; another basic difference is that the measure of the set of points such that $|\mathbf{x}_i - \mathbf{x}_j| < \sigma$ $(i \neq j)$ is not negligible and the fact that $P^{(s+1)}$ is zero there must be taken into account when passing to the limit $\sigma \to 0$. The effect of this shielding of one molecule by another is to reduce the probability of collisions. The total volume of the molecules is, however, comparable with the volume occupied by the gas; the effect of the latter factor is to reduce the volume in which the center of any one molecule can lie, and so to increase the probability of a collision. A detailed discussion of these effects would take us too far.

5. Details of the collision term

In order to specify completely the right-hand side of the Boltzmann equation:

$$\frac{\partial f}{\partial t} + \boldsymbol{\xi} \cdot \frac{\partial f}{\partial \mathbf{x}} = \frac{1}{m} \int (f'f_*' - ff_*) B(\theta, V)\, d\theta\, d\varepsilon\, d\boldsymbol{\xi}_* \tag{5.1}$$

we have to find the expression of $B(\theta, V)$ defined by Eq. (4.14). To this end it is necessary to study the two-body problem for a given potential $U(\rho)$. Let m, m_* be the masses of the two molecules; then it is well known that the relative motion takes place as if one of the molecules (the "target" molecule) were at rest and the other (the "bullet" molecule) had a mass equal to the reduced mass

$$\mu = \frac{mm_*}{m + m_*} \tag{5.2}$$

(in particular, if $m = m_*$, $\mu = m/2$). If ρ, φ are the radial and angular coordinates in the plane of motion (see Fig. 12), then conservation of energy and angular momentum (with respect to a pole located at the position of the target molecule) give:

$$\begin{gathered} \tfrac{1}{2}\mu(\dot{\rho}^2 + \rho^2\dot{\varphi}^2) + U(\rho) = \tfrac{1}{2}\mu V^2 + U(\sigma) \qquad (\rho \leq \sigma) \\ \rho^2\dot{\varphi} = rV \end{gathered} \tag{5.3}$$

where r is the impact parameter and V the relative speed; the right hand sides of these equations are evaluated when the bullet molecule is outside the sphere of interaction and the kinetic energy is constant and equal to $\mu V^2/2$, while the potential energy is also constant and equal to $U(\sigma)$ and the angular momentum equals the product of the linear momentum and the impact parameter. We could omit $U(\sigma)$ in Eq. (5.3) by stipulating $U(\sigma) = 0$ as is always possible; it is more instructive, however, to retain the constant explicitly. Also, we shall restrict our considerations to repulsive potentials, which is the important case for close interactions between molecules in a gas, as we saw before.

Now we can easily integrate the above equations (one can eliminate time derivatives by using φ as the independent variable); the orbit is, as we anticipated in Section 4, symmetric with respect to the apse line. The angle θ can be easily evaluated since it is the angle between $\mathbf{V}$ and the apse line (directed along $\mathbf{n}$) and the solution of Eqs. (5.3) gives:

$$\theta = \left(\frac{\mu}{2}\right)^{\frac{1}{2}} Vr \int_{\rho_0}^{\sigma} \rho^{-2}\left[\frac{\mu}{2} V^2\left(1 - \frac{r^2}{\rho^2}\right) - U(\rho) + U(\sigma)\right]^{-\frac{1}{2}} d\rho + \sin^{-1}\left(\frac{r}{\sigma}\right) \tag{5.4}$$

where ρ_0 is the distance of closest approach which satisfies the equation:

$$\frac{\mu}{2} V^2\left(1 - \frac{r^2}{\rho_0^{\,2}}\right) = U(\rho_0) - U(\sigma) \tag{5.5}$$

We note that $\rho_0 \leqslant \sigma$ (otherwise no deflection arises, since the molecules do not enter into an interaction); it is also clear that $r \leqslant \sigma$, as follows from $\rho_0 \leqslant \sigma$ and the assumption of a repulsive potential [which implies $U(\rho_0) - U(\sigma) \geqslant 0$].

What should now be done is to invert Eq. (5.4) to give $r = r(\theta)$ (the assumption of a repulsive potential guarantees that $\theta = \theta(r)$ is a monotonic function) and to insert it into Eq. (4.14) to obtain $B(\theta, V)$. If there is more than one species, one has to compute $B_{ij}(\theta, V)$ for all the possible pairs ($n(n+1)/2$ in all for n species).

In the case of a mixture it is necessary to obtain the relation between $\boldsymbol{\xi}'$, $\boldsymbol{\xi}_*'$ and $\boldsymbol{\xi}$, $\boldsymbol{\xi}_*$ since Eqs. (3.11) are valid only if the two molecules have equal masses (see Chapter I, Section 4). In the general case we have to write the equations of momentum and energy conservation in the form:

$$\begin{aligned} m\boldsymbol{\xi}' + m_*\boldsymbol{\xi}_*' &= m\boldsymbol{\xi} + m_*\boldsymbol{\xi}_* \\ m\xi'^2 + m_*\xi_*'^2 &= m\xi^2 + m_*\xi_*^2 \end{aligned} \tag{5.6}$$

We identically satisfy the first of these equations by putting

$$\begin{aligned} \boldsymbol{\xi}' &= \boldsymbol{\xi} + \frac{C}{m}\mathbf{n} \\ \boldsymbol{\xi}_*' &= \boldsymbol{\xi}_* - \frac{C}{m_*}\mathbf{n} \end{aligned} \tag{5.7}$$

where $\mathbf{n}$ is a unit vector and C is a scalar to be determined. If we then insert these equations into the second of Eqs. (5.6), we find:

$$2C\mathbf{n}\cdot\boldsymbol{\xi} + \frac{C^2}{m} - 2C\mathbf{n}\cdot\boldsymbol{\xi}_* + \frac{C^2}{m_*} = 0 \tag{5.8}$$

Hence, if $C \neq 0$,

$$C = -2\mu\mathbf{n}\cdot(\boldsymbol{\xi} - \boldsymbol{\xi}_*) = -2\mu\mathbf{n}\cdot\mathbf{V} \tag{5.9}$$

where $\mathbf{V}$ is the relative velocity $\boldsymbol{\xi} - \boldsymbol{\xi}_*$ and μ the reduced mass defined by Eq. (5.2). Inserting Eq. (5.9) into Eq. (5.7) gives:

$$\begin{aligned} \boldsymbol{\xi}' &= \boldsymbol{\xi} - \frac{2\mu}{m}\mathbf{n}(\mathbf{n}\cdot\mathbf{V}) \\ \boldsymbol{\xi}_*' &= \boldsymbol{\xi}_* + \frac{2\mu}{m_*}\mathbf{n}(\mathbf{n}\cdot\mathbf{V}) \end{aligned} \qquad (\mathbf{V} = \boldsymbol{\xi} - \boldsymbol{\xi}_*) \tag{5.10}$$

In order to find the meaning of $\mathbf{n}$, we compute $\mathbf{V}' = \boldsymbol{\xi}' - \boldsymbol{\xi}_*'$, the relative velocity before the impact:

$$\mathbf{V}' = \mathbf{V} - 2\mathbf{n}(\mathbf{n}\cdot\mathbf{V}) \tag{5.11}$$

Hence $\mathbf{n}$ bisects the angle between the straight lines directed along $-\mathbf{V}'$ and $\mathbf{V}$ and is directed along the apse line. In terms of the angles θ and ε we have, of course:

$$\begin{aligned} \mathbf{n}\cdot\mathbf{V} &= V\cos\theta \\ \mathbf{n} &= (\sin\theta\cos\varepsilon, \sin\theta\sin\varepsilon, \cos\theta) \end{aligned} \tag{5.12}$$

since θ is the angle between $\mathbf{n}$ and $\mathbf{V}$ and ε is the azimuth angle in a plane orthogonal to $\mathbf{V}$. The angle ε ranges from 0 to 2π and θ from 0 (head-on collisions, $r = 0$) to $\pi/2$ (grazing collisions, $r = \sigma$). We observe that the Jacobian of $(\boldsymbol{\xi}_*', \boldsymbol{\xi}_*)$ with respect to $(\boldsymbol{\xi}', \boldsymbol{\xi})$ is -1, since the argument of Chapter I, Section 5 can be repeated.

It is seen that all the complicated details of the two-body interactions are summarized by the quantity $B(\theta, V)$ (or the quantities $B_{ij}(\theta, V)$) giving the (unnormalized) probability density of a relative deflection equal to $\pi - 2\theta$ for a pair of molecules having relative speed V. $B(\theta, V)$ cannot be expressed in terms of elementary functions even for such simple potentials as inverse power potentials ($U = k\rho^{1-n}$; $n \neq 2, 3$); the cases of inverse-square and inverse-cube force laws are amenable to an analytic treatment, but describe too soft an interaction at small distances to be realistic for a neutral gas. In spite of these negative remarks, it is worthwhile considering the case of power law potentials in more detail. Eq. (5.4) becomes:

$$\theta = \left(\frac{\mu}{2}\right)^{\frac{1}{2}} Vr\int_{\rho_0}^{\sigma}\rho^{-2}\left[\frac{\mu}{2}V^2\left(1 - \frac{r^2}{\rho^2}\right) - \frac{k}{\rho^{n-1}} + \frac{k}{\sigma^{n-1}}\right]^{-\frac{1}{2}} d\rho + \sin^{-1}\left(\frac{r}{\sigma}\right) \tag{5.13}$$

where ρ_0 satisfies

$$\frac{\mu}{2} V^2\left(1 - \frac{r^2}{\rho_0^{\,2}}\right) - \frac{k}{\rho_0^{n-1}} + \frac{k}{\sigma^{n-1}} = 0 \tag{5.14}$$

If we now put

$$\begin{aligned} b &= r\left(\frac{\mu}{2k} V^2 + \frac{k}{\sigma^{n-1}}\right)^{1/n-1} \\ x &= \frac{r}{\rho}\left(1 + \frac{2k}{\mu V^2 \sigma^{n-1}}\right)^{\frac{1}{2}} \\ \lambda &= \frac{r}{\sigma}\left(1 + \frac{2k}{\mu V^2 \sigma^{n-1}}\right)^{\frac{1}{2}} \end{aligned} \tag{5.15}$$

Eq. (5.13) becomes

$$\theta = \int_\lambda^{x_0} \frac{dx}{\sqrt{1 - x^2 - (x/b)^{n-1}}} + \sin^{-1}\left(\frac{r}{\sigma}\right) \tag{5.16}$$

where x_0 satisfies

$$1 - x_0^{\,2} - (x_0/b)^{n-1} = 0 \tag{5.17}$$

It is clear that the computation of $\theta = \theta(r, V)$ is a rather complicated task. An essential simplification occurs in the limiting case $\sigma \to \infty$ which occurs when we analyse a many-body interaction as a sequence of grazing binary collisions (see the discussion in Section 9). Since all the standard work on the computation of viscosity and heat conduction coefficients [1, 2] is based on this assumption, we give the relevant formulas:

$$\theta = \int_0^{x_0} \frac{dx}{\sqrt{1 - x^2 - (x/b)^{n-1}}} \tag{5.18}$$

$$b = r\left(\frac{\mu}{2k}\right)^{1/n-1} V^{2/n-1} \tag{5.19}$$

where x_0 satisfies Eq. (5.17). Taken together, Eq. (5.18) and Eq. (5.17) give $\theta = \theta(b)$ and, inverting, $b = b(\theta)$. Hence Eq. (5.19) gives

$$r = \left(\frac{2k}{\mu}\right)^{1/n-1} V^{-2/n-1} b(\theta) \tag{5.20}$$

Eq. (5.20) shows that the dependence of r upon V and θ factorizes and consequently, Eq. (4.14) gives:

$$B(\theta, V) = V^{n-5/n-1}\left(\frac{2k}{\mu}\right)^{2/n-1} b\frac{db}{d\theta} = V^\gamma \beta(\theta) \tag{5.21}$$

where $\gamma = (n - 5)/(n - 1)$ and

$$\beta(\theta) = \left(\frac{2k}{\mu}\right)^{2/n-1} b\,\frac{db}{d\theta} \tag{5.22}$$

The relevant simplification for inverse-power laws without a cutoff distance is therefore that $B(\theta, V)$ becomes the product of a function of θ alone times a fractional power of V. A significant simplification arises when $n = 5$, because then V disappears. This simplification was discovered by Maxwell [17] and the fictitious molecules interacting in this way are usually called Maxwellian molecules. Although actual molecules are not Maxwellian molecules, yet the concept is useful because the assumption of an inverse fifth-power law frequently simplifies the calculations in a striking fashion and gives satisfactory answers or, at least, first approximations to satisfactory answers.

We note that $\beta(\theta)$ has the following behaviour:

$$\begin{aligned} \beta(\theta) &= O(\theta) \qquad (\theta \to 0) \\ \beta(\theta) &= O\left[\left(\frac{\pi}{2} - \theta\right)^{-(n+1/n-1)}\right] \qquad \left(\theta \to \frac{\pi}{2}\right) \end{aligned} \tag{5.23}$$

where $O(x)$ denotes a quantity of the order of x. The first of these relations is easily obtained by noting that when $\theta \to 0$, $x_0 \to 0$ from Eq. (5.18), and hence $b \simeq x_0 \to 0$ from Eq. (5.17), while Eq. (5.18) becomes

$$\theta \simeq \int_0^b \frac{dx}{\sqrt{1 - \left(\frac{x}{b}\right)^{n-1}}} = b\int_0^1 \frac{du}{\sqrt{1 - u^{n-1}}}$$

and so $b \simeq K\theta$ where $K \neq 0$, $bdb \simeq k^2\theta\, d\theta$, $\beta(\theta) = O(\theta)$. When $\theta \to \pi/2$, $b \to \infty$ (as is seen by letting $b \to \infty$ in Eq. (5.18)). Hence $x_0^2 \simeq 1 - b^{1-n}$, as follows from Eq. (5.17) and:

$$\begin{aligned} \theta &= x_0\int_0^1 \frac{dy}{\sqrt{1 - x_0^2y^2 - \left(\frac{x_0}{b}\right)^{n-1} y^{n-1}}} \simeq x_0\int_0^1 \frac{dy}{\sqrt{1 - y^2 + \frac{1}{b^{n-1}}(y^2 - y^{n-1})}} \\ &\simeq x_0\int_0^1 \frac{dy}{\sqrt{1-y^2}}\left[1 - \frac{1}{2}\frac{1}{b^{n-1}}\left(\frac{y^2 - y^{n-1}}{1 - y^2}\right)\right] \\ &\simeq \frac{\pi}{2} - \frac{\pi}{2}\frac{1}{2}\frac{1}{b^{n-1}} - \frac{1}{2}\frac{1}{b^{n-1}}\int_0^1 \frac{y^2(1 - y^{n-3})}{(1 - y^2)^{\frac{3}{2}}}\, dy \\ &= \frac{\pi}{2} + O\left(\frac{1}{b^{n-1}}\right) \qquad (n > 3) \end{aligned} \tag{5.25}$$

Hence

$$\begin{aligned} b &= O\left[\left(\frac{\pi}{2} - \theta\right)^{-(1/n-1)}\right] \\ b\frac{db}{d\theta} &= O\left[\left(\frac{\pi}{2} - \theta\right)^{-(n+1/n-1)}\right] \end{aligned} \tag{5.26}$$

and Eq. (5.23) follows.

6. Elementary properties of the collision operator. Collision invariants

The right hand side of Eq. (5.1) contains a quadratic expression $Q(f,f)$ defined by:

$$Q(f,f) = \frac{1}{m}\int (f'f_*' - ff_*)B(\theta, V)\, d\boldsymbol{\xi}_*\, d\varepsilon\, d\theta \tag{6.1}$$

The operator Q acts on the velocity-dependence of f; it describes the effect of interactions, and is accordingly called the collision operator. $Q(f,f)$, that is the integral in Eq. (6.1), is called the collision integral or, simply, the collision term. In this section we shall study some properties which make the manipulation of Q possible in many problems of basic character in spite of its complicated form. Actually, we shall study here a slightly more general expression, the bilinear quantity

$$Q(f, g) = \frac{1}{2m}\int (f'g_*' + f_*'g' - fg_* - f_*g)B(\theta, V)\, d\boldsymbol{\xi}_*\, d\varepsilon\, d\theta \tag{6.2}$$

It is clear that when $g = f$, Eq. (6.2) reduces to Eq. (6.1); in addition,

$$Q(f, g) = Q(g, f) \tag{6.3}$$

Our first aim is to study some manipulations of the eightfold integral

$$\int Q(f, g)\varphi(\boldsymbol{\xi})\, d\boldsymbol{\xi} = \frac{1}{2m}\int (f'g_*' + f_*'g' - fg_* - f_*g)\varphi(\boldsymbol{\xi})B(\theta, V)\, d\boldsymbol{\xi}\, d\boldsymbol{\xi}_*\, d\theta\, d\varepsilon \tag{6.4}$$

where the integrals with respect to $\boldsymbol{\xi}$ are extended to the whole velocity space and $\varphi(\boldsymbol{\xi})$ is any function of $\boldsymbol{\xi}$ such that the indicated integrals exist.

We now perform the interchange of variables $\boldsymbol{\xi} \to \boldsymbol{\xi}_*$, $\boldsymbol{\xi}_* \to \boldsymbol{\xi}$ (which implies also $\boldsymbol{\xi}' \to \boldsymbol{\xi}_*'$, $\boldsymbol{\xi}_*' \to \boldsymbol{\xi}'$ because of Eqs. (3.11)). Then, since both $B(\theta, V)$ and the quantity within parentheses transform into themselves, and the Jacobian of the transformation is obviously unity, we have

$$\int Q(f, g)\varphi(\boldsymbol{\xi})\, d\boldsymbol{\xi} = \frac{1}{2m}\int (f'g_*' + f_*'g' - fg_* - f_*g)\varphi(\boldsymbol{\xi}_*)B(\theta, V)\, d\boldsymbol{\xi}\, d\boldsymbol{\xi}_*\, d\theta\, d\varepsilon \tag{6.5}$$

This equation is identical to Eq. (6.4) except for having $\varphi(\boldsymbol{\xi}_*)$ in place of $\varphi(\boldsymbol{\xi})$. Now we consider another transformation of variables in Eq. (6.4): $\boldsymbol{\xi} \to \boldsymbol{\xi}'$ and $\boldsymbol{\xi}_* \to \boldsymbol{\xi}_*'$ (here, as above, the unit vector $\mathbf{n}$ in Eq. (3.11) is considered as fixed). As we know (Chapter I, Section 4), the absolute value of the Jacobian of this transformation is unity, so that $d\boldsymbol{\xi}\, d\boldsymbol{\xi}_* = d\boldsymbol{\xi}'\, d\boldsymbol{\xi}'_*$ and Eq. (6.5) becomes

$$\int Q(f, g)\varphi(\boldsymbol{\xi})\, d\boldsymbol{\xi} = \frac{1}{2m}\int (f'g_*' + f_*'g' - fg_* - f_*g)\varphi(\boldsymbol{\xi})B(\theta, V)\, d\boldsymbol{\xi}'\, d\boldsymbol{\xi}_*'\, d\theta\, d\varepsilon \quad (6.6)$$

where now, since $\boldsymbol{\xi}'$ and $\boldsymbol{\xi}_*'$ are integration variables, we must express $\boldsymbol{\xi}$ and $\boldsymbol{\xi}_*$ by means of the relations inverting Eqs. (3.11), which are (see Eq. (I.4.3))

$$\begin{aligned} \boldsymbol{\xi} &= \boldsymbol{\xi}' - \mathbf{n}(\mathbf{n}\cdot\mathbf{V}') \\ \boldsymbol{\xi}_* &= \boldsymbol{\xi}_*' + \mathbf{n}(\mathbf{n}\cdot\mathbf{V}') \end{aligned} \quad (6.7)$$

where $\mathbf{V}' = \boldsymbol{\xi}' - \boldsymbol{\xi}_*'$ is related to $\mathbf{V} = \boldsymbol{\xi} - \boldsymbol{\xi}_*$ by Eq. (5.1), and consequently

$$\mathbf{V}'\cdot\mathbf{n} = -\mathbf{V}\cdot\mathbf{n} \quad (6.8)$$

The hemisphere $\mathbf{V}\cdot\mathbf{n} > 0$ corresponds to $\mathbf{V}'\cdot\mathbf{n} < 0$; we may change, however, $\mathbf{n}$ into $-\mathbf{n}$, without altering the expressions of $\boldsymbol{\xi}$, $\boldsymbol{\xi}_*$ and integrate over the hemisphere $\mathbf{V}'\cdot\mathbf{n} > 0$. We can also change the names of integration variables and call $\boldsymbol{\xi}$, $\boldsymbol{\xi}_*$ what we called $\boldsymbol{\xi}'$, $\boldsymbol{\xi}_*'$ before. Then, because of Eqs. (3.11) and (6.7), we can consistently call $\boldsymbol{\xi}'$ and $\boldsymbol{\xi}_*'$ what we called $\boldsymbol{\xi}$ and $\boldsymbol{\xi}_*$ before, and write Eq. (6.6) as follows:

$$\int Q(f, g)\varphi(\boldsymbol{\xi})\, d\boldsymbol{\xi} = \frac{1}{2m}\int (fg_* + f_*g - f'g_*' - f_*'g')\varphi(\boldsymbol{\xi}')B(\theta, V)\, d\boldsymbol{\xi}\, d\boldsymbol{\xi}_*\, d\theta\, d\varepsilon \quad (6.9)$$

where $B(\theta, V)$ is not affected by the change, since Eq. (5.11) implies $V' = V$. We can rewrite Eq. (6.9) as follows:

$$\int Q(f, g)\varphi(\boldsymbol{\xi})\, d\boldsymbol{\xi} = -\frac{1}{2m}\int (f'g_*' + f_*'g' - fg_* - f_*g)\varphi(\boldsymbol{\xi}')B(\theta, V)\, d\boldsymbol{\xi}\, d\boldsymbol{\xi}_*\, d\theta\, d\varepsilon \quad (6.10)$$

This equation is identical to Eq. (6.4) except for a minus sign and having $\varphi(\boldsymbol{\xi}')$ in place of $\varphi(\boldsymbol{\xi})$.

Finally, let us interchange $\boldsymbol{\xi}$ and $\boldsymbol{\xi}_*$ in Eq. (6.10) as we did in Eq. (6.4) to obtain Eq. (6.5). The result is:

$$\int Q(f, g)\varphi(\boldsymbol{\xi})\, d\boldsymbol{\xi} = -\frac{1}{2m}\int (f'g_*' + f_*'g' - fg_* - f_*g) \times \\ \times \varphi(\boldsymbol{\xi}_*')B(\theta, V)\, d\boldsymbol{\xi}\, d\boldsymbol{\xi}_*\, d\theta\, d\varepsilon \quad (6.11)$$

which is identical to Eq. (6.4) except for a minus sign and for having $\varphi(\boldsymbol{\xi}_*')$ in place of $\varphi(\boldsymbol{\xi})$.

We have thus obtained four different expressions for the same quantity: Eqs. (6.4), (6.5), (6.10), (6.11). We can now obtain more expressions by taking appropriate linear combinations of the four basic ones; we are particularly interested in the combination which is obtained by adding the above four expressions and dividing by four. The result is:

$$\int Q(f, g)\varphi(\boldsymbol{\xi})\, d\boldsymbol{\xi} = \frac{1}{8m} \int (f'g_*' + f_*'g' - fg_* - f_*g) \times$$
$$\times (\varphi + \varphi_* - \varphi' - \varphi_*')B(\theta, V)\, d\boldsymbol{\xi}\, d\boldsymbol{\xi}_*\, d\theta\, d\varepsilon \quad (6.12)$$

This equation expresses a basic property of the collision term which will be frequently used in the following. In the particular case of $g = f$, Eq. (6.12) reads:

$$\int Q(f, f)\varphi(\boldsymbol{\xi})\, d\boldsymbol{\xi} = \frac{1}{4m} \int (f'f_*' - ff_*) \times$$
$$\times (\varphi + \varphi_* - \varphi' - \varphi_*')B(\theta, V)\, d\boldsymbol{\xi}\, d\boldsymbol{\xi}_*\, d\theta\, d\varepsilon \quad (6.13)$$

We now observe that the integral appearing in Eq. (6.12) is zero, independent of the particular f and g, if

$$\varphi + \varphi_* = \varphi' + \varphi_*' \quad (6.14)$$

is valid almost everywhere in velocity space. Since the integral appearing on the left hand side of Eq. (6.11) is the average change of the function $\varphi(\boldsymbol{\xi})$ in unit time due to the effect of the collisions, the functions satisfying Eq. (6.14) are usually called "collision invariants". We now have the property that, if $\varphi(\boldsymbol{\xi})$ is assumed to be continuous, then Eq. (6.14) is satisfied if and only if

$$\varphi(\boldsymbol{\xi}) = a + \mathbf{b} \cdot \boldsymbol{\xi} + c\xi^2 \quad (6.15)$$

where a and c are constant scalars and $\mathbf{b}$ is a constant vector. The functions $\psi_0 = 1$, $(\psi_1, \psi_2, \psi_3) = \boldsymbol{\xi}$, $\psi_4 = \xi^2$ are usually called the elementary collision invariants; thus a general collision invariant is a linear combination of the five quantities ψ.

In order to prove the above statement that Eq. (6.14) is satisfied if and only if $\varphi(\boldsymbol{\xi})$ has the form shown in Eq. (6.15), we need the following:

LEMMA *Let* $\mathbf{x}$ *be a vector in an n-dimensional space* E_n *and* $f(\mathbf{x})$ *be a function continuous at at least one point and satisfying*

$$f(\mathbf{x}) + f(\mathbf{x}_1) = f(\mathbf{x} + \mathbf{x}_1) \quad (6.16)$$

for any $\mathbf{x}, \mathbf{x}_1 \in E_n$, *then* $f(\mathbf{x}) = \mathbf{A} \cdot \mathbf{x}$ *where* $\mathbf{A}$ *is a constant vector.*

In fact if f is continuous at a point say $\mathbf{x}_0$, it will be continuous everywhere because

$$f(\mathbf{x}+\mathbf{h}) - f(\mathbf{x}) = f(\mathbf{h}) = f(\mathbf{x}_0+\mathbf{h}) - f(\mathbf{x}_0) \tag{6.17}$$

on account of Eq. (6.16) with $\mathbf{x}_1 = \mathbf{h}$. Eq. (6.16) gives, for any integer p, by induction:

$$f\left(\sum_{i=1}^{p} \mathbf{x}_i\right) = \sum_{i=1}^{p} f(\mathbf{x}_i) \tag{6.18}$$

and, in particular, for $\mathbf{x}_i = \mathbf{x}$ $(i = 1, \ldots, p)$

$$f(p\mathbf{x}) = pf(\mathbf{x}) \tag{6.19}$$

and replacing p by q, $\mathbf{x}$ by $\mathbf{x}/q$:

$$f(\mathbf{x}/q) = \frac{1}{q} f(\mathbf{x}) \tag{6.20}$$

Then by means of Eq. (6.19) (with $\mathbf{x}/q$ in place of $\mathbf{x}$) and Eq. (6.20):

$$f\left(\frac{p}{q}\mathbf{x}\right) = pf(\mathbf{x}/q) = \frac{p}{q} f(\mathbf{x}) \tag{6.21}$$

which is simply

$$f(\alpha\mathbf{x}) = \alpha f(\mathbf{x}) \tag{6.22}$$

for any rational $\alpha > 0$. By the continuity of $f(\mathbf{x})$, Eq. (6.21) is valid for any real $\alpha > 0$. Eq. (6.16) implies $f(0) = 0$ (by letting $\mathbf{x} = \mathbf{x}_1 = 0$) and, by letting $\mathbf{x}_1 = -\mathbf{x}$, $f(-\mathbf{x}) = -f(\mathbf{x})$; hence Eq. (6.22) is valid for any real α. The result then follows because if $\mathbf{a}_k$ $(k = 1, \ldots, n)$ are n orthogonal vectors, then any $\mathbf{x}$ can be written in the form $\sum_{k=1}^{n} (\mathbf{x}\cdot\mathbf{a}_k)\mathbf{a}_k$ and consequently, by means of Eqs. (6.18)

$$\begin{aligned} f(\mathbf{x}) &= f\left(\sum_{k=1}^{n} (\mathbf{x}\cdot\mathbf{a}_k)\mathbf{a}_k\right) = \sum_{k=1}^{n} f((\mathbf{x}\cdot\mathbf{a}_k)\mathbf{a}_k) = \sum_{k=1}^{n} \mathbf{x}\cdot\mathbf{a}_k f(\mathbf{a}_k) \\ &= \mathbf{x}\cdot\left(\sum_{k=1}^{n} f(\mathbf{a}_k)\mathbf{a}_k\right) = \mathbf{A}\cdot\mathbf{x} \qquad \left(\mathbf{A} = \sum_{k=1}^{n} f(\mathbf{a}_k)\mathbf{a}_k\right) \end{aligned} \tag{6.23}$$

and the proof of the lemma is achieved.

Now, Eq. (6.14) implies that $\varphi + \varphi_*$ has the same value for all the pairs of vectors $(\boldsymbol{\xi}, \boldsymbol{\xi}_*)$ which satisfy the conservation equations; that is, $\varphi + \varphi_*$ is constant whenever $\boldsymbol{\xi} + \boldsymbol{\xi}_*$ and $\xi^2 + \xi_*{}^2$ are.

In other words, $\varphi + \varphi_*$ is a function of the latter quantities alone:

$$\varphi(\boldsymbol{\xi}) + \varphi(\boldsymbol{\xi}_*) = \Phi(\xi^2 + \xi_*{}^2, \boldsymbol{\xi} + \boldsymbol{\xi}_*) \tag{6.24}$$

Let us define:

$$\varphi_\pm(\boldsymbol{\xi}) = \varphi(\boldsymbol{\xi}) \pm \varphi(-\boldsymbol{\xi})$$

$$\Phi_\pm(\xi^2 + \xi_*{}^2, \boldsymbol{\xi} + \boldsymbol{\xi}_*) = \Phi(\xi^2 + \xi_*{}^2, \boldsymbol{\xi} + \boldsymbol{\xi}_*) \pm \Phi(\xi^2 + \xi_*{}^2, -\boldsymbol{\xi} - \boldsymbol{\xi}_*) \tag{6.25}$$

and add Eq. (6.24) to the equation obtained from it by the change $(\boldsymbol{\xi}, \boldsymbol{\xi}_*) \to (-\boldsymbol{\xi}, -\boldsymbol{\xi}_*)$ and then subtract the latter equation from Eq. (6.24). The result is

$$\varphi_\pm(\boldsymbol{\xi}) + \varphi_\pm(\boldsymbol{\xi}_*) = \Phi_\pm(\xi^2 + \xi_*{}^2, \boldsymbol{\xi} + \boldsymbol{\xi}_*) \tag{6.26}$$

If we put $\boldsymbol{\xi}_* = -\boldsymbol{\xi}$ in the equation for $\varphi_\pm$ we obtain $(\varphi_\pm(-\boldsymbol{\xi}) = \varphi_\pm(\boldsymbol{\xi}))$:

$$2\varphi_+(\boldsymbol{\xi}) = \Phi_+(2\xi^2, 0) \tag{6.27}$$

which shows that φ_+ depends on ξ^2 alone, $\varphi_+ = \psi(\xi^2)$. Then Eq. (6.26) shows that Φ_+ depends on $\xi^2 + \xi_*{}^2$ alone, because no function of $\boldsymbol{\xi} + \boldsymbol{\xi}_*$ can be constructed from ξ^2 and $\xi_*{}^2$ ($f(\boldsymbol{\xi} + \boldsymbol{\xi}_*) = g(\xi^2, \xi_*{}^2)$ implies, by letting $\boldsymbol{\xi}_* = 0$, $f(\boldsymbol{\xi}) = h(\xi^2)$; hence $h(\xi^2 + \xi_*{}^2 + 2\boldsymbol{\xi} \cdot \boldsymbol{\xi}_*) = g(\xi^2, \xi_*{}^2)$ which is impossible, unless h is constant, because the left hand side takes on different values for $\boldsymbol{\xi} \cdot \boldsymbol{\xi}_* = 0$, $\xi_*{}^2 = \lambda^2\xi^2$ and $\boldsymbol{\xi}_* = \lambda\boldsymbol{\xi}$, while the right hand side takes on equal values for these two choices of the argument of h). We conclude that Eq. (6.26) can be rewritten as follows

$$\psi(\xi^2) + \psi(\xi_*{}^2) = \Phi_+(\xi^2 + \xi_*{}^2) = \psi(\xi^2 + \xi_*{}^2) + \psi(0) \tag{6.28}$$

where the last expression of Φ_+ comes from letting $\boldsymbol{\xi}_* = 0$. If we put $f(\xi^2) = \psi(\xi^2) - \psi(0)$, Eq. (6.28) becomes Eq. (6.16) in the one-dimensional case (with $x = \xi^2$, $x_1 = \xi_*{}^2$) and, by applying the lemma, we conclude that $f(\xi^2) = 2c\xi^2$ for some constant c. Hence:

$$\varphi_+(\xi^2) = \psi(\xi^2) = \psi(0) + f(\xi^2) = 2a + 2c\xi^2 \tag{6.29}$$

where $2a = \psi(0)$ is a constant.

Let us consider the equation for φ_- in Eq. (6.26). If we take $\boldsymbol{\xi}$ and $\boldsymbol{\xi}_*$ orthogonal, we conclude that Φ_- can be considered to depend on the second argument alone, because $\xi^2 + \xi_*{}^2 = (\boldsymbol{\xi} + \boldsymbol{\xi}_*)^2$. Hence we may write:

$$\varphi_-(\boldsymbol{\xi}) + \varphi_-(\boldsymbol{\xi}_*) = h(\boldsymbol{\xi} + \boldsymbol{\xi}_*) = \varphi_-(\boldsymbol{\xi} + \boldsymbol{\xi}_*) \tag{6.30}$$

where the last expression for $h(\boldsymbol{\xi} + \boldsymbol{\xi}_*)$ comes from letting $\boldsymbol{\xi}_* = 0$ and taking into account the fact that $\varphi_-(0) = 0$ by Eq. (6.25). In order to show that the condition $\boldsymbol{\xi} \cdot \boldsymbol{\xi}_* = 0$ can be avoided in Eq. (6.30), we take now $\boldsymbol{\xi}$ and $\boldsymbol{\xi}_*$ arbitrary and a third vector $\boldsymbol{\rho}$ orthogonal to both of them

$$\boldsymbol{\rho} \cdot \boldsymbol{\xi} = \boldsymbol{\rho} \cdot \boldsymbol{\xi}_* = 0 \tag{6.31}$$

while the magnitude ρ of $\boldsymbol{\rho}$ is fixed by

$$\rho^2 = |\boldsymbol{\xi} \cdot \boldsymbol{\xi}_*| \geqslant 0 \tag{6.32}$$

We have, by applying Eq. (6.30) to the orthogonal vectors $(\boldsymbol{\xi}, \boldsymbol{\rho})$ and $(\boldsymbol{\xi}_*, \mp\boldsymbol{\rho})$:

$$\begin{aligned} \varphi_-(\boldsymbol{\xi} + \boldsymbol{\rho}) &= \varphi_-(\boldsymbol{\xi}) + \varphi_-(\boldsymbol{\rho}) \\ \varphi_-(\boldsymbol{\xi}_* \mp \boldsymbol{\rho}) &= \varphi_-(\boldsymbol{\xi}_*) + \varphi_-(\mp\boldsymbol{\rho}) = \varphi_-(\boldsymbol{\xi}_*) \mp \varphi_-(\boldsymbol{\rho}) \end{aligned} \tag{6.33}$$

where the minus or plus sign in the second equation is taken according to $\boldsymbol{\xi} \cdot \boldsymbol{\xi}_* \gtrless 0$. Now, because of this choice and Eqs. (6.31), (6.32)

$$(\boldsymbol{\xi} + \boldsymbol{\rho}) \cdot (\boldsymbol{\xi}_* \mp \boldsymbol{\rho}) = \boldsymbol{\xi} \cdot \boldsymbol{\xi}_* \mp \boldsymbol{\xi} \cdot \boldsymbol{\rho} + \boldsymbol{\rho} \cdot \boldsymbol{\xi}_* \mp \rho^2 = 0 \tag{6.34}$$

and we can apply Eq. (6.30) to $\boldsymbol{\xi} + \boldsymbol{\rho}$ and $\boldsymbol{\xi}_* \mp \boldsymbol{\rho}$, and to $\boldsymbol{\xi} + \boldsymbol{\xi}_*$, $\boldsymbol{\rho} \mp \boldsymbol{\rho}$ to deduce:

$$\begin{aligned} \varphi_-(\boldsymbol{\xi} + \boldsymbol{\rho}) + \varphi_-(\boldsymbol{\xi}_* \mp \boldsymbol{\rho}) &= \varphi_-(\boldsymbol{\xi} + \boldsymbol{\xi}_* + \boldsymbol{\rho} \mp \boldsymbol{\rho}) \\ &= \varphi_-(\boldsymbol{\xi} + \boldsymbol{\xi}_*) + \varphi_-(\boldsymbol{\rho} \mp \boldsymbol{\rho}) \end{aligned} \tag{6.35}$$

Inserting in the left hand side the expressions given by Eq. (6.33), we obtain:

$$\varphi_-(\boldsymbol{\xi}) + \varphi_-(\boldsymbol{\xi}_*) + \varphi_-(\boldsymbol{\rho}) \mp \varphi_-(\boldsymbol{\rho}) = \varphi_-(\boldsymbol{\xi} + \boldsymbol{\xi}\boldsymbol{\rho}) + \varphi_-(\boldsymbol{\rho} \mp \boldsymbol{\rho} \tag{6.36}$$

If $\boldsymbol{\xi} \cdot \boldsymbol{\xi}_* > 0$, the minus sign holds and we obtain:

$$\varphi_-(\boldsymbol{\xi}) + \varphi_-(\boldsymbol{\xi}_*) = \varphi_-(\boldsymbol{\xi} + \boldsymbol{\xi}_*) \tag{6.37}$$

In particular, if we put $\boldsymbol{\xi} = \boldsymbol{\xi}_* = \boldsymbol{\rho}$ $(\boldsymbol{\rho} \cdot \boldsymbol{\rho} = \boldsymbol{\rho}^2 > 0)$, Eq. (6.37) gives:

$$2\varphi_-(\boldsymbol{\rho}) = \varphi_-(2\boldsymbol{\rho}) \tag{6.38}$$

Using the result in Eq. (6.36) when $\boldsymbol{\xi} \cdot \boldsymbol{\xi}_* < 0$ and the positive sign holds, we finally prove that Eq. (6.37) is identical with Eq. (6.16) provided we identify $\boldsymbol{\xi}$, $\boldsymbol{\xi}_*$, φ_- with $\mathbf{x}$, $\mathbf{x}_1$, f and the lemma can be applied to conclude that

$$\varphi_-(\boldsymbol{\xi}) = 2\mathbf{b} \cdot \boldsymbol{\xi} \tag{6.39}$$

for some constant vector $\mathbf{b}$. Since Eq. (6.25) shows that $\varphi = (\varphi_+ + \varphi_-)/2$, Eq. (6.15) is proved by adding Eqs. (6.29) and (6.39).

Summarizing, if φ is a collision invariant, given by Eq. (6.15), then

$$\int \varphi(\boldsymbol{\xi}) Q(f, g)\, d\boldsymbol{\xi} = 0 \tag{6.40}$$

In the case of a mixture of gases, the above treatment can be extended to show that, if $Q_{ij}(f_i, f_j)$ denotes the collision term for the interaction between a particle of the j-th species and a particle of the i-th species as shown in the right hand side of Eq. (4.17) (which is $\sum_i Q_{ij}(f_i, f_j)$), then

$$\int \varphi \sum_i Q_{ij}(f_i, f_j)\, d\boldsymbol{\xi} = 0 \tag{6.41}$$

for any f_i, f_j if φ is a constant (conservation of the mass of particles of the j-th species) and

$$\sum_j \int \varphi_j \sum_i Q_{ij}(f_i, f_j)\, d\boldsymbol{\xi} = 0 \tag{6.42}$$

if $\varphi_j = \text{const.}$, $\varphi_j = \boldsymbol{\xi}$ or $\varphi_j = \xi^2$ (conservation of the total mass, momentum, energy). This, of course, is correct in the absence of reactions; if the latter occur, Eq. (6.41) is never satisfied in general and Eq. (6.42) is satisfied by $\varphi_j = 1$, $\varphi_i = \boldsymbol{\xi}$, $\varphi_j = \xi^2 + 2E_j/m_j$ where E_j is the internal energy of the j-th species (this applies, in particular, to the treatment of a polyatomic gas, according to the remarks of Section 4).

7. Solution of the equation $Q(f,f) = 0$

In this section we investigate the existence of positive functions f which give a vanishing collision integral:

$$Q(f,f) = \int (f'f_*' - ff_*)B(\theta, V)\, d\boldsymbol{\xi}_*\, d\theta\, d\varepsilon = 0 \tag{7.1}$$

We want to show that such functions exist and are all given by

$$f(\boldsymbol{\xi}) = \exp(a + \mathbf{b}\cdot\boldsymbol{\xi} + c\xi^2) \tag{7.2}$$

where a, $\mathbf{b}$, c have the same meaning as in Eq. (6.15). In order to show that this statement is true, we prove a preliminary result which will also be important later, this is that no matter what the distribution function is, the following inequality (Boltzmann's inequality) holds:

$$\int \log f Q(f,f)\, d\boldsymbol{\xi} \leqslant 0 \tag{7.3}$$

and the equality sign applies if, and only if, f is given by Eq. (7.2). Now it is seen that the first statement is a simple corollary of the second one: in fact, if Eq. (7.1) is satisfied then multiplying it by $\log f$ and integrating gives Eq. (7.3) with the equality sign, which implies Eq. (7.2) if the second statement applies. The other way round, if Eq. (7.2) holds, then, because of the results of the previous section applied to $\varphi = \log f$, $f'f_*' = ff_*$ and Eq. (7.1) is satisfied.

Let us prove, therefore, that Eq. (7.3) always holds for $f > 0$ and that the equality sign implies, and is implied by, Eq. (7.2). If we use Eq. (6.13) with $\varphi = \log f$ we have:

$$\int \log f Q(f,f)\, d\boldsymbol{\xi} = \frac{1}{4m}\int (f'f_*' - ff_*)\log(ff_*/f'f_*')B(\theta, V)\, d\boldsymbol{\xi}\, d\boldsymbol{\xi}_*\, d\theta\, d\varepsilon$$

$$= \frac{1}{4m}\int f'f_*'(1-\lambda)\log\lambda B(\theta, V)\, d\boldsymbol{\xi}\, d\boldsymbol{\xi}_*\, d\theta\, d\varepsilon \tag{7.4}$$

where

$$\lambda = ff_*/(f'f_*') \tag{7.5}$$

Now $f'f_*' > 0$, $B \geqslant 0$ (the equality sign applying only at $\theta = 0$); also, for any $\lambda \geqslant 0$ we have

$$(1-\lambda)\log\lambda \leqslant 0 \tag{7.6}$$

and the equality sign applies if, and only if, $\lambda = 1$ (note that $(1 - \lambda)$ and $-\log \lambda$ are negative and positive together and are both zero if and only if $\lambda = 1$). If we use Eq. (7.6), Eq. (7.4) implies Eq. (7.3) and the equality sign applies if, and only if, $\lambda = 1$, that is:

$$ff_* = f'f_*' \tag{7.7}$$

applies almost everywhere. But taking the logarithms of both sides of this equation, we find that $\varphi = \log f$ satisfies Eq. (6.14), so that $\varphi = \log f$ is given by Eq. (6.15); hence f is given by Eq. (7.2), as was to be shown.

We note that in Eq. (7.2) c must be negative, since f must be integrable over the whole velocity space. If we put $c = -\alpha$, $\mathbf{b} = 2\alpha\mathbf{v}$, where $\mathbf{v}$ is another constant vector, Eq. (7.2) can be written as follows:

$$f(\boldsymbol{\xi}) = A \exp[-\alpha(\boldsymbol{\xi} - \mathbf{v})^2] \tag{7.8}$$

where A is a constant related to a, α, v^2 (α, $\mathbf{v}$, A constitute a new set of arbitrary constants). Eq. (7.8) is the familiar Maxwellian distribution; it is different from Eq. (I.6.25) because Eq. (7.8) describes a gas which is not at rest (for $\mathbf{v} \neq 0$). However, Eq. (7.8) reduces to Eq. (I.6.25) (apart from trivial changes) if we change the reference frame to one moving with velocity $\mathbf{v}$ with respect to the frame for which Eq. (7.8) holds, and express the quantities A and α suitably in terms of internal energy and mass density. This interpretation will be shown to be correct in the next section.

In the case of a gas mixture, the above treatment can be modified to show that:

$$\sum_j \int \log f_j \sum_i Q_{ij}(f_i, f_j)\, d\boldsymbol{\xi} \leq 0 \tag{7.9}$$

the equality sign holding if and only if:

$$f_j = A_j \exp[-\alpha m_j(\boldsymbol{\xi} - \mathbf{v})^2] \tag{7.10}$$

where A_j, $\mathbf{v}$, α are constants. In the case of a polyatomic gas, the internal energy is to be taken into account and:

$$f_j = A_j \exp[-\alpha m_j(\boldsymbol{\xi} - \mathbf{v})^2 - 2\alpha E_j] \tag{7.11}$$

8. Connection between the microscopic description and the macroscopic description of gas dynamics

In this section we shall consider the problem of evaluating the macroscopic quantities once the distribution function is given.

We have occasionally used the density in physical space $\rho(\mathbf{x}, t)$, which is nothing else than the integral of the density in the one-particle phase space

$f(\mathbf{x}, \boldsymbol{\xi}, t)$ with respect to all possible velocities

$$\rho(\mathbf{x}, t) = \int f \, d\boldsymbol{\xi} \tag{8.1}$$

Because of the probabilistic meaning of f, the density ρ is the expected mass per unit volume at $(\mathbf{x}, t)$ or the product of the molecular mass m by the probability density of finding a molecule at $(\mathbf{x}, t)$, that is the (expected) number density $n(\mathbf{x}, t)$

$$n(\mathbf{x}, t) = \rho(\mathbf{x}, t)/m = \int P \, d\boldsymbol{\xi} \tag{8.2}$$

where P is the probability density related to f by Eq. (3.12).

The mass velocity $\mathbf{v}$ is given by the average of the molecular velocity $\boldsymbol{\xi}$

$$\mathbf{v} = \frac{\int \boldsymbol{\xi} P \, d\boldsymbol{\xi}}{\int P \, d\boldsymbol{\xi}} = \frac{\int \boldsymbol{\xi} f \, d\boldsymbol{\xi}}{\int f \, d\boldsymbol{\xi}} \tag{8.3}$$

where the integral in the denominator is due to the fact that P is not normalized to unity when we consider $\mathbf{x}$ as fixed and we integrate only with respect to $\boldsymbol{\xi}$. Because of Eq. (8.1), Eq. (8.3) can also be written as follows

$$\rho \mathbf{v} = \int \boldsymbol{\xi} f \, d\boldsymbol{\xi} \tag{8.4}$$

or, using components:

$$\rho v_i = \int \xi_i f \, d\boldsymbol{\xi} \tag{8.5}$$

The mass velocity $\mathbf{v}$ is what we can directly perceive of the molecular motion by means of macroscopic observations; it is zero for the steady state of a gas enclosed in a specularly reflecting box at rest. Each molecule has its own velocity $\boldsymbol{\xi}$ which can be decomposed into the sum of $\mathbf{v}$ and another velocity

$$\mathbf{c} = \boldsymbol{\xi} - \mathbf{v} \tag{8.6}$$

which describes the random deviation of the molecular velocity from the ordered motion with velocity $\mathbf{v}$. The velocity $\mathbf{c}$ is usually called the peculiar velocity or the random velocity; it coincides with $\boldsymbol{\xi}$ when the gas is macroscopically at rest. We note that, because of Eqs. (8.6), (8.5) and (8.1), we have:

$$\int c_i f \, d\boldsymbol{\xi} = \int \xi_i f \, d\boldsymbol{\xi} - v_i \int f \, d\boldsymbol{\xi} = \rho v_i - \rho v_i = 0 \tag{8.7}$$

The quantity ρv_i which appears in Eq. (8.5) can be interpreted as the momentum density or, alternatively, as the mass flow (in the i-th direction).

Other quantities which will be needed in the following are the momentum flow, the energy density and the energy flow. Since momentum is a vector quantity, we have to consider the flow of the j-th component of momentum in the i-th direction; this is given by:

$$\int \xi_i(\xi_j f)\, d\boldsymbol{\xi} = \int \xi_i \xi_j f\, d\boldsymbol{\xi} \tag{8.8}$$

where we use the general fact that if a quantity has a density G in phase space (in this case $G = \xi_j f$), the associated flow through a surface S (i.e. the amount of that quantity that goes through S per unit surface and unit time) is given by $\int G\xi_n\, dt\, dS\, d\boldsymbol{\xi}/(dS\, dt) = \int G\xi_n\, d\boldsymbol{\xi}$, where integration is extended to all the possible velocities, dS denotes the area of a surface element and ξ_n is the component of $\boldsymbol{\xi}$ along the normal to such an element. Eq. (8.8) shows that the momentum flow is described by a symmetric tensor of second order. It is to be expected that in a macroscopic description only a part of the microscopically evaluated momentum flow will be identified as such, because the integral in Eq. (8.8) will be in general different from zero even if the gas is macroscopically at rest (absence of macroscopic momentum flow). In order to find out how the above momentum flow will appear in a macroscopic description, we have to use the splitting of $\boldsymbol{\xi}$ into mass velocity $\mathbf{v}$ and peculiar velocity $\mathbf{c}$, according to Eq. (8.6). We have:

$$\int \xi_i \xi_j f\, d\boldsymbol{\xi} = \int (v_i + c_i)(v_j + c_j) f\, d\boldsymbol{\xi} = v_i v_j \int f\, d\boldsymbol{\xi} + v_i \int c_j f\, d\boldsymbol{\xi} +$$
$$+ v_j \int c_i f\, d\boldsymbol{\xi} + \int c_i c_j f\, d\boldsymbol{\xi} = \rho v_i v_j + \int c_i c_j f\, d\boldsymbol{\xi} \tag{8.9}$$

where Eqs. (8.1) and (8.7) have been used. Thus the momentum flow decomposes into two parts, one of which is recognized as the macroscopic momentum flow (momentum density times velocity), while the second part is a hidden momentum flow due to the random motion of the molecules. How will this second part manifest itself in a macroscopic description? If we take a fixed region of the gas and observe the change of momentum inside it, we find that (in the absence of external body forces) the change can only in part be attributed to the matter which enters and leaves the region, leaving a second part which has no macroscopic explanation unless we attribute it to the action of a force exerted on the boundary of the region of interest by the contiguous regions of the gas. In other words the integral of $\int c_i c_j f\, d\boldsymbol{\xi}$ appears as a contribution to the stress tensor (and, indeed, the only contribution to the stress tensor if the gas is a Boltzmann gas, for which the actual actions exerted by the molecules of a region on the molecules of another are neglected). We shall therefore write

$$p_{ij} = \int c_i c_j f\, d\boldsymbol{\xi} \tag{8.10}$$

(a complete identification is correctly justified by the fact that, as we shall see later, p_{ij} plays, in the macroscopic equations to be derived from the Boltzmann equation, the same role as the stress tensor in the conservation equations derived from macroscopic considerations).

An analogous decomposition is to be introduced for the energy density and energy flow. The energy density is given by $\frac{1}{2}\int \xi^2 f\, d\boldsymbol{\xi}$ and we have only to take $j = i$ and sum from $i = 1$ to $i = 3$ in Eq. (8.9) to deduce

$$\tfrac{1}{2}\int \xi^2 f\, d\boldsymbol{\xi} = \tfrac{1}{2}\rho v^2 + \tfrac{1}{2}\int c^2 f\, d\boldsymbol{\xi} \tag{8.11}$$

Again the first term on the right-hand side will be macroscopically identified with the kinetic energy density, while the second term will be ascribed to the "internal energy" of the gas. Therefore, if we introduce the internal energy per unit mass e we have for the density of internal energy per unit volume ρe:

$$\rho e = \tfrac{1}{2}\int c^2 f\, d\boldsymbol{\xi} \tag{8.12}$$

We note that a relation exists between the internal energy density and the spur or trace (i.e. the sum of the three diagonal terms) of the stress tensor. In fact, Eqs. (8.12) and (8.10) give

$$p_{ii} = \int c^2 f\, d\boldsymbol{\xi} = 2\rho e \tag{8.13}$$

(Here and in the following, unless otherwise stated, we shall use the convention of summing over repeated subscripts from 1 to 3; in other words $p_{ii} \equiv \sum_{i=1}^{3} p_{ii}$). The spur divided by 3 gives the isotropic part of the stress tensor; it is therefore convenient to identify $p = p_{ii}/3$ with the gas pressure, at least in the case of equilibrium. The identification is also correct for nonequilibrium situations in the case of a monatomic perfect gas, but is generally incorrect. Therefore

$$p = \tfrac{2}{3}\rho e \tag{8.14}$$

Eq. (8.14) is called the state equation of the gas and allows us to express any of the three quantities p, ρ, e in terms of the remaining two. As we saw in Chapter I, Section 6, for a monatomic perfect gas, e is a function of temperature; that is, of an index which has the property of taking the same value for two systems in contact with each other in a state of equilibrium. Eq. (8.14) shows that p/ρ is constant at constant temperature for rarefied monatomic gases. It is this property which identifies such gases as the perfect gases obeying Boyle's law:

$$p = \rho RT \tag{8.15}$$

where T is the absolute temperature and R a constant (depending on the molecular mass, in the way shown by Eq. (I.6.27), where $m = \bar{m}$ for a single gas). Eqs. (8.14) and (8.15) give the identification

$$e = \tfrac{3}{2}RT \tag{8.16}$$

which was anticipated in Eq. (I.6.26).

We now have to investigate the energy flow; the total energy flow is obviously given by

$$\int \xi_i(\tfrac{1}{2}\xi^2 f)\, d\boldsymbol{\xi} = \tfrac{1}{2}\int \xi_i \xi^2 f\, d\boldsymbol{\xi} \tag{8.17}$$

Using Eq. (8.6) gives:

$$\begin{aligned} \tfrac{1}{2}\int \xi_i \xi^2 f\, d\boldsymbol{\xi} &= \tfrac{1}{2}\int (v_i + c_i)(v^2 + 2c_j v_j + c^2) f\, d\boldsymbol{\xi} \\ &= \tfrac{1}{2}v_i v^2 \int f\, d\boldsymbol{\xi} + v_i v_j \int c_j f\, d\boldsymbol{\xi} + \tfrac{1}{2}v_i \int c^2 f\, d\boldsymbol{\xi} + \\ &\quad + \tfrac{1}{2}v^2 \int c_i f\, d\boldsymbol{\xi} + v_j \int c_i c_j f\, d\boldsymbol{\xi} + \tfrac{1}{2}\int c_i c^2 f\, d\boldsymbol{\xi} \end{aligned} \tag{8.18}$$

or, using Eqs. (8.1), (8.7), (8.12), (8.10),

$$\tfrac{1}{2}\int \xi_i \xi^2 f\, d\boldsymbol{\xi} = v_i(\tfrac{1}{2}\rho v^2 + e) + v_j p_{ij} + \tfrac{1}{2}\int c_i c^2 f\, d\boldsymbol{\xi} \tag{8.19}$$

We now have three terms: the first of which is obviously the energy flow due to macroscopic convection and the second is to be macroscopically interpreted as due to the work done by the stresses in unit time.

The third term represents another kind of energy flow; the additional term is usually called the heat flux vector, and is denoted by $\mathbf{q}$:

$$q_i = \tfrac{1}{2}\int c_i c^2 f\, d\boldsymbol{\xi} \tag{8.20}$$

As for the case of the stress tensor the identification is justified, as we shall see later, by the fact that $\mathbf{q}$ plays the same role as the heat flux vector in the macroscopic equations. However, the name "heat flux" is somewhat misleading, because there are situations when $q_i \neq 0$ and the temperature is practically constant everywhere; in this case one has to speak of a heat flux at constant temperature. The name "nonconvective energy flow" would be more appropriate for $\mathbf{q}$ but it is not used.

The above discussion links the distribution function with the quantities used in the macroscopic description; in particular, for example, p_{ij} can be used to evaluate the drag on a body moving inside the gas and $\mathbf{q}$ to evaluate the heat transfer from a hot body to a colder one when they are separated by a region filled by the gas.

In order to complete the connection, as a simple mathematical consequence of the Boltzmann equation, we now derive five differential equations satisfied by the macroscopic quantities considered above: these equations are usually called the conservation equations, since they can be physically interpreted as expressing conservation of mass, momentum and energy.

In order to obtain these equations we consider the Boltzmann equation:

$$\frac{\partial f}{\partial t} + \xi_i \frac{\partial f}{\partial x_i} + X_i \frac{\partial f}{\partial \xi_i} = Q(f, f) \tag{8.21}$$

where, for the sake of generality, we have introduced the body force term which is usually left out. We multiply both sides of Eq. (8.21) by the five collision invariants ψ_α ($\alpha = 0, 1, 2, 3, 4$) defined in Section 6 and integrate with respect to $\boldsymbol{\xi}$ in accordance with the results of Section 6 Eq. (6.40) with $g = f$, $\varphi = \psi_\alpha$:

$$\int \psi_\alpha Q(f, f)\, d\boldsymbol{\xi} = 0 \qquad (\alpha = 0, 1, 2, 3, 4) \tag{8.22}$$

for any f. Therefore for any f satisfying Eq. (8.21):

$$\frac{\partial}{\partial t}\int \psi_\alpha f\, d\boldsymbol{\xi} + \frac{\partial}{\partial x_i}\int \xi_i \psi_\alpha f\, d\boldsymbol{\xi} + X_i \int \psi_\alpha \frac{\partial f}{\partial \xi_i} = 0 \qquad (\alpha = 0, 1, 2, 3, 4) \tag{8.23}$$

provided X_i does not depend upon $\boldsymbol{\xi}$.

If we take successively $\alpha = 0, 1, 2, 3, 4$ and use Eqs. (8.1), (8.4), (8.9) to (8.12), (8.19) and (8.20), we obtain

$$\begin{gathered}
\frac{\partial \rho}{\partial t} + \frac{\partial}{\partial x_i}(\rho v_i) = 0 \\
\frac{\partial}{\partial t}(\rho v_j) + \frac{\partial}{\partial x_i}(\rho v_i v_j + p_{ij}) = \rho X_j \\
\frac{\partial}{\partial t}[\rho(\tfrac{1}{2}v^2 + e)] + \frac{\partial}{\partial x_i}[\rho v_i(\tfrac{1}{2}v^2 + e) + p_{ij}v_j + q_i] = \rho X_i v_i
\end{gathered} \tag{8.24}$$

where we have also used the following relations:

$$\begin{gathered}
\int \partial f/\partial \xi_i\, d\boldsymbol{\xi} = 0; \qquad \int \xi_j\, \partial f/\partial \xi_i\, d\boldsymbol{\xi} = -\int \delta_{ij} f\, d\boldsymbol{\xi} = -\rho\, \delta_{ij}; \\
\tfrac{1}{2}\int \xi^2\, \partial f/\partial \xi_i\, d\boldsymbol{\xi} = -\int \xi_i f\, d\boldsymbol{\xi} = -\rho v_i.
\end{gathered} \tag{8.25}$$

which follow by partial integration and the conditions $\lim_{\xi\to\infty}(\psi_\alpha f) = 0$ which are required in order that all the integrals considered in the equations of this section exist. In the above $\delta_{ij} = 1$ for $i = j$, $\delta_{ij} = 0$ for $i \neq j$. Eqs. (8.24) are the basic equations of continuum mechanics, in particular of macroscopic gas dynamics; as they stand, however, they constitute an empty scheme, since there are five equations for 13 quantities (if Eq. (8.13) is taken into account). In order to have useful equations, one must have some expressions for p_{ij} and q_i in terms of ρ, v_i, e. Otherwise, one has to go back to the Boltzmann equation and solve it; and once this has been done, everything is done, and Eqs. (8.24) are useless!

In any macroscopic approach to fluid dynamics, one has to postulate, either on the basic of experiments or by plausible arguments, some phenomenological relations (the so-called "constitutive equations") between p_{ij}, q_i on one hand and ρ, v_i, e on the other. In the case of a gas, or, more generally, a fluid, there are two models which are well known: the Euler (or ideal) fluid:

$$p_{ij} = p\,\delta_{ij}; \qquad q_i = 0 \tag{8.26}$$

and the Navier-Stokes-Fourier (or viscous and thermally conducting) fluid:

$$\begin{aligned} p_{ij} &= p\,\delta_{ij} - \mu\left(\frac{\partial v_i}{\partial x_j} + \frac{\partial v_j}{\partial x_i}\right) - \lambda\frac{\partial v_k}{\partial x_k}\delta_{ij} \\ q_i &= -\kappa\,\partial T/\partial x_i \end{aligned} \tag{8.27}$$

where μ and λ are the viscosity coefficients (usually one neglects the so called bulk viscosity; then $\lambda = -(2\mu)/3)$) and κ is the heat conduction coefficient (μ, λ and κ can be functions of the density ρ and temperature T).

No such relations are to be introduced in the microscopic description; the single unknown f contains all the information about density, velocity, temperature, stresses and heat flux! Of course this is possible because f is a function of seven variables instead of four; the macroscopic approach (five functions of four variables) is simpler than the microscopic one (one function of seven variables) and is to be preferred whenever it can be applied. Therefore one of the tasks of a theory based on the Boltzmann equation is to deduce, for a gas in ordinary conditions, some approximate macroscopic model' (in particular Eqs. (8.27) with μ, λ, κ expressed in terms of molecular constants) and find out what the limits of application of this model are. We shall consider this part of the theory in Chapter V.

There are, however, regimes of such rarefaction that no general macroscopic theory in the usual sense is possible (constitutive equations such as Eq. (8.26) and (8.27) are not valid); in this case the Boltzmann equation must be solved and not used only to justify the macroscopic equations.

We note that if we apply Eqs. (8.1), (8.3), (8.12) to the Maxwellian given by Eq. (7.8), we find that the constant $\mathbf{v}$ appearing in the latter equation is actually the mass velocity, while

$$\alpha = 3(4e)^{-1} = (2RT)^{-1}, \qquad A = \rho(\tfrac{4}{3}\pi e)^{-\frac{3}{2}} = \rho(2\pi RT)^{-\frac{3}{2}} \tag{8.28}$$

Furthermore,

$$p_{ij} = p\,\delta_{ij} = \tfrac{2}{3}\rho e\,\delta_{ij} \qquad q_i = 0 \tag{8.29}$$

that is, a gas with a Maxwellian distribution satisfies the constitutive equations of the Euler fluid, Eq. (8.26).

The above discussion can be extended to the case of a mixture of gases. Of course now we have, for each component, all the quantities defined before and we may also consider total or average quantities for the whole mixture. In the case of polyatomic gases the total internal energy per unit mass, e, is the sum of $\frac{3}{2}RT$ (translational energy) and the average internal energy of the molecules, say ε (rotational and vibrational energy).

In the absence of reactions, conservation of mass holds for each species, but the equations of momentum and energy conservation are valid only for the total momentum and energy. In a reacting gas, even the mass conservation equation is not valid for the single species.

9. Non-cutoff potentials and grazing collisions. Fokker-Planck equation

In this section we give a brief discussion of some topics related to the use of a noncutoff potential and, in particular, the effect of grazing collisions. The question arises when we want to include the effect of long range interactions in the Boltzmann equation. We might modify the arguments of Section 4 by allowing $\partial U/\partial r$ to be different from zero for $r > \sigma$; then additional terms would arise, exactly as in Eqs. (4.2), (4.3), (4.4) (with a non-zero right-hand side, however). The integral defining $\bar{\mathbf{X}}$ is now extended to $|\mathbf{x} - \mathbf{x}_*| > \sigma$:

$$\bar{\mathbf{X}}(\mathbf{x}) = \frac{1}{m}\int_{|\mathbf{x}-\mathbf{x}_*|>\sigma} \rho(\mathbf{x}_*, t)\mathbf{X}(\mathbf{x}, \mathbf{x}_*)\, d\mathbf{x}_* \tag{9.1}$$

where $\rho = mn$ is the mass density defined by Eq. (8.1). If $\mathbf{X}$ is a central force, whose magnitude varies as $|\mathbf{x} - \mathbf{x}_*|^{-n}$ for $|\mathbf{x} - \mathbf{x}_*| \to \infty$ then the above integral is negligible for a Boltzmann gas, provided $n > 4$. In fact if $|X(\mathbf{x}, \mathbf{x}_*)| \leqslant \alpha\sigma^{n-1}/|\mathbf{x} - \mathbf{x}_*|^n$ for $|\mathbf{x} - \mathbf{x}_*| > \sigma$, where α is bounded when $\sigma \to 0$, and $A = \max_{x,x_*}(|\rho(\mathbf{x}) - \rho(\mathbf{x}_*)|/|\mathbf{x} - \mathbf{x}_*|)$, $r = |\mathbf{x} - \mathbf{x}_*|$, then:

$$|\bar{\mathbf{X}}| \leqslant \frac{1}{m}\rho(\mathbf{x})\left|\int_{r>\sigma} \mathbf{X}\, d\mathbf{x}_*\right| + \frac{1}{m}\left|\int_{r>\sigma} (\rho_* - \rho)\mathbf{X}\, d\mathbf{x}_*\right|$$

$$\leqslant \frac{1}{m} A\alpha\sigma^{n-1}4\pi\int_\sigma^\infty \frac{r}{r^n} r^2\, dr = \frac{4\pi\alpha A}{(n-4)M} N\sigma^3$$

where $M = Nm$ is the total mass of the gas and $\int \mathbf{X}\, d\mathbf{x}_* = 0$ for a central force. Eq. (9.2) shows that $|\bar{\mathbf{X}}| \to 0$ for $N\sigma^3 \to 0$, provided $n > 4$. Note that the latter conditions is sufficient but could turn out not to be necessary, because our crude estimates prevented possible cancellations. In any case

$n > 2$ appears to be a necessary condition and this excludes, for example, Coulomb forces between charged particles. In the latter case, as we mentioned before, the collective behaviour described by the Vlasov term $\bar{\mathbf{X}} \cdot \partial P/\partial \boldsymbol{\xi}$ is very important. Actually, the Vlasov term is sufficient to describe the effect of long range interactions for short times, but is inadequate for a study of the long time as well as the large distance behavior. A similar remark holds for the case of a neutral gas, because, although large distance interactions are formally negligible, they could be nonuniformly small and produce a non-negligible effect, perhaps a dominating one, when times and distances tend to infinity (as in $\varepsilon + e^{-t}$, where $\varepsilon \ll 1$ and $t \to \infty$).

A better way to include the effect of long range forces is obtained by treating their influence in a statistical way, on the basis of the fact that they produce a continuous sequence of small and almost random changes of velocity. This can be done as follows. In the equation of motion of a particle with position $\mathbf{x}$ and velocity $\boldsymbol{\xi}$, the force per unit mass is replaced by a randomly varying force, whose average properties reflect the influence exerted by the other particles. We write:

$$\frac{d\boldsymbol{\xi}}{dt} = \mathbf{X}(t, \mathbf{x}, \boldsymbol{\xi}) \tag{9.3}$$

and assume that if $\mathbf{R} = \overline{\mathbf{X}(t)}$ is the average force at time t then there is no correlation between $\mathbf{X}(t_1) - \mathbf{R}(t_1)$ and $\mathbf{X}(t_2) - \mathbf{R}(t_2)$ unless $t_1 = t_2$. That is, we concentrate all the correlation (which extends over a certain time interval) into a time instant, in such a way that

$$\overline{[\mathbf{X}(t_1) - \bar{\mathbf{X}}(t_1)][\mathbf{X}(t_2) - \bar{\mathbf{X}}(t_2)]} = 2\mathsf{D}\,\delta(t_1 - t_2) \tag{9.4}$$

where the bar denotes an average, and D is a tensor with components D_{ij} called the diffusion tensor for reasons that will become clear later. $\mathbf{R} = \bar{\mathbf{X}}$ should be zero for a particle with velocity equal to the mass velocity of the neighboring particles. If the latter condition is not satisfied, however, we expect an average change of $\boldsymbol{\xi}$, hence $\mathbf{R} = \mathbf{R}(\boldsymbol{\xi}) \neq 0$. Eq. (9.3) gives for the change of $\boldsymbol{\xi}$ in a time interval Δt:

$$\Delta\boldsymbol{\xi} = \int_t^{t+\Delta t} \mathbf{X}(\tau, \mathbf{x}, \boldsymbol{\xi})\, d\tau \tag{9.5}$$

Hence

$$\lim_{\Delta t \to 0} \frac{\overline{\Delta\boldsymbol{\xi}}}{\Delta t} = \lim_{\Delta t \to 0} \frac{1}{\Delta t} \int_t^{t+\Delta t} \overline{\mathbf{X}(\tau, \mathbf{x}, \boldsymbol{\xi})}\, d\tau = \mathbf{R}(\mathbf{x}, \boldsymbol{\xi}) \tag{9.6}$$

and

$$\lim_{\Delta t \to 0} \frac{\overline{\Delta\boldsymbol{\xi}\,\Delta\boldsymbol{\xi}}}{\Delta t} = \lim_{\Delta t \to 0} \frac{1}{\Delta t} \iint^{t+\Delta t} \overline{\mathbf{X}(\tau_1, \mathbf{x}, \boldsymbol{\xi})\mathbf{X}(\tau_2, \mathbf{x}, \boldsymbol{\xi})}\, d\tau_1\, d\tau_2 = 2\mathsf{D}(\mathbf{x}, \boldsymbol{\xi}) \tag{9.7}$$

Higher order correlations such as $\overline{\Delta\boldsymbol{\xi}\,\Delta\boldsymbol{\xi}\,\Delta\boldsymbol{\xi}\,\Delta\boldsymbol{\xi}}$ are of the order of higher powers of Δt, at least under the assumption of a Gaussian distribution of $\mathbf{X}(t, \mathbf{x}, \boldsymbol{\xi})$ about its mean value $\mathbf{R}(\mathbf{x}, \boldsymbol{\xi})$.

Let us denote by $T(\mathbf{x}, \mathbf{x}', \boldsymbol{\xi}, \boldsymbol{\xi}', \Delta t)$ the probability density that a molecule with velocity $\boldsymbol{\xi}'$ at $\mathbf{x}'$ will, after a time interval Δt, be at $\mathbf{x}$ with velocity $\boldsymbol{\xi}$ as a consequence of the random motion considered above (accordingly, T is determined by Eq. (9.3), at least in principle). Then the distribution function f at time $t + \Delta t$ is given in terms of f at time t by:

$$f(\mathbf{x}, \boldsymbol{\xi}, t + \Delta t) = \int T(\mathbf{x}, \mathbf{x}', \boldsymbol{\xi}, \boldsymbol{\xi}', \Delta t) f(\mathbf{x}', \boldsymbol{\xi}', t)\, d\mathbf{x}'\, d\boldsymbol{\xi}' \tag{9.8}$$

Eqs. (9.5) and (9.7) are equivalent to:

$$\begin{aligned}
&\lim_{\Delta t \to 0} \frac{1}{\Delta t} \int (\boldsymbol{\xi} - \boldsymbol{\xi}') T(\mathbf{x}, \mathbf{x}', \boldsymbol{\xi}, \boldsymbol{\xi}', \Delta t)\, d\boldsymbol{\xi}\, d\mathbf{x} = R(\mathbf{x}', \boldsymbol{\xi}') \\
&\lim_{\Delta t \to 0} \frac{1}{\Delta t} \int (\boldsymbol{\xi} - \boldsymbol{\xi}')(\boldsymbol{\xi} - \boldsymbol{\xi}') T(\mathbf{x}, \mathbf{x}', \boldsymbol{\xi}, \boldsymbol{\xi}', \Delta t)\, d\boldsymbol{\xi}\, d\mathbf{x} = 2\mathsf{D}(\mathbf{x}', \boldsymbol{\xi}') \\
&\lim_{\Delta t \to 0} \frac{1}{\Delta t} \int \underbrace{(\boldsymbol{\xi} - \boldsymbol{\xi}')(\boldsymbol{\xi} - \boldsymbol{\xi}') \cdots (\boldsymbol{\xi} - \boldsymbol{\xi}')}_{n \text{ times}} \times \\
&\qquad \times\, T(\mathbf{x}, \mathbf{x}', \boldsymbol{\xi}, \boldsymbol{\xi}', \Delta t)\, d\boldsymbol{\xi}\, d\mathbf{x} = 0 \qquad (n > 2)
\end{aligned} \tag{9.9}$$

Besides, since $d\mathbf{x}/dt = \boldsymbol{\xi}$:

$$\begin{aligned}
&\lim_{\Delta t \to 0} \frac{1}{\Delta t} \int (\mathbf{x} - \mathbf{x}') T(\mathbf{x}, \mathbf{x}', \boldsymbol{\xi}, \boldsymbol{\xi}', \Delta t)\, d\boldsymbol{\xi}\, d\mathbf{x} = \boldsymbol{\xi}' \\
&\lim_{\Delta t \to 0} \frac{1}{\Delta t} \int \underbrace{(\mathbf{x} - \mathbf{x}')(\mathbf{x} - \mathbf{x}') \cdots (\mathbf{x} - \mathbf{x}')}_{n \text{ times}} \times \\
&\qquad \times\, T(\mathbf{x}, \mathbf{x}', \boldsymbol{\xi}, \boldsymbol{\xi}', \Delta t)\, d\boldsymbol{\xi}\, d\mathbf{x} = 0 \qquad (n > 1)
\end{aligned} \tag{9.10}$$

Also, on account of the meaning of T

$$\int T(\mathbf{x}, \mathbf{x}', \boldsymbol{\xi}, \boldsymbol{\xi}', \Delta t)\, d\mathbf{x}\, d\boldsymbol{\xi} = 1 \tag{9.11}$$

If $\varphi(\mathbf{x}, \boldsymbol{\xi})$ is an arbitrary smooth function, then:

$$\begin{aligned}
&\int \varphi(\mathbf{x}, \boldsymbol{\xi}) \frac{\partial f}{\partial t}(\mathbf{x}, \boldsymbol{\xi}, t)\, d\mathbf{x}\, d\boldsymbol{\xi} \\
&\quad = \lim_{\Delta t \to 0} \frac{1}{\Delta t} \left[\int \varphi(\mathbf{x}, \boldsymbol{\xi}) f(\mathbf{x}, \boldsymbol{\xi}, t + \Delta t)\, d\mathbf{x}\, d\boldsymbol{\xi} - \int \varphi(\mathbf{x}, \boldsymbol{\xi}) f(\mathbf{x}, \boldsymbol{\xi}, t)\, d\mathbf{x}\, d\boldsymbol{\xi} \right]
\end{aligned} \tag{9.12}$$

By means of Eq. (9.8) and a Taylor expansion about $\mathbf{x}'$, $\boldsymbol{\xi}'$, we find:

$$\int \varphi(\mathbf{x}, \boldsymbol{\xi}) f(\mathbf{x}, \boldsymbol{\xi}, t + \Delta t)\, d\mathbf{x}\, d\boldsymbol{\xi}$$

$$= \int \varphi(\mathbf{x}, \boldsymbol{\xi}) T(\mathbf{x}, \mathbf{x}', \boldsymbol{\xi}, \boldsymbol{\xi}', \Delta t) f(\mathbf{x}', \boldsymbol{\xi}', t)\, d\mathbf{x}'\, d\boldsymbol{\xi}'\, d\mathbf{x}\, d\boldsymbol{\xi}$$

$$= \int \varphi(\mathbf{x}', \boldsymbol{\xi}') f(\mathbf{x}', \boldsymbol{\xi}', t)\, d\mathbf{x}'\, d\boldsymbol{\xi}' +$$

$$+ \Delta t \int \left[\frac{\partial \varphi}{\partial \mathbf{x}'} \cdot \boldsymbol{\xi}' + \frac{\partial \varphi}{\partial \boldsymbol{\xi}'} \cdot \mathbf{R}(\boldsymbol{\xi}') + \frac{\partial^2 \varphi}{\partial \boldsymbol{\xi}'\, \partial \boldsymbol{\xi}'} \cdot \mathsf{D}(\boldsymbol{\xi}')\right] f(\mathbf{x}', \boldsymbol{\xi}', t)\, d\mathbf{x}'\, d\boldsymbol{\xi}' +$$

$$+ o(\Delta t) \tag{9.13}$$

where $\mathsf{A} \cdot \mathsf{B} \equiv A_{ij} B_{ji}$ for any two tensors A and B, and $o(\Delta t)$ denotes terms which tend to zero faster than Δt. Eq. (9.12) then becomes, if the change of names $\mathbf{x}', \boldsymbol{\xi}' \to \mathbf{x}, \boldsymbol{\xi}$ is made:

$$\int \varphi(\mathbf{x}, \boldsymbol{\xi}) \frac{\partial f}{\partial t}\, d\mathbf{x}\, d\boldsymbol{\xi}$$

$$= \int \left[\frac{\partial \varphi}{\partial x} \cdot \boldsymbol{\xi} + \frac{\partial \varphi}{\partial \boldsymbol{\xi}} \cdot R(\mathbf{x}, \boldsymbol{\xi}) + \frac{\partial^2 \varphi}{\partial \boldsymbol{\xi}\, \partial \boldsymbol{\xi}} \cdot \mathsf{D}(\mathbf{x}, \boldsymbol{\xi})\right] f(\mathbf{x}, \boldsymbol{\xi}, t)\, d\mathbf{x}\, d\boldsymbol{\xi}$$

$$= \int \left[-\boldsymbol{\xi} \cdot \frac{\partial f}{\partial \mathbf{x}} - \frac{\partial}{\partial \boldsymbol{\xi}} \cdot (\mathbf{R} f) + \frac{\partial^2 \varphi}{\partial \boldsymbol{\xi}\, \partial \boldsymbol{\xi}} \cdot (\mathsf{D} f)\right] \varphi(\mathbf{x}, \boldsymbol{\xi})\, d\mathbf{x}\, d\boldsymbol{\xi} \tag{9.14}$$

where the last form comes from a suitable partial integration. Because of the arbitrariness of $\varphi(\mathbf{x}, \boldsymbol{\xi})$, Eq. (9.14) implies that f satisfies the following equation (Generalized Fokker-Planck equation in velocity space):

$$\frac{\partial f}{\partial t} + \boldsymbol{\xi} \cdot \frac{\partial f}{\partial \mathbf{x}} = \frac{\partial^2 f}{\partial \boldsymbol{\xi}\, \partial \boldsymbol{\xi}} \cdot (\mathsf{D} f) - \frac{\partial}{\partial \boldsymbol{\xi}} \cdot (\mathbf{R} f) \tag{9.15}$$

The first term describes a diffusion in velocity space and this is the reason why D is called the diffusion tensor.

Since the particle considered so far is just a generic particle, and the force $\mathbf{X}$ is produced by the other particles, there must not be any loss of momentum and energy on average; that is, the following relations must hold:

$$\begin{aligned} &\int \boldsymbol{\xi} \left[\frac{\partial^2}{\partial \boldsymbol{\xi}\, \partial \boldsymbol{\xi}} \cdot (\mathsf{D} f) - \frac{\partial}{\partial \boldsymbol{\xi}} \cdot (\mathbf{R} f)\right] d\boldsymbol{\xi} = 0 \\ &\int \xi^2 \left[\frac{\partial^2}{\partial \boldsymbol{\xi}\, \partial \boldsymbol{\xi}} \cdot (\mathsf{D} f) - \frac{\partial}{\partial \boldsymbol{\xi}} \cdot (\mathbf{R} f)\right] d\boldsymbol{\xi} = 0 \end{aligned} \tag{9.16}$$

which, after partial integration, become:

$$\int \mathbf{R} f \, d\boldsymbol{\xi} = 0$$
$$\int (Tr\,\mathsf{D}) f \, d\boldsymbol{\xi} + \int \boldsymbol{\xi} \cdot \mathbf{R} f \, d\boldsymbol{\xi} = 0 \tag{9.17}$$

where $\mathrm{Tr}(\mathsf{D}) = D_{ii} = D_{11} + D_{22} + D_{33}$ is the trace of D. Eqs. (9.17) cannot be satisfied for all f, unless $\mathbf{R}$ and D depend upon f itself (the trivial case $\mathbf{R} = \mathrm{Tr}\,\mathsf{D} = 0$ being excluded). This is not only acceptable but also necessary, because the random force $\mathbf{X}(t)$ depends upon the distribution of the particles.

Isotropy considerations suggest that

$$\mathsf{D} = D\mathsf{I}, \qquad \mathbf{R} = -F(\boldsymbol{\xi} - \mathbf{v}) \tag{9.18}$$

where I is the fundamental tensor (unit matrix), $\mathbf{v}$ the mass velocity of the gas, defined by Eq. (8.3) and D and F appropriate functions of $\mathbf{x}$ and $|\boldsymbol{\xi} - \mathbf{v}|$. In this case Eq. (9.15) becomes:

$$\frac{\partial f}{\partial t} + \boldsymbol{\xi} \cdot \frac{\partial f}{\partial \mathbf{x}} = \Delta_\xi(Df) + \frac{\partial}{\partial \boldsymbol{\xi}} \cdot [(\boldsymbol{\xi} - \mathbf{v})Ff] \tag{9.19}$$

where Δ_ξ is the Laplace operator in velocity space. A particularly simple case is obtained by taking D and F independent of $\boldsymbol{\xi}$. In this case the first of Eqs. (9.17) is trivially satisfied while the second one becomes:

$$D = RTF \tag{9.20}$$

provided use is made of the definition of temperature T (see Section 8).

When we derive the Boltzmann equation for molecules interacting with a central force, we can repeat the arguments of Sections 2, 3 and 4, except for the fact that the evolution of $P_N = \sum_{i=1}^{N} [f(\mathbf{x}_i, \boldsymbol{\xi}_i, t)/m]$ between two subsequent collisions involves a diffusion and a friction in the velocity space, as described by Eq. (9.15). This means that a term of the kind indicated on the right hand side of the latter equation should be added to the right hand side of the Boltzmann equation, Eq. (4.16).

Another way of treating interactions between widely separated molecules is to let $\sigma \to \infty$ in the Boltzmann collision operator, which amounts to extending the binary collision analysis to distances where it is not strictly applicable. This seems very odd at first sight, because σ, as defined above, is of the order 10^{-8} cm and we used the limit $\sigma \to 0$ when deriving the Boltzmann equation. However, there is a justification in putting $\sigma = \infty$; as a matter of fact, σ enters in Eq. (4.16) only through $B(\theta, V)$, and increasing σ means that we take more and more grazing collisions into account. Now these added collisions are so grazing that they hardly deflect the molecules from their initial paths; accordingly, a molecule which enters into such a grazing collision in a certain state of motion emerges in practically the same state and therefore, it is argued, the contribution from such molecules to the integral in Eq. (4.16) is practically zero ($f'f_*' \approx ff_*$). In other words, if we accept this argument, we should say that, if we arbitrarily enlarge the value of σ and, in particular, let $\sigma \to \infty$ we do not change anything, since we just add the same large number to each of the two terms involved in the difference:

$$\left(\iiint f'f_*'B(\theta, V)\, d\boldsymbol{\xi}_*\, d\varepsilon\, d\theta\right) - \left(\iiint ff_*B(\theta, V)\, d\boldsymbol{\xi}_*\, d\varepsilon\, d\theta\right) \quad (9.21)$$

How large these numbers are, one can easily see, since for any potential extending to infinity the above two integrals diverge! This is obvious if one remembers that, for example, the second integral is also given by (see Eq. (4.14)):

$$\int\int_0^\sigma\int_0^{2\pi} ff_*V\, d\boldsymbol{\xi}_* r\, dr\, d\varepsilon = f(\boldsymbol{\xi})\left[\int f(\boldsymbol{\xi}_*)\, |\boldsymbol{\xi}_* - \boldsymbol{\xi}|\, d\boldsymbol{\xi}_*\right]\pi\sigma^2 \quad (9.22)$$

which is plainly divergent when $\sigma \to \infty$.

It is true that, although both integrals diverge, if one does not separate them and writes the collision term as in Eq. (4.16), the final result is a finite (for a reasonably smooth f). However, this is not a justification for taking $\sigma = \infty$; such a justification should be based upon a proof that the grazing collisions corresponding to very large values of the impact parameter correctly describe the effect of a large number of simultaneous grazing collisions. In this case the collision integral should embody the effects which were previously described by means of the Fokker-Planck term. This can be shown in a formal way by noticing that for small deflections ($\theta \to \pi/2$), $\boldsymbol{\xi}'$ is close to $\boldsymbol{\xi}$ and $\boldsymbol{\xi}_*'$ to $\boldsymbol{\xi}_*$, and by using the fact that:

$$\begin{aligned}
&\int d\boldsymbol{\xi}\varphi(\boldsymbol{\xi})\iiint (f'f_*' - ff_*)B(\theta, V)\, d\boldsymbol{\xi}_*\, d\theta\, d\varepsilon \\
&= \iiiint \varphi(f'f_*' - ff_*)B(\theta, V)\, d\boldsymbol{\xi}\, d\boldsymbol{\xi}_*\, d\theta\, d\varepsilon \\
&= \iint d\theta\, d\varepsilon\left[\iint \varphi f'f_*'B(\theta, V)\, d\boldsymbol{\xi}\, d\boldsymbol{\xi}_* - \iint \varphi ff_*B(\theta, V)\, d\boldsymbol{\xi}\, d\boldsymbol{\xi}_*\right] \\
&= \iint d\theta\, d\varepsilon\left[\iint \varphi' ff_*B(\theta, V)\, d\boldsymbol{\xi}\, d\boldsymbol{\xi}_* - \iint \varphi ff_*B(\theta, V)\, d\boldsymbol{\xi}\, d\boldsymbol{\xi}_*\right] \\
&= \iiiint (\varphi' - \varphi)ff_*B(\theta, V)\, d\boldsymbol{\xi}\, d\boldsymbol{\xi}_*\, d\theta\, d\varepsilon
\end{aligned} \quad (9.23)$$

where $\varphi = \varphi(\boldsymbol{\xi})$ is a sufficiently smooth test function, $\varphi' = \varphi(\boldsymbol{\xi}')$ and the integration variables in the integral involving $f'f_*'$, $\boldsymbol{\xi}$ and $\boldsymbol{\xi}_*$, have been changed to $\boldsymbol{\xi}'$, $\boldsymbol{\xi}_*'$ with $d\boldsymbol{\xi}'\,d\boldsymbol{\xi}_*' = d\boldsymbol{\xi}\,d\boldsymbol{\xi}_*$ (see Chapter I, Section 4; also, Section 6 of this Chapter). A Taylor series expansion gives:

$$\varphi' - \varphi = (\boldsymbol{\xi}' - \boldsymbol{\xi})\cdot\frac{\partial f}{\partial \boldsymbol{\xi}} + \tfrac{1}{2}(\boldsymbol{\xi}' - \boldsymbol{\xi})(\boldsymbol{\xi}' - \boldsymbol{\xi})\cdot\frac{\partial^2 f}{\partial \boldsymbol{\xi}\,\partial \boldsymbol{\xi}} + o(|\boldsymbol{\xi}' - \boldsymbol{\xi}|^2) \tag{9.24}$$

where $o(|\boldsymbol{\xi}' - \boldsymbol{\xi}|^2)$ denotes terms which, for a twice differentiable f, are of order higher than second with respect to $|\boldsymbol{\xi}' - \boldsymbol{\xi}|$ when the latter tends to zero. When we consider grazing collisions, $|\boldsymbol{\xi}' - \boldsymbol{\xi}|$ is very small and we can neglect these higher order terms. If we insert Eq. (9.24) into Eq. (9.23) we find

$$\begin{aligned}\int \varphi(\boldsymbol{\xi}) \iiint (f'f_*' - ff_*)B(\theta, V)\, d\boldsymbol{\xi}_*\, d\theta\, d\varepsilon &= \int \left[\mathbf{R}(\boldsymbol{\xi})\cdot\frac{\partial \varphi}{\partial \boldsymbol{\xi}} + \mathsf{D}(\boldsymbol{\xi})\cdot\frac{\partial^2 \varphi}{\partial \boldsymbol{\xi}\,\partial \boldsymbol{\xi}}\right] f\, d\boldsymbol{\xi} \\ &= \int \varphi \left[\frac{\partial^2}{\partial \boldsymbol{\xi}\,\partial \boldsymbol{\xi}}\cdot(\mathsf{D}f) - \frac{\partial}{\partial \boldsymbol{\xi}}\cdot(\mathbf{R}f)\right] d\boldsymbol{\xi}\end{aligned} \tag{9.25}$$

where integration with respect to θ refers only to the grazing collisions (i.e. it runs from $\pi/2 - \eta$ to $\pi/2$, $\eta \ll \pi/2$) and:

$$\begin{aligned}\mathbf{R}(\boldsymbol{\xi}) &= \iiint (\boldsymbol{\xi}' - \boldsymbol{\xi}) f_* B(\theta, V)\, d\boldsymbol{\xi}_*\, d\theta\, d\varepsilon \\ 2\mathsf{D}(\boldsymbol{\xi}) &= \iiint (\boldsymbol{\xi}' - \boldsymbol{\xi})(\boldsymbol{\xi}' - \boldsymbol{\xi}) f_* B(\theta, V)\, d\boldsymbol{\xi}_*\, d\theta\, d\varepsilon\end{aligned} \tag{9.26}$$

Because of the arbitrariness of φ, Eq. (9.25) shows that the contribution of grazing collisions has exactly the form of the Fokker-Planck term appearing in Eq. (9.15). The present calculation also gives the exact expressions for $\mathbf{R}$ and D which turn out to be dependent upon f, in agreement with what was said before. The same expressions could be obtained from Eqs. (9.6) and (9.7), by noticing that the frequency of collisions of a given molecule with a molecule having velocity between $\boldsymbol{\xi}_*$ and $\boldsymbol{\xi}_* + d\boldsymbol{\xi}_*$ at impact angles between θ and $\theta + d\theta$, ε and $\varepsilon + d\varepsilon$, is $f_* B(\theta, |\boldsymbol{\xi} - \boldsymbol{\xi}_*|)\, d\boldsymbol{\xi}_*\, d\theta\, d\varepsilon$.

The above expressions for $\mathbf{R}$ and D may be simplified by noticing that, in a reference frame with the z-axis directed along $\mathbf{V}$:

$$\boldsymbol{\xi}' - \boldsymbol{\xi} = -\mathbf{n}(\mathbf{n}\cdot\mathbf{V}) = -V(\sin\theta\cos\theta\cos\varepsilon,\ \sin\theta\cos\theta\sin\varepsilon,\ \cos^2\theta) \tag{9.27}$$

Consequently:

$$\int_{(\pi/2)-\eta}^{\pi/2} d\theta \int_0^{2\pi} d\varepsilon B(\theta, V)(\boldsymbol{\xi}' - \boldsymbol{\xi})$$

$$= -V \int_{(\pi/2)-\eta}^{\pi} B(\theta, V)(0, 0, \cos^2\theta)\, d\theta = -\mathbf{V}F(V) \quad (9.28)$$

where

$$F(V) = \int_{(\pi/2)-\eta}^{\pi/2} \cos^2\theta B(\theta, V)\, d\theta \quad (9.29)$$

Hence:

$$\mathbf{R} = -\int \mathbf{V}F(V) f(\boldsymbol{\xi}_*)\, d\boldsymbol{\xi}_* \qquad (\mathbf{V} = \boldsymbol{\xi} - \boldsymbol{\xi}_*) \quad (9.30)$$

Analogously

$$\int_{(\pi/2)-\eta}^{\pi/2} \int_0^{2\pi} (\boldsymbol{\xi}' - \boldsymbol{\xi})(\boldsymbol{\xi}' - \boldsymbol{\xi}) B(\theta, V)\, d\theta\, d\varepsilon$$

$$= V^2 \int_{(\pi/2)-\eta}^{\pi/2} \begin{pmatrix} \frac{1}{2}\sin^2\theta & 0 & 0 \\ 0 & \frac{1}{2}\sin^2\theta & 0 \\ 0 & 0 & \cos^2\theta \end{pmatrix} \cos^2\theta B(\theta, V)\, d\theta$$

$$= (V^2\mathsf{I} - \mathbf{VV})\tfrac{1}{2} \int_{(\pi/2)-\eta}^{\pi/2} \sin^2\theta \cos^2\theta B(\theta, V)\, d\theta$$

$$+ \mathbf{VV} \int_{(\pi/2)-\eta}^{\pi/2} \cos^4\theta B(\theta, V)\, d\theta \quad (9.31)$$

The integral involving $\cos^4\theta$ is of the order of already neglected quantities, while the integral involving $\sin^2\theta\cos^2\theta = \cos^2\theta - \cos^4\theta$ differs from $F(V)$ by the same negligible integral. Hence, within the adopted approximation, the second of Eqs. (9.26) becomes:

$$2\mathsf{D}(\boldsymbol{\xi}) = \tfrac{1}{2}\int (V^2\mathsf{I} - \mathbf{VV}) F(V) f(\boldsymbol{\xi}_*)\, d\boldsymbol{\xi}_* \qquad (\mathbf{V} = \boldsymbol{\xi} - \boldsymbol{\xi}_*) \quad (9.32)$$

Note that $\mathbf{R}$ and D as given by Eqs. (9.30) and (9.32) satisfy Eqs. (9.17). In fact we have

$$\int \mathbf{R} f\, d\boldsymbol{\xi} = -\iint (\boldsymbol{\xi} - \boldsymbol{\xi}_*) F(|\boldsymbol{\xi} - \boldsymbol{\xi}_*|) f(\boldsymbol{\xi}) f(\boldsymbol{\xi}_*)\, d\boldsymbol{\xi}\, d\boldsymbol{\xi}_* = 0 \quad (9.33)$$

because the integral changes its sign when we effect the immaterial interchange between $\boldsymbol{\xi}$ and $\boldsymbol{\xi}_*$.

Also:

$$\int \xi \cdot \mathbf{R} f \, d\xi = -\int \xi \cdot (\xi - \xi_*) F(|\xi - \xi_*|) f(\xi) f(\xi_*) \, d\xi \, d\xi_*$$

$$= -\tfrac{1}{2} \int (\xi - \xi_*) \cdot (\xi - \xi_*) f(\xi) f(\xi_*) F(|\xi - \xi_*|) \, d\xi \, d\xi_*$$

$$= -Tr(\mathsf{D}) \tag{9.34}$$

where the interchange between ξ and ξ_* has been used again to symmetrize the integral, and the relation $\mathbf{V} \cdot \mathbf{V} = V^2 = \mathrm{Tr}(V^2 \mathsf{I} - \mathbf{VV})/2$ used to compute $\mathrm{Tr}(\mathsf{D})$.

For power law potentials, Eqs. (9.29), (5.21) and (5.23) give ($n > 2$)

$$F(V) = V^{(n-5)/(n-1)} O(\eta^{2-(2/n-1)}) \tag{9.35}$$

which shows that D and $\mathbf{R}$ are small for $n > 2$ ($\eta \ll \pi/2$). Hence grazing collisions are formally negligible (for $n > 2$), *provided* f is sufficiently smooth and sufficiently fast decreasing as $\xi \to \infty$. Nevertheless, the long time and long distance behavior can be affected by the grazing collisions. The length and time scales involved here are of the order of $l\eta^{2-(2/n-1)}$ and $l\eta^{2-(2/n-1)}/\bar{c}$ respectively, where l is the mean free path and $\bar{c}$ a suitable average velocity; the experimental verification of effects on these scales is probably very difficult and, as a consequence, the grazing collisions can be left out, unless we want to discuss these unusual effects. In this connection it is also to be pointed out that the expansion used above to deduce the Fokker-Planck term from the Boltzmann equation is not uniformly valid.

We conclude that in most problems the inclusion of the long range part of the potential (provided the exponent of the force for $r \to \infty$ is larger than 2; this excludes Coulomb forces) should not make much difference. The physical meaning of the corresponding part of the collision operator, however, is not the same as in the derivation of the Boltzmann equation, because it is to be interpreted as a description of many simultaneous deflections due to a many body interaction rather than a description of two body collisions. The strict binary collision analysis, in fact, is meaningless for distances larger than $n^{-\frac{1}{3}}$ where n is the number density. Consistency forces us to accept the restriction $\sigma < \sigma_0 \simeq n^{-\frac{1}{3}}$ [18]. It is the smallness of the deflections which makes the binary collision analysis significant, because we can apply the superposition principle and describe the result of a many-body interaction as a linear combination of the results of several two-body interactions.

For a different approach to the subject matter of this section, as well as for further details on the case of Coulomb forces see Refs. [19–21].

10. Model equations

One of the major shortcomings in dealing with the Boltzmann equation is the complicated structure of the collision integral, Eq. (6.1).

It is therefore not surprising that alternative, simpler expressions have been proposed for the collision term; they are known as collision models, and any Boltzmann-like equation where the Boltzmann collision integral is replaced by a collision model is called a model equation or a kinetic model.

The idea behind this replacement is that a large amount of detail of the two-body interaction (which is contained in the collision term) is not likely to influence significantly the values of many experimentally measured quantities. That is, unless very refined experiments are devised, it is expected that the fine structure of the collision operator $Q(f,f)$ can be replaced by a blurred image, based upon a simpler operator $J(f)$ which retains only the qualitative and average properties of the true collision operator.

The most widely known collision model is usually called the Bhatnagar, Gross and Krook (BGK) model, although Welander proposed it independently at about the same time as the abovementioned authors [22, 23]. The idea behind the BGK model (retained by more sophisticated models) is that the essential features of a collision operator are:

(1) The true collision term $Q(f,f)$ satisfies Eq. (8.22); hence the collision model $J(f)$ must satisfy

$$\int \psi_\alpha J(f)\, d\boldsymbol{\xi} = 0 \qquad (\alpha = 0, 1, 2, 3, 4) \tag{10.1}$$

(2) The collision term satisfies Eq. (7.3). Hence $J(f)$ must satisfy

$$\int \log f J(f)\, d\boldsymbol{\xi} \leqslant 0 \tag{10.2}$$

with equality holding if, and only if, f is a Maxwellian.

As we shall see in Chapter III, this second property expresses the tendency of the gas to a Maxwellian distribution. The simplest way of taking this feature into account seems to assume that the average effect of collisions is to change the distribution function $f(\boldsymbol{\xi})$ by an amount proportional to the departure of f from a Maxwellian $\Phi(\boldsymbol{\xi})$. So, if ν is a constant with respect to $\boldsymbol{\xi}$, we introduce the following collision model

$$J(f) = \nu[\Phi(\boldsymbol{\xi}) - f(\boldsymbol{\xi})] \tag{10.3}$$

The Maxwellian $\Phi(\boldsymbol{\xi})$ has five disposable scalar parameters $(\rho, \mathbf{v}, T)$ according to Eqs. (7.8) and (8.28); however, these are fixed by Eq. (10.1) which implies

$$\int \psi_\alpha \Phi(\boldsymbol{\xi})\, d\boldsymbol{\xi} = \int \psi_\alpha f(\boldsymbol{\xi})\, d\boldsymbol{\xi} \tag{10.4}$$

so that at any space point and time instant $\Phi(\boldsymbol{\xi})$ must have exactly the density, velocity and temperature of the gas, given by the distribution function $f(\boldsymbol{\xi})$ (see Eqs. (8.1), (8.3), (8.12), (8.16)). Since the latter will, in general, vary with both time and space coordinates, the same will be true for the parameters of $\Phi(\boldsymbol{\xi})$ which is accordingly called the local Maxwellian. The "collision frequency" ν is not restricted at this level and has to be fixed by means of additional considerations; we note, however, that ν can be a function of the local state of the gas and hence vary with both time and space coordinates.

We still have to prove that the BGK model satisfies Eq. (10.2) and that equality applies if, and only if, f is a Maxwellian. We have

$$\begin{aligned}\int \log f J(f)\, d\boldsymbol{\xi} &= \int \log(f/\Phi) J(f)\, d\boldsymbol{\xi} + \int \log \Phi J(f)\, d\boldsymbol{\xi} \\ &= \int \nu\Phi\left[\left(1 - \frac{f}{\Phi}\right)\log\left(\frac{f}{\Phi}\right)\right] d\boldsymbol{\xi}\end{aligned} \tag{10.5}$$

where the integral involving $\log \Phi$ is zero because the latter quantity is a linear combination of the ψ_α's and Eq. (10.1) applies. Eq. (7.6) with $\lambda = f/\Phi$ then shows that the last integral in Eq. (10.5) is nonpositive and equal to zero if and only if $f = \Phi$, that is if and only if f is a Maxwellian, as required.

We observe that the nonlinearity of the proposed $J(f)$ is much worse than the nonlinearity of the collision term $Q(f, f)$; in fact, the latter is simply quadratic in f, while the former contains f in both the numerator and the denominator of an exponential (the $\mathbf{v}$ and T appearing in Φ are functionals of f, defined by Eqs. (8.3), (8.12), (8.16)).

The main advantage in using the BGK collision term is that for any given problem one can deduce integral equations for the macroscopic variables ρ, $\mathbf{v}$, T (see Chapter VII); these equations are strongly nonlinear, but simplify some iteration procedures and make the treatment of interesting problems feasible on a high speed computer. Another advantage of the BGK model is offered by its linearized form (see Chapter IV).

The BGK model contains the most basic features of the Boltzmann collision integral, but has some shortcomings. Some of them can be avoided by suitable modifications, at the expense, however, of the simplicity of the model. A first modification can be introduced in order to allow the collision frequency to depend on the molecular velocity instead of being locally constant; this modification is suggested by the circumstance that a computation of the collision frequency for physical models of the molecules (rigid spheres, finite range potentials) shows that ν varies with the molecular velocity and this variation is expected to be important at high molecular velocities. Formally the modification is quite simple [24]; we have only to allow ν to depend on $\boldsymbol{\xi}$ (more precisely on the magnitude c of the random velocity $\mathbf{c}$, defined by Eq. (8.6)), while requiring that Eq. (10.1) still holds. All the basic formal properties, including Eq. (10.2), are retained, but the density, velocity and temperature which now appear in the Maxwellian Φ are not the local density, velocity and temperature of the gas, but some fictitious local parameters related to five functionals of f different from ρ, $\mathbf{v}$, T; this follows from the fact that Eq. (10.1) now gives

$$\int \nu(c)\psi_\alpha \Phi \, d\boldsymbol{\xi} = \int \nu(c)\psi_\alpha f \, d\boldsymbol{\xi} \tag{10.6}$$

instead of Eq. (10.4).

A different kind of correction to the BGK model is obtained when we want to adjust the model to give the same Navier-Stokes equations as the full Boltzmann equation; in fact, as we shall see in Chapter V, the BGK model gives the value $Pr = 1$ for the Prandtl number, a value which is not in agreement with both the true Boltzmann equation and the experimental data for a monatomic gas (which agree in giving $Pr \simeq \frac{2}{3}$). In order to have a correct Prandtl number, a further adjustable parameter is required beside the already available ν; accordingly one is led [25, 26] to generalize the BGK model by substituting a local anisotropic threedimensional Gaussian in place of the local Maxwellian (which is an isotropic Gaussian):

$$\begin{aligned} \Phi &= \rho\pi^{-\frac{3}{2}}(\det \mathsf{A})^{\frac{1}{2}} \exp\left[-\sum_{i,j=1}^{3} A_{ij}(\xi_j - v_j)(\xi_i - v_i)\right] \\ \mathsf{A} &= \|A_{ij}\| = \|(2RT/Pr)\,\delta_{ij} - 2(1 - Pr)p_{ij}/(\rho Pr)\|^{-1} \end{aligned} \tag{10.7}$$

If we let $Pr = 1$, we recover the BGK model. A disadvantage of this model (called the ES, or ellipsoidal statistical model) is that it has not been possible to prove (or disprove) that Eq. (10.2) holds. Other models with different choices of Φ have been proposed [27, 28] but they are not interesting, except for linearized problems (see Chapter IV), because they are extremely complicated from the point of view of obtaining solutions.

Another model is offered by a Fokker-Planck collision term with the choice (9.18), made simpler by taking a velocity independent D; then F is also velocity independent because of Eq. (9.20) and the collision model can be written as follows:

$$J(f) = D\sum_{k=1}^{3}\left\{\frac{\partial^2 f}{\partial \xi_k^{\,2}} + \frac{1}{RT}\frac{\partial}{\partial \xi_k}[(\xi_k - v_k)f]\right\} \tag{10.8}$$

where v_k ($k = 1, 2, 3$) are the components of the (local) mass velocity and T is the (local) temperature of the gas. This model seems to have been proposed for the first time by Frisch, Helfand, Lebowitz [29] in connection with the kinetic theory of liquids; but is also a good description of the grazing collisions in a gas, as was shown in Section 9. It is interesting to notice that if D is taken to be proportional to the pressure $p = \rho RT$, Eq. (10.8) has the same kind of nonlinearity (i.e. quadratic) as the true Boltzmann equation, an advantage over the BGK model.

The idea of kinetic models can be naturally extended to mixtures and polyatomic gases [27, 30, 31]. A typical collision term of the BGK type will take the form

$$J_i(f_r) = \sum_{i=1}^{n} J_{ij}(f_r) = \sum_{i=1}^{n} \nu_{ji}(\Phi_{ji} - f_j) \tag{10.9}$$

where the ν_{ij} are collision frequencies and Φ_{ij} is a Maxwellian to be determined by suitable conditions which generalize Eq. (10.1).

Appendix

In this appendix we show that if P is a probability density, then:

$$H(P) = -\overline{\log P} = -\int P \log P \, d\mu \qquad \left(\int P \, d\mu = 1\right) \tag{A.1}$$

(where $d\mu$ is the volume element in the space M of the events, whose probability density is P) is a suitable measure of the likelihood of P. In other words, if we take several P's "at random", provided positive and normalized, most of them will be close to the probability density P for which $H(P)$ is maximum. In order to give a meaning to the words "at random", we assume that the space M is subdivided into n cells Ω_i of volume μ_i and the probability density P is replaced by a set of numbers P_i which are the averages of P over the cells:

$$P_i = \frac{1}{\mu_i} \int_{\Omega_i} P \, d\mu \qquad \left(\sum_i P_i \mu_i = 1\right) \tag{A.2}$$

We then take N objects and distribute them at random in the cells. If N_i of them are in Ω_i $(0 \leqslant N_i \leqslant N)$, we take the number $P_i = N_i/(N\mu_i)$ to give the probability for the cell Ω_i. Conversely, given a probability density P, we can represent it as closely as we wish by taking n and N sufficiently large, by means of a distribution of N objects in n cells. For a *given order of approximation*, however, we have only a finite, albeit huge, number of possible distributions. If we distribute the objects at random, there are $W(P) = N!/N_1!N_2! \cdots N_n!$ ways of realizing the distribution $P = (P_1, P_2, \ldots, P_n)$.

In fact, we can take the first set of N_1 objects at random from the set of N objects in

$$\binom{N}{N_1} = \frac{N!}{N_1!\,(N - N_1)!}$$

ways, the second set of N_2 objects at random from the remaining $N - N_1$ objects in $\binom{N - N_1}{N_2}$ ways, etc. so that the distribution $(P_1, \ldots, P_n)$ can be realized in

$$W(P) = \binom{N}{N_1}\binom{N - N_1}{N_2}\binom{N - (N_1 + N_2)}{N_3}\cdots\binom{N - (N_1 + \cdots + N_{n-1})}{N_n}$$
$$= \frac{N!}{N_1!\,N_2!\cdots N_n!} \tag{A.3}$$

ways, as stated. The total number of possible distributions is, by the formula which gives the N-th power of a polynomial:

$$\sum W(P) = \sum_{N,\ldots,N_n} \frac{N!}{N_1!\,N_2!\cdots N_n!} = (\underbrace{1 + 1 + \cdots + 1}_{n \text{ terms}})^N = n^N \tag{A.4}$$

Hence we are entitled to take, as a measure of the likelihood for a discrete distribution $(P_1, P_2, \ldots, P_n)$, the quantity $W(P)/n^N$ or its natural logarithm divided by N (this in order to obtain a finite limit for $N \to \infty$):

$$H(P) = \frac{1}{N}\log\left(\frac{N!}{N_1!\,N_2!\cdots N_n!\,n^N}\right) \tag{A.5}$$

Now we compute the limit for $N \to \infty$ which should give the appropriate expression for $H(P)$ when P_i takes all the possible real values. When $N \to \infty$ the following estimate holds:

$$\log N! = N\log N - N + o(N) \tag{A.6}$$

where $o(N)$ denotes a quantity such that $o(N)/N$ tends to zero when $N \to \infty$. Eq. (A.6) follows from Stirling's formula [32–33] or from the following inequality

$$2 < \frac{N!\,e^N}{N^N} < 3N \tag{A.7}$$

where $e = 2.71828\ldots$ is the basis of natural logarithms. In turn, Eq. (A.7) follows from the elementary inequality

$$\left(1 + \frac{1}{N}\right)^N < e < \left(1 + \frac{1}{N}\right)^{N+1}; \tag{A.8}$$

in fact, if we let $a_N = N!e^N/N^N$, then $a_{N+1} > a_N$ and $a_{N+1}/(N+1) < a_N/N$, because of Eq. (A.8), and Eq. (A.7) follows from the recurrence relation since $a_1 = e > 2$ and $a_1/1 = e < 3$. If we use Eq. (A.6) in Eq. (A.5) we find:

$$H(P) = \log N - 1 - \sum_{i=1}^{n} \frac{N_i}{N}(\log N_i - 1) - \log n + \frac{o(N)}{N}$$
$$= -\sum_{i=1}^{n} \frac{N_i}{N}\log\left(\frac{N_i}{N}\right) - \log n + \frac{o(N)}{N}$$
$$= -\sum_{i=1}^{n} P_i\mu_i \log(P_i n\mu_i) + \frac{o(N)}{N} \tag{A.9}$$

where the relations $\sum N_i = N$ and $P_i = N_i/(N\mu_i)$ have been taken into account. If we let $N \to \infty$ the last term in Eq. (A.9) disappears.

If we let $n \to \infty$, $\mu_i \to 0$ we can always *arrange* that $n\mu_i \to \bar{\mu}$ (total volume of the space of events, assumed to be finite for the sake of simplicity); it is sufficient to take $\mu_i = \bar{\mu}/n$. Then $(P_1, P_2, \ldots, P_n)$ tends to a continuous probability density P and Eq. (A.9) takes the form:

$$H(P) = -\int P \log(P\bar{\mu})\, d\mu = -\int P \log P\, d\mu - \log \bar{\mu} \int P\, d\mu$$

$$= -\int P \log P\, d\mu - \log \bar{\mu} \tag{A.10}$$

This is exactly Eq. (A.1) except for a nonessential additive constant (which, moreover, is zero if the measure $d\mu$ is normalized so that $\bar{\mu} = 1$).

We note a difficulty. If we change the variables describing the space of events, then $d\mu$ will become $d\mu = J\, d\mu'$ where $d\mu'$ is the volume element with respect to the new variables and J is the Jacobian determinant of the old variables with respect to the new ones. Then the appropriate probability density in the new variables is $P' = PJ$ (such that $P'\, d\mu' = P\, d\mu$ and consequently $\int P'\, d\mu' = 1$) and $H(P)$ becomes:

$$H(P) = -\int P' \log(P'/J)\, d\mu' = -\int P' \log P'\, d\mu' + \int P' \log J\, d\mu'$$

$$= H'(P') + \int P' \log J\, d\mu' \tag{A.11}$$

If J is constant, then $H'(P')$ differs from $H(P)$ by a constant, and hence is equivalent to it as a measure of likelihood (in particular, if $J = 1$, $H'(P') = H(P)$). If J is not a constant, then $H'(P')$ and $H(P)$ are not equivalent. Hence Eq. (A.1) is significant only when we use a particular class of variables such that two sets of variables of the class are related to each other by a transformation with a constant Jacobian. This class privileges a measure, which must have some physical property in order to be chosen for such purpose. In the case of a dynamical system of particles interacting with velocity independent forces, it is appropriate to use the physical variables, that is, the Cartesian components of the position vectors and velocities of the particles or variables related to them by a transformation with a constant Jacobian (in particular, for conservative forces, the so called canonical variables related to the Cartesian components of position vectors and momenta by a transformation with a unit Jacobian). In fact, the volume element $\prod_{k=1} d\mathbf{x}_k\, d\boldsymbol{\xi}_k$ is invariant during the evolution of the system because of the Liouville theorem, a property which selects this volume element from amongst the possible measures.

An application of Eq. (A.1) was given in the main text to show that if $P^{(1)}$ is given, then the most probable s-particle distribution $P^{(s)}$ is the factored one, $P^{(s)} = \prod_{i=1}^{s} P^{(1)}(\mathbf{x}_i, \boldsymbol{\xi}_i)$. Another possible application is to show that the most probable one-particle distribution for a perfect gas of given total energy is a Maxwellian. In fact, if we maximize

$$H(P) = -\int P \log P \, d\boldsymbol{\xi} \, d\mathbf{x} \tag{A.12}$$

with the constraints (e = energy per unit mass)

$$\int P \, d\boldsymbol{\xi} \, d\mathbf{x} = 1 \qquad \tfrac{1}{2}\int \xi^2 P \, d\boldsymbol{\xi} \, d\mathbf{x} = e \tag{A.13}$$

we find

$$-(1 + \log P) + \mu + \frac{\lambda}{2}\xi^2 = 0 \tag{A.14}$$

where λ and μ are (constant) Lagrange multipliers. Eq. (A.14) gives:

$$P = A \exp\left(-\frac{\lambda}{2}\xi^2\right) \qquad (A = \exp(1 - \mu)) \tag{A.15}$$

Inserting Eq. (A.15) into Eqs. (A.13) gives the appropriate values of A and λ; Eq. (I.6.25) is thus recovered.

A subtler argument is required to obtain the most likely N-particle distribution function. If we use the constraints (we assume equal masses for simplicity):

$$\int P_N \prod_{k=1}^{N} d\mathbf{x}_k \, d\boldsymbol{\xi}_k = 1 \qquad \int \left(\sum_{i=1}^{N} \tfrac{1}{2}\xi_i^2\right) P_N \prod_{k=1}^{N} d\mathbf{x}_k \, d\boldsymbol{\xi}_k = Ne \tag{A.16}$$

when maximising

$$H(P_N) = -\int P_N \log P_N \prod_{k=1}^{N} d\mathbf{x}_k \, d\boldsymbol{\xi}_k \tag{A.17}$$

we obtain

$$-(1 + \log P_N) + \mu + \frac{\lambda}{2}\sum_{i=1}^{N} \xi_i^2 = 0 \tag{A.18}$$

or

$$P_N = A \exp\left(-\frac{\lambda}{2}\sum_{i=1}^{N} \xi_i^2\right) \tag{A.19}$$

so that we do not recover the appropriate distribution function, given by Eq. (I.6.12). The reason is that we have imposed the weaker constraint given by the second of Eqs. (A.16) rather than

$$P_N = 0 \quad \text{if} \quad \sum_{k=1}^{N} \tfrac{1}{2}\xi_k^2 \neq Ne \tag{A.20}$$

which is the correct one for the N-particle distribution function. Eq. (A.21) can be taken into account in two ways: either we switch to a modified $H(P_N)$ where integration extends to the hypersphere $\sum_{i=1}^{N} \xi_i^2 = 2Ne$ with respect to the measure given by the surface element (in this case Eq. (A.20) is automatically satisfied and $P_N = \text{const.}$ over the surface maximizes $H(P_N)$ as required by Eq. (I.6.12)), or we assume that $P_N = 0$ for $|\sum_{i=1}^{N} \xi_i^2 - 2Ne)| > \varepsilon$ and then let $\varepsilon \to 0$ (in this case $P_N^{(\varepsilon)} = A_N H(\varepsilon - |\sum_{i=1}^{N} \frac{1}{2}\xi_i^2 - 2Ne|)$ and A_N turns out to be proportional to $1/\varepsilon$ for the normalization of P_N to be valid; then, when $\varepsilon \to 0$, $P_N^{(\varepsilon)}$ tends to a multiple of the delta function, as is checked by letting $\varepsilon = 1/m$, $m \to \infty$ and recalling Eq. (I.2.8)).

References

[1] S. Chapman and T. G. Cowling, "The Mathematical Theory of Nonuniform Gases," Cambridge University Press, Cambridge (1952).

[2] J. O. Hirschfelder, C. F. Curtiss and R. B. Bird, "Molecular Theory of Gases and Liquids," Wiley, New York (1954).

[3] L. Waldmann, in "Handbuch der Physik", vol. XII, p. 384, Springer Verlag, Berlin (1958).

[4] L. Boltzmann, "Lectures on Gas Theory" (English translation from the German), University of California Press, Berkeley (1964).

[5] T. Carleman, "Problemes Mathematiques dans la Theorie Cinetique des Gas", Almqvist and Wiksells, Uppsala (1957).

[6] J. Jeans, "Dynamical Theory of Gases," Dover, New York (1954).

[7] M. N. Kogan, "Rarefied Gas Dynamics," Plenum Press, New York (1969).

[8] H. Grad, "Comm. Pure Appl. Math.", **14,** 323 (1961).

[9] C. Cercignani, "Transport Theory and Statistical Physics", **2,** 211 (1972).

[10] N. N. Bogoliubov, "Problems of a Dynamical Theory in Statistical Physics," (English translation from the Russian) in "Studies in Statistical Mechanics", J. de Boer and G. E. Uhlenbeck, Eds., Vol. I, p. 5, North-Holland Publishing Company, Amsterdam (1962).

[11] M. Born and H. S. Green, "A General Kinetic Theory of Fluids", Cambridge University Press, Cambridge (1949).

[12] J. G. Kirkwood, "J. Chem. Phys.", **14,** 180 (1946).

[13] J. G. Kirkwood, "J. Chem. Phys.", **15,** 72 (1947).

[14] J. Yvon, "La Theorie Statistique des Fluides", Actualites Scientifiques et Industrielles, no. 203, Hermann, Paris (1935).

[15] H. Grad, in "Handbuch der Physik", vol. XII, Springer, Berlin (1958).

[16] C. S. Wang Chang, G. E. Uhlenbeck and J. de Boer, in "Studies in Statistical Mechanics", J. de Boer and G. E. Uhlenbeck, Eds. vol. II, Part c, North Holland Publishing Company, Amsterdam (1964).

[17] J. C. MAXWELL, "Phil. Trans. Roy. Soc.", 157 (1866); reprinted in: "The Scientific Papers by J. C. Maxwell", Dover, New York (1965).
[18] C. CERCIGNANI, "Phys. Fluids", **10,** 2097 (1967).
[19] I. PRIGOGINE, "Non-Equilibrium Statistical Mechanics", Interscience, New York (1962).
[20] R. BALESCU, "Statistical Mechanics of Charged Particles", Interscience, New York (1962).
[21] R. L. LIBOFF, "Introduction to the Theory of Kinetic Equations", Wiley, New York (1969).
[22] P. L. BHATNAGAR, E. P. GROSS and M. KROOK, "Phys. Rev.", **94,** 511 (1954).
[23] P. WELANDER, "Arkiv Fysik", **7,** 507 (1954).
[24] M. KROOK, "J. Fluid Mech.", **6,** 523 (1959).
[25] L. H. HOLWAY, JR., Ph.D. Thesis, Harvard (1963).
[26] C. CERCIGNANI and G. TIRONI, "in Rarefied Gas Dynamics" (C. L. Brundin, ed.), vol. I, Academic Press, New York (1967), p. 441.
[27] L. SIROVICH, "Phys. Fluids", **5,** 908 (1962).
[28] C. CERCIGNANI, "Mathematical Methods in Kinetic Theory", Plenum Press, New York (1969).
[29] J. L. LEBOWITZ, H. L. FRISCH, and E. HELFAND, "Phys. Fluids", **3,** 325 (1960).
[30] T. F. MORSE, "Phys. Fluids", **7,** 2012 (1964).
[31] F. B. HANSON and T. F. MORSE, "Phys. Fluids", **10,** 345 (1967).
[32] H. JEFFREYS, "Asymptotic Approximations", Oxford University Press, Oxford (1962).
[33] E. ARTIN, "The Gamma Function", Holt, Rinehart and Winston, New York (1964).

III GAS-SURFACE INTERACTION AND THE H-THEOREM

1. Boundary conditions and the gas-surface interaction

In this and the following sections we shall investigate the conditions which must be satisfied by the distribution function f at the boundary ∂R of the region R where the motion of the particles under study takes place. This subject was briefly touched upon when we derived the Boltzmann equation for a gas of rigid spheres (see Chapter II, Section 2). There the discussion of the boundary conditions satisfied by the N-particle distribution function was needed in order to discard some surface integrals. It is clear that such conditions are also required in order to solve the Boltzmann equation, since the latter contains the space derivatives of f.

In the case of a gas flowing past a solid body or within a region bounded by one or more solid bodies, the boundary conditions describe the interaction of the gas molecules with the solid walls. It is to this interaction that one can trace the origin of the drag and lift exerted by the gas on the body and the heat transfer between the gas and the solid boundary. Unfortunately, both theoretical and experimental information on gas-surface interactions is rather scanty.

The difficulties of a theoretical investigation are due, mainly, to our lack of knowledge of the structure of the surface layers of solid bodies and hence of the interaction potential of the gas molecules with molecules of the solid. When a molecule impinges upon a surface, it is adsorbed and may form chemical bonds, dissociate, become ionized or displace surface molecules. The state of the surface layer depends not only on the surface temperature, but also on its roughness and cleanliness. The latter, in turn, may vary with time because outgassing or preliminary heating promote purification. In general, adsorbed layers may be present; in this case, the interaction of a given molecule with the surface also depends on the distribution of molecules incident on the surface element.

The main source of experimental data is given by the patterns of re-emitted molecules obtained when a molecular beam is shone upon a surface (see Section 7).

The simplest possible model of the gas-surface interaction is to assume that the molecules are specularly reflected at the solid boundary

$$f(\mathbf{x}, \boldsymbol{\xi}, t) = f(\mathbf{x}, \boldsymbol{\xi} - 2\mathbf{n}(\mathbf{n} \cdot \boldsymbol{\xi}), t), \qquad (\mathbf{x} \in \partial R, \boldsymbol{\xi} \cdot \mathbf{n} > 0) \qquad (1.1)$$

where $\mathbf{n}$ is the unit vector normal to the surface at $\mathbf{x}$ (see Fig. 14). This assumption is extremely unrealistic in general and can be used only in particular cases. In general, a molecule striking the surface at a velocity $\boldsymbol{\xi}'$ reflects from it at a velocity $\boldsymbol{\xi}$ which is strictly determined only if the path of the molecule within the wall can be computed exactly. This computation is impossible because it depends upon a great number of details, such as the locations and velocities of all the molecules of the wall. Hence we may only hope to compute the probability density $R(\boldsymbol{\xi}' \to \boldsymbol{\xi}; \mathbf{x}, t; \tau)$ that a molecule striking the surface with velocity between $\boldsymbol{\xi}'$ and $\boldsymbol{\xi}' + d\boldsymbol{\xi}'$ at the point $\mathbf{x}$ and time t re-emerges at practically the same point with velocity between $\boldsymbol{\xi}$ and $\boldsymbol{\xi} + d\boldsymbol{\xi}$ after a time interval τ (adsorption or sitting time). If R is known, then we can easily write the boundary condition for f by the following argument, in which we assume the wall to be at rest (otherwise $\boldsymbol{\xi}$, $\boldsymbol{\xi}'$ must be replaced by $\boldsymbol{\xi} - \mathbf{u}_0$, $\boldsymbol{\xi}' - \mathbf{u}_0$ throughout, $\mathbf{u}_0$ denoting the velocity of the wall).

The mass of molecules emerging with velocity between $\boldsymbol{\xi}$ and $\boldsymbol{\xi} + d\boldsymbol{\xi}$ from a surface element dA about $\mathbf{x}$ in the time interval between t and $t + dt$ is (see the analogous argument when we derived Eq. (II, 2.23))

$$d^*\mathcal{M} = f(\mathbf{x}, \boldsymbol{\xi}, t)\, |\boldsymbol{\xi} \cdot \mathbf{n}|\, dt\, dA\, d\boldsymbol{\xi} \qquad (\mathbf{x} \in \partial R,\ \boldsymbol{\xi} \cdot \mathbf{n} > 0) \tag{1.2}$$

Analogously, the probability that a molecule impinges upon the same surface element with velocity between $\boldsymbol{\xi}'$ and $\boldsymbol{\xi}' + d\boldsymbol{\xi}'$ in the time interval

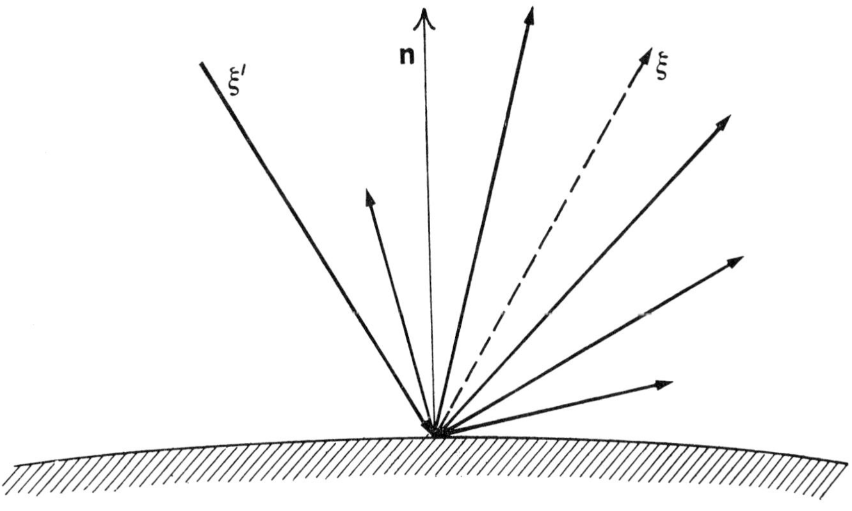

Fig. 14. The velocity $\boldsymbol{\xi}$ of a reemerging molecule is not strictly determined by the velocity possessed by the same molecule before hitting the wall, unless specular reflection applies (dashed line).

between $t - \tau$ and $t - \tau + dt$ is

$$d^*\mathscr{M}' = f(\mathbf{x}, \boldsymbol{\xi}', t - \tau)\, |\boldsymbol{\xi}' \cdot \mathbf{n}|\, dt\, dA\, d\boldsymbol{\xi} \qquad (\mathbf{x} \in \partial R, \boldsymbol{\xi}' \cdot \mathbf{n} < 0) \quad (1.3)$$

If we multiply $d^*\mathscr{M}'$ by the probability of a scattering from velocity $\boldsymbol{\xi}'$ to a velocity between $\boldsymbol{\xi}$ and $\boldsymbol{\xi} + d\boldsymbol{\xi}$ with an adsorption time between τ and $\tau + d\tau$; that is, consider $R(\boldsymbol{\xi}' \to \boldsymbol{\xi}; \mathbf{x}, t; \tau)\, d\boldsymbol{\xi}\, d\tau$ and "sum" (integrate) over all the possible values of $\boldsymbol{\xi}'$ and τ, we must obtain $d^*\mathscr{M}$ (here we assume that each molecule re-emerges from the surface element into which entered, which is unrealistic when τ is not small):

$$d^*\mathscr{M} = d\boldsymbol{\xi} \int_0^\infty d\tau \int_{\boldsymbol{\xi}' \cdot \mathbf{n} < 0} R(\boldsymbol{\xi}' \to \boldsymbol{\xi}; \mathbf{x}, t; \tau)\, d^*\mathscr{M}' \qquad (\mathbf{x} \in \partial R, \boldsymbol{\xi} \cdot \mathbf{n} > 0) \quad (1.4)$$

Using Eqs. (1.2) and (1.3) and cancelling the common factor $dA\, d\boldsymbol{\xi}\, dt$, we obtain

$$|\boldsymbol{\xi} \cdot \mathbf{n}|\, f(\mathbf{x}, \boldsymbol{\xi}, t)$$
$$= \int_0^\infty d\tau \int_{\boldsymbol{\xi}' \cdot \mathbf{n} < 0} R(\boldsymbol{\xi}' \to \boldsymbol{\xi}; \mathbf{x}, t; \tau) f(\mathbf{x}, \boldsymbol{\xi}', t - \tau) \quad (\mathbf{x} \in \partial R, \boldsymbol{\xi} \cdot \mathbf{n} > 0) \quad (1.5)$$

If $\bar{\tau}$ is an average adsorption time, $\bar{v}$ the average normal velocity with which the gas molecules impinge upon the surface and n, as usual, the number density of the gas, $n\bar{v}\, dA$ molecules will impinge, per unit time, on a surface element of area dA and stay there an average time $\bar{\tau}$; if σ_0 is an effective range of the gas surface interaction each molecule will occupy an area $\pi\sigma_0^2$ and the total area occupied by the adsorbed molecules will be $n\bar{v}\bar{\tau}\pi\sigma_0^2\, dA$, which is to say, a fraction $n\bar{v}\bar{\tau}\pi\sigma_0^2$ of the surface will be occupied. This argument, of course, is just a rough order-of-magnitude argument, because the molecules may penetrate somewhat into the solid wall and not necessarily remain at the surface of it.

If $n\sigma_0^2\bar{v}\bar{\tau}$ is not close to zero, the nature of the interaction of each incident molecule depends on the total number and energy of the incident molecules. Under conditions of extremely low density (for example, for an orbiting satellite $n\sigma_0^2\bar{v}\bar{\tau} \ll 1$) each incident molecule interacts with the surface independently of the others. The same independence may show up in the other limiting case $n\sigma_0^2\bar{v}\bar{\tau} \simeq 1$ (for example, in chemical adsorption when $\bar{\tau}$ may be very large, or in a dense gas when n can be extremely large); in this case an adsorbed layer is formed and the molecule interacts directly with this layer rather than with the atoms of the solid surface (and the adsorption time for this interaction can be much shorter).

Whenever the effective adsorption time $\bar{\tau}$ and number density n are such that $n\sigma_0^2\bar{v}\bar{\tau} \ll 1$, we can assume that $R(\boldsymbol{\xi}' \to \boldsymbol{\xi}; \mathbf{x}, t; \tau)$ does not depend on the distribution function $f(\mathbf{x}, \boldsymbol{\xi}', t)$; hence we can compute $R(\boldsymbol{\xi}' \to \boldsymbol{\xi}; \mathbf{x}, t; \tau)$ under the assumption that one gas molecule of given velocity $\boldsymbol{\xi}'$ impinges upon the wall. If, in addition, $\bar{\tau}$ is small compared with any characteristic time of interest in the evolution of f we can let $\tau = 0$ in the argument of f appearing on the right hand side of Eqs. (1.5); in this case the latter becomes

$$|\boldsymbol{\xi} \cdot \mathbf{n}| f(\mathbf{x}, \boldsymbol{\xi}, t)$$
$$= \int_{\boldsymbol{\xi}' \cdot \mathbf{n} < 0} R(\boldsymbol{\xi}' \to \boldsymbol{\xi}; \mathbf{x}, t) f(\mathbf{x}, \boldsymbol{\xi}', t) \, |\boldsymbol{\xi}' \cdot \mathbf{n}| \, d\boldsymbol{\xi}' \qquad (\mathbf{x} \in \partial R; \boldsymbol{\xi} \cdot \mathbf{n} > 0) \quad (1.6)$$

where

$$R(\boldsymbol{\xi}' \to \boldsymbol{\xi}; \mathbf{x}, t) = \int_0^\infty R(\boldsymbol{\xi}' \to \boldsymbol{\xi}; \mathbf{x}, t; \tau) \, d\tau \qquad (1.7)$$

Eq. (1.6) is, in particular, valid for steady problems.

If the wall restitutes all the gas molecules (i.e. it is nonporous and non-adsorbing), the total probability for an impinging molecule to be re-emitted, with no matter what velocity $\boldsymbol{\xi}$, is unity:

$$\int_{\boldsymbol{\xi}' \cdot \mathbf{n} > 0} R(\boldsymbol{\xi}' \to \boldsymbol{\xi}; \mathbf{x}, t) \, d\boldsymbol{\xi} = 1 \qquad (\boldsymbol{\xi}' \cdot \mathbf{n} < 0) \qquad (1.8)$$

Eq. (1.8) has a simple consequence: the normal component v_n of the mass velocity at the wall must be zero. This is physically obvious in view of the meaning of Eq. (1.8), but the formal proof is of some interest:

$$v_n = \int \boldsymbol{\xi} \cdot \mathbf{n} f(\boldsymbol{\xi}) \, d\boldsymbol{\xi} = \int_{\boldsymbol{\xi} \cdot \mathbf{n} < 0} |\boldsymbol{\xi} \cdot \mathbf{n}| f(\boldsymbol{\xi}) \, d\boldsymbol{\xi} - \int_{\boldsymbol{\xi} \cdot \mathbf{n} < 0} |\boldsymbol{\xi} \cdot \mathbf{n}| f(\boldsymbol{\xi}) \, d\boldsymbol{\xi}$$
$$= \int_{\boldsymbol{\xi}' \cdot \mathbf{n} < 0} d\boldsymbol{\xi}' \int_{\boldsymbol{\xi} \cdot \mathbf{n} > 0} d\boldsymbol{\xi} R(\boldsymbol{\xi}' \to \boldsymbol{\xi}) f(\boldsymbol{\xi}') \, |\boldsymbol{\xi}' \cdot \mathbf{n}| - \int_{\boldsymbol{\xi}' \cdot \mathbf{n} < 0} |\boldsymbol{\xi}' \cdot \mathbf{n}| f(\boldsymbol{\xi}') \, d\boldsymbol{\xi}'$$
$$= \int_{\boldsymbol{\xi}' \cdot \mathbf{n} < 0} |\boldsymbol{\xi}' \cdot \mathbf{n}| f(\boldsymbol{\xi}') \, d\boldsymbol{\xi}' - \int_{\boldsymbol{\xi}' \cdot \mathbf{n} < 0} |\boldsymbol{\xi}' \cdot \mathbf{n}| f(\boldsymbol{\xi}') \, d\boldsymbol{\xi}' = 0 \qquad (1.9)$$

Here we omitted the inessential arguments $\mathbf{x}$, t and used Eqs. (1.6) and (1.8).

An obvious property of the kernel $R(\boldsymbol{\xi}' \to \boldsymbol{\xi}; \mathbf{x}, t)$ is that it cannot assume negative values

$$R(\boldsymbol{\xi}' \to \boldsymbol{\xi}; \mathbf{x}, t) \geqslant 0 \qquad (1.10)$$

In the case of a specularly reflecting boundary, Eqs. (1.1) and (1.6) show that

$$R(\boldsymbol{\xi}' \to \boldsymbol{\xi}; \mathbf{x}, t) = \delta(\boldsymbol{\xi} - \boldsymbol{\xi}' + 2\mathbf{n}[\mathbf{n} \cdot \boldsymbol{\xi}']) \qquad (1.11)$$

Situations may arise in which the wall does not restitute all the particles, and Eq. (1.8) is violated. An extreme case is when no particles enter the region R through ∂R (then the kernel $R(\boldsymbol{\xi}' \to \boldsymbol{\xi}; \mathbf{x}, t) = 0$); such is the case for a gas of neutrons at the boundary of a nuclear reactor, a gas of photons at the surface of completely adsorbing medium or a gas of electrons at a positively charged plate.

Other phenomena which occur in practice are chemical reactions at the wall and the sputtering of ions or atoms of the solid as a consequence of a collision with the molecules of the gas. In this case new species of molecules are present in the gas and we have to introduce kernels $R_i^{\,j}(\boldsymbol{\xi}' \to \boldsymbol{\xi}; \mathbf{x}, t; \tau)$ specifying the transition probabilities from a particle of species i with velocity $\boldsymbol{\xi}'$ to a particle of species j with velocity $\boldsymbol{\xi}$.

2. Computation of scattering kernels

The problem of computing the scattering kernel $R(\boldsymbol{\xi}' \to \boldsymbol{\xi}; \mathbf{x}, t)$ defined by Eq. (1.7) is amenable to further mathematical elaboration if we add the assumption that, although the molecules of the gas disturb the single atoms of the wall, these atoms are found, by an impinging molecule, to be in local thermal equilibrium with each other at a temperature T_0 which may vary from point to point on a macroscopic scale. In this case the dependence of $R(\boldsymbol{\xi}' \to \boldsymbol{\xi}; \mathbf{x}, t)$ on $\mathbf{x}$ and t can be consistently assumed to take place only through the temperature T_0 and the chemical composition of the wall. The latter may be taken to be known and hence, henceforth, we shall omit writing $\mathbf{x}$ and t as arguments of $R(\boldsymbol{\xi}' \to \boldsymbol{\xi})$.

Let us set up the mathematical problem of computing $R(\boldsymbol{\xi}' \to \boldsymbol{\xi})$ for a given wall of known temperature T_0, under the assumption that $R(\boldsymbol{\xi}' \to \boldsymbol{\xi})$ does not depend on $\mathbf{x}$, t (i.e. $n\sigma_0^2\bar{v}\tau \ll 1$). Then a molecule of the gas captured by the wall and the N atoms of the solid form a mechanical system, described by a $(N+1)$-particle distribution function P_{N+1} satisfying the Liouville equation Eq. (I.3.8). If we want to compute $R(\boldsymbol{\xi}' \to \boldsymbol{\xi})$ as given by Eq. (1.7), it is sufficient to study the steady Liouville equation (Eq. (I.6.6) with $N+1$ in place of N):

$$\sum_{k=0}^{N} \boldsymbol{\xi}_k \cdot \frac{\partial P_{N+1}}{\partial \mathbf{x}_k} + \sum_{\substack{j,k=0\\ j\neq k}}^{N} \mathbf{X}_{jk} \cdot \frac{\partial P_{N+1}}{\partial \boldsymbol{\xi}_k} = 0 \tag{2.1}$$

where the index zero refers to the gas molecule and $\mathbf{X}_{jk} = \mathbf{X}_{jk}(\mathbf{x}_l)$ denotes the interaction force (per unit mass) between the j-th and k-th molecule. The latter forces are such as to keep the solid body together, but in order to have a sharply defined problem, we shall assume that at a well defined geometrical boundary (the boundary of the solid) particles other than the gas molecule are specularly reflected (this excludes sputtering from our considerations). Since we assume that the kernel depends only on the local state of the wall, we shall consider a flat wall and a solid occupying the half space $\mathbf{x} \cdot \mathbf{n} < 0$.

Then:

$$P_{N+1}(\mathbf{x}_0, \boldsymbol{\xi}_0, \mathbf{x}_1, \boldsymbol{\xi}_1, \ldots, \mathbf{x}_i, \boldsymbol{\xi}_i, \ldots, \mathbf{x}_N, \boldsymbol{\xi}_N)$$
$$= P_{N+1}(\mathbf{x}_0, \boldsymbol{\xi}_0, \mathbf{x}_1, \boldsymbol{\xi}_1, \ldots, \mathbf{x}_i, \boldsymbol{\xi}_i - 2\mathbf{n}_i(\mathbf{n}_i \cdot \boldsymbol{\xi}_i), \ldots, \mathbf{x}_N, \boldsymbol{\xi}_N)$$
$$(1 \leqslant i \leqslant N; \mathbf{x}_i \cdot \mathbf{n} = 0; \boldsymbol{\xi}_i \cdot \mathbf{n} < 0) \qquad (2.2)$$

If we denote by $\mathbf{y}_0$ the two-dimensional position vector in the plane $\mathbf{x}_0 \cdot \mathbf{n} = 0$ the boundary condition at this plane is

$$P_{N+1}(\mathbf{x}_k, \boldsymbol{\xi}_k) = \delta(\boldsymbol{\xi}_0 - \boldsymbol{\xi}')\, \delta(\mathbf{y}_0 - \mathbf{y}')\bar{P}_N(\mathbf{x}_l, \boldsymbol{\xi}_l)$$
$$(\mathbf{x}_0 \cdot \mathbf{n} = 0; \boldsymbol{\xi}_0 \cdot \mathbf{n} < 0) \quad (2.3)$$

where k ranges from 0 to N, l from 1 to N and $\bar{P}_N$ is the N-particle equilibrium distribution for the atoms of the solid wall. Eq. (2.3) expresses the condition that a gas molecule entering the surface at $\mathbf{y}'$ with velocity $\boldsymbol{\xi}'$ finds a system of atoms in thermal equilibrium. If we assume that Eq. (I.6.11) holds for the solid, then $\bar{P}_N$ should be replaced by a delta function; since the gas molecule interacts with a comparatively small number of atoms of the solid, and we shall eventually integrate with respect to positions and velocities of all the atoms, we may write

$$\bar{P}_N = Z_N^{-1} \exp\{-[\textstyle\sum \frac{1}{2} m_l \xi_l^2 + V(\mathbf{x}_j)]/(kT_0)\} \qquad (2.4)$$

where $V(\mathbf{x}_j)$ is the potential energy of the system of atoms, m_l is the mass of the l-th atom, k is the Boltzmann constant and Z_N is a normalizing factor (the so-called partition function). Eq. (2.4) is equivalent to the delta function $\delta(\frac{1}{2}\sum_l m_l\, \xi_l^2 + V(\mathbf{x}_j) - E_0)$ when integrated over the majority of coordinates and velocities in the limit $N \to \infty$ as shown by the expressions of $P_N^{(1)}$ and $P_N^{(2)}$ discussed in Chapter I, Section 6.

If we were able to solve Eq. (2.1) with the boundary conditions (2.2) and (2.3), then we could compute $R(\boldsymbol{\xi}' \to \boldsymbol{\xi})$ as follows:

$$R(\boldsymbol{\xi}' \to \boldsymbol{\xi}) = \frac{|\boldsymbol{\xi} \cdot \mathbf{n}|}{|\boldsymbol{\xi}' \cdot \mathbf{n}|} \int [P_{N+1}^{(1)}(\mathbf{x}_0, \boldsymbol{\xi})]_{\mathbf{x}_0 \cdot \mathbf{n}=0}\, dA_0$$
$$= \frac{|\boldsymbol{\xi} \cdot \mathbf{n}|}{|\boldsymbol{\xi}' \cdot \mathbf{n}|} \int [P_{N+1}(\mathbf{x}_k, \boldsymbol{\xi}_k)]_{\substack{\mathbf{x}_0 \cdot \mathbf{n}=0 \\ \xi_0 = \xi}}\, dA_0 \prod_{j=1}^{N} d\mathbf{x}_j\, d\xi_j \qquad (2.5)$$

where dA_0 is the surface element in the plane $\mathbf{x}_0 \cdot \mathbf{n} = 0$. The first expression in Eq. (2.5) follows from Eq. (1.6) when we note that, in our problem, f is proportional to $\delta(\boldsymbol{\xi} - \boldsymbol{\xi}')$ when $\boldsymbol{\xi} \cdot \mathbf{n} < 0$ and to the one-particle probability density $P_{N+1}^{(1)}(\mathbf{x}_0, \boldsymbol{\xi}_0)$ when $\boldsymbol{\xi}_0 \cdot \mathbf{n} > 0$; the integration with respect to dA_0 depends upon the fact that we do not care about the exact point of remission (provided it is not too far from the point where the molecule entered the wall). The second expression in Eq. (2.5) comes from the definition of $P_{N+1}^{(1)}$ in terms of P_{N+1}.

The problem in the stated form is extremely difficult to handle except, perhaps, in the case of harmonic forces. In general it is convenient to pass to an equation involving only the one-particle distribution function $P = P_{N+1}^{(1)}$ given by

$$P(\mathbf{x}_0, \boldsymbol{\xi}_0) = \int P_{N+1}(\mathbf{x}_k, \boldsymbol{\xi}_k) \prod_{k=1}^{N} d\mathbf{x}_k \, d\boldsymbol{\xi}_k \tag{2.6}$$

For this purpose it is necessary to take into account separately long range interactions and short range interactions (collisions) treating the former in the same way as in the Vlasov equation (see Chapter II, Section 4) and the latter as in the Boltzmann equation; if one wants to take into account the intermediate range interactions, a Fokker-Planck term will also arise (see Chapter II, Section 9). Accordingly, we obtain the equation

$$\boldsymbol{\xi} \cdot \frac{\partial P}{\partial \mathbf{x}} + \frac{\partial P}{\partial \boldsymbol{\xi}} \cdot \int n(\mathbf{x}_*) \mathbf{X}(\mathbf{x}_*, \mathbf{x}) \, d\mathbf{x}_* = LP \tag{2.7}$$

where

$$LP = \int (P_{0*}'P' - P_{0*}P) B(\theta, V) \, d\boldsymbol{\xi}_* \, d\theta \, d\varepsilon + \\ + \sum_{i,j=1}^{3} \left[\frac{\partial}{\partial \xi_i} \left(D_{ij} \frac{\partial P}{\partial \xi_j} \right) + \frac{\partial}{\partial \xi_i} (F_{ij} \xi_j P) \right] \tag{2.8}$$

in which $\mathbf{X}$ denotes the long range force (per unit mass) exerted by a representative atom of the solid (located at $\mathbf{x}_*$) upon a gas molecule (located at $\mathbf{x} = (x, y, z)$ with velocity $\boldsymbol{\xi} = (\xi, \eta, \zeta)$), while $n(\mathbf{x}_*)$ denotes the number density of solid atoms; $P_{0*} = P_0(\boldsymbol{\xi}_*)$, $P_{0*}' = P_0(\boldsymbol{\xi}_*')$, where P_0 denotes the equilibrium Maxwellian of the solid atoms, while D_{ij} and F_{ij} are the diffusion and friction tensors in velocity space (see Chapter II, Section 9). $B(\theta, V)$ has the same meaning as in the Boltzmann equation, except for the fact that it has to be computed from the details of a collision between a gas molecule and a solid atom. The integral term in Eq. (2.8) is the collision term relative to gas-solid collisions: gas-gas collisions have been neglected and the solid atoms have been assumed to be in thermal equilibrium (see also Chapter IV, Section 3).

If the solid is considered to be completely homogeneous and nonpolar and we take the x-axis along $\mathbf{n}$, Eq. (2.7) reduces to

$$\boldsymbol{\xi} \cdot \frac{\partial P}{\partial \mathbf{x}} + \frac{\partial P}{\partial \xi} X(x) = LP \tag{2.9}$$

where $\mathbf{X}(x)$ is the average of $\mathbf{X}(\mathbf{x}_*, \mathbf{x})$ which, by symmetry, has only the x-component and this component depends simply on x. Our assumptions exclude the important case of a crystal for which X depends on y and z as well (in a periodic fashion).

Eq. (2.9) has to be solved in connection with the boundary condition

$$P = \delta(\boldsymbol{\xi} - \boldsymbol{\xi}') \qquad x = 0,\ \xi = \boldsymbol{\xi}\cdot\mathbf{n} < 0 \tag{2.10}$$

and the vanishing of P at infinity ($P \to 0$ when $x \to -\infty$). By symmetry, again, P will not depend upon y and z and we can write Eq. (2.9) in the form

$$\xi\frac{\partial P}{\partial x} + \frac{\partial P}{\partial \xi}X(x) = LP \tag{2.11}$$

which has to be solved with the boundary conditions specified above. In this case, Eq. (2.5) reduces to

$$R(\boldsymbol{\xi}' \to \boldsymbol{\xi}) = \frac{|\boldsymbol{\xi}\cdot\mathbf{n}|}{|\boldsymbol{\xi}'\cdot\mathbf{n}|}[P(\mathbf{x}, \boldsymbol{\xi})]_{x=0} \tag{2.12}$$

3. Reciprocity

As was said before, the basic problem of a gas-surface interaction, Eq. (2.1) with the boundary conditions (2.2) and (2.3), appears to be hopeless and one has to use Eq. (2.11) with the boundary condition (2.10) in order to compute $R(\boldsymbol{\xi}' \to \boldsymbol{\xi})$. We can derive, however, a significant result from Eq. (2.1) when the two body force $\mathbf{X}_{jk}$ does not depend upon the particle velocities.

Let us denote by $G_{N+1}(\mathbf{x}_k, \boldsymbol{\xi}_k; \mathbf{x}_{1k}, \boldsymbol{\xi}_{1k})$ the Green function of Eq. (2.1) for the half space problem described by Eqs. (2.2) and (2.3): in other words, let G_{N+1} be the function (which we assume to exist) satisfying

$$\sum_{k=0}^{N}\boldsymbol{\xi}_k\cdot\frac{\partial G_{N+1}}{\partial \mathbf{x}_k} + \sum_{\substack{j,k=0\\ j\neq k}}^{N}\mathbf{X}_{jk}\cdot\frac{\partial G_{N+1}}{\partial \boldsymbol{\xi}_k} = \prod_{k=0}^{N}\delta(\mathbf{x}_k - \mathbf{x}_{1k})\,\delta(\boldsymbol{\xi}_k - \boldsymbol{\xi}_{1k}) \tag{3.1}$$

and

$$G_{N+1} = 0 \qquad (\boldsymbol{\xi}_0\cdot\mathbf{n} < 0,\ \mathbf{x}_0\cdot\mathbf{n} = 0) \tag{3.2}$$

$$\begin{aligned} G_{N+1}(\mathbf{x}_0, \boldsymbol{\xi}_0, \mathbf{x}_1, \boldsymbol{\xi}_1, \ldots, \mathbf{x}_i, \boldsymbol{\xi}_i, \ldots, \mathbf{x}_N, \boldsymbol{\xi}_N) =&\\ = G_{N+1}(\mathbf{x}_0, \boldsymbol{\xi}_0, \mathbf{x}_1, \boldsymbol{\xi}_1, \ldots, \mathbf{x}_i, \boldsymbol{\xi}_i - 2\mathbf{n}_i(\mathbf{n}_i\cdot\boldsymbol{\xi}_i), \ldots, \mathbf{x}_N, \boldsymbol{\xi}_N)&\\ (1 \leqslant i \leqslant N,\ \mathbf{x}_i\cdot\mathbf{n} = 0,\ \boldsymbol{\xi}_i\cdot\mathbf{n} > 0)& \end{aligned} \tag{3.3}$$

G_{N+1} also has to vanish at infinity in both the velocity and physical space. In Eq. (3.3) we omitted part of the arguments upon which

$$G = G(\mathbf{x}_k, \boldsymbol{\xi}_k; \mathbf{x}_{1k}, \boldsymbol{\xi}_{1k})$$

depends.

We have then, by using the basic property of the delta function, Eq. (I.2.10):

$$G_{N+1}(\mathbf{x}_k, \boldsymbol{\xi}_k; \mathbf{x}_{1k}, \boldsymbol{\xi}_{1k})$$

$$= \int G_{N+1}(\mathbf{x}_{2k}, \boldsymbol{\xi}_{2k}; \mathbf{x}_{1k}, \boldsymbol{\xi}_{1k}) \prod_{k=0}^{N} \delta(\mathbf{x}_{2k} - \mathbf{x}_{1k})\, \delta(\boldsymbol{\xi}_{2k} - \boldsymbol{\xi}_k)\, d\mathbf{x}_{2k}\, d\boldsymbol{\xi}_{2k}$$

$$= \int G_{N+1}(\mathbf{x}_{2k}, \boldsymbol{\xi}_{2k}; \mathbf{x}_{1k}, \boldsymbol{\xi}_{1k}) \left[-\sum_{k=0}^{N} \boldsymbol{\xi}_{2k} \cdot \frac{\partial}{\partial \mathbf{x}_{2k}} - \sum_{\substack{j,k=0 \\ j \neq k}}^{N} \mathbf{X}_{jk} \cdot \frac{\partial}{\partial \boldsymbol{\xi}_{2k}} \right] \times$$

$$\times\, G_{N+1}(\mathbf{x}_{2k}, -\boldsymbol{\xi}_{2k}; \mathbf{x}_k, -\boldsymbol{\xi}_k) \prod_{j=0}^{N} d\mathbf{x}_{2j}\, d\boldsymbol{\xi}_{2j} \tag{3.4}$$

where integration extends to $\mathbf{x}_k \cdot \mathbf{n} < 0$ and we have used Eq. (3.1) (with $\mathbf{x}_{2k}, \mathbf{x}_{1k}, -\boldsymbol{\xi}_{2k}, -\boldsymbol{\xi}_{1k}$ in place of $\mathbf{x}_k, \mathbf{x}_{1k}, \boldsymbol{\xi}_k, \boldsymbol{\xi}_{1k}$) to eliminate the product of delta functions. Let us now integrate by parts and observe that the integrated terms are zero if the boundary conditions are satisfied. Eq. (3.4) then becomes:

$$G_{N+1}(\mathbf{x}_k, \boldsymbol{\xi}_k; \mathbf{x}_{1k}, \boldsymbol{\xi}_{1k})$$

$$= \int \left[\sum_{k=0}^{N} \boldsymbol{\xi}_{2k} \cdot \frac{\partial}{\partial \mathbf{x}_{2k}} + \sum_{\substack{j \neq k \\ j,k=0}}^{N} \mathbf{X}_{jk} \cdot \frac{\partial}{\partial \boldsymbol{\xi}_{2k}} \right] G_{N+1}(\mathbf{x}_{2k}, \boldsymbol{\xi}_{2k}; \mathbf{x}_{1k}, \boldsymbol{\xi}_{1k}) \times$$

$$\times\, G_{N+1}(\mathbf{x}_{2k}, -\boldsymbol{\xi}_{2k}; \mathbf{x}_k, -\boldsymbol{\xi}_k) \prod_{j=0}^{N} d\mathbf{x}_{2j}\, d\boldsymbol{\xi}_{2j}$$

$$= \int \prod_{k=0}^{N} \delta(\mathbf{x}_{2k} - \mathbf{x}_{1k})\, \delta(\boldsymbol{\xi}_{2k} - \boldsymbol{\xi}_{1k}) G_{N+1}(\mathbf{x}_{2k}, -\boldsymbol{\xi}_{2k}; \mathbf{x}_k, -\boldsymbol{\xi}_k) \prod_{j=0}^{N} d\mathbf{x}_{2j}\, d\boldsymbol{\xi}_{2j} \tag{3.5}$$

where we have used Eq. (3.1) with $\boldsymbol{\xi}_{2k}, \mathbf{x}_{2k}$ in place of $\boldsymbol{\xi}_k, \mathbf{x}_k$. By using the basic property of the delta function again, Eq. (3.5) gives

$$G_{N+1}(\mathbf{x}_k, \boldsymbol{\xi}_k; \mathbf{x}_{1k}, \boldsymbol{\xi}_{1k}) = G_{N+1}(\mathbf{x}_{1k}, -\boldsymbol{\xi}_{1k}; \mathbf{x}_k, -\boldsymbol{\xi}_k) \tag{3.6}$$

which expresses a basic property of the Green function of the Liouville equation with the boundary conditions (3.2) and (3.3). This property is crucially dependent on the time reversibility of the dynamics (when forces are velocity independent); as a consequence the Liouville operator appearing in Eq. (3.1) just changes its sign when $\boldsymbol{\xi}$ is changed into $-\boldsymbol{\xi}$.

The knowledge of G_{N+1} is sufficient for the computation of $R(\boldsymbol{\xi}' \to \boldsymbol{\xi})$.

In fact, if P_{N+1} satisfies Eqs. (2.1), (2.2) and (2.3) we have:

$$P_{N+1}(\mathbf{x}_k, \boldsymbol{\xi}_k) = \int P_{N+1}(\mathbf{x}_k', \boldsymbol{\xi}_k') \prod_{k=0}^{N} \delta(\mathbf{x}_k - \mathbf{x}_k')\, \delta(\boldsymbol{\xi}_k' - \boldsymbol{\xi}_k)\, d\mathbf{x}_k'\, d\boldsymbol{\xi}_k'$$

$$= \int P_{N+1}(\mathbf{x}_k', \boldsymbol{\xi}_k') \left[-\sum_{k=0}^{N} \boldsymbol{\xi}_k' \cdot \frac{\partial}{\partial \mathbf{x}_k'} - \sum_{\substack{j,k=0 \\ j \neq k}}^{N} \mathbf{X}_{jk}' \cdot \frac{\partial}{\partial \boldsymbol{\xi}_k'} \right] \times$$

$$\times\, G(\mathbf{x}_k', -\boldsymbol{\xi}_k'; \mathbf{x}_k, -\boldsymbol{\xi}_k) \prod_{k=0}^{N} d\mathbf{x}_k'\, d\boldsymbol{\xi}_k'$$

$$= -\int \boldsymbol{\xi}_0' \cdot \mathbf{n}_0 [G(\mathbf{x}_k', -\boldsymbol{\xi}_k'; \mathbf{x}_k, -\boldsymbol{\xi}_k) P_{N+1}(\mathbf{x}_k', \boldsymbol{\xi}_k')]_{\mathbf{x}_0' \cdot \mathbf{n}_0 = 0} \times$$

$$\times\, dA_0'\, d\boldsymbol{\xi}_0' \prod_{k=1}^{N} d\mathbf{x}_k'\, d\boldsymbol{\xi}_k' + \int G_{N+1}(\mathbf{x}_k', \boldsymbol{\xi}_k'; \mathbf{x}_k, \boldsymbol{\xi}_k) \times$$

$$\times \left[\sum_{k=0}^{N} \boldsymbol{\xi}_k' \cdot \frac{\partial P_{N+1}'}{\partial \mathbf{x}_k'} + \sum_{\substack{j,k=0 \\ j \neq k}}^{N} \mathbf{X}_{jk}' \cdot \frac{\partial P_{N+1}'}{\partial \boldsymbol{\xi}_k'} \right] \prod_{k=0}^{N} d\mathbf{x}_k'\, d\boldsymbol{\xi}_k'$$

$$= -\boldsymbol{\xi}' \cdot \mathbf{n} \int [G_{N+1}(\mathbf{x}_k', -\boldsymbol{\xi}_k'; \mathbf{x}_k, -\boldsymbol{\xi}_k)]_{\substack{\mathbf{x}_0' \cdot \mathbf{n} = 0 \\ \boldsymbol{\xi}_0' = \boldsymbol{\xi}'}} \times$$

$$\times\, \delta(\mathbf{y}_0 - \mathbf{y}_0') \bar{P}_N(\mathbf{x}_l, \boldsymbol{\xi}_l) \prod_{k=1}^{N} d\mathbf{x}_k'\, d\boldsymbol{\xi}_k' \tag{3.7}$$

where we proceeded as above with the only difference being that P_{N+1} satisfies Eq. (2.1) and (2.3) rather than (3.1) and (3.2). dA_0', of course, denotes the surface element in the plane $\mathbf{x}_0' \cdot \mathbf{n} = 0$. Using Eq. (2.5) and denoting by $f_0(\boldsymbol{\xi})$ a Maxwellian corresponding to a gas in equilibrium at the temperature and velocity of the wall, we have:

$$|\boldsymbol{\xi}' \cdot \mathbf{n}|\, R(\boldsymbol{\xi}' \to \boldsymbol{\xi}) f_0(\boldsymbol{\xi}')$$

$$= |\boldsymbol{\xi} \cdot \mathbf{n}|\, |\boldsymbol{\xi}' \cdot \mathbf{n}|\, Z_N^{-1} \int G_{N+1}^0(\mathbf{x}_k', -\boldsymbol{\xi}_k'; \mathbf{x}_k, -\boldsymbol{\xi}_k) \times$$

$$\times\, \delta(\mathbf{y}_0 - \mathbf{y}_0') \exp\left[-\frac{\sum_{k=0}^{N} (\frac{1}{2} m_k \boldsymbol{\xi}_k'^2) + V(\mathbf{x}_j')}{kT_0} \right] \times$$

$$\times\, dA_0\, dA_0' \prod_{k=1}^{N} d\mathbf{x}_k'\, d\boldsymbol{\xi}_k'\, d\mathbf{x}_k\, d\boldsymbol{\xi}_k \tag{3.8}$$

where the zero superscript in G_{N+1}^0 signifies that G_{N+1} has to be evaluated at $\mathbf{x}_0 \cdot \mathbf{n} = \mathbf{x}_0' \cdot \mathbf{n} = 0$, $\boldsymbol{\xi}_0' = \boldsymbol{\xi}'$, $\boldsymbol{\xi}_0 = \boldsymbol{\xi}$. Now G_{N+1} is zero unless

$$\sum_{k=0}^{N} (\tfrac{1}{2} m_k \xi_k'^2) + V(\mathbf{x}_j') = \sum_{k=0}^{N} (\tfrac{1}{2} m_k \xi_k^2) + V(\mathbf{x}_j)$$

because it is the Green function of the Liouville equation (3.1), which conserves total energy. Hence the integral on the right hand side of Eq. (3.8) is invariant with respect to the interchange

$$(\mathbf{x}_k, \boldsymbol{\xi}_k, \mathbf{x}_k', \boldsymbol{\xi}_k') \to (\mathbf{x}_k', -\boldsymbol{\xi}_k', \mathbf{x}_k, -\boldsymbol{\xi}_k)$$

accordingly the left hand side is invariant with respect to the interchange $(\boldsymbol{\xi}, \boldsymbol{\xi}') \to (-\boldsymbol{\xi}', -\boldsymbol{\xi})$ and we obtain ($f_0(-\boldsymbol{\xi}) = f_0(\boldsymbol{\xi})$ because the wall is assumed to be at rest):

$$|\boldsymbol{\xi}' \cdot \mathbf{n}| f_0(\boldsymbol{\xi}') R(\boldsymbol{\xi}' \to \boldsymbol{\xi}) = |\boldsymbol{\xi} \cdot \mathbf{n}| f_0(\boldsymbol{\xi}) R(-\boldsymbol{\xi} \to -\boldsymbol{\xi}') \qquad (\boldsymbol{\xi} \cdot \mathbf{n} > 0, \boldsymbol{\xi}' \cdot \mathbf{n} < 0) \quad (3.9)$$

Eq. (3.9) is a basic property of the kernel $R(\boldsymbol{\xi}' \to \boldsymbol{\xi})$ which can be called the "reciprocity law" or the "detailed balance". One physical interpretation of Eq. (3.9) is that if a gas is at equilibrium at the temperature T_0 of the wall, and hence has distribution function f_0, the number of molecules scattered from a velocity range $(\boldsymbol{\xi}', \boldsymbol{\xi}' + d\boldsymbol{\xi}')$ to a velocity range $(\boldsymbol{\xi}, \boldsymbol{\xi} + d\boldsymbol{\xi})$ (per unit time and unit surface) is equal to the number of molecules scattered from any velocity between $-\boldsymbol{\xi}$ and $-\boldsymbol{\xi} - d\boldsymbol{\xi}$ to any velocity between $-\boldsymbol{\xi}'$ and $-\boldsymbol{\xi}' - d\boldsymbol{\xi}'$. Eq. (3.9) was suggested by the author [1, 2] on the basis of the latter intuitive meaning, except for the fact that $\boldsymbol{\xi} - 2\mathbf{n}(\mathbf{n} \cdot \boldsymbol{\xi})$, $\boldsymbol{\xi}' - 2\mathbf{n}(\mathbf{n} \cdot \boldsymbol{\xi}')$ are replaced $-\boldsymbol{\xi}$ and $-\boldsymbol{\xi}'$; this circumstance makes no difference for usual walls, for which there is rotational symmetry in the tangential plane. Eq. (3.9), as it stands, was first written down by Kuščer [3]. Proofs of Eq. (3.9), having various degree of rigor have been given in subsequent papers [4–6]. The proof given above is substantially taken from the paper of the author [6]. It is to be noted that Kuščer [5] proved Eq. (3.9) in the case of quantum mechanical systems and generalized it to molecules with internal degrees of freedom. Relations similar to Eq. (3.9) had previously appeared in a series of papers on nonequilibrium processes, where the gas surface interaction was described by means of impulsive terms contained in the Liouville equation [7–9].

We note a simple consequence of reciprocity; if the impinging distribution is the wall Maxwellian f_0, *and* mass is conserved at the wall according to Eq. (1.8), then the distribution function of the emerging molecules is again f_0 or, in other words, the wall Maxwellian satisfies the boundary conditions. In fact, if we integrate Eq. (3.9) with respect to $\boldsymbol{\xi}'$ and use Eq. (1.8) we obtain

$$\int_{\boldsymbol{\xi}' \cdot \mathbf{n} < 0} |\boldsymbol{\xi}' \cdot \mathbf{n}| f_0(\boldsymbol{\xi}') R(\boldsymbol{\xi}' \to \boldsymbol{\xi}) \, d\boldsymbol{\xi}' = |\boldsymbol{\xi} \cdot \mathbf{n}| f_0(\boldsymbol{\xi}) \qquad (\boldsymbol{\xi} \cdot \mathbf{n} > 0) \quad (3.10)$$

and this equation proves our statement according to Eq. (1.6). It is to be remarked that Eq. (3.10), although a consequence of Eq. (1.8) (when Eq. (3.9) holds), is less restrictive than Eq. (3.10) and could be satisfied even if Eq. (3.9) fails.

4. A remarkable inequality

We want to prove the following

THEOREM *Let $C(g)$ be a strictly convex continuous function (see Fig.* 15) *of its argument g. Then for any scattering kernel $R(\xi' \to \xi)$ satisfying Eqs.* (1.8), (1.10), (3.10), *the following inequality holds*

$$\int f_0 \xi \cdot \mathbf{n} C(g)\, d\xi \leqslant 0 \tag{4.1}$$

where f_0 is the wall Maxwellian, $g = f/f_0$ and integration extends to the full ranges of values of the components of ξ, the values of f for $\xi \cdot \mathbf{n} > 0$ being related to those for $\xi \cdot \mathbf{n} < 0$ through Eq. (1.6). *Equality in Eq.* (4.1) *holds if and only if $g =$ const. almost everywhere, unless $R(\xi' \to \xi)$ is proportional to a delta function.*

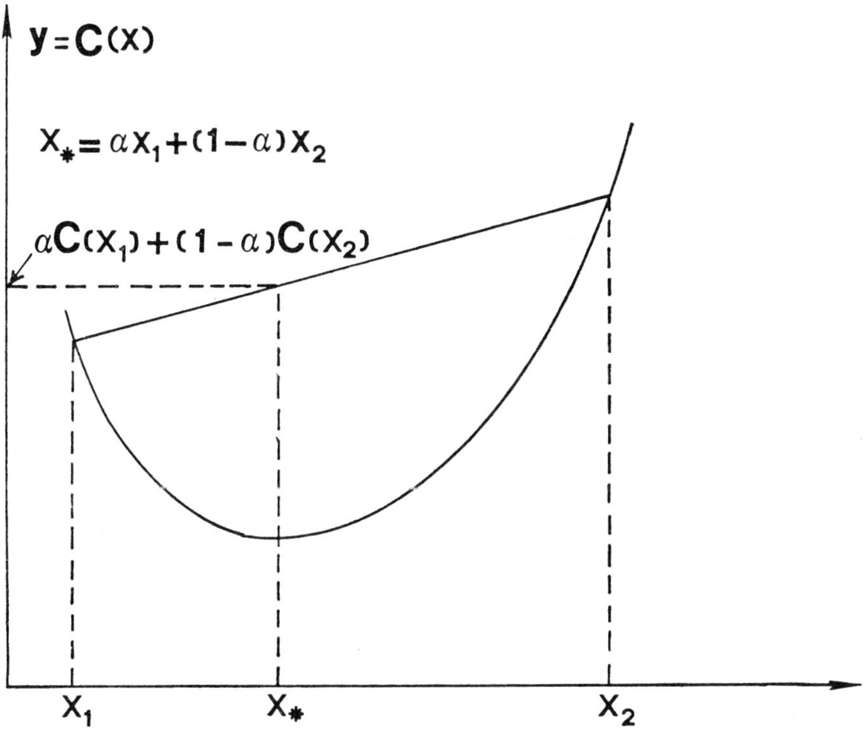

Fig. 15. A convex function.

In fact the convexity of $C(g)$ implies:

$$C\left(\sum_{i=1}^{n}\alpha_i x_i\right) \leqslant \sum_{i=1}^{n}\alpha_i C(x_i) \qquad \left(\sum_{i=1}^{n}\alpha_i = 1\right) \tag{4.2}$$

and putting $\alpha_i = w_i\,\Delta\boldsymbol{\xi}_i$, $x_i = g(\boldsymbol{\xi}_i)$ and letting the volume $\Delta\boldsymbol{\xi}_i$ tend to zero, gives also:

$$C\left(\int W(\boldsymbol{\xi}')g(\boldsymbol{\xi}')\,d\boldsymbol{\xi}'\right) \leqslant \int W(\boldsymbol{\xi}')C(g(\boldsymbol{\xi}'))\,d\boldsymbol{\xi}' \tag{4.3}$$

for any set of positive weights satisfying:

$$\int W(\boldsymbol{\xi}')\,d\boldsymbol{\xi}' = 1 \tag{4.4}$$

Eq. (4.3) is known as Jensen's inequality [39–41]. In particular, if we take (for any fixed $\boldsymbol{\xi}$):

$$W(\boldsymbol{\xi}') = \frac{R(\boldsymbol{\xi}' \to \boldsymbol{\xi})\,|\boldsymbol{\xi}' \cdot \mathbf{n}|\,f_0(\boldsymbol{\xi}')}{|\boldsymbol{\xi} \cdot \mathbf{n}|\,f_0(\boldsymbol{\xi})}, \tag{4.5}$$

Eq. (4.4) is satisfied because of Eq. (3.10) and Eqs. (1.6) and (4.3) give:

$$\begin{aligned} C(g(\boldsymbol{\xi})) &= C\left(\int_{\boldsymbol{\xi}'\cdot\mathbf{n}<0} \frac{R(\boldsymbol{\xi}' \to \boldsymbol{\xi})\,|\boldsymbol{\xi}' \cdot \mathbf{n}|\,f_0(\boldsymbol{\xi}')}{|\boldsymbol{\xi} \cdot \mathbf{n}|\,f_0(\boldsymbol{\xi})}\, g(\boldsymbol{\xi}')\,d\boldsymbol{\xi}'\right) \\ &\leqslant \int_{\boldsymbol{\xi}'\cdot\mathbf{n}<0} \frac{R(\boldsymbol{\xi}' \to \boldsymbol{\xi})\,|\boldsymbol{\xi}' \cdot \mathbf{n}|\,f_0(\boldsymbol{\xi}')}{|\boldsymbol{\xi} \cdot \mathbf{n}|\,f_0(\boldsymbol{\xi})}\, C(g(\boldsymbol{\xi}'))\,d\boldsymbol{\xi}' \qquad (\boldsymbol{\xi} \cdot \mathbf{n} > 0) \end{aligned} \tag{4.6}$$

Multiplying by $|\boldsymbol{\xi} \cdot \mathbf{n}|\,f_0(\boldsymbol{\xi})$, integrating with respect to $\boldsymbol{\xi}$ and using Eq. (1.8) gives:

$$\int |\boldsymbol{\xi} \cdot \mathbf{n}|\,f_0(\boldsymbol{\xi})C(g(\boldsymbol{\xi}))\,d\boldsymbol{\xi} \leqslant \int |\boldsymbol{\xi}' \cdot \mathbf{n}|\,f_0(\boldsymbol{\xi}')C(g(\boldsymbol{\xi}'))\,d\boldsymbol{\xi}' \tag{4.7}$$

which is Eq. (4.1). Equality applies if, and only if, it applies (except for a zero measure set) in Eq. (4.6); this can happen only if just one value of g appears in both sides; that is, if either g = constant or

$$\frac{R(\boldsymbol{\xi}' \to \boldsymbol{\xi})f_0(\boldsymbol{\xi}')\,|\boldsymbol{\xi}' \cdot \mathbf{n}|}{f_0(\boldsymbol{\xi})\,|\boldsymbol{\xi} \cdot \mathbf{n}|} = \delta(\boldsymbol{\xi}' - \boldsymbol{\xi}_*) \tag{4.8}$$

according to which $g(\boldsymbol{\xi}) = g(\boldsymbol{\xi}^*)$ (here $\boldsymbol{\xi}^*$ can be any function of $\boldsymbol{\xi}$ provided Eq. (1.8) is satisfied). This completes the proof.

The above theorem was given by Darrozes and Guiraud [10] in a particular case; the above proof is taken from a paper of the author [6]. We state the result of Darrozes and Guiraud in the form of a corollary:

COROLLARY I *For any scattering kernel satisfying Eqs.* (1.8), (1.10), (3.10), *the following inequality holds:*

$$\int \boldsymbol{\xi} \cdot \mathbf{n} f \log f \, d\boldsymbol{\xi} \leqslant -\frac{1}{RT_0}(q_i + p_{ij}v_j)n_i \tag{4.9}$$

where q_i is the heat flux vector and p_{ij} the stress tensor of the gas, v_j the velocity of the gas relative to the wall (f, p_{ij}, v_j are all evaluated at the wall, of course) and T_0 is the temperature of the wall (R, as usual, is the gas constant). Equality in Eq. (4.9) *holds if and only if $f = f_0$ (unless the kernel $R(\boldsymbol{\xi}' \to \boldsymbol{\xi})$ is proportional to a delta function).*

In fact if we take

$$C(g) = g \log g \qquad (g > 0)$$
$$C(0) = 0 \tag{4.10}$$

Eq. (4.1) gives:

$$\int \boldsymbol{\xi} \cdot \mathbf{n} g \log g \, f_0 \, d\boldsymbol{\xi} \leqslant 0 \tag{4.11}$$

In terms of $f = f_0 g$ we obtain:

$$\int \boldsymbol{\xi} \cdot \mathbf{n} f \log f \, d\boldsymbol{\xi} \leqslant \int \boldsymbol{\xi} \cdot \mathbf{n} f \log f_0 \, d\boldsymbol{\xi}$$
$$= -\frac{1}{2RT_0} \int \boldsymbol{\xi} \cdot \mathbf{n} \xi^2 f \, d\boldsymbol{\xi} = -\frac{1}{RT_0}(q_i + p_{ij}v_j)n_i \tag{4.12}$$

where Eq. (1.8) has been used. Eq. (4.12) is just Eq. (4.9).

We note that, even if the gas is moving relatively to the wall, the right hand side of Eq. (4.9) is proportional to the heat flux from the solid wall to the gas. In fact, conservation of energy implies the following relation (in a reference frame at rest with respect to the wall):

$$[(q_i + p_{ij}v_j)n_i]_{\text{gas}} = [(q_i + p_{ij}v_j)n_i]_{\text{solid}} = [\mathbf{q} \cdot \mathbf{n}]_{\text{solid}} \tag{4.13}$$

Eq. (4.13) expresses the fact that, if the gas slips over the wall, a discontinuity of normal heat flux is present.

Let us now introduce a linear operator defined as follows

$$Ag = [f_0(\boldsymbol{\xi}) \, |\boldsymbol{\xi} \cdot \mathbf{n}|]^{-1} \int_{\boldsymbol{\xi}' \cdot \mathbf{n} > 0} f_0(\boldsymbol{\xi}') R(-\boldsymbol{\xi}' \to \boldsymbol{\xi}) g(\boldsymbol{\xi}') \, |\boldsymbol{\xi}' \cdot \mathbf{n}| \, d\boldsymbol{\xi}'$$
$$= \int_{\boldsymbol{\xi}' \cdot \mathbf{n} > 0} H(\boldsymbol{\xi}, \boldsymbol{\xi}') g(\boldsymbol{\xi}') \, d\mu' \tag{4.14}$$

where

$$H(\boldsymbol{\xi}, \boldsymbol{\xi}') = R(-\boldsymbol{\xi}' \to \boldsymbol{\xi})[f_0(\boldsymbol{\xi})\, |\boldsymbol{\xi} \cdot \mathbf{n}|]^{-1} \tag{4.15}$$

$$d\mu = |\boldsymbol{\xi} \cdot \mathbf{n}|\, f_0(\boldsymbol{\xi})\, d\boldsymbol{\xi} \tag{4.16}$$

Then we have the following

COROLLARY II *The operator A defines a contraction mapping in the function space* $L_p (p > 1)$ *of the functions of* $\boldsymbol{\xi}$ $(\boldsymbol{\xi} \cdot \mathbf{n} > 0)$ *with integrable* p-th *power with respect to the measure defined by Eq.* (4.16). *That is:*

$$\|Ag\|_p \leqslant \|g\|_p \tag{4.17}$$

where

$$\|g\|_p = \left[\int_{\boldsymbol{\xi}\cdot\mathbf{n}>0} |g(\boldsymbol{\xi})|^p\, d\mu\right]^{1/p} \tag{4.18}$$

and the equality sign holds if, and only if, g is a constant almost everywhere (unless $R(\boldsymbol{\xi}' \to \boldsymbol{\xi})$ *is proportional to a delta function).*

To prove this corollary it is sufficient to let $C(g) = |g|^p$ in the above theorem.

5. Maxwell's boundary conditions. Accommodation coefficients

The treatment given in Sections 1 and 2 shows that the theoretical investigations of the gas-wall interaction are exceedingly difficult because of the complexity of phenomena taking place at the wall. Even if we restrict our attention to the linear boundary conditions expressed by Eq. (1.6) with a kernel independent of f, we are still faced with the problem of finding $R(\boldsymbol{\xi}' \to \boldsymbol{\xi})$. General considerations can only succeed in placing restrictions upon this kernel, such as the reciprocity law expressed by Eq. (3.9); otherwise, we have to construct a physical model of the surface and try to evaluate the corresponding kernel $R(\boldsymbol{\xi}' \to \boldsymbol{\xi})$.

The first attempt at solving such problem is due to Maxwell. In an appendix to a paper published in 1879 [11] he discusses the problem of finding a boundary condition for the distribution function. As a first hypothesis he supposes the surface of a physical wall to be a perfectly elastic smooth fixed surface, having the apparent shape of the solid without any minute deviations. In this case the gas molecules are specularly reflected; therefore the gas cannot exert any stress on the surface, except in the direction of the normal. Then Maxwell points out that since gases can actually exert oblique stresses against real surfaces, such surfaces cannot be represented as perfectly reflecting.

As a second model for a real wall Maxwell considers a stratum in which fixed elastic spheres are placed so far apart from one another that any one sphere is not sensibly protected by any other sphere from the impact of molecules. He also assumes the stratum to be so deep that every molecule which comes from the gas toward such wall must strike one or more of the spheres; when at last the molecule leaves the stratum of spheres and returns into the gas, its velocity must of course be directed from the surface toward the gas, but the probability of any particular magnitude and direction of the velocity will be the same as in a gas in thermal and mechanical equilibrium with the solid.

Then Maxwell considers more complicated models of physical walls, and finally concludes by saying that he prefers to treat the surface as something intermediate between a perfectly reflecting and perfectly absorbing surface, and, in particular, to suppose that a portion of every surface element absorbs all the incident molecules, and afterwards allows them to re-evaporate with the velocities corresponding to those in a still gas at the temperature of the solid wall, while the remaining portion perfectly reflects all the molecules incident upon it.

If we call α the fraction of evaporated molecules, Maxwell's assumption is equivalent to choosing

$$R(\boldsymbol{\xi}' \to \boldsymbol{\xi}; \mathbf{x}, t) = (1 - \alpha)\, \delta(\boldsymbol{\xi}' - \boldsymbol{\xi} + 2\mathbf{n}[\mathbf{n} \cdot \boldsymbol{\xi}]) + \alpha f_0(\boldsymbol{\xi})\, |\boldsymbol{\xi} \cdot \mathbf{n}| \qquad (\boldsymbol{\xi} \cdot \mathbf{n} > 0;\ \boldsymbol{\xi}' \cdot \mathbf{n} < 0) \tag{5.1}$$

where, if T_0 is the temperature of the wall, f_0 is given by:

$$f_0 = [2\pi(RT_0)^2]^{-1} \exp[-\xi^2(2RT_0)^{-1}] \tag{5.2}$$

the normalization being chosen in such a way that Eq. (1.8) is satisfied. Eq. (4.8) refers to the case of a wall at rest; otherwise, $\boldsymbol{\xi}$ must be replaced by $\boldsymbol{\xi} - \mathbf{u}_0$, $\mathbf{u}_0$ being the velocity of the wall. α, $\mathbf{u}_0$, T_0 may vary from point to point and with time.

It is easily seen that Maxwell's kernel, Eq. (5.1), satisfies Eq. (3.9) and hence also Eq. (3.10), Eq. (1.8) being satisfied by construction. It is to be stressed, however, that Maxwell was not able to master his problem completely and had to resort to qualitative arguments and introduce a phenomenological parameter α ($0 \leqslant \alpha \leqslant 1$) which is not directly related to the structure of the surface.

We note that according to Maxwell's boundary conditions the tangential momentum and the random kinetic energy of the re-evaporated molecules are influenced partially by the velocity and temperature of the boundary and partially by the momentum and random kinetic energy of the incoming stream; if $\alpha = 0$ (specular reflection) the re-emitted stream does not feel the boundary (as far as tangential momentum and kinetic energy are concerned), while if $\alpha = 1$ (completely diffuse reevaporation), the reemitted stream has completely lost its memory of the incoming stream (except for conservation of the number of molecules). For this reason the coefficient α (originally defined as the fraction of the diffusively evaporated molecules) is sometimes called the "accommodation coefficient" because it expresses the tendency of the gas to accommodate to the state of the wall. It is to be stressed, however, that momentum and energy accommodate differently in physical interactions, momentum being lost or gained much faster than energy; this points out a basic inaccuracy of Maxwell's boundary conditions.

In general, one may define an accommodation coefficient $\alpha(\varphi)$ for any function of molecular velocity $\varphi(\boldsymbol{\xi})$ as follows:

$$\alpha(\varphi) = \frac{\int_{\boldsymbol{\xi}\cdot\mathbf{n}<0} \varphi(\boldsymbol{\xi})\,|\boldsymbol{\xi}\cdot\mathbf{n}|\,f(\boldsymbol{\xi})\,d\boldsymbol{\xi} - \int_{\boldsymbol{\xi}\cdot\mathbf{n}>0} \varphi(\boldsymbol{\xi})\,|\boldsymbol{\xi}\cdot\mathbf{n}|\,f(\boldsymbol{\xi})\,d\boldsymbol{\xi}}{\int_{\boldsymbol{\xi}\cdot\mathbf{n}<0} \varphi(\boldsymbol{\xi})\,|\boldsymbol{\xi}\cdot\mathbf{n}|\,f(\boldsymbol{\xi})\,d\boldsymbol{\xi} - J_0\int_{\boldsymbol{\xi}\cdot\mathbf{n}>0} \varphi(\boldsymbol{\xi})\,|\boldsymbol{\xi}\cdot\mathbf{n}|\,f_0(\boldsymbol{\xi})\,d\boldsymbol{\xi}} \tag{5.3}$$

where $f_0(\boldsymbol{\xi})$ is given by Eq. (5.2), and so is the wall Maxwellian, and J_0 is a normalizing factor chosen in such a way that $J_0 f_0$ gives the same mass flow as f. The numerator of Eq. (5.3) is the difference between the impinging and emerging flow of the quantity $\varphi(\boldsymbol{\xi})$, the denominator is the same quantity when the molecules are completely accommodated ($f = J_0 f_0$ for $\boldsymbol{\xi}\cdot\mathbf{n} > 0$) In particular, if we let $\varphi(\boldsymbol{\xi}) = \boldsymbol{\xi} \wedge \mathbf{n}$ (where $\wedge$ denotes vector product), then we obtain the accommodation coefficient for tangential momentum; analogously, if we let $\varphi = \boldsymbol{\xi}\cdot\mathbf{n}$, or $\xi^2/2$, then we obtain the accommodation coefficient for normal momentum and energy. Note that the numerator of Eq. (5.3) is $\int \varphi(\boldsymbol{\xi})\boldsymbol{\xi}\cdot\mathbf{n} f(\boldsymbol{\xi})\,d\boldsymbol{\xi}$.

It is convenient to restrict the definition in Eq. (5.3) to functions enjoying the property $\varphi(\boldsymbol{\xi}) = \varphi(\boldsymbol{\xi} - 2\mathbf{n}[\mathbf{n}\cdot\boldsymbol{\xi}])$, which are even functions of $\boldsymbol{\xi}\cdot\mathbf{n}$. This condition is not satisfied by $\varphi = \boldsymbol{\xi}\cdot\mathbf{n}$; accordingly, if one wants to define an accommodation coefficient for normal momentum, one has to take $\varphi(\boldsymbol{\xi}) = |\boldsymbol{\xi}\cdot\mathbf{n}|$.

In general, $\alpha(\varphi)$ turns out to be dependent upon the distribution function of the impinging molecules; accordingly, the definition (5.3) is not very useful in general. The notion of an accommodation coefficient becomes more meaningful if we restrict f somewhat. To this end, we put

$$(\varphi, \psi) = (\psi, \varphi) = \int_{\boldsymbol{\xi}\cdot\mathbf{n}>0} \varphi(\boldsymbol{\xi})\psi(\boldsymbol{\xi})\,d\mu = \int_{\boldsymbol{\xi}\cdot\mathbf{n}>0} \varphi(\boldsymbol{\xi})\psi(\boldsymbol{\xi})f_0(\boldsymbol{\xi})\,|\boldsymbol{\xi}\cdot\mathbf{n}|\,d\boldsymbol{\xi} \tag{5.5}$$

where $d\mu$ is defined by Eq. (4.17). (φ, ψ) is a scalar product in the Hilbert space of functions defined for $\boldsymbol{\xi}\cdot\mathbf{n} > 0$, whose square is integrable with respect to the measure $d\mu$. The corresponding norm is given by Eq. (4.19) with $p = 2$. If reciprocity holds, the operator A, defined by Eq. (4.14), is symmetric with respect to the scalar product in Eq. (5.5):

$$\begin{aligned}(\varphi, A\psi) &= \iint \varphi(\boldsymbol{\xi}) f_0(\boldsymbol{\xi}') R(-\boldsymbol{\xi}' \to \boldsymbol{\xi}) \psi(\boldsymbol{\xi}') \, |\boldsymbol{\xi}' \cdot \mathbf{n}| \, d\boldsymbol{\xi}' \, d\boldsymbol{\xi} \\ &= \iint \varphi(\boldsymbol{\xi}) f_0(\boldsymbol{\xi}) R(-\boldsymbol{\xi} \to \boldsymbol{\xi}') \psi(\boldsymbol{\xi}') \, |\boldsymbol{\xi}' \cdot \mathbf{n}| \, d\boldsymbol{\xi}' \, d\boldsymbol{\xi} \\ &= (A\varphi, \psi)\end{aligned} \tag{5.6}$$

We now define the reflection or parity operators in velocity space as follows:

$$\begin{aligned}P\varphi &= \varphi(-\boldsymbol{\xi}) \\ P_n\varphi &= \varphi(\boldsymbol{\xi} - 2\mathbf{n}[\mathbf{n} \cdot \boldsymbol{\xi}]) \\ P_t\varphi &= \varphi(-\boldsymbol{\xi} + 2\mathbf{n}[\mathbf{n} \cdot \boldsymbol{\xi}])\end{aligned} \tag{5.7}$$

and observe that

$$P^2 = P_t^2 = P_n^2 = I, \qquad P_tP_n = P_nP_t = P, \qquad (\varphi, P_t\psi) = (P_t\varphi, \psi) \tag{5.8}$$

where I is the identity operator. As stated above, the function φ in Eq. (5.3) will be taken to satisfy the restriction $P_n\varphi = \varphi$.

Let us define $\psi(\boldsymbol{\xi})$ by

$$f(\boldsymbol{\xi}) = J_0 f_0(\boldsymbol{\xi})[1 + \psi(-\boldsymbol{\xi})] = J_0 f_0(1 + P\psi) \qquad (\boldsymbol{\xi} \cdot \mathbf{n} < 0) \tag{5.9}$$

where $f_0(\boldsymbol{\xi})$ and J_0 are as in Eq. (5.3). Then $(1, \psi) = 0$.

Eq. (5.3) then becomes:

$$\begin{aligned}\alpha(\varphi, \psi) &= \frac{(P\varphi, \psi) - (\varphi, A\psi)}{(P\varphi, \psi)} = 1 - \frac{(\varphi, A\psi)}{(P\varphi, \psi)} \\ &= 1 - \frac{(\varphi, A\psi)}{(P_t\varphi, \psi)} = 1 - \frac{(\varphi, A\psi)}{(\varphi, P_t\psi)} \\ &= 1 - \frac{(A\varphi, \psi)}{(P_t\varphi, \psi)}\end{aligned} \tag{5.10}$$

Here we have wr:tten $\alpha(\varphi, \psi)$ in place of $\alpha(\varphi)$ in order to emphasize that the accommodation coefficient depends on ψ as well.

Eq. (5.10) shows that if φ is taken to be a solution φ_j of the equation

$$A\varphi = \lambda P_t\varphi \tag{5.11}$$

corresponding to the eigenvalue $\lambda = \lambda_j$ ($j = 0, 1, 2, \ldots$) then the accommodation coefficient

$$\alpha_j \equiv \alpha(\varphi_j, \psi) = 1 - \frac{(A\varphi_j, \psi)}{(P_t\varphi_j, \psi)} = 1 - \lambda_j \tag{5.12}$$

is independent of ψ, and hence of the distribution function of the impinging molecules. This is an obvious advantage of this definition of α_j which is due to the present author [1, 2]; a disadvantage is that these accommodation coefficients correspond to the eigenfunctions φ_j, which do not possess, in general, a simple physical significance. Another possibility was considered by Shen and Kuščer [12, 5]. Let us take a set of physically meaningful quantities $\{\varphi_j(\boldsymbol{\xi})\}$ and let $\varphi = \varphi_i$, $\psi = \varphi_j$ (thus restricting the distribution function). We obtain a matrix of accommodation coefficients, whose elements are:

$$\alpha_{ij} = 1 - \frac{(\varphi_i, A\varphi_j)}{(\varphi_i, P_t\varphi_j)} = 1 - \frac{(\varphi_j, A\varphi_i)}{(\varphi_j, P_t\varphi_i)} = \alpha_{ji} \tag{5.13}$$

These coefficients have a clear physical meaning, but only for the special distribution function $f = f_0(1 + P\varphi_j)$. They are more easily computed, in general, than those defined above, because they do not require the solution of Eq. (5.11) as a prerequisite.

6. Mathematical models for gas-surface interaction

In view of the difficulty of computing the kernel $R(\boldsymbol{\xi}' \to \boldsymbol{\xi})$ from a physical model of the wall (see Section 7), we shall presently discuss a different procedure, which is less physical in nature and is similar to the approach to the collision term discussed in Section 10 of Chapter II. The idea is to construct a mathematical model in the form of a kernel $R(\boldsymbol{\xi}' \to \boldsymbol{\xi})$ which satisfies the basic requirements expressed by Eqs. (1.8), (1.10), (3.9) and which is not otherwise restricted except by the condition of not being too complicated.

A possible approach to the construction of such models is through the eigenvalue equation (5.11). Let the λ_j's form a discrete set and let φ_j be the corresponding eigenfunctions ($j = 0, 1, \ldots$). We assume that $\lambda_i \neq \lambda_j$ for $i \neq j$ in order to simplify the treatment. Then

$$\lambda_j(\varphi_i, P_t\varphi_j) = (\varphi_i, A\varphi_j) = (A\varphi_i, \varphi_j) = \lambda_i(P_t\varphi_i, \varphi_j) = \lambda_i(\varphi_i, P_t\varphi_j) \tag{6.1}$$

If $\lambda_i \neq \lambda_j$, i.e. $\varphi_i \neq \varphi_j$, Eq. (6.1) implies

$$(\varphi_i, P_t\varphi_j) = 0 \qquad (i \neq j) \tag{6.2}$$

We assume that $(\varphi_i, P_t\varphi_i) \neq 0$ (this could be easily shown to be true for nonpolar walls) and normalize φ_i in such a way that $(\varphi_i, P_t\varphi_i) = 1$; then

$$(\varphi_i, P_t\varphi_j) = \delta_{ij} \tag{6.3}$$

We further assume that the φ_j's form a complete set so that any sufficiently regular function ψ can be expanded into a series

$$\psi = \sum_i c_i\varphi_i \tag{6.4}$$

Then Eq. (6.3) shows that the coefficients of the expansion are given by

$$c_j = (\psi, P_t\varphi_j) \tag{6.5}$$

Inserting Eq. (6.5) into Eq. (6.4) we obtain

$$\begin{aligned}\psi(\boldsymbol{\xi}) &= \sum_j (\psi, P_t\varphi_j)\varphi_j(\boldsymbol{\xi}) = \sum_j \int \psi(\boldsymbol{\xi}')P_t[\varphi_j(\boldsymbol{\xi}')]\varphi_j(\boldsymbol{\xi})\, d\mu' \\ &= \int \psi(\boldsymbol{\xi}')\left\{\sum_{j=0}^{\infty} P_t[\varphi_j(\boldsymbol{\xi}')]\varphi_j(\boldsymbol{\xi})\right\} f_0(\boldsymbol{\xi}')\, |\boldsymbol{\xi}'\cdot\mathbf{n}|\, d\boldsymbol{\xi}' \end{aligned} \tag{6.6}$$

where the last step is valid provided the equations are now interpreted in the sense of the theory of generalized functions (see Section 2 of Chapter I). Hence completeness is equivalent to

$$f_0(\boldsymbol{\xi}')\, |\boldsymbol{\xi}'\cdot\mathbf{n}| \sum_{j=0}^{\infty} P_t[\varphi_j(\boldsymbol{\xi}')]\varphi_j(\boldsymbol{\xi}) = \delta(\boldsymbol{\xi}-\boldsymbol{\xi}') \tag{6.7}$$

Thus, using Eqs. (5.11), (5.5), (5.6), (6.7), (4.16):

$$\begin{aligned}&\int \psi(\boldsymbol{\xi}')\left\{\sum_{j=0}^{\infty} \lambda_j P_t[\varphi_j(\boldsymbol{\xi}')]\varphi_j(\boldsymbol{\xi})\right\} d\mu' \\ &\quad= \int \psi(\boldsymbol{\xi}')\left\{\sum_{j=0}^{\infty} A\varphi_j(\boldsymbol{\xi}')\varphi_j(\boldsymbol{\xi})\right\} d\mu' \\ &\quad= \sum_{j=0}^{\infty} (\psi, A\varphi_j)\varphi_j(\boldsymbol{\xi}) = \sum_{j=0}^{\infty} (A\psi, \varphi_j)\varphi_j(\boldsymbol{\xi}) \\ &\quad= \sum_{j=0}^{\infty} (P_tA\psi, P_t\varphi_j)\varphi_j(\boldsymbol{\xi}) = P_tA\psi \\ &\quad= [f_0(\boldsymbol{\xi})\, |\boldsymbol{\xi}\cdot\mathbf{n}|]^{-1} \int_{\boldsymbol{\xi}'\cdot\mathbf{n}>0} R(-\boldsymbol{\xi}' \to -\boldsymbol{\xi} + 2\mathbf{n}[\mathbf{n}\cdot\boldsymbol{\xi}])\psi(\boldsymbol{\xi}')\, d\mu' \end{aligned} \tag{6.8}$$

where the penultimate step is based upon Eqs. (6.4) and (6.5) with $A\psi$ in place of ψ. Since $\psi(\boldsymbol{\xi})$ is essentially an arbitrary function of $\boldsymbol{\xi}$, Eq. (6.8) implies

$$R(-\boldsymbol{\xi}' \to -\boldsymbol{\xi} + 2\mathbf{n}[\mathbf{n}\cdot\boldsymbol{\xi}]) = f_0(\boldsymbol{\xi})\, |\boldsymbol{\xi}\cdot\mathbf{n}| \sum_{j=0}^{\infty} \lambda_j P_t[\varphi_j(\xi')]\varphi_j(\boldsymbol{\xi}) \tag{6.9}$$

Changing $\boldsymbol{\xi}'$ into $-\boldsymbol{\xi}'$ and $\boldsymbol{\xi}$ into $-\boldsymbol{\xi} + 2\mathbf{n}(\mathbf{n}\cdot\boldsymbol{\xi})$, we obtain:

$$\begin{aligned}R(\boldsymbol{\xi}' \to \boldsymbol{\xi}) &= f_0(\boldsymbol{\xi})\, |\boldsymbol{\xi}\cdot\mathbf{n}| \sum_{j=0}^{\infty} \lambda_j P_t[\varphi_j(-\boldsymbol{\xi}')]P_t[\varphi_j(\boldsymbol{\xi})] \\ &= f_0(\boldsymbol{\xi})\, |\boldsymbol{\xi}\cdot\mathbf{n}| \sum_{j=0}^{\infty} \lambda_j \psi_j(-\boldsymbol{\xi}')\psi_j(\boldsymbol{\xi}) \end{aligned} \tag{6.10}$$

where

$$\psi_j(\boldsymbol{\xi}) = P_t[\varphi_j(\boldsymbol{\xi})] = \varphi_j(-\boldsymbol{\xi} + 2\mathbf{n}[\mathbf{n}\cdot\boldsymbol{\xi}]) \tag{6.11}$$

Eq. (6.10) is an expansion of the kernel $R(\boldsymbol{\xi}' \to \boldsymbol{\xi})$; if λ_j does not tend to zero sufficiently fast when $j \to \infty$, Eq. (6.10) is to be understood in the sense of the theory of generalized functions.

If we now make special choices for the λ_j's and ψ_j's in Eq. (6.10), we generate kernels $R(\boldsymbol{\xi}' \to \boldsymbol{\xi})$. The condition of mass conservation, Eq. (1.8), reduces to taking

$$\lambda_0 = 1, \qquad \psi_0 = 1 \tag{6.12}$$

provided f_0, the wall Maxwellian, is normalized as in Eq. (5.2). Reciprocity is automatically satisfied. Hence the only restriction which remains to be satisfied is Eq. (1.10). This condition gives some trouble if one tries to truncate the sum in Eq. (6.10) as was done by the author [1, 2]. The resulting kernels can be of some use in linearized problems (see Chapter IV) but are unacceptable in general. The only satisfactory procedure is to try to sum the series for special choices of infinitely many λ's and ψ's and to hope to obtain a positive result. The simplest possible assumption is complete degeneracy, when $\lambda_j = 1$ $(j = 0, 1, \ldots)$. Then Eqs. (6.10) and (6.7) show that $R(\boldsymbol{\xi}' \to \boldsymbol{\xi}) = \delta(\boldsymbol{\xi}' - \boldsymbol{\xi} + 2\mathbf{n}[\mathbf{n} \cdot \boldsymbol{\xi}])$ and we obtain the boundary condition for a purely specular reflection.

The next possible assumption is to let $\lambda_0 = 1$, $\lambda_j = 1 - \alpha$ $(j = 1, 2, \ldots)$; then

$$\begin{aligned} R(\boldsymbol{\xi}' \to \boldsymbol{\xi}) &= f_0(\boldsymbol{\xi})\,|\boldsymbol{\xi} \cdot \mathbf{n}|\,[\psi_0(-\boldsymbol{\xi}')\psi_0(\boldsymbol{\xi}) + (1-\alpha)\sum_{j=1}^{\infty} \psi_j(-\boldsymbol{\xi}')\psi_j(\boldsymbol{\xi})] \\ &= f_0(\boldsymbol{\xi})\,|\boldsymbol{\xi} \cdot \mathbf{n}|\left[\alpha\psi_0(-\boldsymbol{\xi}')\psi_0(\boldsymbol{\xi}) + (1-\alpha)\sum_{j=0}^{\infty} \psi_j(-\boldsymbol{\xi}')\psi_j(\boldsymbol{\xi})\right] \\ &= \alpha f_0(\boldsymbol{\xi})\,|\boldsymbol{\xi} \cdot \mathbf{n}| + (1-\alpha)\,\delta(\boldsymbol{\xi}' - \boldsymbol{\xi} + 2\mathbf{n}[\mathbf{n} \cdot \boldsymbol{\xi}]) \end{aligned} \tag{6.13}$$

where we have taken into account Eq. (6.7) again and noticed that $\psi_0 = 1$. Eq. (6.13) coincides with Eq. (15.1), and thus Maxwell's boundary conditions are obtained. Note that $R \geqslant 0$ implies $0 \leqslant \alpha \leqslant 1$.

It is clear that in order to obtain more sophisticated kernels, we have either to make more special assumptions about the eigenfunctions ψ_j, or to use a completely different procedure. If we follow the first approach, the simplest assumption is to let the eigenfunctions be polynomials, orthogonal to each other in the sense that Eq. (6.2) holds. If the polynomials are taken to have a definite parity with respect to $\boldsymbol{\xi}_t$, so that $R_t\psi_j = \pm\psi_j$, then the above requirement reduces to standard orthogonality with respect to the scalar product (5.5).

We can choose the ψ's in two ways, by assuming that they are polynomials in either of the three components of $\boldsymbol{\xi}$, or in the two tangential components ξ_{t_1}, ξ_{t_2} and the square ξ_n^2 of the normal component (we assume, as usual, that $\boldsymbol{\xi}$ is the velocity of a molecule with respect to an observer at rest with respect to the wall; otherwise $\boldsymbol{\xi}$ must be replaced by $\boldsymbol{\xi} - \mathbf{u}_0$, $\mathbf{u}_0$ being the wall velocity). The first procedure yields the half range polynomials introduced by Gross et al. [13] for different purposes, while the second one has the advantage of yielding classical polynomials. In fact, the second method, which we shall presently follow, suggests eigenfunctions of the following form:

$$\psi_{lmn}(\boldsymbol{\xi}) = (m!\, n!\, 2^{m+n})^{-\frac{1}{2}} L_l\left(\frac{\xi_n^{\,2}}{2RT_0}\right) H_m\left(\frac{\xi_{t_1}}{\sqrt{2RT_0}}\right) H_n\left(\frac{\xi_{t_2}}{\sqrt{2RT_0}}\right) \tag{6.14}$$

where L_l is the l-th Laguerre polynomial and H_n is the n-th Hermite polynomial (see Appendix). The numerical coefficient in Eq. (6.14) is chosen in such a way that the ψ's are not only orthogonal but also normalized with respect to the measure $d\mu$ given by Eq. (4.16) (see Appendix). Also, the following formula holds (see Appendix):

$$\begin{aligned}
&\sum_{l,m,n} p^l q^m s^n \psi_{lmn}(\boldsymbol{\xi})\psi_{lmn}(\boldsymbol{\xi}') \\
&\quad = \frac{1}{(1-p)\sqrt{(1-q^2)(1-s^2)}} \exp\left[-\frac{p}{1-p}\,\frac{\xi_n^{\,2} + \xi_n'^{\,2}}{2RT_0} + \right. \\
&\quad + \frac{q}{2(1+q)}\,\frac{(\xi_{t_1} + \xi_{t_1}')^2}{2RT_0} - \frac{q}{2(1-q)}\,\frac{(\xi_{t_1} - \xi_{t_1}')^2}{2RT_0} + \\
&\quad \left. + \frac{s}{2(1+s)}\,\frac{(\xi_{t_2} + \xi_{t_2}')^2}{2RT_0} - \frac{s}{2(1-s)}\,\frac{(\xi_{t_2} - \xi_{t_2}')^2}{2RT_0}\right] \times \\
&\quad \times I_0\left(\frac{\sqrt{p}}{1-p}\,\frac{\xi_n \xi_n'}{RT_0}\right) \qquad (|p|, |q|, |s|, <1)
\end{aligned} \tag{6.15}$$

where $I_0(x)$ denotes the modified Bessel function of first kind and zeroth order defined by

$$I_0(x) = (2\pi)^{-1} \int_0^{2\pi} \exp(x \cos \varphi)\, d\varphi \tag{6.16}$$

If we compare Eq. (6.15) with Eq. (6.10), we can immediately find kernels in a closed form by taking the eigenvalues λ_{lmn} corresponding to ψ_{lmn} to be of the form $p^l q^m s^n$, because the series appearing in Eq. (6.10) can then be summed by means of Eq. (6.15). Since we are not interested in anisotropies in the tangential plane, and in order to simplify the equations, we take

$s = q$ in Eq. (6.15). Then we find a particular kernel of the form:

$$R(\boldsymbol{\xi}' \to \boldsymbol{\xi}) = f_0(\boldsymbol{\xi})\, |\boldsymbol{\xi} \cdot \mathbf{n}| \sum_{l,m,n} p^l q^{m+n} \psi_{lmn}(-\boldsymbol{\xi}')\psi_{lmn}(\boldsymbol{\xi})$$

$$= \xi_n f_0(\boldsymbol{\xi})(1-p)^{-1}(1-q^2)^{-1} \exp\left[-\frac{p}{1-p}\frac{\xi_n{}^2 + \xi_n{}'^2}{2RT_0} + \right.$$

$$\left. + \frac{q}{2(1+q)}\frac{(\boldsymbol{\xi}_t + \boldsymbol{\xi}_t')^2}{2RT_0} - \frac{q}{2(1-q)}\frac{(\boldsymbol{\xi}_t - \boldsymbol{\xi}_t')^2}{2RT_0}\right] \times$$

$$\times I_0\left(\frac{\sqrt{p}}{1-p}\frac{\xi_n \xi_n'}{RT_0}\right)$$

$$= f_0(\boldsymbol{\xi})\xi_n(1-p)^{-1}(1-q^2)^{-1} \times$$

$$\times \exp\left[-\frac{p}{1-p}\frac{\xi_n{}^2 + \xi_n{}'^2}{2RT_0} - \frac{q^2}{1-q^2}\frac{\xi_t^2 + \xi_t'^2}{2RT_0} + \right.$$

$$\left. + \frac{q}{1-q^2}\frac{\boldsymbol{\xi}_t \boldsymbol{\xi}_t'}{2RT_0}\right] I_0\left(\frac{\sqrt{p}}{1-p}\frac{\xi_n \xi_n'}{RT_0}\right)$$

$$= \frac{(1-p)^{-1}(1-q^2)^{-1}}{2\pi(RT_0)^2}\,\xi_n \times$$

$$\times \exp\left[-\frac{\xi_n{}^2 + p\xi_n{}'^2}{2RT_0(1-p)} - \frac{1}{1-q^2}\frac{(\boldsymbol{\xi}_t - q\boldsymbol{\xi}_t')^2}{2RT_0}\right] \times$$

$$\times I_0\left(\frac{\sqrt{p}}{1-p}\frac{\xi_n \xi_n'}{RT_0}\right) \qquad (\boldsymbol{\xi}' \cdot \mathbf{n} < 0, \boldsymbol{\xi} \cdot \mathbf{n} > 0) \tag{6.17}$$

where $\boldsymbol{\xi}_t = \boldsymbol{\xi} - \mathbf{n}\xi_n$ is the two-dimensional vector whose components were previously denoted by ξ_{t_1} and ξ_{t_2}. The kernel given by Eq. (6.17) satisfies the conditions of reciprocity and normalization by construction if $|p| < 1$, $|q| < 1$ and further satisfies $R \geqslant 0$ if and only if q is real and p real non-negative (if $p < 0$ the argument of I_0 becomes imaginary and I_0 takes on negative as well as positive values). Hence Eq. (6.17) provides us with a two-parameter family of kernels, which has all the required properties provided the two parameters p and q are real and satisfy $0 \leqslant p \leqslant 1$, $-1 \leqslant q \leqslant 1$. The values $p = 1$, $q = \pm 1$ are limiting values because the kernel becomes peaked and tends to a product of delta functions in the limit. If Eqs. (A.11) and (A.26) of the Appendix are taken into account, then it is easily seen that the kernel given by Eq. (6.17) degenerates into the kernel corresponding to specular reflection, Eq. (1.11).

What is the physical meaning of the two parameters p, q? From what we said above it is clear that p and q are related to the accommodation properties of the wall. In particular, Eq. (5.12) shows that $q = \lambda_{010}$ is related to the accommodation coefficient of the quantity $\psi_{010}(\boldsymbol{\xi}) = \xi_{t_1}(RT_0)^{-\frac{1}{2}}$ (where we have used Eq. (6.14) and the fact that $L_0(x) = H_0(x) = 1$, $H_1(x) = 2x$, according to Eq. (A.3) of the Appendix). The factor $(RT_0)^{-\frac{1}{2}}$ does not matter because it cancels in Eq. (5.12) and we conclude that q is very simply related to the accommodation coefficient of tangential momentum α_t:

$$\alpha_t = 1 - q \tag{6.18}$$

Analogously p is related to the accommodation coefficient of the quantity $\psi_{100}(\boldsymbol{\xi}) = 1 - \xi_n{}^2(2RT_0)^{-1}$. Since the latter quantity does not have a simple physical meaning, let us investigate the accommodation coefficient of $\xi_n{}^2$, or what is the same, of $\xi_n{}^2(2RT_0)^{-1} = 1 - \psi_{100}(\boldsymbol{\xi})$. According to Eq. (5.10) this coefficient, which will be denoted by α_n, is given by

$$\alpha_n = 1 - \frac{(1 - p\psi_{100}, \psi)}{(1 - \psi_{100}, \psi)} = 1 - \frac{p(\psi_{100}, \psi)}{(\psi_{100}, \psi)} = 1 - p \tag{6.19}$$

where we made use of $A(1) = 1$, $A(\psi_{100}) = p\psi_{100}$, $R_t(\psi_{100}) = \psi_{100}$ and $(1, \psi) = 0$, the latter relation following from the definition of ψ, Eq. (5.9). Hence $p = 1 - \alpha_n$, where α_n is the accommodation coefficient for the part of the kinetic energy corresponding to the motion normal to the wall.

If we try to compute the accommodation coefficients for normal momentum and total kinetic energy, we immediately realize that they are *not* constant according to the considered model, but depend on the distribution function of the impinging molecules.

Eqs. (6.18) and (6.19) can be used to express the kernel (5.31) in terms of the physically meaningful parameters α_n and α_t. The result is:

$$\begin{aligned} &R(\boldsymbol{\xi}' \to \boldsymbol{\xi}; \alpha_n, \alpha_t) \\ &\quad = \frac{[\alpha_n\alpha_t(2 - \alpha_t)]^{-1}}{2\pi(RT_0)^2}\, \xi_n \times \\ &\quad\quad \times \exp\left\{-\frac{\xi_n{}^2 + (1-\alpha_n)\xi_n'^2}{2RT_0\alpha_n} - \frac{1}{\alpha_t(2-\alpha_t)}\frac{[\boldsymbol{\xi}_t - (1-\alpha_t)\boldsymbol{\xi}_t']^2}{2RT_0}\right\} \times \\ &\quad\quad \times I_0\left(\frac{\sqrt{1-\alpha_n}\,\xi_n\xi_n'}{\alpha_nRT_0}\right) \\ &\quad\quad\quad (\xi_n' < 0, \xi_n > 0; 0 \leqslant \alpha_t \leqslant 2; 0 \leqslant \alpha_n \leqslant 1) \end{aligned} \tag{6.20}$$

The two parameters α_n and α_t depend, of course, upon the physical nature of the gas and the wall as well as on the temperature of the latter. We note that we allowed α_t to be larger than 1, though smaller than 2. An extreme case is when $\alpha_n \to 0$ and $\alpha_t \to 2$; then $R(\boldsymbol{\xi}' \to \boldsymbol{\xi}) = \delta(\boldsymbol{\xi} + \boldsymbol{\xi}')$, so that each molecule exactly reverses its own velocity. For rough physical surfaces a value of α_t rather close to 1, possibly slightly larger than unity, is to be expected.

The kernel given by Eq. (6.17) or, equivalently, by Eq. (6.20) is by no means the only kernel of the form shown in Eq. (6.10) with the ψ's given by Eq. (6.14). More general kernels can be obtained by linearly combining several kernels of the form (6.17) with different values of p and q; the weights of the linear combination must sum to 1 (in order to ensure normalization) and be positive (in order to ensure $R \geqslant 0$). The distribution of values of p and q can be a continuous one; then we obtain the kernel

$$R(\boldsymbol{\xi}' \to \boldsymbol{\xi}) = \int_0^1 \int_{-1}^1 \frac{w(p, q)\xi_n}{2\pi(RT_0)^2} \times$$
$$\times \exp\left[-\frac{\xi_n{}^2 + p\xi_n{}'^2}{2RT_0(1 - p)} - \frac{1}{1 - q^2}\frac{(\boldsymbol{\xi}_t - q\boldsymbol{\xi}_t{}')^2}{2RT_0}\right] \times$$
$$\times I_0\left(\frac{\sqrt{p}}{1 - p}\frac{\xi_n\xi_n{}'}{RT_0}\right) dp\, dq \tag{6.21}$$

where

$$w(p, q) \geqslant 0 \qquad \int_0^1 \int_{-1}^1 w(p, q)(1 - p)(1 - q^2)\, dp\, dq = 1 \tag{6.22}$$

Eq. (6.21) gives the most general kernel having the polynomial eigenfunctions given by Eq. (6.14) and satisfying all the physical conditions (including absence of privileged directions in the tangential plane), provided we allow the weight function $w(p, q)$ to be a generalized function. Eq. (6.20), for example, is obtained by letting

$$w(p, q) = [\alpha_n \alpha_t (2 - \alpha_t)]^{-1}\, \delta(p + \alpha_n - 1)\, \delta(q + \alpha_t - 1) \tag{6.23}$$

All the kernels of the form (6.21) possess accommodation coefficients for tangential momentum and normal kinetic energy which are independent of the impinging distribution. In order to obtain less particular kernels, we may use the polynomials ψ_{lmn} given by Eq. (6.14) as a vector base in Hilbert space, even if we do not assume them to be eigenfunctions of A in the sense of Eq. (5.11). In this case Eq. (6.10) is replaced by:

$$R(\boldsymbol{\xi}' \to \boldsymbol{\xi}) = |\boldsymbol{\xi} \cdot \mathbf{n}|\, f_0(\boldsymbol{\xi}) \sum_{i,j=0}^{\infty} \lambda_{ij} \psi_i(-\boldsymbol{\xi}') \psi_j(\boldsymbol{\xi}) \tag{6.24}$$

where the indexes i and j stand for the sets (l, m, n), (l', m', n').

Thanks to reciprocity, $((\lambda_{ij}))$ is a symmetric matrix. The matrix $((\delta_{ij} - \lambda_{ij}))$ can be regarded as an "accommodation matrix" [1, 2]. Models of this kind have been independently considered by Cercignani [1, 2] and

Shen [14]. Again the trouble is that one cannot truncate the sum without violating $R \geqslant 0$, and very cumbersome algebra is expected in any attempt to sum series of the form (6.24) with $\lambda_{ij} \neq 0$ for infinitely many $i \neq j$.

Different approaches seem worth exploring in order to find simple but not too special scattering kernels. Recently, it has been shown [15] that in Eq. (6.20) one can let either α_t depend symmetrically on ξ_n and $-\xi_n'$ or α_n depend symmetrically on $\boldsymbol{\xi}_t$ and $-\boldsymbol{\xi}_t'$ without violating Eqs. (1.8), (1.10) and (3.9). In this way, kernels with nonconstant accommodation coefficients for tangential momentum or normal energy, respectively, are obtained.

A very general method for obtaining kernels, although somewhat artificial, is the following. Take any arbitrary *symmetric* positive function of $\boldsymbol{\xi}$, $\boldsymbol{\xi}'$, $K(\boldsymbol{\xi}, \boldsymbol{\xi}')$, and put

$$R(\boldsymbol{\xi}' \to \boldsymbol{\xi}) = |\boldsymbol{\xi} \cdot \mathbf{n}| f_0(\boldsymbol{\xi})[H(\boldsymbol{\xi})H(-\boldsymbol{\xi}') + K(\boldsymbol{\xi}, -\boldsymbol{\xi}')] \qquad (\boldsymbol{\xi}' \cdot \mathbf{n} < 0,\ \boldsymbol{\xi} \cdot \mathbf{n} > 0) \tag{6.25}$$

where

$$H(\boldsymbol{\xi}) = \frac{1 - \int_{\boldsymbol{\xi}' \cdot \mathbf{n} > 0} K(\boldsymbol{\xi}', \boldsymbol{\xi})\, d\mu'}{\sqrt{1 - \iint_{\substack{\boldsymbol{\xi} \cdot \mathbf{n} > 0 \\ \boldsymbol{\xi}' \cdot \mathbf{n} > 0}} K(\boldsymbol{\xi}', \boldsymbol{\xi})\, d\mu'\, d\mu}} \tag{6.26}$$

Reciprocity is trivially satisfied and so is normalization thanks to Eq. (6.26). $R \geqslant 0$ is ensured provided $H(\boldsymbol{\xi})$ turns out to be positive. As an example we take

$$K(\boldsymbol{\xi}, \boldsymbol{\xi}') = B(\boldsymbol{\xi})\, \delta(\boldsymbol{\xi} + \boldsymbol{\xi}' - 2\mathbf{n}[\mathbf{n} \,.\, \boldsymbol{\xi}']) \tag{6.27}$$

where $B(\boldsymbol{\xi})$ is an arbitrary positive even function of the molecular speed. Then:

$$H(\boldsymbol{\xi}) = \frac{1 - B(\boldsymbol{\xi})\xi_n f_0(\boldsymbol{\xi})}{\sqrt{1 - \int_{\xi_n > 0} B(\boldsymbol{\xi})\xi_n f_0(\boldsymbol{\xi})\, d\mu}} \tag{6.28}$$

and Eq. (6.25) becomes:

$$R(\boldsymbol{\xi}' \to \boldsymbol{\xi}) = |\boldsymbol{\xi} \cdot \mathbf{n}| f_0(\boldsymbol{\xi}) \left[\frac{\alpha(\boldsymbol{\xi})\alpha(\boldsymbol{\xi}')}{\int_{\xi_n > 0} \alpha(\boldsymbol{\xi})\, d\mu} + [1 - \alpha(\boldsymbol{\xi})]\, \delta(\boldsymbol{\xi} - \boldsymbol{\xi}' + 2\mathbf{n}\xi_n') \right] \tag{6.29}$$

where

$$\alpha(\boldsymbol{\xi}) = 1 - \xi_n f_0(\boldsymbol{\xi}) B(\boldsymbol{\xi}) \tag{6.30}$$

is an arbitrary even function of $\boldsymbol{\xi}$ satisfying $0 \leqslant \alpha(\boldsymbol{\xi}) \leqslant 1$.

The kernel in Eq. (6.29) was proposed by Epstein in 1967 [16] as a generalization of Maxwell's boundary conditions; in fact Eq. (6.30) reduces to Eq. (5.1) when α is a constant. Epstein chose his model on the basis of mass conservation, Eq. (1.8), and preservation of the wall Maxwellian, Eq. (3.10). It is interesting to note that reciprocity, although not mentioned at all in Epstein's paper, turns out to be automatically satisfied as a consequence of these constraints; this, of course, is due to the specific form chosen by Epstein for the kernel.

7. Physical models for gas-surface interaction

The approach presented in Sect. 6 can be very useful for obtaining relatively simple expressions for families of scattering kernels involving several parameters. The disadvantage is that these parameters are not related to basic physical quantities, such as the masses of the gas and solid atoms, the strength and range of interaction force, the temperature and number density of the wall atoms. In particular, for example, the parameters α_n and α_t appearing in Eq. (6.20) are left completely unspecified.

In order to obtain more detailed information, it is necessary to consider a physical model of the wall and to compute the scattering kernel from the dynamics of a molecule captured by the wall itself. It is clear that such computations must be either very schematic or very complicated [17–24]. This approach can be combined with the approach of Section 5; the latter provides analytical expressions for the distribution function and the former numerical values or simple expressions for the parameters left unspecified in such expressions.

A more useful approach, though a largely undeveloped one at present, would seem to use Eq. (2.9) with LP given by Eq. (2.8), where $B(\theta, V)$, D_{ij}, F_{ij}, $\bar{X}$ are to be computed from dynamical considerations. The major problem seems to be that of solving Eq. (2.9) for realistic, or at least not so unrealistic, expressions of B, F_{ij}, D_{ij}, $\bar{X}$. Here we consider an extremely simple case [6], in which $B(\theta, V) = 0$ (i.e. hard collisions are negligible compared to the effect of simultaneously grazing collisions) while

$$\begin{aligned} &D_{ij} = 0 \quad (i \neq j), && D_{33} = \frac{2RT_0}{l_n}\,|\xi_n|, \\ &D_{11} = D_{22} = \frac{2RT_0}{l_t}\,|\xi_n| && F_{ij} = D_{ij}/(RT_0) \end{aligned} \tag{7.1}$$

where l_t and l_n are constants, with the physical significance of diffusion lengths in the normal and tangential directions (here $i = 1, 2$ correspond to two mutually orthogonal directions in the tangential plane). Further we assume that the body force is a delta function centered at $x = -d (x \equiv x_3)$, so that the continuous interaction is replaced by a (ficticious) rigid wall placed inside the physical wall (see Fig. 16). Then we have to solve the equation:

$$\xi_n \frac{\partial P}{\partial x} = \frac{2RT_0}{l_n}\left[\frac{\partial}{\partial \xi_n}\left(|\xi_n| \frac{\partial P}{\partial \xi_n} + \frac{|\xi_n|\,\xi_n}{RT_0} P\right)\right] +$$
$$+ \frac{2RT_0}{l_t}|\xi_n|\left[\frac{\partial^2 P}{\partial \xi_1^2} + \frac{\partial}{\partial \xi_1}\left(\frac{\xi_1}{RT_0} P\right) + \frac{\partial^2 P}{\partial \xi_2^2} + \frac{\partial}{\partial \xi_2}\left(\frac{\xi_2}{RT_0} P\right)\right] \quad (7.2)$$

in the strip $-d < x < 0$ (see Fig. 16) with the boundary conditions:

$$\begin{aligned} P(0, \boldsymbol{\xi}) &= \delta(\boldsymbol{\xi} - \boldsymbol{\xi}') \qquad (\xi_n < 0) \\ P(-d, \boldsymbol{\xi}) &= P(-d, \boldsymbol{\xi} - 2\mathbf{n}[\mathbf{n} \cdot \boldsymbol{\xi}]) \end{aligned} \quad (7.3)$$

which is equivalent to solving the same equation in the strip $-2d \leqslant x \leqslant 0$ (see Fig. 16) with boundary conditions

$$\begin{aligned} P(0, \boldsymbol{\xi}) &= \delta(\boldsymbol{\xi} - \boldsymbol{\xi}') && (\xi_n < 0) \\ P(-2d, \boldsymbol{\xi}) &= \delta(\boldsymbol{\xi} - \boldsymbol{\xi}' + 2\mathbf{n}[\mathbf{n} \cdot \boldsymbol{\xi}]) && (\xi_n > 0) \end{aligned} \quad (7.4)$$

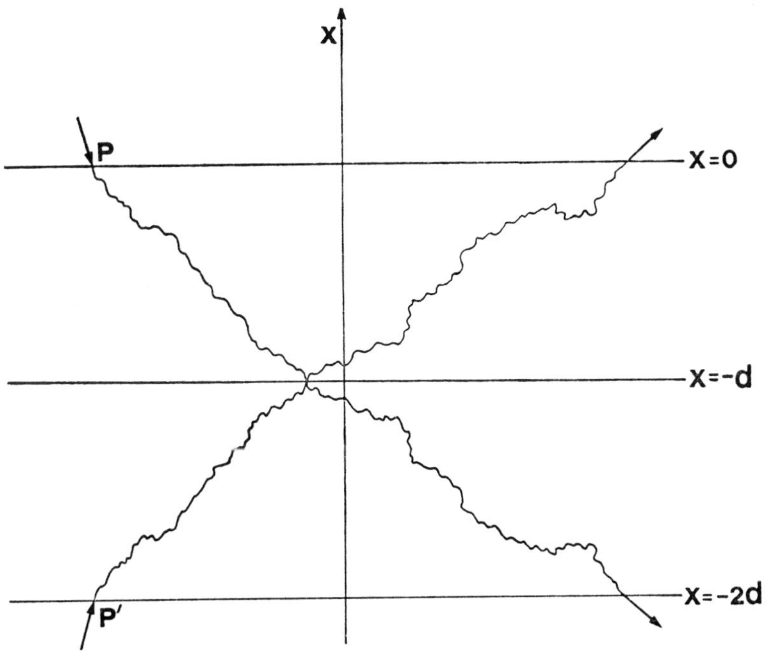

Fig. 16. Motion of a gas molecule inside a solid wall. A ficticious rigid wall is placed at $x = -d$ to simulate the body force. The region $-2d < x < -d$ is simply the specular image of the physical region $-d < x < 0$.

In addition, P and $\xi_n\,\partial P/\partial\xi_n$ must be continuous for the derivatives in Eq. (7.2) to exist.

In order to solve our problem we introduce the following device [6]; we interpret $|\xi_n|$ as the radial coordinate in a fictitious plane, so that we introduce fictitious velocity components:

$$\begin{aligned} \xi_3 &= |\xi_n| \cos\varphi & \xi_3' &= |\xi_n'| \\ \xi_4 &= |\xi_n| \sin\varphi & \xi_4' &= 0 \end{aligned} \tag{7.5}$$

where φ is a fictitious angle. Let $Q = Q(x, \xi_1, \xi_2, \xi_3, \xi_4)$ be continuous and continuously differentiable, and satisfy

$$-\frac{\partial Q}{\partial x} = \frac{2RT_0}{l_t}\left[\frac{\partial^2 Q}{\partial \xi_1{}^2} + \frac{\partial}{\partial \xi_1}\left(\frac{\xi_1}{RT_0}Q\right) + \frac{\partial^2 Q}{\partial \xi_2{}^2} + \frac{\partial}{\partial \xi_2}\left(\frac{\xi_2}{RT_0}Q\right)\right] + \\ + \frac{2RT_0}{l_n}\left[\frac{\partial^2 Q}{\partial \xi_3{}^2} + \frac{\partial}{\partial \xi_3}\left(\frac{\xi_3 Q}{RT_0}\right) + \frac{\partial^2 Q}{\partial \xi_4{}^2} + \frac{\partial}{\partial \xi_4}\left(\frac{\xi_4 Q}{RT_0}\right)\right] \tag{7.6}$$

$$Q(0, \xi_1, \xi_2, \xi_3, \xi_4) = \prod_{k=1}^{4} \delta(\xi_k - \xi_k') \tag{7.7}$$

Then

$$P = |\xi_n'| \int_{-\pi}^{\pi} Q(\xi_1, \xi_2, |\xi_n| \cos\varphi, |\xi_n| \sin\varphi)\, d\varphi \tag{7.8}$$

satisfies Eq. (7.2) and (7.4) for $\xi_n < 0$, as is seen by transforming from ξ_3, ξ_4 to $|\xi_n|$, φ in Eqs. (7.6) and (7.7) and integrating with respect to φ. (Note that

$$|\xi_n'| \int_{-\pi}^{\pi} \delta(\xi_3 - \xi_3')\, \delta(\xi_4 - \xi_4')\, d\varphi = \int_{-\pi}^{\pi} \delta(\varphi)\, \delta(|\xi_n| - |\xi_n'|)\, d\varphi \\ = \delta(\xi_n - \xi_n') \qquad (\xi_n < 0, \xi_n' < 0) \tag{7.9}$$

as follows by a transformation of the delta function from Cartesian to polar coordinates).

Eq. (7.6) is separable, and accordingly the solution can be factored as follows (note that the boundary condition, Eq. (7.7), is also factored):

$$Q = \frac{1}{(RT_0)^2} \prod_{k=1}^{4} G\left(\frac{\xi_k}{\sqrt{RT_0}}, \frac{\xi_k'}{\sqrt{RT_0}}, \frac{|x|}{l_k}\right) \qquad (l_1 = l_2 = l_t;\ l_3 = l_4 = l_n) \tag{7.10}$$

where $G(y, y', t)$ satisfies (as is verified by direct substitution):

$$\frac{\partial G}{\partial t} = \frac{\partial^2 G}{\partial y^2} + \frac{\partial}{\partial y}(yG); \lim_{t\to 0} G = \delta(y - y') \tag{7.11}$$

Now, if we introduce

$$z = ye^t, \tag{7.12}$$

Eq. (7.11) becomes ($\partial G/\partial t \to \partial G/\partial t + z\,\partial G/\partial z$, $\partial G/\partial y \to e^t\,\partial G/\partial z$):

$$\frac{\partial G}{\partial t} = e^{2t}\frac{\partial^2 G}{\partial z^2} + G \tag{7.13}$$

If we let

$$W = Ge^{-t} \tag{7.14}$$

Eq. (7.13) becomes

$$\frac{\partial W}{\partial t} = e^{2t}\frac{\partial^2 W}{\partial z^2} \tag{7.15}$$

Finally if we put

$$\tau = \tfrac{1}{2}(e^{2t} - 1) \tag{7.16}$$

W satisfies:

$$\frac{\partial W}{\partial \tau} = \frac{\partial^2 W}{\partial z^2} \tag{7.17}$$

which is the heat equation which has the well known fundamental solution [25]

$$\begin{aligned} W &= (4\pi\tau)^{-\frac{1}{2}}\exp[-(z - z')^2/(4\tau)] \\ &= [2\pi(e^{2t} - 1)]^{-\frac{1}{2}}\exp\{-(z - z')^2/[2(e^{2t} - 1)]\} \end{aligned} \tag{7.18}$$

which gives ($z_0 = y_0$ according to Eq. (7.12) with $t = 0$):

$$\begin{aligned} G &= [2\pi(1 - e^{-2t})]^{-\frac{1}{2}}\exp\{-(z - z')^2/[2(e^{2t} - 1)]\} \\ &= [2\pi(1 - e^{-2t})]^{-\frac{1}{2}}\exp\{-(y - y'e^{-t})^2/[2(1 - e^{-2t})]\} \end{aligned} \tag{7.18}$$

Hence

$$\begin{aligned} Q &= \frac{\exp\left[-\sum_{i=1}^{2}\frac{(\xi_i - \xi_i' e^{-|x|/l_t})^2}{2RT_0(1 - e^{-2|x|/l_t})} - \sum_{i=3}^{4}\frac{(\xi_i - \xi_i' e^{-|x|/l_n})^2}{2RT_0(1 - e^{-2|x|/l_n})}\right]}{(2\pi RT_0)^2(1 - e^{-2|x|/l_t})(1 - e^{-2|x|/l_n})} \\ &= (2\pi RT_0)^{-2}(1 - e^{-2|x|/l_t})^{-1}(1 - e^{-2|x|/l_n})^{-1}\exp\left[-\frac{(\boldsymbol{\xi}_t - \boldsymbol{\xi}_t' e^{-|x|/l_t})^2}{2RT_0(1 - e^{-2|x|/l_t})} - \right. \\ &\quad \left. - \frac{(\xi_n^{\,2} + \xi_n'^2 e^{-2|x|/l_n}) - 2\xi_n\xi_n' e^{-|x|/l_n}\cos\varphi}{2RT_0(1 - e^{-2|x|/l_n})}\right] \end{aligned} \tag{7.20}$$

and because of Eqs. (7.8) and (6.16):

$$\begin{aligned} P(x, \boldsymbol{\xi}) &= \xi_n'(2\pi)^{-1}(RT_0)^{-2}(1 - e^{-2|x|/l_t})^{-1}(1 - e^{-2|x|/l_n})^{-1} \times \\ &\times \exp\left[-\frac{(\boldsymbol{\xi}_t - \boldsymbol{\xi}_t' e^{-|x|/l_t})^2}{2RT_0(1 - e^{-2|x|/l_t})} - \frac{(\xi_n^{\,2} + \xi_n'^2 e^{-2|x|/l_n})}{2RT_0(1 - e^{-2|x|/l_t})}\right] \times \\ &\times I_0\left[\frac{\xi_n\xi_n'}{RT_0}e^{-|x|/l_n}(1 - e^{-2|x|/l_n})^{-1}\right] \end{aligned} \tag{7.21}$$

In order to compute $R(\boldsymbol{\xi}' \to \boldsymbol{\xi})$ we must compute $P(0, \boldsymbol{\xi})$ for $\xi_n > 0$. But because of symmetry of the boundary conditions (7.4) $P(0, \boldsymbol{\xi}) = P(-2d, \boldsymbol{\xi} - 2\mathbf{n}[\mathbf{n} \cdot \boldsymbol{\xi}])$, and Eq. (2.12) and (7.21) give

$$R(\boldsymbol{\xi}' \to \boldsymbol{\xi}) = \xi_n(2\pi)^{-1}(RT_0)^{-2}(1 - e^{-4d/l_n})^{-1}(1 - e^{-4d/l_t}) \times$$
$$\times \exp\left[-\frac{(\boldsymbol{\xi}_t - \boldsymbol{\xi}_t' e^{-2d/l_t})^2}{2RT_0(1 - e^{-4d/l_t})} - \frac{(\xi_n^{\,2} + \xi_n'^2 e^{-4d/l_n})}{2RT_0(1 - e^{-4d/l_n})}\right] \times$$
$$\times I_0\left[\frac{\xi_n \xi_n'}{RT_0} e^{-2d/l_n}(1 - e^{-4d/l_n})^{-1}\right] \qquad (\xi_n' < 0, \xi_n > 0) \quad (7.22)$$

Eq. (7.22) is identical with the kernel given by Eq. (6.20) provided we put

$$\alpha_n = 1 - e^{-4d/l_n} \qquad \alpha_t = 1 - e^{-2d/l_t} \qquad (7.23)$$

Here the accommodation coefficients are related to the ratios d/l_n, d/l_t, where d is the depth of penetration, and l_n and l_t are two characteristic lengths related to diffusion in velocity space.

It is noticeable that by solving a Fokker-Planck equation, the kernel postulated in Section 4 has been recovered. It is to be noted that the same kernel was found by Williams [26] by means of an analogy with scattering of electromagnetic waves by a wall, while the corresponding one-dimensional kernel involving ξ_n and ξ_n', but not $\boldsymbol{\xi}_t$ and $\boldsymbol{\xi}_t'$, was obtained by Kuščer et al. [27] on the basis of a similar Fokker-Planck equation and a completely different interpretation.

It is perhaps surprising that Eq. (7.2) has been solved for $\xi_n < 0$ without any reference to what happens for $\xi_n > 0$. In general one would expect a rather complicated interplay between the solution for $\xi_n > 0$ and $\xi_n < 0$. This interplay occurs because a boundary condition at $\xi_n = 0$ is required to determine the solution for $\xi_n > 0$, and another boundary condition at $\xi_n = 0$ is required to determine the solution for $\xi_n < 0$ when one considers the two regions separately. These conditions are none other than the continuity of P and $\sum_{i=1}^{3} D_{3i}\, \partial P/\partial \xi_i$ at $\xi_n = 0$ and, for a general Fokker-Planck equation, give rise to a set of two coupled integral equations [6]. An enormous simplification arises if $D_{ij} = 0$ at $\xi_n = 0$, because then the conditions at the boundary $\xi_n = 0$ are automatically satisfied provided $\partial P/\partial \xi_i$ is finite there. Physically this is due to the fact that the gas molecule undergoes a diffusion process in velocity space in which the sign of ξ_n cannot be reversed because the diffusion coefficient vanishes at $\xi_n = 0$; hence the molecule can turn back only at the artificial boundary $x = -d$ simulating the body force.

8. Scattering of molecular beams

As was said before, scattering of molecular beams from solid surfaces is the main source of experimental data on the interaction of molecules with a solid wall. In these experiments, a beam of molecules with a given distribution function impinges upon a wall and the distribution function of the emerging molecules is determined by counting the molecules. Experimental results on scattering patterns are usually presented for scattering in the plane of incidence (i.e., the plane containing the incident velocity vector **U** and the surface normal **n**) only. Further, results usually refer only to angular distributions. What is actually measured is the ratio of the number of molecules $N(\theta, \varphi)$ scattered per unit solid angle to the total number of scattered molecules, without distinguishing the molecules according to their speed:

$$N(\theta, \varphi) = \frac{1}{N_0}\int_0^\infty f(\boldsymbol{\xi})\,|\boldsymbol{\xi}\cdot\mathbf{n}|\,\xi^2\,d\xi \qquad (\boldsymbol{\xi}\cdot\mathbf{n} > 0) \tag{8.1}$$

where θ and φ are the polar angles in velocity space with the polar axis along the normal and

$$N_0 = \int_{\boldsymbol{\xi}\cdot\mathbf{n}>0} f\,|\boldsymbol{\xi}\cdot\mathbf{n}|\,d\boldsymbol{\xi} \tag{8.2}$$

The ideal distribution function for the incident beam is the monoenergetic collimated beam:

$$f = \frac{N_0}{|\boldsymbol{\xi}\cdot\mathbf{n}|}\,\delta(\boldsymbol{\xi} - \mathbf{U}) = \frac{N_0}{m \sin\theta_0 \cos\theta_0}\,\delta(E - E_0)\,\delta(\theta - \theta_0)\,\delta(\varphi - \varphi_0) \tag{8.3}$$

where $E = \xi^2/(2m)$ is the energy of the beam (the molecules are assumed to be monatomic). Such beams, however, are difficult to obtain in the laboratory and another type of incident beam, the "collimated thermal beam", is commonly used. The latter is obtained by collimating the gas atoms effusing from an equilibrium oven source, and hence is of the form

$$\begin{aligned} f &= 2N_0 m^{\frac{3}{2}}(RT_s)^{-2}\,|\cos\theta_0 \sin\theta_0|^{-1} \times \\ &\quad \times \exp[-\xi^2/(2RT_s)]\,\delta(\theta - \theta_0)\,\delta(\varphi - \varphi_0) \\ &= N_0(RT_s)^{-2}\,|\cos\theta_0 \sin\theta_0|^{-1} \exp[-E/(kT_s)]\,\delta(\theta - \theta_0)\,\delta(\varphi - \varphi_0) \end{aligned} \tag{8.4}$$

where T_s is the source temperature.

The computation of the total flux per solid angle, is particularly simple if the assumption (8.3) is made and Eq. (1.6) is used; then:

$$\begin{aligned} N(\theta, \varphi) &= \int_0^\infty \left[\int_{\boldsymbol{\xi}'\cdot\mathbf{n}<0} R(\boldsymbol{\xi}' \to \boldsymbol{\xi})\,\delta(\boldsymbol{\xi}' - \mathbf{U})\right] |\boldsymbol{\xi}\cdot\mathbf{n}|\,\xi^2\,d\xi \\ &= \int_0^\infty R(\mathbf{U} \to \boldsymbol{\xi})\,|\boldsymbol{\xi}\cdot\mathbf{n}|\,\xi^2\,d\xi = \cos\theta \int_0^\infty R(\mathbf{U} \to \boldsymbol{\xi})\xi^3\,d\xi \end{aligned} \tag{8.5}$$

Eq. (8.5) shows that a simple quadrature yields $N(\theta, \varphi)$ once $R(\boldsymbol{\xi}' \to \boldsymbol{\xi})$ is known and Eq. (8.3) holds. If the latter is replaced by Eq. (8.4) an additional integration with respect to U is required.

In particular, Eq. (8.5) can be used in connection with the kernel given by Eq. (6.20). The results can be compared with experiments [4, 28]: the agreement is satisfactory. Here we present a single case (Fig. 17); the experimental data are taken from an experiment with a 295°K argon beam on platinum at 1081° [29]. The four polar diagrams refer to four different incidence angles and represent the pattern of scattered molecules in the

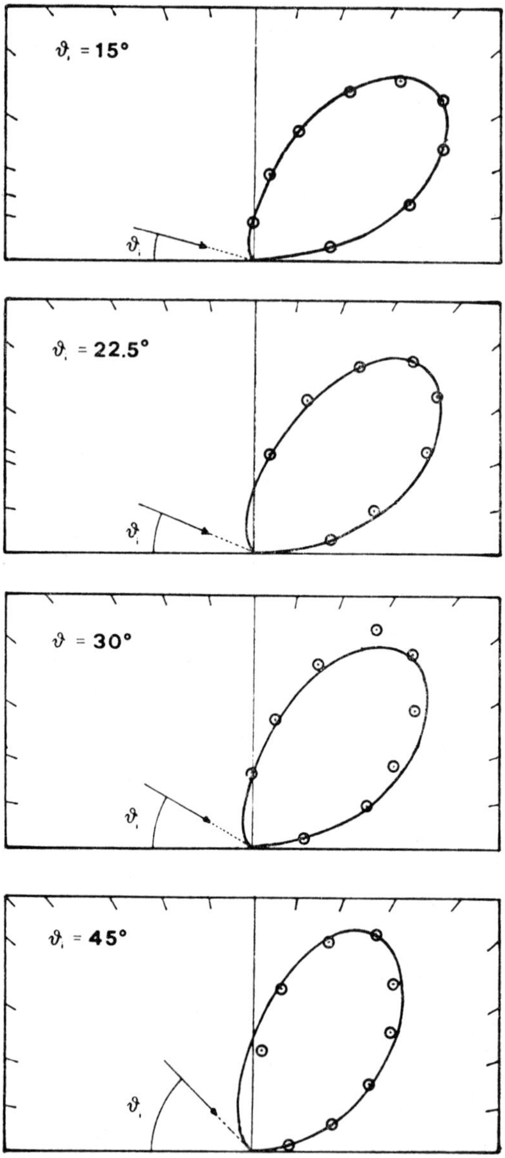

Fig. 17. Comparison of the model based on the scattering kernel given by Eq. (6.20) with the experimental data of Hinchen and Foley.

incidence plane. The small circles represent experimental data while the curve is computed by means of the kernel given by Eq. (6.20) with $\alpha_n = 0.3$, $\alpha_t = 0.1$; note that the agreement is not simply due to a best fitting, since the values of α_n and α_t are the same in the four cases. Notice that the experimental data are widely different from the patterns corresponding to specular and diffuse re-emission.

Before the method of modelling the scattering kernel was developed and used, the best way of representing the experimental data was to assume that the distribution function of emerging molecules had some specific expression containing available parameters to be adjusted to fit experimental data. Pioneering work in this direction was done by Schamberg [30] and Nocilla [31, 32]. Schamberg's model assumes uniform re-emission speed and is rather cumbersome. Nocilla's model is based on a comparatively simple form for the distribution function of the emerging molecules. The latter were assumed to be Maxwellian with a mass velocity $\mathbf{U}_r$ and a temperature T_r different from those of the wall:

$$f = \rho_r(2\pi RT_r)^{-\frac{3}{2}}\exp[-(\boldsymbol{\xi} - \mathbf{U}_r)^2/(2RT_r)] \tag{8.6}$$

where ρ_r is determined by mass conservation. Particular cases of this model had been previously considered by Knudsen [33] ($\mathbf{U}_r = 0$) and Grad [34] ($T_r = T_0$). The calculations based on Eq. (8.6) are tractable but the model suffers from the serious limitation of the absence of any correlation between the distributions of the impinging and emerging molecules. The latter difficulty can be avoided by a re-interpretation [35, 27, 4] of Eq. (8.6) as rigorously giving the emerging distribution for the case of any impinging monocromatic beam, Eq. (8.3). In this case Eq. (8.6) implies a scattering kernel which is also Maxwellian in $\boldsymbol{\xi}$; reciprocity, Eq. (3.9) and normalization, Eq. (1.8), can then be used to reduce the possible variety of kernels, with the consequential destruction of the flexibility of the model [4, 27].

We conclude by remarking that scattering experiments with crystals at comparatively low temperature show a new phenomenon, which is the presence of more than one maximum in the patterns of re-emitted molecules. This is probably due to the periodicity of the crystal, a feature which was left out from the models described in this section.

9. The H-theorem. Irreversibility

We want to show that if we define

$$\mathscr{H} = \int f \log f \, d\boldsymbol{\xi} \tag{9.1}$$

$$\mathscr{H}_i = \int \xi_i f \log f \, d\boldsymbol{\xi} \tag{9.2}$$

where f is any function which satisfies the Boltzmann equation, Eq. (II.8.21), then

$$\frac{\partial \mathscr{H}}{\partial t} + \frac{\partial \mathscr{H}_i}{\partial x_i} \leqslant 0 \tag{9.3}$$

Eq. (9.3) is an obvious consequence of the inequality in Eq. (II.7.3) and of the fact that f satisfies the Boltzmann equation; it is sufficient to multiply both sides of Eq. (II.8.21) by $1 + \log f$ and integrate over all possible velocities $\boldsymbol{\xi}$, taking into account $d(f \log f) = (1 + \log f)\, df$, Eq. (II.7.3) and the fact that 1 is a collision invariant (vanishing of f for $\boldsymbol{\xi} \to \infty$ is also understood).

We can now interpret Eq. (9.3) by rewriting it as follows:

$$\frac{\partial \mathscr{H}}{\partial t} + \frac{\partial \mathscr{H}_i}{\partial x_i} = \mathscr{I}; \; \mathscr{I} = \int \log f Q(f, f)\, d\boldsymbol{\xi} \leqslant 0 \tag{9.4}$$

and introducing

$$H = \int_R \mathscr{H}\, d\mathbf{x} \tag{9.5}$$

where R is a region filled by gas. If we had $\mathscr{I} = 0$, then H would be a conserved quantity like mass, energy and momentum, since we can interpret the vector $\boldsymbol{\mathscr{H}} = (\mathscr{H}_1, \mathscr{H}_2, \mathscr{H}_3)$ as the flow of H (note that here $\mathscr{H}$ is not the magnitude of the vector $\boldsymbol{\mathscr{H}}$ but the density corresponding to the flow $\boldsymbol{\mathscr{H}}$). Since $\mathscr{I} \neq 0$ in general, but $\mathscr{I} \leqslant 0$, we can say that molecular collisions act as a negative source for the quantity H. We also know that the source $\mathscr{I}$ is zero if and only if the distribution function is Maxwellian. Finally, we note that, as we did for momentum flow (Chapter II, Section 8), $\boldsymbol{\mathscr{H}}$ can be split into a macroscopic (convective) flow of H, $\mathscr{H}\mathbf{u}$ and a microscopic flow of H, $\boldsymbol{\mathscr{H}} - \mathscr{H}\mathbf{u}$.

If we integrate both sides of Eq. (9.4) with respect to $\mathbf{x}$ over the region R we have, if the boundary ∂R of R moves with velocity $\mathbf{u}_0$.

$$\frac{dH}{dt} - \int_{\partial R} (\boldsymbol{\mathscr{H}} \cdot \mathbf{n} - \mathscr{H}\mathbf{u}_0 \cdot \mathbf{n})\, dS = \int_R \mathscr{I}\, d\mathbf{x} \leqslant 0 \tag{9.6}$$

where dS is a surface element of the boundary ∂R and $\mathbf{n}$ the inward normal. The second term in the integral comes from the fact that, if the boundary is moving, when forming the time derivative of H we have to take into account that the region of integration changes with time.

From Eq. (9.4) and (9.6) we deduce two classical forms of the celebrated H-theorem of Boltzmann:

(a) If the gas is homogeneous ($\partial f/\partial \mathbf{x} = 0$ and hence $\partial \mathscr{H}_i/\partial x_i = 0$ in Eq. (9.3)) the quantity $\mathscr{H}$ (which is a function of time only) never increases with time and is steady if and only if the distribution function is Maxwellian (in fact $d\mathscr{H}/dt = 0$ implies $\mathscr{I} = 0$ and because of the results of Chapter II, Section 7, f is Maxwellian).

(b) If the gas is enclosed in a region such that the integral appearing on the left hand side of Eq. (9.6) is nonpositive:

$$\int_{\partial R} (\boldsymbol{\mathscr{H}} - \mathscr{H}\mathbf{u}_0) \cdot \mathbf{n}\, dS \leqslant 0 \tag{9.7}$$

the quantity H (which is a function of time only) never decreases with time and is steady if and only if the distribution function is Maxwellian.

The second form of the H-theorem covers more general situations but is somewhat unsatisfactory, because it can be applied only after having checked that Eq. (9.7) is satisfied. For example, Eq. (9.7) is satisfied (with the equality sign), if the molecules are specularly reflected at the boundary ∂R (in the reference frame in which ∂R is at rest); this is the only case considered in the traditional treatments [36, 35].

We can however obtain a sharp and significant form of the H-theorem if the boundary ∂R is a solid wall at which Eq. (1.6) holds. It is sufficient to note that

$$\begin{aligned}\int_{\partial R} (\boldsymbol{\mathscr{H}} - \mathscr{H}\mathbf{u}_0) \cdot \mathbf{n}\, dS &= \int (\boldsymbol{\xi} - \mathbf{u}_0) \cdot \mathbf{n} f \log f\, d\boldsymbol{\xi}\, dS \\ &= \int \boldsymbol{\xi}_r \cdot \mathbf{n} f \log f\, d\boldsymbol{\xi}_r\, dS \end{aligned} \tag{9.8}$$

where $\boldsymbol{\xi}_r = \boldsymbol{\xi} - \mathbf{u}_0$ is the molecular velocity in a reference frame moving with the wall. In this reference frame, Eqs. (4.9) and (4.13) hold. Hence

$$\begin{aligned}\int_{\partial R} (\boldsymbol{\mathscr{H}} - \mathscr{H}\mathbf{u}_0) \cdot \mathbf{n}\, dS &\leqslant -\int \frac{1}{RT_0} [(q_i + p_{ij}v_j)n_i]_{\text{gas}}\, dS \\ &= -\frac{1}{R} \int_{\partial R} \frac{(\mathbf{q} \cdot \mathbf{n})_{\text{solid}}}{T_0}\, dS \end{aligned} \tag{9.9}$$

and Eq. (9.6) gives

$$\frac{dH}{dt} \leqslant -\frac{1}{R} \int \frac{(\mathbf{q} \cdot \mathbf{n})_{\text{solid}}\, dS}{T_0} = -\frac{1}{R} \int \frac{d^*Q}{T_0} \tag{9.10}$$

where T_0 may vary from point to point along ∂R and d^*Q is the heat transferred from the body to the gas at temperature T_0.

Hence we obtain the following

THEOREM (Generalized H-theorem). *If the gas is bounded by nonporous solid walls at which a linear boundary condition*, Eq. (1.6), *holds, the quantity H given by Eq.* (9.5) *satisfies Eq.* (9.10). *In particular, if at no point of the boundary heat is flowing* from *the gas to the solid* (*i.e. at no point of ∂R is $d^*Q < 0$*) *then*

H always decreases with time and can be constant only if the distribution function is Maxwellian.

Boltzmann's H-theorem is of basic importance because it shows that the Boltzmann equation has a basic feature of irreversibility: the quantity H always decreases even when it is not released to the surroundings [equality of sign in Eq. (9.7)] or [Eq. (9.10)] when no energy exchange takes place between the gas and the surroundings.

These circumstances seems to be in conflict with the fact that the molecules constituting the gas follow the laws of classical mechanics which are time reversible. Accordingly, given a motion at $t = t_0$ with velocities $\mathbf{v}_1, \ldots, \mathbf{v}_N$, we can always consider the motion with velocities $-\mathbf{v}_1, \ldots, -\mathbf{v}_N$ (and the same positions as before) at $t = t_0$; the backward evolution of the latter state will be equal to the forward evolution of the original one. Therefore if $dH/dt < 0$ in the first case, we shall have $dH/d(-t) < 0$ in the second case; i.e. $dH/dt > 0$, which contradicts Boltzmann's H theorem. This is Loschmidt's paradox (to simplify, the gas is assumed to be enclosed in a specularly reflecting box; otherwise the same objection applies to Eq. (4.9)).

The answer to this paradox is roughly as follows: If one obeys the laws of mechanics, then "before" and "after" have no strict meaning, so that one can use the equations of motion to "predict" either the future or the past. However, we passed from this strictly dynamical description to a statistical one based on the Boltzmann equation. When deriving the Boltzmann equation (Chapter II, Sections 2 and 3) we observed that we had to use Eq. (II.2.14) to express the distribution functions corresponding to an after-collision state in terms of the distribution functions corresponding to the state before the collision, rather than the latter in terms of the former. It is obvious that the first way is the right one to follow if the equation are to be used to predict the future from the past and not *vice versa;* it is clear, however, that this choice introduced a connection with the everyday concepts of past and future which are extraneous to molecular dynamics. In other words, we prepared the way to a definition of these concepts on the basis of the statistical behavior of many-particle systems. When we took the Boltzmann limit ($N \to \infty$, $\sigma \to 0$, $N\sigma^2$ finite) we obtained an equation, the Boltzmann equation, which describes the statistical behavior of the gas molecules: our basic choice has the striking consequence that the equation describes only motions for which the quantity H decreases, while the opposite choice would have led to an equation having a negative sign in front of the collision term, and hence describing only motions with increasing H! The latter motions are not impossible, but only extremely improbable; they correspond to initial data in which the molecular velocities of the molecules which are about to collide show an unusual correlation (this situation can be simulated by studying the dynamics of many interacting particles on a computer and leads to an evolution in which there is an increasing H, as expected, while "randomly" chosen initial data invariably lead to evolutions with decreasing H). In other words, the fact that H decreases is not an intrinsic property of the dynamical system but a property of the level of description; in turn the choice of the latter is dictated by the intention of describing a well defined set of measurements on the system (the "macroscopic" measurements).

The fact that H shows a decreasing rather than increasing behaviour is matter of definition; i.e. we can say that $t_2 > t_1$, *by definition*, if $H(t_2) > H(t_1)$ and this fixes a time arrow pointing into the future in agreement with the ordinary concepts of past and future. This can be done since all the physical systems of the Universe are interacting with each other, and therefore it is unusual (an exception is given, for example, by the phenomenon of *spin echo*) that what is "earlier" for one system is "later" for another (the propagation of electromagnetic radiation would play an important role in a complete discussion of this point).

The fact that a system is never really isolated is to be kept in mind also in connection with another paradoxical objection to any mechanical interpretation of irreversibility; this objection is much subtler than the argument based upon reversing the molecular velocities. The objection is based upon a theorem of Poincaré, which says that any finite mechanical system obeying the laws of classical mechanics will return arbitrarily close to its initial state, for almost any choice of the latter, provided we wait long enough. This follows, for a gas of molecules repelling each other and enclosed in a specularly reflecting box, from the conservation of energy, which implies that the representative point in phase space moves on a bounded surface S (the surface of constant energy). These facts imply that a measure $\mu(A)$ (the "area" of A) is associated with each subset A of S in such a way that if A_t is the set of points into which the points of A are transformed at time t by the motion, $\mu(A_t) = \mu(A)$ and $\mu(S) < \infty$. To prove Poincaré's theorem, let us consider a subset A of S such that all the points of A will never come back to A. We choose A small enough and τ large enough so that A_τ and A do not overlap (if this is impossible, the theorem is trivially true); then none of the sets $A_{2\tau}, A_{3\tau}, \ldots$ overlap, because, if $A_{n\tau}$ and $A_{(n+k)\tau}$ had points in common, then, by tracing the motion backwards and using the uniqueness of the motion through any given phase space point, it would follow that A and $A_{k\tau}$ must have points in common and this would contradict the definition of A. If A, A_τ, $A_{2\tau}, \ldots$ do not overlap, then, since $\mu(A) = \mu(A_\tau) = \mu(A_{2\tau}) = \ldots$, the total measure of the union of these sets would be infinite (which is impossible because $\mu(S) < \infty$), unless $\mu(A) = 0$, and Poincaré's theorem is proved.

This theorem implies that our molecules can have, after a "recurrence time", positions and velocities so close to the initial ones that the one particle distribution function f should also be practically the same; therefore H should also be practically the same, and if it increased initially, then must have decreased at some later time. The answer to this objection (Zermelo's paradox) is that the recurrence time is so large that, practically speaking, one would never observe a significant portion of the recurrence cycle; in fact, according to approximate calculations, the recurrence time for a typical amount of gas is a huge number even if the estimated age of the Universe is taken as the time unit. It is clear that, on such an enormous scale, we need not feel uneasy in conceding that irreversibility disappears (again, however, the theorem should be applied to the whole Universe, including radiation; the interesting possibility of a connection between irreversibility and the expansion of the Universe then arises).

What is the meaning of H? There are two meanings, one with respect to the microscopic description, the other with respect to the macroscopic one. The first meaning follows from the fact that (see Appendix to Chapter II) $-H$ can be interpreted as the likelihood of a microscopic state; Eq. (9.6) then states that in an isolated system (absence of the surface integral) the evolution is towards the more probable states. Alternatively, one can say that the more likely is a microscopic state, the larger is the number of the states with the same f, and hence the knowledge of f gives little information on the microscopic state; hence H, as a measure of unlikelihood, is also a measure of the information which f contains about the microscopic state, and this information decreases with time, because the Boltzmann equation describes an evolution toward more likely states. The second interpretation of H is at a macroscopic level and is suggested by Eq. (9.10) which has the form of the Clausius-Duhem inequality, well known in thermodynamics, provided H is related to the entropy η of the gas as follows

$$H = -\eta/R \tag{9.11}$$

In Sect. 10 we shall check that Eq. (9.11) is valid for equilibrium states, when we take for η the ordinary expression for a perfect gas; for nonequilibrium states, Eq. (9.11) can be regarded as the definition of η for a Boltzmann gas. With this interpretation of H, we realize that the H-theorem is none other than a proof of the second principle of thermodynamics (for a Boltzmann gas). In this respect, the second principle is not a strict consequence of mechanics (this would be untenable in view of Loschmidt's and Zermelo's paradoxes) but requires statistical arguments, asymptotic estimates (for $N \to \infty$, $\sigma \to 0$, $N\sigma^2$ finite, see Sections 2 and 3 of Chapter II) and the definition of the future as the direction of time for which there is a statistical tendency to pass from unprobable to probable states.

10. Equilibrium states and Maxwellian distributions.

The decreasing of H in the absence of energy exchange with the surroundings ($d^*Q = 0$ in Eq. (9.10)) shows that the Boltzmann equation describes an evolution towards a state of minimum H (compatible with the assigned constraints, such as the volume of the region where the gas is contained, number of molecules, temperature of the solid bodies surrounding the gas, etc.), provided no additional amount of H flows in from the exterior. The final state (to be reached as $t \to \infty$) will presumably be a steady state provided such a state is compatible with the boundary conditions and is stable. If $d^*Q \geqslant 0$ in Eq. (9.10), then a steady state can be reached if and only if the equality sign holds in the same equation and this, as we know, implies that the distribution function is Maxwellian almost everywhere in velocity space. If Maxwellians are not compatible with the boundary conditions, no steady state is reached. Analogously, if we consider a steady problem which is *not* solved by a Maxwellian distribution then $\int d^*Q/T_0$ must necessarily be negative; this is only natural, because if f is not Maxwellian, then entropy is produced by collisions and this entropy must be released by a suitable heat exchange with the surroundings. Thus, if we have a gas between two parallel plates at temperatures T_{0_1} and T_{0_2} with $T_{0_1} > T_{0_2}$, there must be a heat flow from plate 1 to plate 2, in such a way that $(d^*Q)_1 > 0, (d^*Q)_2 = -(d^*Q)_1 < 0$ and $\int d^*Q/T_0 = \int (d^*Q)_1(T_{0_2} - T_{0_1})/(T_{0_1}T_{0_2}) < 0$; this is, of course, one of the elementary illustrations of the second principle of thermodynamics.

More particular than steady states are the equilibrium states which we define as follows; steady states with no energy exchange with the surroundings. The H-theorem with $d^*Q = 0$ yields the following.

COROLLARY I *The distribution function must be Maxwellian in an equilibrium state.*

In particular, energy exchange is always zero at a wall, if specular reflection applies. Accordingly we have, as a special case, the following

COROLLARY II *No steady solutions of the Boltzmann equation exist when the gas is bounded by specularly reflecting walls except for Maxwellians (describing equilibrium states).*

The latter result, of course, is not valid for more realistic boundary conditions.

Let us now look for the most general Maxwellian which satisfies the Boltzmann equation. We insert the force term as in Eq. (II.8.21) for the sake of generality and write:

$$\frac{\partial f}{\partial t} + \boldsymbol{\xi} \cdot \frac{\partial f}{\partial \mathbf{x}} + \mathbf{X} \cdot \frac{\partial f}{\partial \boldsymbol{\xi}} = Q(f, f) \tag{10.1}$$

If f is Maxwellian, then $Q(f,f) = 0$ and (Eq. (II.7.2)):

$$\log f = a + \mathbf{b} \cdot \boldsymbol{\xi} + c\xi^2 = a + b_k\xi_k + c\xi^2 \tag{10.2}$$

where a, b_k, c are functions of t and $\mathbf{x}$ and are to be determined. Inserting Eq. (10.2) into Eq. (9.1) gives:

$$\frac{\partial a}{\partial t} + \xi_k \frac{\partial b_k}{\partial t} + \xi^2 \frac{\partial c}{\partial t} + \xi_i \frac{\partial a}{\partial x_i} + \xi_i \xi_k \frac{\partial b_k}{\partial x_i} + \xi^2 \xi_i \frac{\partial c}{\partial x_i} + \\ + X_i b_i + 2cX_i \xi_i = 0 \quad (10.3)$$

where the sum from 1 to 3 with respect to repeated indexes is understood, as usual.

If we equate the coefficients of the various powers of ξ, we obtain

$$\frac{\partial a}{\partial t} + X_i b_i = 0 \quad (10.4)$$

$$\frac{\partial b_k}{\partial t} + \frac{\partial a}{\partial x_k} + 2cX_k = 0 \quad (10.5)$$

$$\frac{\partial c}{\partial t} \delta_{ik} + \frac{1}{2}\left(\frac{\partial b_k}{\partial x_i} + \frac{\partial b_i}{\partial x_k}\right) = 0 \quad (10.6)$$

$$\frac{\partial c}{\partial x_i} = 0 \quad (10.7)$$

If we differentiate Eq. (10.6) with respect to x_j we obtain

$$\frac{\partial^2 b_k}{\partial x_i \, \partial x_j} + \frac{\partial^2 b_i}{\partial x_k \, \partial x_j} = 0 \quad (10.8)$$

where use of Eq. (10.7) has been made.

By permuting indexes in Eq. (10.8) we also obtain:

$$\begin{aligned} \frac{\partial^2 b_i}{\partial x_j \, \partial x_k} + \frac{\partial^2 b_j}{\partial x_i \, \partial x_k} &= 0 \\ \frac{\partial^2 b_j}{\partial x_k \, \partial x_i} + \frac{\partial^2 b_k}{\partial x_j \, \partial x_i} &= 0 \end{aligned} \quad (10.9)$$

If we add Eqs. (10.9) to each other and subtract Eq. (10.8) from the result, we obtain, except for a factor 2:

$$\frac{\partial^2 b_j}{\partial x_i \, \partial x_k} = 0 \quad (10.10)$$

Since all the second derivatives of each b_j vanish, the b_j have the form

$$b_j(t, x) = \alpha_j(t) + \beta_{jl}(t) x_l \quad (10.11)$$

Eq. (10.6) then becomes (c does not depend upon the space variables according to Eq. (10.7)):

$$\frac{dc}{dt}\delta_{ik} + \tfrac{1}{2}(\beta_{ki} + \beta_{ik}) = 0 \tag{10.12}$$

Hence

$$\beta_{ik} = -\frac{dc}{dt}\delta_{ik} + \omega_{ik}(t) \tag{10.13}$$

where $\omega_{ik} = -\omega_{ki}$ is an antisymmetric Cartesian tensor.

Using Eqs. (10.11) and (10.13) in Eq. (10.5) gives:

$$\frac{d\alpha_k}{dt} - \frac{d^2c}{dt^2}x_k + \frac{d\omega_{kl}}{dt}x_l + \frac{\partial a}{\partial x_k} + 2cX_k = 0 \tag{10.14}$$

Differentiating with respect to x_i and subtracting the resulting equation from the same equation with interchanged i and k, we obtain (α_k, c and ω_{kl} do not depend on space variables):

$$\frac{d\omega_{ki}}{dt} + c(X_{k/i} - X_{i/k}) = 0 \tag{10.15}$$

Hence

$$X_k = -\frac{1}{2c}\frac{d\omega_{kj}}{dt}x_j - \frac{\partial\varphi}{\partial x_k} \tag{10.16}$$

where $\varphi = \varphi(t, \mathbf{x})$ is a potential. In other words, the external force is not completely arbitrary: it must be the sum of a conservative force and of a very special nonconservative force linear in the space variables (Cartesian coordinates).

Inserting Eq. (10.16) into Eq. (10.14) we obtain

$$\frac{d\alpha_k}{dt} - \frac{d^2c}{dt^2}x_k + \frac{\partial a}{\partial x_k} - 2c\frac{\partial\varphi}{\partial x_k} = 0 \tag{10.17}$$

or, equivalently,

$$a = 2c\varphi(t, x) + \frac{1}{2}\frac{d^2c}{dt^2}x^2 - \frac{d\alpha_k}{dt}x_k + \gamma(t) \tag{10.18}$$

where $\gamma(t)$ is an integration constant, which is an arbitrary function of time, and $x^2 = x_i x_i$.

Finally, Eq. (10.4) is to be satisfied. If we insert Eqs. (10.18), (10.11), (10.13) into Eq. (10.4), we find:

$$2c\frac{\partial\varphi}{\partial t} + 2\varphi\frac{dc}{dt} - \frac{\partial\varphi}{\partial x_i}b_i = -\frac{1}{2}\frac{d^3c}{dt^3}x^2 + \frac{d^2\alpha_k}{dt^2}x_k - \frac{d\gamma}{dt} + \frac{1}{2c}\frac{d\omega_{ij}}{dt}x_j b_i \tag{10.19}$$

Eq. (10.18) can be regarded either as a partial differential equation for the potentials compatible with a local Maxwellian distribution and arbitrarily assigned $c = c(t)$, $\omega_{ik} = \omega_{ik}(t) = -\omega_{ik}$, $\alpha_k = \alpha_k(t)$, $\gamma = \gamma(t)$ or as a constraint for the latter quantities once the potential is given.

An important case is afforded by external forces admitting a steady potential. Eq. (10.16) then shows that $\omega_{kj} =$ constant, when Eq. (10.19) reduces to:

$$2\varphi\frac{dc}{dt} - \frac{\partial\varphi}{\partial x_i}b_i = -\frac{1}{2}\frac{d^3c}{dt^3}x^2 + \frac{d^2\alpha_k}{dt^2}x_k - \frac{d\gamma}{dt} \tag{10.20}$$

Suppose now that the potential φ is four times differentiable in some neighborhood. Differentiating Eq. (10.20) subsequently with respect to x_l, x_m, x_n gives

$$5\frac{\partial^2\varphi}{\partial x_l\,\partial x_m\,\partial x_n}\frac{dc}{dt} - \frac{\partial^3\varphi}{\partial x_i\,\partial x_l\,\partial x_m}\omega_{in} - \frac{\partial^3\varphi}{\partial x_i\,\partial x_l\,\partial x_k}\omega_{im} -$$

$$-\frac{\partial^3\varphi}{\partial x_i\,\partial x_m\,\partial x_n}\omega_{il} - \frac{\partial^4\varphi}{\partial x_i\,\partial x_l\,\partial x_m\,\partial x_n}\left(\alpha_i - \frac{dc}{dt}x_i + \omega_{ik}x_k\right) = 0 \tag{10.21}$$

This is a system of ten homogeneous linear equations in the seven unknowns dc/dt, ω_{ik} (3 distinct components), α_i (3 components). The determinant will in general be nonzero (note that the coefficients vary from place to place, while the unknowns do not depend on space variables). Hence we conclude that, unless the potential is special (such as to have, for example, third derivatives which vanish everywhere), $\omega_{ik} = 0$, $\alpha_i = 0$, $c =$ constant, and because of Eqs. (10.20) and (10.18), $a = 2c\varphi(\mathbf{x}) + \gamma$, where c and γ are constants. Accordingly, for sufficiently general steady potentials, the only Maxwellians solving the Boltzmann equation are steady; if, in addition, no energy exchange takes place with the surroundings, any given initial state will tend to an equilibrium state with the Maxwellian distribution (the so-called barometric distribution or Maxwell-Boltzmann distribution):

$$f = \exp[2c\varphi(\mathbf{x}) + \gamma + c\xi^2] = \rho_0\exp\left[-\frac{\xi^2}{2RT} - \frac{\varphi(\mathbf{x})}{RT}\right]$$

where $T = -(2Rc)^{-1}$ is the (constant) temperature of the gas and $\rho_0 = e^\gamma$ is the density at the point where $\varphi = 0$.

Potentials with vanishing third derivatives are now to be considered (this includes the absence of external forces). Then

$$\varphi = A_l x_l + \tfrac{1}{2}B_{lm}x_l x_m \tag{10.23}$$

and Eq. (10.20) gives, by equating the coefficients of corresponding powers:

$$A_i\alpha_i = \frac{d\gamma}{dt}$$

$$2A_i\frac{dc}{dt} - A_l\omega_{li} - B_{li}\alpha_l = \frac{d^2\alpha_i}{dt^2}$$

$$2B_{lm}\frac{dc}{dt} - \tfrac{1}{2}(B_{li}\omega_{im} + B_{mi}\omega_{il}) = -\frac{1}{2}\frac{d^3c}{dt^3}\delta_{lm} \tag{10.24}$$

These are ten differential equations for the eight unknowns γ, α_i, c, ω_{im}. In spite of this, the system has nontrivial solutions in the case in which second order terms are isotropic, $B_{li} = B\,\delta_{li}$. In this case, the last of Eqs. (10.24) reduces to the single condition:

$$2B\frac{dc}{dt} + \frac{1}{2}\frac{d^3c}{dt^3} = 0 \tag{10.25}$$

which gives

$$c = c_0 + c_1\cos(2\sqrt{B}t) + c_2\sin(2\sqrt{B}t) \tag{10.26}$$

where c_0, c_1, c_2 are constants. The second of Eqs. (10.24) becomes:

$$\frac{d^2\alpha_i}{dt^2} + B\alpha_i = 2A_i\frac{dc}{dt} - A_l\omega_{li} = -\frac{2}{3}\frac{A_i}{B}\left(\frac{d^3c}{dt^3} + B\frac{dc}{dt}\right) - A_l\omega_{li} \tag{10.27}$$

where the last form follows from a suitable use of Eq. (10.25). Eq. (10.27) is easily solved to give

$$\alpha_i = -\frac{2}{3}\frac{A_i}{B}\frac{dc}{dt} - \frac{A_l\omega_{li}}{B} + c_3\cos(\sqrt{B}t) + c_4\sin(\sqrt{B}t) \tag{10.28}$$

where c_3 and c_4 are constants while $c(t)$ is given by Eq. (10.26).

The first of Eqs. (10.24) becomes

$$\frac{d\gamma}{dt} = A_i\alpha_i = \frac{1}{B}A_i\left(-\frac{d^2\alpha_i}{dt^2} + 2A_i\frac{dc}{dt}\right) \tag{10.29}$$

and consequently

$$\gamma = -\frac{1}{B}A_i\frac{d\alpha_i}{dt} + \frac{2}{B}A_iA_ic + c_5 \tag{10.30}$$

where c_5 is a constant and α_i and c are given by Eqs. (10.26) and (10.28).

The above result, due to Boltzmann [37], shows that equilibrium is not necessarily achieved in an harmonic field. According to Eqs. (10.26), (10.29), (10.28), density, velocity and temperature oscillate with the natural frequency of the field or with twice such a frequency.

Another particular case is obtained when the field is absent, $B_{lm} = A_i = 0$. Eqs. (10.24) then give:

$$\frac{d\gamma}{dt} = \frac{d^2\alpha_i}{dt^2} = \frac{d^3c}{dt^3} = 0 \tag{10.31}$$

so that

$$\gamma = \gamma_0, \qquad \alpha_i = \alpha_{i0} + \alpha_{i1}t, \qquad c = c_0 + c_1t + c_2t^2 \tag{10.32}$$

where γ_0, α_{i0}, c_0, α_{i1}, c_1, c_2 are constants. A particularly interesting case is when the solution is also required to be steady; then $c_1 = c_2 = \alpha_{i1} = 0$ and Eqs. (10.2), (10.11), (10.18) give

$$\begin{aligned} f &= \exp(\gamma_0 + \alpha_0 \cdot \boldsymbol{\xi} + \omega_{jl}x_l\xi_j + c\xi^2) \\ &= \rho(2\pi RT)^{-\frac{3}{2}} \exp\left[-\frac{(\boldsymbol{\xi} - \mathbf{v})^2}{2RT}\right] \end{aligned} \tag{10.33}$$

where the temperature $T = -(2Rc)^{-1}$ is constant and, if $\mathbf{v}_0 = RT\alpha_0$, $\boldsymbol{\omega} = RT(\omega_{23}, \omega_{31}, \omega_{12})$:

$$\begin{aligned} \mathbf{v} &= \mathbf{v}_0 + \boldsymbol{\omega} \wedge \mathbf{x} \\ \rho &= \rho_0 \exp\left(\frac{\omega^2 r^2}{2RT}\right) \end{aligned} \tag{10.34}$$

where r is the distance from the rotation axis and ρ_0 is the density at $r = 0$ (in order to obtain the last expression in Eq. (10.33), take $\mathbf{v}_0$ to be the velocity of the rotation axis taken, for example, as the z-axis).

Therefore the most general steady Maxwellian, in the absence of body forces, describes a gas with constant temperature, rigid motion and density given by Eq. (10.34); the space variation of density reflects a centrifugation of the gas as a consequence of rotation.

The present results, together with the fact that the only steady solutions in the presence of specularly reflecting walls are Maxwellians (Corollary II), shows how small is the class of steady problems to which the assumption of specular reflection can be applied. For more general boundary conditions, further constraints will be imposed upon the Maxwellian; hence the latter distribution is exceptional in the presence of realistic boundary conditions, even in steady (nonequilibrium) states.

We end this section with the evaluation of H in an equilibrium state, in order to show that Eq. (9.11) applies. As we know, f is given by Eq. (10.31) in an equilibrium state. Eq. (9.5) and (9.1) then give (assuming ρ and T to be constant)

$$\begin{aligned} H &= \log[\rho(2\pi RT)^{-\frac{3}{2}}]\int f\, d\boldsymbol{\xi}\, d\mathbf{x} - (2RT)^{-1}\int c^2 f\, d\boldsymbol{\xi}\, d\mathbf{x} \\ &= M \log[\rho(2\pi RT)^{-\frac{3}{2}}] - \tfrac{3}{2}M = -M\log(T^{\frac{3}{2}}/\rho) + \\ &\quad + [-\tfrac{3}{2}M\log(2\pi R) - \tfrac{3}{2}M] \end{aligned} \tag{10.35}$$

where M is the total mass of gas. The entropy η of a monatomic gas is

$$\eta = RM\log(T^{\frac{3}{2}}/\rho) + \text{const.} \tag{10.36}$$

where the constant can depend on the total mass but not on the state of the gas. With a suitable and permissible identification of the additive constants, Eq. (9.11) follows, as was to be shown.

Appendix

In this Appendix the basic results about Hermite and Laguerre polynomials are briefly reviewed. These polynomials are best introduced by expanding into a power series the so called generating functions:

$$\exp(-z^2 + 2xz) = \sum_{n=0}^{\infty} \frac{z^n}{n!} H_n(x) \tag{A.1}$$

$$(1 - z)^{-1} \exp[-(xz)/(1 - z)] = \sum_{n=0}^{\infty} z^n L_n(x) \tag{A.2}$$

In other words, the n-th Hermite polynomial, $H_n(x)$, is by, definition, $n!$ times the coefficient of z^n in the power series expansion of $\exp(-x^2 + 2xz)$ and the n-th Laguerre polynomial, $L_n(x)$, is the coefficient of z^n in the power series expansion of $(1 - z)^{-1}\exp[-(xz)/(1 - z)]$. The fact that $H_n(x)$ and $L_n(x)$ are indeed polynomials of n-th degree is very easily checked by means of Eqs. (A.1) and (A.2), from which the expressions for H_n and L_n can also be computed. In particular:

$$H_0(x) = 1, \qquad H_1(x) = 2x, \qquad H_2 = 2x^2 - 1$$
$$L_0(x) = 1, \qquad L_1(x) = 1 - x \tag{A.3}$$

The series in Eq. (A.1) converges for any complex z and the one in Eq. (A.2) for $|z| < 1$ (as a quick glance at the singularities of the left hand sides of Eq. (A.1) and (A.2) shows). Note that the left hand side of Eq. (A.1) is invariant with respect to the change $x, z \to -x, -z$. Hence $H_n(-x) = (-1)^n H_n(x)$, so that a Hermite polynomial has the parity of its own index.

If we multiply Eq. (A.1) by the same equation with z' in place of z and then multiply by $\exp(-x^2)$ and integrate with respect to x from $-\infty$ to ∞, we obtain:

$$\begin{aligned}
&\sum_{n=0}^{\infty} \sum_{n'=0}^{\infty} \frac{z^n z'^{n'}}{n!\, n'!} \int_{-\infty}^{\infty} H_n(x) H_{n'}(x) e^{-x^2}\, dx \\
&= \int_{-\infty}^{\infty} \exp(-z^2 - z'^2 + 2xz + 2xz' - x^2)\, dx \\
&= e^{2zz'} \int_{-\infty}^{\infty} \exp[-(x - z - z')^2]\, dx \\
&= \sqrt{\pi} e^{2zz'} = \sqrt{\pi} \sum_{n=0}^{\infty} \frac{2^n z^n z'^n}{n!}
\end{aligned} \tag{A.4}$$

Comparison of the two power series appearing in Eq. (A.4) shows that

$$\int_{-\infty}^{\infty} H_n(x)H_{n'}(x)e^{-x^2}\,dx = \sqrt{\pi}n!\,2^n\,\delta_{nn'} \tag{A.5}$$

Eq. (A.5) shows that the Hermite polynomials are orthogonal with respect to the measure $\exp(-x^2)\,dx$.

If we now take Eq. (A.1) with $z = \rho\exp(i\varphi)$, multiply it by the same equation with y in place of x and with $\bar{z} = \rho\exp(-i\varphi)$ in place of z and then integrate from $\varphi = 0$ to 2π, and then from $\rho = 0$ to ∞ after multiplying by $\rho\exp(-\rho^2/s)$, we obtain

$$\begin{aligned}
\mathscr{I} &\equiv \int_0^{\infty}\int_0^{2\pi} \exp[-\rho^2/s - \rho^2(e^{2i\varphi} + e^{-2i\varphi}) + 2\rho(xe^{i\varphi} + ye^{-i\varphi})]\rho\,d\rho\,d\varphi \\
&= \sum_{n,n'=0}^{\infty} \frac{1}{n!\,n'!} H_n(x)H_{n'}(y) \int_0^{2\pi} e^{i(n-n')\varphi}\,d\varphi \int_0^{\infty} \rho^{n+n'-1}e^{-\rho^2/s}\,d\rho \\
&= \sum_{n,n'=0}^{\infty} \frac{1}{n!\,n'!} H_n(x)H_{n'}(y)2\pi\,\delta_{nn'} \int_0^{\infty} \rho^{2n+1}e^{-\rho^2/s}\,d\rho \\
&= \pi\sum_{n=0}^{\infty}\frac{s^{n+1}}{n!} H_n(x)H_n(y)
\end{aligned} \tag{A.6}$$

The integral $\mathscr{I}$ appearing in Eq. (A.6) can be easily evaluated if we pass from the polar coordinates ρ, φ to the Cartesian coordinates $u = \rho\cos\varphi$, $v = \rho\sin\varphi$, for then

$$\begin{aligned}
\mathscr{I} &= \int_{-\infty}^{\infty}\int_{-\infty}^{\infty} \exp\left[-u^2\left(\frac{1}{s}+2\right) - v^2\left(\frac{1}{s}-2\right) + 2(x+y)u + 2i(x-y)v\right]du\,dv \\
&= \int_{-\infty}^{\infty} \exp\left\{-\left(\frac{1}{s}+2\right)\left[u - \frac{(x+y)s}{1+2s}\right]^2\right\} du \\
&\quad\times \int_{-\infty}^{\infty} \exp\left\{-\left(\frac{1}{s}-2\right)\left[v - i\frac{(x-y)s}{1-2s}\right]^2\right\} dv \times \\
&\quad\times \exp\left[(x+y)^2\frac{s}{1+2s} - (x-y)^2\frac{s}{1-2s}\right] \\
&= \frac{\pi s}{\sqrt{1-4s^2}}\exp\left[\frac{s(x+y)^2}{1+2s} - \frac{(x-y)^2 s}{1-2s}\right]
\end{aligned} \tag{A.7}$$

where use has been made of the well known integral:

$$\int_{-\infty}^{\infty} e^{-\alpha(u-\beta)^2}\,du = \sqrt{\pi/\alpha} \tag{A.8}$$

Comparing the two expressions of $\mathscr{I}$ and cancelling the common factor πs, we obtain

$$\sum_{n=0}^{\infty} \frac{s^n}{n!} H_n(x)H_n(y) = \frac{1}{\sqrt{1-4s^2}} \exp\left[\frac{s(x+y)^2}{2s+1} - \frac{s(x-y)^2}{1-2s}\right] \tag{A.9}$$

This is a very important formula when dealing with operators whose eigenfunctions are Hermite polynomials. It is expedient to replace s by $s/2$ and rewrite Eq. (A.9) as follows:

$$\sum_{n=0}^{\infty} \frac{s^n}{2^n n!} H_n(x)H_n(y) = \frac{1}{\sqrt{1-s^2}} \exp\left[\frac{s(x+y)^2}{2(1+s)} - \frac{s(x-y)^2}{2(1-s)}\right] \quad (|s| < 1) \tag{A.10}$$

In particular, by letting $s \to 1$, we see that the right hand side does not converge in the usual sense; convergence in the sense of generalized functions holds, however, and, in the latter sense

$$\sum_{n=0}^{\infty} \frac{1}{2^n n!} H_n(x)H_n(y) = \sqrt{\pi}\, e^{-x^2} \delta(x-y) \tag{A.11}$$

which proves the completeness of Hermite polynomials.

If we now multiply Eq. (A.2) by the same equation with z' in place of z, and then multiply by e^{-x} and integrate with respect to x from 0 to ∞ we obtain:

$$\sum_{n=0}^{\infty} \sum_{m=0}^{\infty} z^n z'^m \int_0^{\infty} L_n(x)L_m(x)e^{-x}\, dx$$

$$= [(1-z)(1-z')]^{-1} \int_0^{\infty} \exp\left[-x\left(\frac{z}{1-z} + \frac{z'}{1-z'}\right) - x\right] dx$$

$$= \frac{1}{1-zz'} = \sum_0^{\infty} z^n z'^n \tag{A.12}$$

Comparison between the two series expansions appearing in this equation shows that

$$\int_0^{\infty} L_n(x)L_m(x)e^{-x^2}\, dx = \delta_{nm} \tag{A.13}$$

which shows that the Laguerre polynomials are orthogonal on the half line $(0, \infty)$ with respect to the measure $e^{-x}\, dx$.

Let us now put $x = \rho \cos \varphi$, $y = \rho_1 \cos \varphi_1$ in Eq. (A.10), multiply it by the same equation with $x = \rho \sin \varphi$, $y = \rho_1 \sin \varphi_1$ and integrate with respect to φ_1. Then:

$$\sum_{n=0}^{\infty}\frac{s^n}{2^n}\sum_{k=0}^{n}[k!\,(n-k)!]^{-1}H_k(\rho\cos\varphi)H_{n-k}(\rho\sin\varphi)\times$$

$$\times\int_0^{2\pi}H_k(\rho_1\cos\varphi_1)H_{n-k}(\rho_1\sin\varphi_1)\,d\varphi_1$$

$$=\frac{1}{1-s^2}\exp\left[-\frac{s^2}{1-s^2}(\rho^2+\rho_1^{\,2})\right]\int_0^{2\pi}\exp\left[\frac{2s\rho\rho_1}{1-s^2}\cos(\varphi_1-\varphi)\right]d\varphi_1$$

$$=\frac{1}{1-s^2}\exp\left[-\frac{s^2}{1-s^2}(\rho^2+\rho_1^{\,2})\right]\int_0^{2\pi}\exp\left[\frac{2s\rho\rho_1}{1-s^2}\cos\varphi_1\right]d\varphi_1 \qquad \text{(A.14)}$$

where the last expression is obtained by changing φ_1 into $\varphi_1+\varphi$ and taking the periodicity of the integrand into account. The right hand side of Eq. (A.14) does not depend upon φ; hence the sums

$$S_n(\rho,\rho_1)=\frac{1}{2^n}\sum_{k=0}^{n}\frac{1}{k!\,(n-k)!}H_k(\rho\cos\varphi)H_{n-k}(\rho\sin\varphi)\times$$

$$\times\int_0^{2\pi}H_k(\rho_1\cos\varphi_1)H_{n-k}(\rho_1\sin\varphi_1)\,d\varphi_1 \qquad \text{(A.15)}$$

which are the coefficients of the expansion of the right hand side of Eq. (A.14) into a power series in s, do not depend on φ. Hence, if we integrate both sides of Eq. (A.15) from 0 to 2π, we find another expression for $S(\rho,\rho_1)$ which is symmetric in ρ, ρ_1:

$$S_n(\rho,\rho_1)=\frac{1}{2^n}\sum_{k=0}^{n}\frac{1}{k!\,(n-k)!}\frac{1}{2\pi}\int_0^{2\pi}\int_0^{2\pi}H_k(\rho\cos\varphi)H_{n-k}(\rho\sin\varphi)\times$$

$$\times H_k(\rho_1\sin\varphi_1)H_{n-k}(\rho_1\cos\varphi_1)\,d\varphi\,d\varphi_1=S_n(\rho_1,\rho) \qquad \text{(A.16)}$$

If we change φ into $\varphi+\pi$, the integrals must remain unchanged because integration is extended over a period of the integrand; on the other hand $H_k(\rho\cos\varphi)H_{n-k}(\rho\sin\varphi)$ becomes $(-1)^nH_k(\rho\cos\varphi)H_{n-k}(\rho\sin\varphi)$ because of the parity property of Hermite polynomials. Hence $S_n(\rho,\rho_1)=-S_n(\rho,\rho_1)$ for odd n giving:

$$S_{2m+1}(\rho,\rho_1)=0 \qquad (m=0,1,\ldots) \qquad \text{(A.17)}$$

We now fix ρ_1 and let ρ vary. From Eq. (A.16), $S_{2m}(\rho,\rho_1)$ is clearly a polynomial containing only even powers of ρ and having degree $2m$ and so is a (nonzero, except for a finite set of values of ρ_1) polynomial of degree m in ρ^2.

We now compute, by means of Eq. (A.15), the following integral

$$\int_0^\infty \int_0^{2\pi} S_{2n}(\rho, \rho_1) S_{2m}(\rho, \rho_1) e^{-\rho^2} \rho \, d\rho \, d\varphi$$
$$= \frac{1}{2^{2n+2m}} \sum_{k=0}^{2n} \sum_{h=0}^{2m} \frac{1}{k!\,(2n-k)!\,h!\,(2m-h)!} \times$$
$$\times \int_{-\infty}^{\infty} \int_{-\infty}^{\infty} H_k(x) H_h(x) H_{2n-k}(y) H_{2m-h}(y) e^{-(x^2+y^2)} \, dx \, dy \times$$
$$\times \int_0^{2\pi} H_k(\rho_1 \cos \varphi_1) H_{2n-k}(\rho_1 \sin \varphi_1) \, d\varphi_1 \times$$
$$\times \int_0^{2\pi} H_h(\rho_1 \cos \varphi_2) H_{2m-h}(\rho_1 \sin \varphi_2) \, d\varphi_2 \tag{A.18}$$

where a change from the polar variables ρ, φ to the Cartesian ones $x = \rho \cos \varphi$, $y = \rho \sin \varphi$ has been made. Let us use the fact that $S_n(\rho, \rho_1) S_m(\rho, \rho_1)$ does not depend upon φ in conjunction with Eq. (A.5); then:

$$2 \int_0^\infty S_{2n}(\rho, \rho_1) S_{2m}(\rho, \rho_1) e^{-\rho^2} \rho \, d\rho = \Phi(\rho_1) \, \delta_{nm} \tag{A.19}$$

where

$$\Phi(\rho_1) = \frac{1}{2_{2n}} \sum_{k=0}^{2n} \frac{\pi}{k!\,(n-k)!} \int_0^{2\pi} \int_0^{2\pi} H_k(\rho_1 \cos \varphi_1) H_{2n-k}(\rho_1 \sin \varphi_1) \times$$
$$\times H_h(\rho_1 \cos \varphi_2) H_{2m-h}(\rho_1 \sin \varphi_2) \, d\varphi_1 \, d\varphi_2 \tag{A.20}$$

and we have used the fact that $\delta_{kh} \, \delta_{2n-k, 2m-h} = \delta_{kh} \, \delta_{nm}$.

Eq. (A.19) shows that for fixed ρ_1 the polynomials in ρ^2, $S_{2n}(\rho, \rho_1)$, are orthogonal with respect to the measure $2\rho e^{-\rho^2} \, d\rho$, so that the polynomials $P_n(x) = S_{2n}(\sqrt{x}, \rho_1)$ are orthogonal with respect to the measure $e^{-x} \, dx$. However, the set of polynomials of all degrees which are orthogonal with respect to a given measure are uniquely determined up to a factor, since they can be easily constructed by recursion (Gram-Schmidt's method). Hence, for any fixed ρ_1, $P_n(x) = S_{2n}(\sqrt{x}, \rho_1)$ must be proportional to $L_n(x)$, so that we have $S_{2n}(\rho, \rho_1) = Q_n(\rho_1) L_n(\rho^2)$. On the other hand, $S_{2n}(\rho, \rho_1) = S_{2n}(\rho_1, \rho)$; hence $Q_n(\rho_1)$ must coincide with $L_n(\rho_1{}^2)$ up to a constant factor k_n and we have

$$S_{2n}(\rho, \rho_1) = k_n L_n(\rho^2) L_n(\rho_1{}^2) \tag{A.21}$$

which establishes a relationship between Hermite and Laguerre polynomials.

Eq. (A.14) then becomes

$$\sum_{n=0}^{\infty} s^{2n} k_n L_n(\rho^2) L_n(\rho_1{}^2)$$
$$= \frac{1}{1-s^2} \exp\left[-\frac{s^2}{1-s^2}(\rho^2+\rho_1{}^2)\right] \int_0^{2\pi} \exp\left(\frac{2s\rho\rho_1 \cos\varphi_1}{1-s^2}\right) d\varphi_1 \quad \text{(A.22)}$$

where Eqs. (A.15) and (A.21) have been used.

In order to compute the constant k_n, we let $\rho_1 = \rho$ and integrate both sides of Eq. (A.22) with respect to ρ after multiplying by $e^{-\rho^2} 2\rho\, d\rho$.

If we use Eq. (A.13) with $m = n$ and $x = \rho^2$ we obtain:

$$\sum_{n=0}^{\infty} s^{2n} k_n = \frac{1}{1-s^2} \int_0^{\infty} \int_0^{2\pi} \exp\left(-\rho^2 - \frac{2s^2\rho^2}{1-s^2} + \frac{2s}{1-s^2}\rho^2 \cos\varphi\right) 2\rho\, d\rho\, d\varphi$$
$$= \frac{1}{1-s^2} \int_0^{2\pi} \frac{d\varphi}{1 + \dfrac{2s^2}{1-s^2} - \dfrac{2s}{1-s^2}\cos\varphi} = \int_0^{2\pi} \frac{d\varphi}{1 + s^2 - 2s\cos\varphi}$$
$$= \frac{2\pi}{1-s^2} = 2\pi \sum_{n=0}^{\infty} s^{2n} \quad \text{(A.23)}$$

where the last expression comes from an expansion of $(1 - s^2)^{-1}$. Comparison of the coefficients in the two series appearing in Eq. (A.23) shows that $k_n = 2\pi$. Hence Eq. (A.22) becomes, if we also replace s^2 by s:

$$\sum_{n=0}^{\infty} s^n L_n(\rho^2) L_n(\rho_1{}^2)$$
$$= \frac{1}{1-s} \exp\left[-\frac{s}{1-s}(\rho^2+\rho_1)\right] \frac{1}{2\pi} \int_0^{2\pi} \exp\left(\frac{2\sqrt{s}\,\rho\rho_1}{1-s}\cos\varphi\right) d\varphi$$
$$= \frac{1}{1-s} \exp\left[-\frac{s}{1-s}(\rho^2+\rho_1{}^2)\right] I_0\left(\frac{2\sqrt{s}\,\rho\rho_1}{1-s}\right) \quad \text{(A.24)}$$

where I_0 is the modified Bessel function of order zero of the first kind [38] which can be defined by

$$I_0(x) = \frac{1}{2\pi} \int_0^{2\pi} \exp(x\cos\varphi)\, d\varphi \quad \text{(A.25)}$$

Eq. (A.24) is a basic equation when dealing with Laguerre polynomials. If we let $s \to 1$ we obtain, in the sense of the theory of generalized functions:

$$\sum_{n=0}^{\infty} L_n(\rho^2)L_n(\rho_1^2) = e^{\rho^2}\,\delta(\rho^2 - \rho_1^2) \tag{A.26}$$

which proves the completeness of the set $L_n(x)$. Eq. (A.26) is verified by noting that the right hand side of Eq. (A.24) can be written in the form

$$I(s, \rho, \rho_1) = \frac{1}{\sqrt{s}} \exp\left[\frac{\sqrt{s}}{1+\sqrt{s}}(\rho^2 + \rho_1^2)\right] \frac{1}{2\pi}\int_0^{2\pi} \frac{\sqrt{s}}{1-s} \times$$
$$\times \exp\left[-\frac{\sqrt{s}}{1-s}(\rho^2 + \rho_1^2 - 2\rho\rho_1\cos\varphi)\right] d\varphi \tag{A.27}$$

where the integrand is positive and tends to zero when $s \to 1$ except when $\rho^2 + \rho_1^2 - 2\rho\rho_1 \cos\varphi = 0$: that is, when $\rho = \rho_1$, $\varphi = 0$. On the other hand, putting $x = \rho\cos\varphi$, $y = \rho \sin\varphi$ gives:

$$\int_0^{\infty}\int_0^{2\pi} \frac{\sqrt{s}}{1-s} \exp\left[-\frac{\sqrt{s}}{1-s}(\rho^2 + \rho_1^2 - 2\rho\rho_1\cos\varphi)\right] d(\rho^2)\, d\varphi$$
$$= 2\int_{-\infty}^{\infty}\int_{-\infty}^{\infty} \frac{\sqrt{s}}{1-s} \exp\left\{-\frac{\sqrt{s}}{1-s}[(x-\rho_1)^2 + y^2]\right\} dx\, dy$$
$$= 2\pi \tag{A.28}$$

Hence the integrand in Eq. (A.27) tends to $2\pi\,\delta(\rho^2 - \rho_1^2)\,\delta(\varphi)$ when $s \to 1$. Since the factor outside the integral in Eq. (A.27) tends to an easily computed finite limit, we obtain

$$\lim_{s\to 1} I(s, \rho, \rho_1) = \exp\left(\frac{\rho^2 + \rho_1^2}{2}\right)\delta(\rho^2 - \rho_1^2) = e^{\rho^2}\,\delta(\rho^2 - \rho_1^2) \tag{A.29}$$

and Eq. (A.26) is verified.

From the properties of Hermite and Laguerre polynomials we infer that the polynomials ψ_{lmn} in the components ξ_n, ξ_{t1}, ξ_{t2} of the velocity vector defined by Eq. (6.14) of the main text:

$$\psi_{lmn}(\boldsymbol{\xi}) = (m!\,n!\,2^m 2^n)^{-\frac{1}{2}} L_l\left(\frac{\xi_n^2}{2RT_0}\right) H_m\left(\frac{\xi_{t1}}{\sqrt{2RT_0}}\right) H_n\left(\frac{\xi_{t2}}{\sqrt{2RT_0}}\right) \tag{A.30}$$

satisfy

$$\int \psi_{lmn}(\boldsymbol{\xi})\psi_{l'm'n'}(\boldsymbol{\xi}) f_0(\boldsymbol{\xi})\, |\boldsymbol{\xi}\cdot\mathbf{n}|\, d\boldsymbol{\xi} = \delta_{ll'}\,\delta_{mm'}\,\delta_{nn'} \tag{A.31}$$

In addition, Eqs. (A.10) and (A.24) give:

$$\sum_{l,m,n} p^l q^m s^n \psi_{lmn}(\boldsymbol{\xi})\psi_{lmn}(\boldsymbol{\xi}')$$

$$= \frac{1}{(1-p)\sqrt{(1-q^2)(1-s^2)}} \exp\left[-\frac{p}{1-p}\frac{\xi_n^2+\xi_n'^2}{2RT_0}+\right.$$

$$+\frac{q}{2(1+q)}\frac{(\xi_{t1}+\xi_{t1}')^2}{2RT_0}-\frac{q}{2(1-q)}\frac{(\xi_{t1}-\xi_{t1}')^2}{2RT_0}+$$

$$\left.+\frac{s}{2(1+s)}\frac{(\xi_{t2}+\xi_{t2}')^2}{2RT_0}-\frac{s}{2(1-s)}\frac{(\xi_{t2}-\xi_{t2}')^2}{2RT_0}\right] I_0\left(\frac{\sqrt{p}}{1-p}\frac{\xi_n\xi_n'}{RT_0}\right)$$

$$(|p|, |q|, |s| < 1) \quad \text{(A.32)}$$

which is Eq. (6.15) of the main text.

References

[1] C. CERCIGNANI, in "Transport Theory", G. Birkhoff et al., eds., SIAM-AMS Proceedings, Vol. I, p. 249, AMS, Providence (1968).

[2] C. CERCIGNANI, "Mathematical Methods in Kinetic Theory", Plenum Press, New York (1969).

[3] I. KUŠČER, in "Transport Theory Conference", AEC Report ORO-3858-1, Blacksburg, Va. (1969).

[4] C. CERCIGNANI and M. LAMPIS, "Transport Theory and Statistical Physics", **1,** 101 (1971).

[5] I. KUŠČER, "Surface Science", **25,** 225 (1971).

[6] C. CERCIGNANI, "Transport Theory and Statistical Physics", **2,** 27 (1972).

[7] J. L. LEBOWITZ and H. L. FRISCH, "Phys. Rev.", **107,** 917 (1957).

[8] P. G. BERGMANN and J. L. LEBOWITZ, "Phys. Rev.", **99,** 56 (1955).

[9] J. L. LEBOWITZ and P. G. BERGMANN, "Ann. Phys." (N.Y.) **1,** 1 (1957).

[10] J. DARROZÈS and J. P. GUIRAUD, "Compt. Rend. Ac. Sci." (Paris), **A262,** 1368 (1966).

[11] J. C. MAXWELL, "Phil. Trans. Royal Soc." I, Appendix (1879); reprinted in "The Scientific Papers of J. C. Maxwell", Dover Publications, New York (1965).

[12] S. F. SHEN and I. KUŠČER, in "Rarefied Gas Dynamics", Dini et al., eds., vol. I, p. 109 (1974).

[13] E. P. GROSS, E. A. JACKSON and S. ZIERING, "Ann. Phys." (N.Y.), **1,** 141 (1957).

[14] S. F. SHEN, "Entropie", **18,** 138 (1967).

[15] C. CERCIGNANI and M. LAMPIS, Presented at the 8th Symposium on Rarefied Gas Dynamics, Stanford, Calif. (1972).

[16] M. EPSTEIN, AIAA Journal, 5, 1797 (1967).
[17] F. O. GOODMAN, "Phys. Chem. Solids", **23,** 1269 (1962).
[18] F. O. GOODMAN, "Phys. Chem. Solids", **23,** 1491 (1962).
[19] F. O. GOODMAN, "Phys. Chem. Solids", **24,** 1451 (1963).
[20] F. O. GOODMAN, "Surface Sci.", **7,** 391 (1967).
[21] L. TRILLING, "Surface Sci.", **21,** 337 (1970).
[22] R. M. LOGAN and R. E. STICKNEY, "J. Chem. Phys." **44,** 195 (1966).
[23] R. M. LOGAN and J. C. KECK, "J. Chem. Phys." **49,** 860 (1968).
[24] R. M. LOGAN and J. C. KECK, "Surface Science", **15,** 387 (1969).
[25] S. L. SOBOLEV, "Partial Differential Equations of Mathematical Physics", Pergamon Press, Oxford (1964).
[26] M. M. R. WILLIAMS, "J. Phys. D.: Appl. Phys.", **4,** 1315 (1971).
[27] I. KUŠČER, J. MOZINA and F. KRIZANIC, in "Rarefied Gas Dynamics", Dini et al., Eds., vol. I, p. 97 (1974).
[28] C. CERCIGNANI, in "Rarefied Gas Dynamics", Dini et al., Eds., vol. I, p. 75 (1974).
[29] J. J. HINCHEN and W. M. FOLEY, in "Rarefied Gas Dynamics", J. H. de Leeuw, Ed., vol. II, p. 505, Academic Press, New York (1966).
[30] R. SCHAMBERG, in "Proc. 1959 Heat Transfer and Fluid Dynamics Institute", p. 1, Stanford University Press (1959).
[31] S. NOCILLA, in "Rarefied Gas Dynamics", L. Talbot, Ed., p. 169, Academic Press, New York (1961).
[32] S. NOCILLA, in "Rarefied Gas Dynamics", J. A. Laurmann, Ed., vol. I, p. 327, Academic Press, New York (1963).
[33] M. KNUDSEN, "Ann. der Phys." **34,** 593 (1911).
[34] H. GRAD, in "Handbuch der Physik", vol. XII, 205, Springer, Berlin (1958).
[35] M. N. KOGAN, "Rarefied Gas Dynamics", Plenum Press, New York (1969).
[36] S. CHAPMAN and T. G. COWLING, "The Mathematical Theory of Non-Uniform Gases", Cambridge University Press, Cambridge (1952).
[37] L. BOLTZMANN, "Wissenschaftliche Abhandlungen," F. Hasenorl, Ed., vol. II, p. 83, J. A. Barth, Leipzig (1909).
[38] G. N. WATSON, "A Treatise on the Theory of Bessel Functions", Cambridge University Press, Cambridge (1958).
[39] J. L. W. V. JENSEN, "Acta Mathematica", **30,** 175 (1906).
[40] A. ZYGMUND, "Trigonometrical Series", p. 67, Dover, New York (1955).
[41] W. RUDIN, "Real and Complex Analysis", McGraw-Hill, London (1970).

IV LINEAR TRANSPORT

1. The linearized collision operator

Because of the nonlinear nature of the collision term, the Boltzmann equation is very difficult to solve and to analyse. In Chapter III, Section 10, we studied a very particular class of solutions; namely, the Maxwellians. The meaning of a Maxwellian distribution is clear: it describes equilibrium states (or slight generalizations of them, characterized by the fact that neither heat flux nor stresses other than isotropic pressure are present). If we want to describe more realistic nonequilibrium situations, when oblique stresses are present and heat transfer takes place, we have to rely upon approximate methods.

Some of the most useful methods of solution are based upon perturbation techniques: we choose a parameter ε which can be small in some situations and expand f in a series of powers of ε [or, more generally, of functions $\sigma_n(\varepsilon)$, such that $\lim_{\varepsilon\to 0} \sigma_{n+1}(\varepsilon)/\sigma_n(\varepsilon) = 0$]. The resulting expansion which in general cannot be expected to be convergent, but only asymptotic to a solution of the Boltzmann equation, gives useful information for a certain range of small values of ε (sometimes larger than would be expected).

There are many different perturbation methods corresponding to different choices of ε; however, in this section we want to study the general features of any perturbation method with respect to the collision operator $Q(f, f)$. We shall restrict ourselves to power series in ε:

$$f = \sum_{n=0}^{\infty} \varepsilon^n f_n \tag{1.1}$$

If we insert this expansion into $Q(f, f)$ and take into account both the quadratic nature of the collision operator and the Cauchy rule for the product of two series, we find

$$Q(f, f) = \sum_{n=0}^{\infty} \varepsilon^n \sum_{k=0}^{n} Q(f_k, f_{n-k}) \tag{1.2}$$

where $Q(f, g)$ is the bilinear operator defined by Eq. (II.6.2). The presence of a symmetrized expression is related to the fact that we can combine the terms with $k = k_0$ and $k = n - k_0$ (for any k_0, $0 \leqslant k_0 \leqslant n$).

Eq. (1.2) shows that expanding f into a power series in the parameter ε implies an analogous expansion of the collision term, the coefficients being as follows

$$Q_n = \sum_{k=0}^{n} Q(f_k, f_{n-k}) \tag{1.3}$$

A significant number of perturbation expansions which are used in connection with the Boltzmann equation have the following feature: either as a consequence of the zero-th order equation, or because of the assumptions underlying the perturbation method, the zero-th order term in the expansion is a Maxwellian. We shall restrict our attention to this case at present: we note, however, that the parameters appearing in the Maxwellian (density, mass velocity, temperature) can depend arbitrarily upon the time and space variables (the Maxwellian is, in general, not required to satisfy the Boltzmann equation), but this will not concern us insofar as we deal with the collision operator which does not act on the space-time dependence of f.

Under the present assumptions we have (Chapter II, Section 7):

$$Q(f_0, f_0) = 0 \quad \text{i.e.} \quad Q_0 = 0 \tag{1.4}$$

We observe now that $Q_n (n \geqslant 1)$ as given by Eq. (1.3) splits as follows:

$$Q_n = 2Q(f_0, f_n) + \sum_{k=1}^{n-1} Q(f_k, f_{n-k}) \qquad (n \geqslant 1) \tag{1.5}$$

where the first term arises from $k = 0, k = n$ in Eq. (1.3) and the second term involves f_k with $k < n$, and is accordingly known at the n-th step of the perturbation procedure (in particular, it is zero for $n = 1$). As a consequence, the relevant operator to be considered is the linear operator $2Q(f_0, f_n)$ acting upon the unknown f_n, while the remainder can be written as a source term, say $f_0 S_n$. It is usual to put $f_n = f_0 h_n$ and to consider h_n as unknown; then we can write Eq. (1.5) as follows:

$$Q_n = f_0 L h_n + f_0 S_n \tag{1.6}$$

where, by definition, the *linearized collision operator* L is given by

$$Lh = 2f_0^{-1} Q(f_0, f_0 h) = \frac{1}{m} \int f_{0*}(h_*' + h' - h_* - h) B(\theta, V) \, d\xi_* \, d\varepsilon \, d\theta \tag{1.7}$$

In order to obtain the second expression of Lh in Eq. (1.7) we used Eq. (II.6.2) and the fact that f_0, being Maxwellian, satisfies Eq. (II.7.7).

We can now use the properties of $Q(f, g)$ and the definition of L to deduce some basic properties of the latter. If we consider Eq. (II.6.12) with $f = f_0$, $g = f_0 h$ and $\varphi = \bar{g}$, we have:

$$\begin{aligned}\int f_0 \bar{g} L h \, d\xi &= 2 \int \bar{g} Q(f_0, f_0 h) \, d\xi \\ &= -\frac{1}{4m} \int f_0 f_{0*} (h_*' + h' - h_* - h) \times \\ &\quad \times (\bar{g}_*' + \bar{g}' - \bar{g} - \bar{g}_*) B(\theta, V) \, d\xi_* \, d\xi \, d\varepsilon \, d\theta\end{aligned} \tag{1.8}$$

The last expression shows that interchange of h and g changes the integral into its complex conjugate (and hence leaves the integral unchanged if g and h are real valued). This suggests the introduction of a Hilbert space $\mathscr{H}$ [1–5] where the scalar product (g, h) and the norm $\|h\|$ are defined by

$$(g, h) = \int f_0(\boldsymbol{\xi})\bar{g}(\boldsymbol{\xi})h(\boldsymbol{\xi})\, d\boldsymbol{\xi}; \qquad \|h\|^2 = (h, h) \tag{1.9}$$

Eq. (1.8) then shows that

$$(g, Lh) = (Lg, h) \tag{1.10}$$

so that L is self-adjoint (here both g and h belong to the domain of the operator L). If we put $g = h$ in Eq. (1.8) we obtain

$$(h, Lh) = -\frac{1}{4m}\int f_0 f_0{}^* |h_*' + h' - h_* - h|^2\, B(\theta, V)\, d\boldsymbol{\xi}\, d\boldsymbol{\xi}_*\, d\varepsilon\, d\theta \tag{1.11}$$

and since $B(\theta, V) > 0$, we have

$$(h, Lh) \leqslant 0 \tag{1.12}$$

and the equality sign holds if and only if the quantity which appears squared in Eq. (1.11) is zero; that is, if h is a collision invariant. Eq. (1.12) says that L is a nonpositive operator in $\mathscr{H}$. We note that when the equality sign holds in Eq. (1.12), so that h is a collision invariant, then Eq. (1.7) gives

$$Lh = 0 \tag{1.13}$$

and, conversely, if we scalarly multiply this equation by h, according to the scalar product in $\mathscr{H}$ defined above, we obtain

$$(h, Lh) = 0 \tag{1.14}$$

which implies that h is a collision invariant. Therefore the collision invariants ψ_α are eigenfunctions of L corresponding to the eigenvalue $\lambda = 0$, and are the only eigenfunctions corresponding to such eigenvalue; all the other eigenvalues, if any, must be negative, because of Eq. (1.12). It can be verified that the latter equation is the linearized version of Eq. (II.7.3); in fact, if we write $f = f_0(1 + h)$ in the latter equation and neglect terms of higher than second order (zero-th and first order terms cancel), we obtain Eq. (1.12). We note a further property of L which is a trivial consequence of the fact that the ψ_α's satisfy:

$$L\psi_\alpha = 0 \qquad (\alpha = 0, 1, 2, 3, 4) \tag{1.15}$$

and of Eq. (1.10) (with $g = \psi_\alpha$), namely that:

$$(\psi_\alpha, Lh) = 0 \tag{1.16}$$

which is the linearized version of Eq. (II.8.22).

2. The linearized Boltzmann equation

As we mentioned in Section 1, a significant number of perturbation expansions which are used in connection with the Boltzmann equation have the form shown in Eq. (1.1). The result of inserting such an expansion into the Boltzmann equation

$$\frac{\partial f}{\partial t} + \boldsymbol{\xi} \cdot \frac{\partial f}{\partial \mathbf{x}} = Q(f, f) \tag{2.1}$$

depends upon the meaning of ε. If ε does not appear directly in Eq. (2.1), then we must equate the coefficients of the various powers to obtain

$$\frac{\partial f_0}{\partial t} + \boldsymbol{\xi} \cdot \frac{\partial f_0}{\partial \mathbf{x}} = Q(f_0, f_0) \tag{2.2}$$

$$\frac{\partial f_n}{\partial t} + \boldsymbol{\xi} \cdot \frac{\partial f_n}{\partial \mathbf{x}} = Q_n \tag{2.3}$$

where Q_n is given by Eq. (1.3).

Eq. (2.3) shows that f_0 must be a solution of the Boltzmann equation. Since we do not know any solutions except Maxwellians (with some irrelevant exceptions), we are practically forced to choose f_0 to be a Maxwellian; otherwise, making the zero-th order step of the approximation procedure would be as hard as solving the original equation. Although there are Maxwellians with variable density, velocity and temperature which solve the Boltzmann equation (see Chapter III, Section 10), they are of limited use; accordingly, we shall choose our Maxwellian f_0 to have constant parameters. This choice is sufficiently broad for our purposes. Physically, it means that we study a situation in which there is little departure from an overall equilibrium.

We can put $f_n = f_0 h_n (n \geqslant 1)$ and write Eq. (2.3) as follows

$$\frac{\partial h_n}{\partial t} + \boldsymbol{\xi} \cdot \frac{\partial h_n}{\partial \mathbf{x}} = Lh_n + S_n \tag{2.4}$$

where, according to Eqs. (1.5) and (1.6):

$$S_1 = 0 \qquad S_n = f_0^{-1} \sum_{k=1}^{n-1} Q(f_0 h_k, f_0 h_{n-k}) \tag{2.5}$$

The sequence of Eqs. (2.4) describes a successive approximation procedure for solving the Boltzmann equation. What is interesting is that at each step we have to solve the same equation, the only change being in the source term, which has to be evaluated in terms of the previous approximations. The equations to be solved involve a complicated integrodifferential operator and have a shape almost as complicated as that of the original Boltzmann equation, except for the fact that we have got rid of the nonlinearity. The fact that the same operator appears at each step allows us to concentrate on the first step, and so to study the following equation:

$$\frac{\partial h}{\partial t} + \boldsymbol{\xi} \cdot \frac{\partial h}{\partial \mathbf{x}} = Lh \tag{2.6}$$

which is called the *linearized Boltzmann equation.*

The presence of a source term in the subsequent steps is hardly a complication in solving the equations, since well-known procedures allow us to solve an inhomogeneous linear equation once we are able to master the corresponding homogeneous equation (see Section 11). In practice, however, one usually makes only the first step and solves the linearized Boltzmann equation in place of the nonlinear one.

The study of the linearized Boltzmann equation is important for at least two reasons:

1. There are conditions (to be specified below) under which the results obtained from the linearized Boltzmann equation can be used to faithfully represent the physical situation.

2. The fact that the linearized equation has the same structure (except for the nonlinearity in the collision term) as the full Boltzmann equation suggests that we can obtain a valuable insight into the features of the solutions of the full Boltzmann equation by studying the linearized one; these features are obviously not those related to nonlinear effects, but are, for example, those related to the behavior in the proximity of boundaries, for which the nonlinear nature of the collisions is expected to have little influence.

We now have to specify the conditions under which one can make use of the linearized Boltzmann equation to obtain physically significant results. Since the parameter ε was assumed not to appear in the Boltzmann equation, we must examine the initial and boundary conditions. Since we look for a solution in the form $f = f_0(1 + h)$ with the condition that h can be regarded, in some sense, as a small quantity with respect to 1, a necessary condition is that h is small for $t = 0$ and at the boundaries.

As a consequence, a first condition is that the initial datum shows little departure from the basic Maxwellian f_0; this does not necessarily mean that h is small everywhere for $t = 0$, but that $\|h\| \ll 1$, where $\|h\|$ is given by Eq. (1.9). In particular, if $\bar{\rho}$, $\bar{v}$, $\bar{T}$ are the initial density, velocity and temperature, ρ_0, v_0, T_0 the corresponding parameters in f_0, then the quantities $|\bar{\rho} - \rho_0|/\rho_0$, $|\bar{T} - T_0|/T_0$, $|\bar{\mathbf{v}} - \mathbf{v}_0|/(RT_0)$ must be small with respect to 1.

The situation is similar, but less obvious, when we examine the boundary conditions. If we put

$$f = f_0(1 + h) \tag{2.7}$$

into the boundary conditions given by Eq. (III.1.6), we obtain

$$h(\mathbf{x}, \boldsymbol{\xi}, t) = h_0(\mathbf{x}, \boldsymbol{\xi}, t) + \int_{\mathbf{c}'\cdot\mathbf{n}<0} B(\boldsymbol{\xi}' \to \boldsymbol{\xi}; \mathbf{x}, t) h(\mathbf{x}, \boldsymbol{\xi}', t)\, d\boldsymbol{\xi}'$$

$$(\mathbf{x} \in \partial R; \mathbf{c}\cdot\mathbf{n} > 0) \quad (2.8)$$

where $\mathbf{c}'\cdot\mathbf{n} < 0$ replaces $\boldsymbol{\xi}'\cdot\mathbf{n} < 0$, in order to take a possible motion of the wall into account, and

$$h_0(\mathbf{x}, \boldsymbol{\xi}, t) = [f_0(\boldsymbol{\xi})\,|\mathbf{c}\cdot\mathbf{n}|]^{-1} \int_{\mathbf{c}'\cdot\mathbf{n}<0} R(\boldsymbol{\xi}' \to \boldsymbol{\xi}; \mathbf{x}, t) f_0(\boldsymbol{\xi}')\,|\mathbf{c}'\cdot\mathbf{n}|\, d\boldsymbol{\xi}' - 1 \quad (2.9)$$

$$B(\boldsymbol{\xi}' \to \boldsymbol{\xi}; \mathbf{x}, t) = [f_0(\boldsymbol{\xi})\,|\mathbf{c}\cdot\mathbf{n}|]^{-1} R(\boldsymbol{\xi}' \to \boldsymbol{\xi}; \mathbf{x}, t) f_0(\boldsymbol{\xi}')\,|\mathbf{c}'\cdot\mathbf{n}| \quad (2.10)$$

It is obvious from Eq. (2.8) that h can be small (in some sense), only if h_0 (the source term) is small (in some related sense). Let $f_w(\boldsymbol{\xi}; \mathbf{x}, t)(\mathbf{x} \in \partial R)$ denote a Maxwellian with velocity and temperature equal to the wall velocity and temperature, respectively; then $R(\boldsymbol{\xi}' \to \boldsymbol{\xi}; \mathbf{x}, t)$ satisfies Eq. (III.3.10) with f_w in place of f_0. Accordingly, Eq. (2.9) can be rewritten as follows

$$h_0(\boldsymbol{\xi}) = |\mathbf{c}\cdot\mathbf{n}|^{-1} \int_{\mathbf{c}'\cdot\mathbf{n}<0} R(\boldsymbol{\xi}' \to \boldsymbol{\xi}) \left[\frac{f_0(\boldsymbol{\xi}')}{f_0(\boldsymbol{\xi})} - \frac{f_w(\boldsymbol{\xi}')}{f_w(\boldsymbol{\xi})}\right] |\mathbf{c}'\cdot\mathbf{n}|\, d\boldsymbol{\xi}' \quad (2.11)$$

where the arguments $\mathbf{x}$ ($\mathbf{x} \in \partial R$) and t have been omitted for conciseness. We see from Eq. (2.11) that if the velocity $\mathbf{u}_w$ and the temperature T_w at the walls are close to the velocity $\mathbf{u}_0$ and the temperature T_0 of the basic Maxwellian f_0, then h_0 will be small. Hence linearization is allowed (formally, at least) if relative velocities and temperature differences are small. We note that h_0 can be small even when f_0 and f_w are not very close to each other; an extreme example is afforded by specular reflection (Eq. (III.1.11) with $\boldsymbol{\xi} - \mathbf{u}_w$, $\boldsymbol{\xi}' - \mathbf{u}_w$ in place of $\boldsymbol{\xi}$, $\boldsymbol{\xi}'$). In this case it is easy to compute

$$h_0 = \exp\left[-2\frac{(\xi_n - u_{wn})(u_{0n} - u_{wn})}{RT_0}\right] - 1 \quad (2.12)$$

where the subscript n denotes the normal component, so that h_0 is (formally) small if

$$\frac{u_{0n} - u_{wn}}{\sqrt{2RT_0}} \ll 1 \quad (2.13)$$

This, in particular, can be achieved if $\mathbf{u}_0 - \mathbf{u}_w$ forms a small angle with the tangential plane at each point of the wall (this situation arises in flow past a thin airfoil at a small angle of attack). It is clear that linearization is still possible if the wall is not specularly reflecting provided that the deviation from specular reflection is sufficiently small [6].

Let us consider in more detail the case in which the re-emission of the molecules at the wall is definitely different from specular reflection. Then

$R(\boldsymbol{\xi}' \to \boldsymbol{\xi}; \mathbf{x}, t)$ will depend, in general, upon T_w and $\mathbf{u}_w$ for reciprocity to be satisfied, and $|T_w - T_0|/T_0$ and $|\mathbf{u}_w - \mathbf{u}_0|(2RT_0)^{-\frac{1}{2}}$ must be small for linearization to be valid. Then it is convenient to take into account a simplification which does not alter the accuracy of a linearized treatment. That is, we linearize $R(\boldsymbol{\xi}' \to \boldsymbol{\xi}; \mathbf{x}, t)$ with respect to the small parameters $|T_w - T_0|/T_0$, $|\mathbf{u}_w - \mathbf{u}_0| \cdot (2RT_0)^{-\frac{1}{2}}$. This linearization procedure can be applied to both Eqs. (2.9) and (2.10); first order terms are to be retained in the former case (zero-th order terms cancel because of reciprocity) while only zero-th order terms are kept in Eq. (2.10) because first order terms in $B(\boldsymbol{\xi}' \to \boldsymbol{\xi}; \mathbf{x}, t)$ become second order when multiplied by h in Eqs. (2.8)–(2.10). The zero-th order approximation to $R(\boldsymbol{\xi}' \to \boldsymbol{\xi}; \mathbf{x}, t)$ is no longer dependent upon $\mathbf{x}$ and t (if these arguments appear only through $\mathbf{u}_w(\mathbf{x}, t)$ and $T_w(\mathbf{x}, t)$, which is the usual case) and satisfies the same relations as the full $R(\boldsymbol{\xi}' \to \boldsymbol{\xi}; \mathbf{x}, t)$ except for the fact that the basic Maxwellian $f_0(\boldsymbol{\xi})$ now replaces the wall Maxwellian. Hence Eqs. (2.8) and (2.9) can be written in the form

$$h^+ = h_0 + APh^- \qquad (\mathbf{x} \in \partial R) \tag{2.14}$$

$$h_0 = \psi_0^+ - AP\psi_0^- \tag{2.15}$$

where h^+ and h^- denote the restrictions of h to $\boldsymbol{\xi} \cdot \mathbf{n} > 0$, $\boldsymbol{\xi} \cdot \mathbf{n} < 0$ and the operators A and P were defined by Eqs. (III.4.14) and (III.5.7), whereas

$$\psi_0 = \frac{(\boldsymbol{\xi} - \mathbf{u}_0) \cdot (\mathbf{u}_w - \mathbf{u}_0)}{RT_0} + \left[\frac{(\boldsymbol{\xi} - \mathbf{u}_0)^2}{2RT_0} - \frac{3}{2}\right]\frac{T_w - T_0}{T_0} \tag{2.16}$$

Eq. (III.5.5) defined a scalar product for functions defined along the boundary. In order to avoid confusion with the scalar product introduced in Section 1, we shall henceforth add a suffix B to the scalar product appearing in Eq. (III.5.5), so that we shall write for any two functions g and h defined for $\mathbf{x} \in \partial R$:

$$(g, h)_B = \int_{\mathbf{c}\cdot\mathbf{n}>0} g(\boldsymbol{\xi})h(\boldsymbol{\xi})\, d\mu = \int_{\mathbf{c}\cdot\mathbf{n}>0} g(\boldsymbol{\xi})h(\boldsymbol{\xi})f_0(\boldsymbol{\xi})\, |\mathbf{c} \cdot \mathbf{n}|\, d\boldsymbol{\xi}$$

$$(\mathbf{c} \equiv \boldsymbol{\xi} - \mathbf{u}_0) \tag{2.17}$$

and analogously for the $\|h\|_B$, related to $(g, h)_B$ by

$$\|g\|_B^2 = (g, g)_B \tag{2.18}$$

Eqs. (III.5.6) and (III.4.17) show that the operator A is symmetric with respect to the considered scalar product:

$$(g, Ah)_B = (Ag, h)_B \tag{2.19}$$

and has a norm not larger than unity

$$\|Ag\|_B \leqslant \|g\|_B \tag{2.20}$$

the equality sign holding if and only if g is a constant (unless the kernel $R(\boldsymbol{\xi}' \to \boldsymbol{\xi})$ is a delta function).

We have found that linearization is formally justified if the inhomogeneous terms in the initial and boundary conditions are small. In order to investigate whether this is a rigorously sufficient condition, it is necessary to investigate the initial and boundary value problems for the linearized Boltzmann equation and to prove that there is one and only one solution to a given boundary problem and that this solution remains small (in some sense) if the above-mentioned inhomogeneous terms are sufficiently small. We have also to prove that the difference between the solutions of the linearized and nonlinear equation is small and of higher order when the formal conditions for linearization are met. These problems will be discussed in Chapter VIII; here we shall accept the formal conditions as sufficient.

3. The linear Boltzmann equation. Neutron transport and radiative transfer

There is no difficulty in extending the treatment of Sections 1–2 to the case of a mixture or a polyatomic gas, starting from Eqs. (II.4.17) and (II.4.22) and writing:

$$f_j = f_{0j}(1 + h_j) \tag{3.1}$$

where the f_{0j} are Maxwellians which differ from each other only in the density.

A completely different and very interesting possibility arises, however, in the case of a two-component mixture, when one of the components has a very small density, so that collisions of particles of this species (to be called species N henceforth) with each other can be neglected in comparison with collisions with particles of the other species (say, M) and the latter collisions, in turn, can be neglected in comparison with collisions of particles M with each other. If this occurs, then the evolution of species M is not influenced by the state of particles N, while the behavior of the latter is influenced by the state of particles M. A particularly interesting case arises when species M is in equilibrium and hence has a Maxwellian distribution, to be denoted by F_0; then the distribution function for species N satisfies:

$$\frac{\partial f}{\partial t} + \boldsymbol{\xi} \cdot \frac{\partial f}{\partial \mathbf{x}} = \frac{1}{m_0} \int (f' F_{0*}' - f F_{0*}) B(\theta, V)\, d\theta\, d\varepsilon\, d\boldsymbol{\xi}_* \tag{3.2}$$

where m_0 is the mass of a particle of species M and $B(\theta, V) = B_{NM}(\theta, V)$ is given by Eq. (II.4.14) where $r = r(\theta)$ is computed through the law of interaction between a particle of species N and a particle of species M. Eq. (3.2) will be called the *linear Boltzmann equation.*

A particular instance of Eq. (3.2) was already met in Section 2 of Chapter III, Eq. (2.7), where it described the behavior of gas molecules (species N) trapped in a solid wall (whose atoms constitute species M). Other important examples are offered by the transport of neutrons in a gas moderator, electron transport in solids and ionized gases and by radiative transfer through a planetary or stellar atmosphere in local thermal equilibrium. In these cases, however, important modifications, to be discussed below, need to be taken into account.

It is convenient to transform Eq. (3.2) into a slightly different (and more general) form. To this end, we separate the right hand side of Eq. (3.2) into two parts:

$$\frac{1}{m_0}\int f' F_{0*}' B(\theta, V)\, d\theta\, d\varepsilon\, d\boldsymbol{\xi}_* - \left[\frac{1}{m_0}\int F_{0*} B(\theta, V)\, d\theta\, d\varepsilon\, d\boldsymbol{\xi}_*\right] f(\boldsymbol{\xi})$$

$$= \int K(\boldsymbol{\xi}', \boldsymbol{\xi}) f(\boldsymbol{\xi}')\, d\boldsymbol{\xi}' - \nu(\boldsymbol{\xi}) f(\boldsymbol{\xi}) \tag{3.3}$$

Here

$$\nu(\boldsymbol{\xi}) = \frac{2\pi}{m_0}\int F_0(\xi_*) B(\theta, V)\, d\theta\, d\boldsymbol{\xi}_* \tag{3.4}$$

and

$$K(\boldsymbol{\xi}', \boldsymbol{\xi}) = m(m + m_0)^2 \int \frac{B(\theta, V)}{2V \cos\theta \sin\theta} F_0(\boldsymbol{\xi}_*') \times$$

$$\times\ \delta(m\boldsymbol{\xi} + m_0\boldsymbol{\xi}_* - m\boldsymbol{\xi}' - m_0\boldsymbol{\xi}_*') \times$$

$$\times\ \delta(m\xi^2 + m_0\xi_*^{\ 2} - m\xi'^2 - m_0\xi_*'^2)\, d\boldsymbol{\xi}_*'\, d\boldsymbol{\xi}_*$$

$$= \frac{(m + m_0)^2}{m_0^{\ 3}}\int \frac{B(\theta, V)}{2V \cos\theta \sin\theta} F_0\left(\boldsymbol{\xi}_* + \frac{m}{m_0}[\boldsymbol{\xi} - \boldsymbol{\xi}']\right) \times$$

$$\times\ \delta\left(2[\boldsymbol{\xi} - \boldsymbol{\xi}'] \cdot [\boldsymbol{\xi} - \boldsymbol{\xi}_*] - \frac{m + m_0}{m_0} |\boldsymbol{\xi} - \boldsymbol{\xi}'|^2\right) d\boldsymbol{\xi}_* \tag{3.5}$$

where m is the mass of a particle of species N and

$$\theta = \cos^{-1}\left[\frac{(\boldsymbol{\xi} - \boldsymbol{\xi}_*) \cdot (\boldsymbol{\xi}' - \boldsymbol{\xi})}{|\boldsymbol{\xi} - \boldsymbol{\xi}_*|\, |\boldsymbol{\xi}' - \boldsymbol{\xi}|}\right] \tag{3.6}$$

Here Eqs. (II.4.18)–(II.4.20) have been used and integration with respect to $\boldsymbol{\xi}_*'$ effected by using the properties of the delta function.

A further integration can be effected by taking the polar components of $\mathbf{V} = \boldsymbol{\xi} - \boldsymbol{\xi}_*$ (with the polar axis along $\boldsymbol{\xi} - \boldsymbol{\xi}'$) as integration variables. Then, putting

$$\mathbf{n} = \mathbf{V}/V = (\sin\theta \cos\varepsilon, \sin\theta \sin\varepsilon, \cos\theta) \tag{3.7}$$

gives:

$$K(\boldsymbol{\xi}', \boldsymbol{\xi}) = \frac{(m + m_0)^2}{m_0^{\,3}} \int \frac{B(\theta, V)}{2 \cos \theta \sin \theta} F_0\left(\boldsymbol{\xi} + \frac{m}{m_0} [\boldsymbol{\xi} - \boldsymbol{\xi}'] - \mathbf{n} V\right) \times$$

$$\times\, \delta\left(2 \,|\boldsymbol{\xi} - \boldsymbol{\xi}'|\, V \cos \theta - \frac{m + m_0}{m_0} |\boldsymbol{\xi} - \boldsymbol{\xi}'|^2\right) V \, dV \, dn$$

$$= \frac{1}{m_0} \left(\frac{m + m_0}{2m_0}\right)^3 \int B\left(\theta, \frac{m + m_0}{2m_0 \cos \theta} |\boldsymbol{\xi} - \boldsymbol{\xi}'| \, (\cos^3\theta)\right)^{-1} \times$$

$$\times\, F_0\left(\boldsymbol{\xi} + \frac{m}{m_0} [\boldsymbol{\xi} - \boldsymbol{\xi}'] - \frac{m + m_0}{2m_0 \cos \theta} \mathbf{n} \, |\boldsymbol{\xi} - \boldsymbol{\xi}'|\right) d\theta \, d\varepsilon \qquad (3.8)$$

According to Eq. (3.3), Eq. (3.2) can be written as follows:

$$\frac{\partial f}{\partial t} + \boldsymbol{\xi} \cdot \frac{\partial f}{\partial \mathbf{x}} = \int K(\boldsymbol{\xi}', \boldsymbol{\xi}) f(\boldsymbol{\xi}') \, d\boldsymbol{\xi}' - \nu(\boldsymbol{\xi}) f(\boldsymbol{\xi}) \qquad (3.9)$$

The main difference between the linear Boltzmann equation and the linearized Boltzmann equation, Eq. (2.6), is that the *linearized* collision operator L given by Eq. (1.7) has five eigenfunctions corresponding to the eigenvalue $\lambda = 0$, Eq. (1.15), whilst the *linear* collision operator appearing on the left hand side of Eq. (3.2), or (3.9), has only one eigenfunction $[f = f_0(\boldsymbol{\xi}) = F_0(\boldsymbol{\xi}[m/m_0]^{\frac{1}{2}})]$. This corresponds to the existence of just one conservation law in a collision (mass conservation); in fact momentum and energy are exchanged with the particles of species M. $f_0(\boldsymbol{\xi})$ is, of course, the equilibrium solution for species N.

Another difference arises from the fact that f in Eqs. (3.2) and (3.9) must be positive, whereas h in Eq. (2.6) could be negative, since it is a small perturbation about unity and only $1 + h$ is required to be positive according to Eq. (2.7). This difference would appear to be inessential because $1 + h$ satisfies Eq. (2.6) if h does; thus the analogy between f and $1 + h$ or $f_0(1 + h)$ should be valid. This is not acceptable, however, because h is not small, in general, with respect to 1 in a local sense in such a way that $|h(\mathbf{x}, \boldsymbol{\xi}, t)| \ll 1$ for any $\mathbf{x}$, $\boldsymbol{\xi}$, t, but only in a global sense (see Chapter VIII). In particular, $1 + h$ is frequently negative for some set of values of $\mathbf{x}$, $\boldsymbol{\xi}$, t.

As we mentioned above, the linear Boltzmann equation has found wide application especially in the field of neutron transport under the impetus of the development of nuclear reactor technology. In this case, the linear Boltzmann equation written above can be extended to cover more general situations, which are of major practical importance. First, the species M, to be called the medium henceforth, need not be a gas; it may be a solid moderator, or the reactor shield or nuclear fuel, or a combination of different materials. However, because of the high dilution (the number of neutrons is

10^{11} times smaller than the number of the atoms of the moderator even in a high flux reactor) and the extremely small range of neutron-neutron interaction (10^5 times smaller than the range of intermolecular forces), neutron-neutron collisions can be safely neglected and neutron-atom interactions described in terms of two body collisions. Of course, the scattering kernel $K(\boldsymbol{\xi}', \boldsymbol{\xi})$ and the collision frequency $\nu(\boldsymbol{\xi})$ in Eq. (3.9) will change with the nature of the medium, and hence from place to place, if different media occupy different parts of the region under study. In addition, such events as inelastic collisions, absorption of a neutron and fission of a nucleus resulting in the creation of new neutrons can occur. In general, Eq. (3.9) is then re-written in terms of a cross section $\sigma(\boldsymbol{\xi}, \mathbf{x})$ and a differential scattering cross section $\sigma(\boldsymbol{\xi}' \to \boldsymbol{\xi}, \mathbf{x})$:

$$\frac{\partial f}{\partial t} + \boldsymbol{\xi} \cdot \frac{\partial f}{\partial \mathbf{x}} + \xi\sigma(\boldsymbol{\xi}, \mathbf{x})f = \int \xi'\sigma(\boldsymbol{\xi}' \to \boldsymbol{\xi}, \mathbf{x})f(\boldsymbol{\xi}')\, d\boldsymbol{\xi}' + S(\mathbf{x}, \boldsymbol{\xi}, t) \quad (3.10)$$

where we have added a source term to take into account the possibility of insertion of neutrons of velocity $\boldsymbol{\xi}$ at point $\mathbf{x}$ and time t. The total cross section $\sigma(\boldsymbol{\xi}, \mathbf{x})$ contains contributions from collisions or scattering (both elastic and inelastic) σ_s and from absorption σ_a:

$$\sigma = \sigma_s + \sigma_a \quad (3.11)$$

The differential cross section (also called the scattering kernel sometimes) usually contains a contribution from scattering (both elastic and inelastic). Thus if scattering is only elastic:

$$\sigma = \xi^{-1}\nu(\boldsymbol{\xi})$$
$$\sigma(\xi' \to \xi) = (\xi')^{-1}K(\boldsymbol{\xi}', \boldsymbol{\xi}) \quad (3.12)$$

where $\nu(\boldsymbol{\xi})$ and $K(\boldsymbol{\xi}', \boldsymbol{\xi})$ are the collision frequency and the scattering kernel appearing in Eq. (3.9). Sometimes the contribution from fission is also included in $\sigma(\boldsymbol{\xi}' \to \boldsymbol{\xi})$ and this can be done when the emission of neutrons can be treated as amounting to simultaneous fission (prompt neutrons). Since however, after a process of fission, some of the fragments remain in a highly excited state and eventually decay to a more stable condition by β-emission, followed immediately by neutron emission [7], it is found convenient to take account of the neutron emission following a fission process in the source term $S(\mathbf{x}, \boldsymbol{\xi}, t)$. The existence of delayed neutrons is of course not essential in steady problems, for which the source is usually written in the form

$$S(\mathbf{x}, \boldsymbol{\xi}) = \int \sigma_f(\boldsymbol{\xi}', \mathbf{x})f(\mathbf{x}, \boldsymbol{\xi}')\bar{\nu}\chi(\boldsymbol{\xi}' \to \boldsymbol{\xi}; \mathbf{x})\, d\boldsymbol{\xi}' \quad (3.13)$$

where σ_f is the fission cross section, $\chi(\boldsymbol{\xi}' \to \boldsymbol{\xi}; \mathbf{x})$ is the probability that at position $\mathbf{x}$ a neutron of velocity $\boldsymbol{\xi}'$ will give rise to a fission neutron with velocity between $\boldsymbol{\xi}$ and $\boldsymbol{\xi} + d\boldsymbol{\xi}$, and $\bar{\nu}$ is the average number of neutrons produced per fission. Frequently it is sufficiently accurate to assume that χ does not depend on $\boldsymbol{\xi}'$ and the direction of $\boldsymbol{\xi}$, so that $\chi = \chi(\xi, \mathbf{x})$.

If we include fission effects in S, then $\sigma(\boldsymbol{\xi}' \to \boldsymbol{\xi}; \mathbf{x})$ describes only the scattering effects. Hence conservation of mass in a scattering process implies:

$$\int \sigma(\boldsymbol{\xi}' \to \boldsymbol{\xi})\, d\boldsymbol{\xi} = \sigma_s(\boldsymbol{\xi}') \tag{3.14}$$

or, if Eqs. (3.12) apply:

$$\int K(\boldsymbol{\xi}', \boldsymbol{\xi})\, d\boldsymbol{\xi} = \nu(\boldsymbol{\xi}') \tag{3.15}$$

Also, for any two functions g and h, in the case of a monatomic gas moderator:

$$\begin{aligned}\int \xi'\sigma(\boldsymbol{\xi}' \to \boldsymbol{\xi})&[f_0(\boldsymbol{\xi})]^{-1}h(\boldsymbol{\xi}')g(\boldsymbol{\xi})\, d\boldsymbol{\xi}'\, d\boldsymbol{\xi}\\ &= \int K(\boldsymbol{\xi}', \boldsymbol{\xi})[f_0(\boldsymbol{\xi})]^{-1}h(\boldsymbol{\xi}')g(\boldsymbol{\xi})\, d\boldsymbol{\xi}'\, d\boldsymbol{\xi}\\ &= \frac{1}{m_0}\int F_0(\boldsymbol{\xi}_*')[f_0(\boldsymbol{\xi})]^{-1}h(\boldsymbol{\xi}')g(\boldsymbol{\xi})B(\theta, |\boldsymbol{\xi} - \boldsymbol{\xi}_*|)\, d\boldsymbol{\xi}\, d\boldsymbol{\xi}_*\end{aligned} \tag{3.16}$$

where $f_0(\boldsymbol{\xi}) = F_0(\boldsymbol{\xi}[m/m_0]^{\frac{1}{2}})$ as above. $\boldsymbol{\xi}'$ and $\boldsymbol{\xi}_*'$ are of course related to $\boldsymbol{\xi}$ and $\boldsymbol{\xi}_*$ by the conservation of momentum and energy; hence $d\boldsymbol{\xi}\, d\boldsymbol{\xi}_* = d\boldsymbol{\xi}'\, d\boldsymbol{\xi}_*'$, $F_0(\xi_*')[f_0(\boldsymbol{\xi})]^{-1} = F_0(\boldsymbol{\xi}_*)[f_0(\boldsymbol{\xi}')]^{-1}$ and $|\boldsymbol{\xi} - \boldsymbol{\xi}_*| = |\boldsymbol{\xi}' - \boldsymbol{\xi}_*'|$. Then the last expression in Eq. (3.16) can be written as follows

$$\begin{aligned}\frac{1}{m_0}\int \frac{F_0(\boldsymbol{\xi}_*)}{f_0(\boldsymbol{\xi}')}& h(\boldsymbol{\xi}')g(\boldsymbol{\xi})B(\theta, |\boldsymbol{\xi}' - \boldsymbol{\xi}_*'|)\, d\boldsymbol{\xi}_*'\, d\boldsymbol{\xi}'\\ &= \frac{1}{m_0}\int \frac{F_0(\boldsymbol{\xi}_*')}{f_0(\boldsymbol{\xi})} h(\boldsymbol{\xi})g(\boldsymbol{\xi}')B(\theta, |\boldsymbol{\xi} - \boldsymbol{\xi}_*|)\, d\boldsymbol{\xi}_*\, d\boldsymbol{\xi}\end{aligned} \tag{3.17}$$

where the last passage is due to an exchange of identity between primed and unprimed variables (see Section 6 of Chapter II). The last expression is the same as the one in Eq. (3.16) except that h and g are interchanged. Hence we can interchange h and g in all the expressions appearing in Eq. (3.16) and obtain:

$$\begin{aligned}\int \xi'\sigma(\boldsymbol{\xi}' \to \boldsymbol{\xi})&[f_0(\boldsymbol{\xi})]^{-1}h(\boldsymbol{\xi}')g(\boldsymbol{\xi})\, d\boldsymbol{\xi}'\, d\boldsymbol{\xi}\\ &= \int \xi'\sigma(\boldsymbol{\xi}' \to \boldsymbol{\xi})[f_0(\boldsymbol{\xi})]^{-1}g(\boldsymbol{\xi}')h(\boldsymbol{\xi})\, d\boldsymbol{\xi}'\, d\boldsymbol{\xi}\\ &= \int \xi\sigma(\boldsymbol{\xi} \to \boldsymbol{\xi}')[f_0(\boldsymbol{\xi}')]^{-1}g(\boldsymbol{\xi})h(\boldsymbol{\xi}')\, d\boldsymbol{\xi}\, d\boldsymbol{\xi}'\end{aligned} \tag{3.18}$$

Since g and h are arbitrary, Eq. (3.18) is equivalent to

$$\xi'\sigma(\boldsymbol{\xi}' \to \boldsymbol{\xi})[f_0(\boldsymbol{\xi})]^{-1} = \xi\sigma(\boldsymbol{\xi} \to \boldsymbol{\xi}')[f_0(\boldsymbol{\xi}')]^{-1} \tag{3.19}$$

or to

$$\xi'\sigma(\boldsymbol{\xi}' \to \boldsymbol{\xi})f_0(\boldsymbol{\xi}') = \xi\sigma(\boldsymbol{\xi} \to \boldsymbol{\xi}')f_0(\boldsymbol{\xi}') \tag{3.20}$$

which is a reciprocity relation (also called detailed balance), similar to the analogous property investigated in Chapter III, Section 3. The extension of Eq. (3.20) to cases in which the medium is not a monatomic gas has been investigated [8, 9]; the only change in Eq. (3.20) is that $\sigma(\boldsymbol{\xi} \to \boldsymbol{\xi}')$ is replaced by $\sigma(-\boldsymbol{\xi} \to -\boldsymbol{\xi}')$ in the right hand side (Eq. (3.20), as it stands, then follows if the neutron-atom interaction is symmetrical with respect to space reflection). The reciprocity property is a further reason for keeping the fission term separated from the scattering term in Eq. (3.11); in fact the kernel describing fission does not satisfy reciprocity.

Another property of the kernel in Eqs. (3.9) is obtained as follows. For any f, by virtue of Eq. (3.20) and (3.12):

$$\begin{aligned}
&\int f(\boldsymbol{\xi})[f_0(\boldsymbol{\xi})]^{-1}\left[\int K(\boldsymbol{\xi}', \boldsymbol{\xi})f(\boldsymbol{\xi}')\,d\boldsymbol{\xi}'\right] d\boldsymbol{\xi}\\
&= \int [f_0(\boldsymbol{\xi})]^{-1}K(\boldsymbol{\xi}', \boldsymbol{\xi})f(\boldsymbol{\xi})f(\boldsymbol{\xi}')\,d\boldsymbol{\xi}'\,d\boldsymbol{\xi}\\
&= \int [K(\boldsymbol{\xi}', \boldsymbol{\xi})/f_0(\boldsymbol{\xi}')]^{\frac{1}{2}}f(\boldsymbol{\xi}')[K(\boldsymbol{\xi}, \boldsymbol{\xi}')/f_0(\boldsymbol{\xi}')]^{\frac{1}{2}}f(\boldsymbol{\xi})\,d\boldsymbol{\xi}\,d\boldsymbol{\xi}'\\
&\leqslant \left\{\int K(\boldsymbol{\xi}', \boldsymbol{\xi})[f_0(\boldsymbol{\xi}')]^{-1}[f(\boldsymbol{\xi}')]^2\,d\boldsymbol{\xi}\,d\boldsymbol{\xi}'\int K(\boldsymbol{\xi}, \boldsymbol{\xi}')[f_0(\boldsymbol{\xi})]^{-1}[f(\boldsymbol{\xi})]^2\,d\boldsymbol{\xi}\,d\boldsymbol{\xi}'\right\}^{\frac{1}{2}}\\
&= \int K(\boldsymbol{\xi}', \boldsymbol{\xi})[f_0(\boldsymbol{\xi}')]^{-1}[f(\boldsymbol{\xi}')]^2\,d\boldsymbol{\xi}\,d\boldsymbol{\xi}'\\
&= \int \nu(\boldsymbol{\xi}')[f_0(\boldsymbol{\xi}')]^{-1}[f(\boldsymbol{\xi}')]^2\,d\boldsymbol{\xi}'\\
&= \int \nu(\boldsymbol{\xi})[f_0(\boldsymbol{\xi})]^{-1}[f(\boldsymbol{\xi})]^2\,d\boldsymbol{\xi}
\end{aligned} \tag{3.21}$$

where the Schwartz inequality [1–5]:

$$\int hg\,d\boldsymbol{\xi}\,d\boldsymbol{\xi}' \leqslant \left[\int h^2\,d\boldsymbol{\xi}\,d\boldsymbol{\xi}'\int g^2\,d\boldsymbol{\xi}\,d\boldsymbol{\xi}'\right]^{\frac{1}{2}} \tag{3.22}$$

and Eq. (3.15) have been used. Eq. (3.21) is analogous to Eq. (1.12).

In conclusion if we let $h = f/f_0$, then the transport of neutrons in a purely scattering medium (no absorption or fission processes) is described by an equation which has all the formal properties of the linearized Boltzmann equation, Eq. (2.6), except for the fact that there is only one collision invariant $\psi_0 = 1$. If there is absorption, but fission is neglected, then there is no collision invariant, but all the other properties hold, including Eq. (1.12) (where the equality sign never applies, however). If fission has to be taken into account, then a truly different situation arises.

In the case of radiative transfer, the particles under study are photons of various frequencies; they are scattered, adsorbed and emitted by the atoms of a medium as in the case of neutrons in a nuclear reactor. The medium is usually a stellar or planetary atmosphere. If the atoms can be assumed to be in thermal equilibrium, at least locally, then the emission coefficient corresponding to frequency ν, j_ν, is given in terms of the absorption coefficient k_ν by Kirchhoff's law which states that at each point we have the relation

$$j_\nu = k_\nu B_\nu(T) \tag{3.23}$$

where T is the local (absolute) temperature and

$$B_\nu(T) = \frac{2h\nu^3}{c_l^2}(e^{h\nu/kT} - 1) \tag{3.24}$$

is the Planck distribution (c_l is the speed of light, k and h are the Boltzmann and Planck constants, respectively). The absorption cross section is given by $k_\nu \rho$, where ρ is the density of atoms, and the emission source is given by $j_\nu \rho$. Hence the transport equation can be written as follows [10]:

$$\frac{\partial f}{\partial t} + c_l \boldsymbol{\Omega} \cdot \frac{\partial f}{\partial \mathbf{x}} = c_l \sigma_a \left(\frac{2h\nu^3 c_l^{-2}}{e^{h\nu/kT} - 1} - f \right) + \\ + c_l \sigma_s \left[\int g(\boldsymbol{\Omega}' \to \boldsymbol{\Omega}) f(\boldsymbol{\Omega}')\, d\boldsymbol{\Omega}' - f \right] \tag{3.25}$$

where σ_s is the scattering cross section, $\sigma_s g(\boldsymbol{\Omega}' \to \boldsymbol{\Omega})$ is the differential cross section and $\boldsymbol{\Omega} = \boldsymbol{\xi}/c_l$. Frequently the assumption of "radiative equilibrium"

$$\int [B_\nu(T) - f]\, d\boldsymbol{\Omega} = 0 \tag{3.26}$$

is made. Eqs. (3.25) and (3.26) form a complicated nonlinear system of two equations for f and T. A simplification arises in the so-called grey case (see Section 9).

Boundary conditions for neutrons and photons are usually much simpler than for gas molecules, but satisfy the same general properties. For example, Eq. (2.14) with $h_0^+ = A = 0$ is frequently applied to neutrons (see Chapter III, Section 1).

We finally remark that the considerations about neutron transport can be transferred to electron transport in solids and ionized gases. There, however, a crucial role is played by the force term (see Eq. II.8.21) which is usually omitted from the Boltzmann equation. This term, in fact, describes the influence of electric and magnetic fields upon the electrons. An additional feature is the smallness of the electron mass with respect to the mass of a scattering atom.

4. Uniqueness of the solution for initial and boundary value problems

In this section we shall briefly examine the question of uniqueness of the solution of the linearized (or linear) Boltzmann equation for a given initial and/or boundary value problem. We assume that the initial datum is given and boundary conditions of the form appearing in Eq. (2.14) are to be satisfied at the solid boundaries. Let us denote by h_1 and h_2 two possible solutions of the problem and let $h = h_1 - h_2$ be their difference.

Then h satisfies Eq. (2.6) and the homogeneous boundary conditions corresponding to Eq. (2.14)

$$h^+ = APh^- \tag{4.1}$$

In order to include the case of a region R extending to infinity, we shall assume that

$$\iint_\Sigma \boldsymbol{\xi} \cdot \mathbf{n} h^2 \, d\boldsymbol{\xi} \, dS \to 0 \tag{4.2}$$

when the points of the surface $\Sigma \subset R$ tend to infinity.

Since h is a function of $\mathbf{x}$ as well as $\boldsymbol{\xi}$, it is convenient to introduce the following scalar products and norms

$$((g, h)) = \int_R (g, h) \, d\mathbf{x} = \iint f_0 gh \, d\boldsymbol{\xi} \, d\mathbf{x} \tag{4.3}$$

$$|||h|||^2 = (h, h) \tag{4.4}$$

$$((g, h))_B = \int_{\partial R} (g, h)_B \, dS = \int_{\partial R} \int_{\boldsymbol{\xi}\cdot\mathbf{n}>0} ghf_0 \, |\boldsymbol{\xi} \cdot \mathbf{n}| \, d\boldsymbol{\xi} \, dS \tag{4.5}$$

$$|||h|||_B{}^2 = ((h, h))_B \tag{4.6}$$

Inequalities (1.12) and (2.20) imply, for any h:

$$((h, Lh)) \leqslant 0 \qquad |||Ah|||_B \leqslant |||h|||_B \tag{4.7}$$

the equality sign holding in the first relation if, and only if, h is almost everywhere a collision invariant $\sum_{\alpha=0}^{4} c_\alpha \psi_\alpha$, and in the second relation if, and only if, h is almost everywhere a constant along the boundary.

Let us assume that $|||h|||$, $|||Ph^-|||_B$, $((h, Lh))$, $((h, \boldsymbol{\xi} \cdot \partial h/\partial \mathbf{x}))$, $((h, \partial h/\partial t))$ exist and are finite. Then if we multiply Eq. (2.6) by h and integrate we obtain:

$$\frac{\partial}{\partial t}(\tfrac{1}{2}\, |||h|||^2) + \left(\left(h, \boldsymbol{\xi} \cdot \frac{\partial h}{\partial \mathbf{x}}\right)\right) = ((h, Lh)) \tag{4.8}$$

The second term on the left hand side can be transformed into a surface term by the divergence theorem. Hence:

$$\frac{\partial}{\partial t}(\tfrac{1}{2}\, |||h|||^2) = |||h^+|||_B{}^2 - |||Ph^-|||_B{}^2 + ((h, Lh)) \tag{4.9}$$

Eq. (4.1) and (4.7) imply

$$\frac{\partial}{\partial t}(\tfrac{1}{2}\, |||h|||^2) \leqslant 0 \tag{4.10}$$

the equality sign holding if and only if h is a collision invariant in R and a constant on ∂R. From Eq. (4.10) it follows that $|||h|||$ cannot increase with time; since it cannot be negative and was zero at $t = 0$ ($h = h_1 - h_2$, where h_1 and h_2 solve the same problem and hence take the same initial value), it follows that $|||h||| = 0$ for $t > 0$ or $h_1 = h_2$ (almost everywhere).

Uniqueness of the solution has thus been established for the initial and boundary value problem. The proof immediately extends to the case of the linear Boltzmann equation (with $h = f/f_0$), when the medium is a purely scattering one or one where absorption dominates emission. The proof, however, holds even when emission of particles dominates over absorption as is the case with nuclear reactors. Then, the first of Eqs. (4.7) fails and is replaced by

$$((h, Lh)) < c|||h|||^2 \tag{4.11}$$

where c is some positive constant. Eq. (4.10) is then replaced by

$$\frac{1}{2}\frac{\partial}{\partial t}\, |||h|||^2 \leqslant c\, |||h|||^2 \tag{4.12}$$

so that

$$\frac{\partial}{\partial t}(|||h|||\, e^{-ct}) \leqslant 0 \tag{4.13}$$

and, by the same argument as above $|||h|||e^{-ct}$, and hence $|||h|||$ must also be zero.

A rather interesting situation arises for steady problems ($\partial h/\partial t = 0$). In this case, there are no initial conditions and Eq. (4.9) reduces to:

$$0 = |||h^+|||_B{}^2 - |||Ph^-|||_B{}^2 + ((h, Lh)) \tag{4.14}$$

Eqs. (4.1) and (4.7) then imply that h must be a collision invariant in R and a constant along the boundary:

$$h = \sum_{\alpha=0}^{4} c_\alpha \psi_\alpha \qquad \mathbf{x} \in R \tag{4.15}$$

$$h = c_0 \qquad \mathbf{x} \in \partial R \tag{4.16}$$

Inserting Eq. (4.15) into Eq. (2.6) with $\partial h/\partial t = 0$ we obtain $c_0 = \text{const.}$, $c_4 = \text{const.}$ and:

$$\frac{\partial c_i}{\partial x_k} + \frac{\partial c_k}{\partial x_i} = 0 \qquad (i, k = 1, 2, 3) \tag{4.17}$$

As we know (cf. Eqs. (III.10.6) and (III.10.11)), this implies

$$c_i = \omega_{ik} x_k + c_i^0 \tag{4.18}$$

where c_i^0 and ω_{ik} are constants and $\omega_{ik} = -\omega_{ki}$. Eqs. (4.18), (4.16) and (4.17) can be simultaneously satisfied only if $c_4 = 0$ and

$$\omega_{ik} x_k + c_i^0 = 0 \qquad (\mathbf{x} \in \partial R) \tag{4.19}$$

Eq. (4.19) can be satisfied only if either the boundary degenerates into a straight line or $\omega_{ik} = 0$, $c_i^0 = 0$. If the boundary is a proper surface, then $c_i = c_4 = 0$ and h reduces to a constant. Hence in the steady case two solutions of the same problem may differ by an additive constant; the latter can be fixed if we assign the total number of particles, by stipulating, for example:

$$((\psi_0, h)) = 0 \tag{4.20}$$

or

$$((\psi_0, h^+))_B = 0 \tag{4.21}$$

In the case of the linear Boltzmann equation, the same result follows if the medium is a purely scattering one; if there is absorption in addition, then the result follows again, even more trivially. The presence of emission events, such as the production of neutrons by fission, changes the picture in a considerable way. For a given shape, there is a critical size of the domain below which uniqueness applies; when the domain has exactly the critical size, then a solution exists in the absence of external sources (self sustaining nuclear reactor). Above the critical size there is no steady solution for which the distribution f is positive, so that there is no physically acceptable steady state solution [11, 12]; in this case the number of neutrons in the assembly increases continuously with time and the nuclear reactor becomes an *atomic bomb*.

5. Further investigation of the linearized collision term

In Section 3 we saw that the linear Boltzmann equation, Eq. (3.2), can be transformed into the form given in Eq. (3.3), which is rather interesting because the collision term is split into two parts. The first part contains an integral operator with kernel $K(\boldsymbol{\xi}', \boldsymbol{\xi})$ given by Eq. (3.8), while the second is simply obtained from the unknown quantity by multiplication by a velocity-dependent factor, which has the physical meaning of a collision frequency. In this section we want to investigate the analogous transformation on the linearized collision operator, given by Eq. (1.7). We can write

$$Lh = K_2 h - K_1 h - \nu(\boldsymbol{\xi})h \tag{5.1}$$

where

$$K_2 h = \frac{1}{m}\iint f_0(\boldsymbol{\xi}_*)h(\boldsymbol{\xi}_*')B(\theta, |\boldsymbol{\xi} - \boldsymbol{\xi}_*|)\, d\boldsymbol{\xi}_*\, d\theta\, d\varepsilon + $$

$$+ \frac{1}{m}\iint f_0(\boldsymbol{\xi}_*)h(\boldsymbol{\xi}')B(\theta, |\boldsymbol{\xi} - \boldsymbol{\xi}_*|)\, d\boldsymbol{\xi}_*\, d\theta\, d\varepsilon \tag{5.2}$$

$$K_1 h = \int\left[\frac{2\pi}{m}\int f_0(\boldsymbol{\xi}_*)B(\theta, |\boldsymbol{\xi} - \boldsymbol{\xi}_*|)\, d\theta\right]h(\boldsymbol{\xi}_*)\, d\boldsymbol{\xi}_* \tag{5.3}$$

$$\nu(\boldsymbol{\xi}) = \frac{2\pi}{m}\int f_0(\boldsymbol{\xi}_*)B(\theta, |\boldsymbol{\xi} - \boldsymbol{\xi}_*|)\, d\boldsymbol{\xi}_*\, d\theta \tag{5.4}$$

provided all the integrals separately exist (see below). The last two terms in Eq. (5.1) are in the desired form (either integral or multiplication operators). The first term must be transformed. According to Eq. (5.2), K_2h is the sum of two integrals differing from each other because of the argument of h which is $\boldsymbol{\xi}_*' = \boldsymbol{\xi}_* + \mathbf{n}(\mathbf{n} \cdot \mathbf{V})$ in the first integral and $\boldsymbol{\xi}' = \boldsymbol{\xi} - \mathbf{n}(\mathbf{n} \cdot \mathbf{V})$ in the second ($\mathbf{V} = \boldsymbol{\xi} - \boldsymbol{\xi}_*$). But

$$\boldsymbol{\xi}_*' = \boldsymbol{\xi} - \mathbf{V} + \mathbf{n}(\mathbf{n} \cdot \mathbf{V}) = \boldsymbol{\xi} - \mathbf{m}(\mathbf{m} \cdot \mathbf{V}) \tag{5.5}$$

where $\mathbf{m}$ is a unit vector in the plane of $\mathbf{V}$ and $\mathbf{n}$, orthogonal to $\mathbf{n}$ [hence $\mathbf{V} = \mathbf{n}(\mathbf{n} \cdot \mathbf{V})) + \mathbf{m}(\mathbf{m} \cdot \mathbf{V})$]. Hence if we rotate $\mathbf{n}$ through $\pi/2$ ($\theta \to (\pi/2) - \theta$, $\varepsilon \to \varepsilon \pm \pi$, amounting to a transformation with unit jacobian), $\boldsymbol{\xi}_*'$ becomes $\boldsymbol{\xi}'$ and Eq. (5.2) can then be written as follows:

$$K_2 h = \frac{2}{m}\int f_0(\boldsymbol{\xi}_*)h(\boldsymbol{\xi}')\bar{B}(\theta, |\boldsymbol{\xi} - \boldsymbol{\xi}_*|)\, d\boldsymbol{\xi}_*\, d\theta\, d\varepsilon \tag{5.6}$$

where

$$\bar{B}(\theta, V) = B(\theta, V) + B\left(\frac{\pi}{2} - \theta, V\right) \tag{5.7}$$

Eq. (5.6) can now be written in the form:

$$K_2 h = \frac{2}{m}\int f_0(\boldsymbol{\xi}_*)h(\boldsymbol{\xi}')\frac{\bar{B}(\theta, |\boldsymbol{\xi} - \boldsymbol{\xi}_*|)}{|\boldsymbol{\xi} - \boldsymbol{\xi}_*|\cos\theta\sin\theta} \times$$

$$\times\, \delta(\boldsymbol{\xi}' + \boldsymbol{\xi}_*' - \boldsymbol{\xi} - \boldsymbol{\xi}_*)\, \delta(\xi'^2 + \xi_*'^2 - \xi^2 - \xi_*^2)\, d\boldsymbol{\xi}'\, d\boldsymbol{\xi}_*\, d\boldsymbol{\xi}_*' \tag{5.8}$$

where θ is given by Eq. (II.4.20), and Eqs. (II.4.18) and (II.4.19) (specialized to the case of a single species) have been used.

If we effect the trivial integration with respect to $\boldsymbol{\xi}_*'$ and pass from $\boldsymbol{\xi}_*$ to the polar variables $|\boldsymbol{\xi}_* - \boldsymbol{\xi}|, \theta, \varepsilon$:

$$K_2h = \frac{2}{m}\int f_0(\boldsymbol{\xi}_*)h(\boldsymbol{\xi}')\bar{B}(\theta, |\boldsymbol{\xi} - \boldsymbol{\xi}_*|) \times$$
$$\times (\cos\theta)^{-1}\,\delta(2\,|\boldsymbol{\xi} - \boldsymbol{\xi}'|\,|\boldsymbol{\xi}_* - \boldsymbol{\xi}|\cos\theta -$$
$$- 2\,|\boldsymbol{\xi} - \boldsymbol{\xi}'|^2)\,d\boldsymbol{\xi}'\,|\boldsymbol{\xi}_* - \boldsymbol{\xi}|\,d(|\boldsymbol{\xi}_* - \boldsymbol{\xi}|)\,d\theta\,d\varepsilon$$
$$= \frac{1}{m}\int h(\boldsymbol{\xi}')f_0(\boldsymbol{\xi} + \mathbf{n}\,|\boldsymbol{\xi} - \boldsymbol{\xi}|'/\cos\theta) \times$$
$$\times (\cos\theta)^{-3}\bar{B}\left(\theta, \frac{|\boldsymbol{\xi} - \boldsymbol{\xi}'|}{\cos\theta}\right)d\boldsymbol{\xi}'\,d\theta\,d\varepsilon \tag{5.9}$$

where $\mathbf{n}$ is the unit vector corresponding to the angles θ and ε. Inserting Eqs. (5.9) and (5.3) into Eq. (5.1) gives

$$Lh = \int K(\boldsymbol{\xi}', \boldsymbol{\xi})h(\boldsymbol{\xi}')\,d\boldsymbol{\xi}' - \nu(\boldsymbol{\xi})h \tag{5.10}$$

where

$$K(\boldsymbol{\xi}', \boldsymbol{\xi}) = K_2(\boldsymbol{\xi}', \boldsymbol{\xi}) - K_1(\boldsymbol{\xi}', \boldsymbol{\xi}) \tag{5.11}$$

$$K_2(\boldsymbol{\xi}', \boldsymbol{\xi}) = \frac{1}{m}\int\left[B(\theta, |\boldsymbol{\xi} - \boldsymbol{\xi}'|/\cos\theta) + B\left(\frac{\pi}{2} - \theta, |\boldsymbol{\xi} - \boldsymbol{\xi}'|/\cos\theta\right)\right] \times$$
$$\times (\cos\theta)^{-3}f_0(\boldsymbol{\xi} + \mathbf{n}\,|\boldsymbol{\xi} - \boldsymbol{\xi}'|/\cos\theta)\,d\theta\,d\varepsilon \tag{5.12}$$

$$K_1(\boldsymbol{\xi}', \boldsymbol{\xi}) = \frac{1}{m}\int f_0(\boldsymbol{\xi}')B(\theta, |\boldsymbol{\xi} - \boldsymbol{\xi}'|)\,d\theta\,d\varepsilon \tag{5.13}$$

Eq. (5.12) can be further elaborated, since

$$f_0(\boldsymbol{\xi} + \mathbf{n}\,|\boldsymbol{\xi} - \boldsymbol{\xi}'|/\cos\theta)$$
$$= f_0(\boldsymbol{\xi}' + [\boldsymbol{\xi} - \boldsymbol{\xi}'] + \mathbf{n}\,|\boldsymbol{\xi} - \boldsymbol{\xi}'|/\cos\theta)$$
$$= \rho_0(2\pi RT_0)^{-\frac{3}{2}}\exp\{-(2RT_0)^{-1}[\xi'^2 + |\boldsymbol{\xi} - \boldsymbol{\xi}'|^2/\cos^2\theta + 2\boldsymbol{\xi}'\cdot(\boldsymbol{\xi} - \boldsymbol{\xi}') +$$
$$+ |\boldsymbol{\xi} - \boldsymbol{\xi}'|^2 + 2\mathbf{n}\cdot\boldsymbol{\xi}'\,|\boldsymbol{\xi} - \boldsymbol{\xi}'|/\cos\theta + 2\mathbf{n}\cdot(\boldsymbol{\xi} - \boldsymbol{\xi}')\,|\boldsymbol{\xi} - \boldsymbol{\xi}'|/\cos\theta]\} \tag{5.14}$$

Here and in the following we assume the mass velocity in the Maxwellian to be zero (otherwise it is sufficient to replace $\boldsymbol{\xi}$ by the peculiar velocity $\mathbf{c}$); accordingly we can write $\nu(\xi)$ in place of $\nu(\boldsymbol{\xi})$.

Provided the plane $\varepsilon = 0$ is suitably chosen and $\alpha(0 \leqslant \alpha \leqslant \pi)$ denotes the angle between $\boldsymbol{\xi}'$ and the polar axis (which is directed as $\boldsymbol{\xi}' - \boldsymbol{\xi}$) the following relations hold:

$$\mathbf{n} \cdot (\boldsymbol{\xi} - \boldsymbol{\xi}') = -|\boldsymbol{\xi} - \boldsymbol{\xi}'| \cos\theta \tag{5.15}$$

$$\begin{aligned}\mathbf{n} \cdot \boldsymbol{\xi}' \, |\boldsymbol{\xi} - \boldsymbol{\xi}'| &= |\boldsymbol{\xi} - \boldsymbol{\xi}'| \, \xi' \cos\theta \cos\alpha + |\boldsymbol{\xi} - \boldsymbol{\xi}'| \, \xi' \sin\theta \sin\alpha \cos\varepsilon \\ &= -(\boldsymbol{\xi} - \boldsymbol{\xi}') \cdot \boldsymbol{\xi}' \cos\theta + |(\boldsymbol{\xi} - \boldsymbol{\xi}') \wedge \boldsymbol{\xi}'| \sin\theta \cos\varepsilon \\ &= -(\boldsymbol{\xi} - \boldsymbol{\xi}') \cdot \boldsymbol{\xi}' \cos\theta + |\boldsymbol{\xi} \wedge \boldsymbol{\xi}'| \sin\theta \cos\varepsilon \end{aligned} \tag{5.16}$$

where $\boldsymbol{\xi} \wedge \boldsymbol{\xi}'$ denotes the vector product between $\boldsymbol{\xi}$ and $\boldsymbol{\xi}'$. Hence Eq. (5.14) becomes

$$\begin{aligned} & f_0(\boldsymbol{\xi} + \mathbf{n} \, |\boldsymbol{\xi} - \boldsymbol{\xi}'|/\cos\theta) \\ &\qquad = \rho_0(2\pi RT)^{-\frac{3}{2}} \exp[-(2RT_0)^{-1}(\xi'^2 + |\boldsymbol{\xi} - \boldsymbol{\xi}'|^2 \operatorname{tg}^2\theta \\ &\qquad\quad + 2\,|\boldsymbol{\xi} \wedge \boldsymbol{\xi}'| \operatorname{tg}\theta \cos\varepsilon)] \end{aligned} \tag{5.17}$$

and Eq. (5.12) takes on the form:

$$\begin{aligned} K_2(\boldsymbol{\xi}', \boldsymbol{\xi}) = \rho_0 m^{-1}(2\pi RT_0)^{-\frac{3}{2}} \int \Big[& B(\theta, |\boldsymbol{\xi} - \boldsymbol{\xi}'|/\cos\theta) + \\ & + B\left(\frac{\pi}{2} - \theta, |\boldsymbol{\xi} - \boldsymbol{\xi}'|/\cos\theta\right)\Big](\cos\theta)^{-3} \times \\ & \times \exp[-(2RT_0)^{-1}(\xi'^2 + |\boldsymbol{\xi} - \boldsymbol{\xi}'|^2 \operatorname{tg}^2\theta + \\ & + 2\,|\boldsymbol{\xi} \wedge \boldsymbol{\xi}'| \operatorname{tg}\theta \cos\varepsilon)] \, d\theta \, d\varepsilon \end{aligned} \tag{5.18}$$

The integration with respect to ε can be handled in terms of the modified Bessel function of first kind and zero-th order (Eq. (III.6.16)), giving

$$\begin{aligned} K_2(\boldsymbol{\xi}, \boldsymbol{\xi}') = \frac{2\pi\rho_0}{m} & (2\pi RT_0)^{-\frac{3}{2}} \times \\ & \times \int_0^{\pi/2} \left[B(\theta, |\boldsymbol{\xi} - \boldsymbol{\xi}'|/\cos\theta) + B\left(\frac{\pi}{2} - \theta, |\boldsymbol{\xi} - \boldsymbol{\xi}'|/\cos\theta\right)\right] \times \\ & \times (\cos\theta)^{-3} \exp[-(2RT_0)^{-1}(\xi'^2 + |\boldsymbol{\xi} - \boldsymbol{\xi}'|^2 \operatorname{tg}^2\theta)] \times \\ & \times I_0(|\boldsymbol{\xi} \wedge \boldsymbol{\xi}'| \, [RT_0]^{-1} \operatorname{tg}\theta) \, d\theta \end{aligned} \tag{5.19}$$

Hence:

$$Lh = Kh - \nu(\xi)h \tag{5.20}$$

where $\nu(\xi)$ is given by Eq. (5.4) and

$$Kh = \int K(\boldsymbol{\xi}', \boldsymbol{\xi})h(\boldsymbol{\xi}')\, d\boldsymbol{\xi}' \tag{5.21}$$

$$K(\boldsymbol{\xi}', \boldsymbol{\xi}) = \int K(\boldsymbol{\xi}', \boldsymbol{\xi}; \theta)\, d\theta \tag{5.22}$$

$$\begin{aligned}
K(\boldsymbol{\xi}', \boldsymbol{\xi}; \theta)& \\
= 2\pi\rho_0 m^{-1}&(2\pi RT_0)^{-\frac{3}{2}} \exp[-(2RT_0)^{-1}\xi'^2] \times \\
&\times \left\{\left[B(\theta, |\boldsymbol{\xi} - \boldsymbol{\xi}'|/\cos\theta) + B\left(\frac{\pi}{2} - \theta, |\boldsymbol{\xi} - \boldsymbol{\xi}'|/\cos\theta\right)\right] \times \right. \\
&\times (\cos\theta)^{-3} \exp[-(2RT_0)^{-1}|\boldsymbol{\xi} - \boldsymbol{\xi}'|^2 \operatorname{tg}^2\theta] I_0\left(|\boldsymbol{\xi} \wedge \boldsymbol{\xi}'| \frac{\operatorname{tg}\theta}{RT_0}\right) \\
&\left. - B(\theta, |\boldsymbol{\xi} - \boldsymbol{\xi}'|)\right\}
\end{aligned} \tag{5.23}$$

The above treatment has a meaning if the range of the potential is finite (in particular, for rigid spheres); if the potential extends to infinity, then the three terms K_1h, K_2h, νh in Eq. (5.1) do not exist separately. In the latter case one can first cut off the angle θ at an angle $\theta_0 < \pi/2$ and then perform all the manipulations described above; the result can be written as follows:

$$Lh = \int \left[\int K(\boldsymbol{\xi}', \boldsymbol{\xi}; \theta)h(\boldsymbol{\xi}')\, d\boldsymbol{\xi}' - \nu(\boldsymbol{\xi}; \theta)h(\boldsymbol{\xi})\right] d\theta \tag{5.24}$$

where $K(\boldsymbol{\xi}', \boldsymbol{\xi}; \theta)$ is given by Eq. (5.23) and

$$\nu(\xi; \theta) = 2\pi\rho_0 m^{-1}(2\pi RT_0)^{-\frac{3}{2}} \int \exp[-\xi'^2(2RT_0)^{-1}] B(\theta, |\boldsymbol{\xi} - \boldsymbol{\xi}'|)\, d\boldsymbol{\xi}' \tag{5.25}$$

The expression in Eq. (5.24) retains a meaning even when we let $\theta_0 \to \pi/2$ and so let the cutoff disappear, and shows that, for potentials extending to infinity, we can write L as an integral, with respect to the parameter θ of the difference between an integral operator K_θ and a multiplication operator ν_θ both depending upon θ. Since, however, the dependence on θ of each operator is non-integrable in the vicinity of $\theta = \pi/2$ we cannot integrate each term to yield Eq. (5.20).

In the case of rigid spheres it is possible to accomplish the integration with respect to θ indicated in Eq. (5.22). In fact, for rigid spheres (see Chapter II, Sections 4), $B(\theta, V) = \sigma^2 V \sin\theta\cos\theta$; hence:

$$\begin{aligned}
K(\boldsymbol{\xi}', \boldsymbol{\xi}; \theta) = 2\pi\rho_0\sigma^2 m^{-1}&(2\pi RT_0)^{-\frac{3}{2}} \exp[-(2RT_0)^{-1}\xi'^2]\{2\,|\boldsymbol{\xi} - \boldsymbol{\xi}'| \times \\
&\times \operatorname{tg}\theta(\cos\theta)^{-2} \exp[-(2RT_0)^{-1}(\boldsymbol{\xi} - \boldsymbol{\xi}')^2 \operatorname{tg}^2\theta] \times \\
&\times I_0(|\boldsymbol{\xi} \wedge \boldsymbol{\xi}'|\,[RT_0]^{-1} \operatorname{tg}\theta) - |\boldsymbol{\xi} - \boldsymbol{\xi}'| \sin\theta\cos\theta\}
\end{aligned} \tag{5.26}$$

To perform the integration, we let $t = tg\theta$ ($dt = (\cos\theta)^{-2}\,d\theta$) and use the easily proved formula [13]

$$\int_0^\infty I_0(at)\exp(-p^2t^2)t\,dt = (2p^2)^{-1}\exp[a^2(2p)^{-2}] \tag{5.27}$$

with $p = |\boldsymbol{\xi} - \boldsymbol{\xi}'|(2RT_0)^{-\frac{1}{2}}$, $a = |\boldsymbol{\xi} \wedge \boldsymbol{\xi}'|(RT_0)^{-1}$ to obtain:

$$K(\boldsymbol{\xi}', \boldsymbol{\xi}) = \rho_0\sigma^2 m^{-1}(2\pi RT_0)^{-\frac{1}{2}}\exp[-(2RT_0)^{-1}\xi'^2]\{2|\boldsymbol{\xi} - \boldsymbol{\xi}'|^{-1}\times$$
$$\times\exp[(2RT_0)^{-1}|\boldsymbol{\xi} - \boldsymbol{\xi}'|^{-2}(\boldsymbol{\xi}\wedge\boldsymbol{\xi}')^2] - |\boldsymbol{\xi} - \boldsymbol{\xi}'|(2RT_0)^{-1}\} \tag{5.28}$$

Finally, the integral giving the collision frequency $\nu(\xi)$ Eq. (5.4), can be easily evaluated for any finite range potential (in particular, for rigid spheres). In fact, if we recall that $B(\theta,V)$ is related to the impact parameter r by Eq. (II.4.14) and r ranges from 0 to σ (the range of the potential, or diameter of the spheres), we obtain

$$\nu(\xi) = \frac{2\pi}{m}\int f_0(\xi_*)\,|\boldsymbol{\xi} - \boldsymbol{\xi}_*|\,d\boldsymbol{\xi}_*\int_0^\sigma r\,dr = \frac{2\pi^2\sigma^2}{m}\rho_0(2\pi RT_0)^{-\frac{3}{2}}\int_0^\infty\int_0^\pi\times$$
$$\times\exp[-(2RT_0)^{-1}(\xi^2 + 2V\xi\cos\theta + V^2)]V^2\,dV\sin\theta\,d\theta$$
$$= \frac{\pi\sigma^2}{m}\rho_0(2\pi RT_0)^{-\frac{1}{2}}\xi^{-1}\int_0^\infty\exp[-(2RT_0)^{-1}(\xi^2 + V^2)]\times$$
$$\times\exp\{[(RT_0)^{-1}V\xi] - \exp[-(RT_0)^{-1}V\xi]\}V^2\,dV$$
$$= \pi\sigma^2\rho_0(m\xi)^{-1}(2\pi RT_0)^{-\frac{1}{2}}\left\{\int_0^\infty\exp[-(2RT_0)^{-1}(V - \xi)^2]V^2\,dV - \right.$$
$$\left. - \int_0^\infty\exp[-(2RT_0)^{-1}(V + \xi)^2]V^2\,dV\right\}$$
$$= \rho_0\sigma^2 m^{-1}(2\pi RT_0)^{\frac{1}{2}}\psi(\xi[2RT_0]^{\frac{1}{2}}) \tag{5.29}$$

where we passed from the integration variable $\boldsymbol{\xi}_*$ to the polar coordinates $V = |\boldsymbol{\xi} - \boldsymbol{\xi}_*|$, θ, φ and let

$$\psi(x) = \frac{1}{x}\left[\int_{-x}^\infty e^{-t^2}(t + x)^2\,dt - \int_x^\infty e^{-t^2}(t - x)^2\,dt\right]$$
$$= \frac{1}{x}\left[\int_{-x}^x e^{-t^2}t^2\,dt + x^2\int_{-x}^x e^{-t^2}\,dt + 2x\int_x^\infty te^{-t^2}\,dt + 2x\int_{-x}^\infty te^{-t^2}\,dt\right]$$
$$= \frac{2}{x}\int_0^x e^{-t^2}t^2\,dt + 2x\int_0^x e^{-t^2}\,dt + 2e^{-x^2}$$
$$= e^{-x^2} + \left(2x + \frac{1}{x}\right)\int_0^x e^{-t^2}\,dt \tag{5.30}$$

where the last expression comes from a partial integration.

A general property of the collision frequency for rigid spheres, finite range potentials and potentials with angular cutoff, is that $\nu(\xi)$ is a monotonic function of ξ. In fact, from Eqs. (5.4) and (II.5.21), we obtain

$$\frac{\partial \nu}{\partial \xi} = \gamma \beta_0 \xi^{-1} \int f_0(\xi_*) \boldsymbol{\xi} \cdot (\boldsymbol{\xi} - \boldsymbol{\xi}_*) \, |\boldsymbol{\xi} - \boldsymbol{\xi}_*|^{\gamma-2} \, d\boldsymbol{\xi}_* \tag{5.31}$$

where

$$\beta_0 = \frac{2\pi}{m} \int_0^{\pi/2} \beta(\theta) \, d\theta \tag{5.32}$$

Since $f_0(\xi_*)$ is proportional to $\exp(-\alpha \xi_*^2)$ where $\alpha > 0$ and $\xi_*^2 = \xi^2 + (\boldsymbol{\xi} - \boldsymbol{\xi}_*)^2 - 2\boldsymbol{\xi} \cdot (\boldsymbol{\xi} - \boldsymbol{\xi}_*)$, the contribution to the integral from the half-space $\boldsymbol{\xi} \cdot (\boldsymbol{\xi} - \boldsymbol{\xi}_*) < 0$ is, in absolute value, smaller than the contribution from the half-space $\boldsymbol{\xi} \cdot (\boldsymbol{\xi} - \boldsymbol{\xi}_*) > 0$; since the former contribution is negative and the latter positive, the total integral is positive and $\partial \nu / \partial \xi$ is greater than or less than zero depending on whether γ is greater than or less than zero. Now $\gamma = (n - 5)/(n - 1)$ for potentials with angular cutoff, $\gamma = 1$ for rigid spheres and finite range potentials. Hence the collision frequency is monotonically increasing for rigid spheres and finite range potentials. For power law potentials with angular cutoff, $\nu(\xi)$ is monotonically increasing for $n > 5$ and monotonically decreasing for $n < 5$. Thus in the last case ($n < 5$), ν is bounded from above

$$0 \leqslant \nu(\xi) \leqslant \nu_0 \tag{5.33}$$

whereas in all the other cases, ν is bounded from below

$$\nu(\xi) \geqslant \nu_0 > 0 \tag{5.34}$$

where $\nu_0 = \nu(0)$ is a finite quantity.

6. The decay to equilibrium and the spectrum of the collision operator

The simplest problem which may be analysed with the linearized Boltzmann equation is the following: assume that the gas has, at $t = 0$, a distribution which is space independent and slightly different from a Maxwellian f_0, $f = f_0(1 + h_0)$. What will happen for $t > 0$? (pure initial value problem). From the H-theorem we expect that f will decay to a Maxwellian distribution; hence, for $t \to \infty$, h should decay to zero if f_0 is chosen to have the same density, velocity and temperature as f. This result is also an immediate consequence of the linearized Boltzmann equation for the perturbation h, which is obtained from Eq. (2.6) by letting $\partial h / \partial \mathbf{x} = 0$:

$$\frac{\partial h}{\partial t} = Lh \tag{6.1}$$

In fact, if h is such that $\|h\|^2$ and (h, Lh) exist, multiplication by h and integration with respect to the velocity variables yields, because of Eq. (1.12):

$$\frac{\partial}{\partial t}[\tfrac{1}{2}\|h\|^2] = (h, Lh) \leqslant 0 \tag{6.2}$$

so that

$$\|h(t_2)\| \leqslant \|h(t_1)\| \qquad (t_2 > t_1) \tag{6.3}$$

and the equality sign holds if and only if h is a collision invariant; but since f and f_0 have been assumed to have the same density, velocity and temperature:

$$(\psi_\alpha, h) = 0 \qquad (\alpha = 0, 1, 2, 3, 4) \tag{6.4}$$

and the only collision invariant which satisfies these relations is $h = 0$. Hence $\|h\|$ always decreases unless it is zero; for $t \to \infty$, $\|h\|$ (being bounded from below and decreasing) must reach a limit with a vanishing derivative. This limit must satisfy Eq. (6.2) with $\partial\|h\|^2/\partial t = 0$; that is, by the previous argument, the limit is zero.

The next problem is: how does h decay to zero? To answer this question we must be able to study the solutions of Eq. (6.1) and, in particular, their behavior for $t \to \infty$. Some general properties may turn out to be sufficient for handling the problem: if, for example, we were able to show that

$$(h, Lh) < -\mu_0 \|h\|^2 \qquad (\mu_0 > 0) \tag{6.5}$$

for any h satisfying Eq. (6.4), then Eq. (6.2) could be replaced by

$$\frac{\partial}{\partial t}[\tfrac{1}{2}\|h\|^2] \leqslant -\mu_0\|h\|^2 \quad \text{or} \quad \frac{\partial}{\partial t}(e^{2\mu_0 t}\|h\|^2) \leqslant 0 \tag{6.6}$$

and hence

$$\|h(t)\| \leqslant \|h(0)\| e^{-\mu_0 t} \tag{6.7}$$

implying an exponential decay to equilibrium for $t \to \infty$. The desire to discuss the decay of h for $t \to \infty$ leads to the study of the eigenvalue problem:

$$Lg = \lambda g \tag{6.8}$$

for the operator L. As we know (Section 1), this problem has five eigensolutions ($g = \psi_\alpha$, $\alpha = 0, 1, 2, 3, 4$) for $\lambda = 0$; all the other values of λ must be negative.

We remark that Eq. (6.8) might even turn out not to have solutions for $\lambda \neq 0$ if we insist that g must be an ordinary function. If we, however, allow g to be a generalized function, then, since L is a self-adjoint operator (Eq. (1.10)), general theorems [14] ensure not only that Eq. (6.8) has solutions, but also that the general solution of Eq. (6.1) can be written in the form

$$h = \int_{-\mu_\infty}^{-\mu_0} e^{\lambda t} g_\lambda(\xi) A_\lambda \, d\lambda + \sum_{k=0}^{4} A_\alpha \psi_\alpha \tag{6.9}$$

where g_λ is the solution corresponding to the "eigenvalue" λ, A_λ an arbitrary function depending upon λ and the integral extends to all the values of $\lambda \neq 0$ for which a $g_\lambda \neq 0$ exists (if all the λ's or part of them form a discrete set, then the corresponding integral has to be replaced by the sum $\sum_k e^{-\lambda_k t} g_k(\boldsymbol{\xi})A_k$). If $\mu_0 \neq 0$ in Eq. (6.9), then h decays exponentially to the collision invariant $\sum A_\alpha \psi_\alpha$ (in particular, to zero, if Eq. (6.4) holds); if $\mu_0 = 0$, the decay to equilibrium is not exponential and depends upon the initial datum (which determines the A_λ).

The above remarks make it clear that it is very important to study the set of values of λ for which Eq. (6.8) has a nonzero solution; that is the *spectrum* of the operator L.

We observe that, according to the usual definition [1–5] the spectrum of an operator is the set of values of λ for which $(L - \lambda I)^{-1}$ does not exist as a bounded operator in $\mathscr{H}$ or is not uniquely determined; in our case (L is self-adjoint and we allow generalized solutions) the two definitions are equivalent. In our case "discrete" and "continuous" spectra correspond exactly to the meaning conveyed by their names.

Before discussing the spectrum of the collision operator in particular cases, it is useful to remark that the order of magnitude of λ in Eq. (6.8) is obtained by noting that, if we make velocities nondimensional by means of an average velocity $\bar{\xi}$ (say $\bar{\xi} = \sqrt{2RT_0}$) and recall that according to Eq. (II.4.14), $B(\theta, V) \simeq V\sigma^2$ where $V \simeq \bar{\xi}$ and σ is the effective molecular radius, then $Lh = \rho_0\sigma^2 m^{-1}\bar{\xi}\mathscr{L}h$ where $\mathscr{L}h$ is nondimensional and may be taken to be of order unity. Hence we conclude that $\lambda \simeq \rho_0\sigma^2\bar{\xi}/m \simeq \bar{\xi}/l$, where l is the mean free path (Eq. (I.4.17)) and $\simeq$ indicates that two quantities are of the same order of magnitude. $\theta = l/\bar{\xi}$ can be called the mean free time since it is of order of magnitude of the time between two subsequent collisions of a given molecule (see also Chapter V, Section 1).

The only case in which a complete solution of Eq. (6.8) has been carried out is the case of Maxwellian molecules, which is to say, molecules repelling each other with a force proportional to the inverse fifth power of the distance. In this case Wang Chang and Uhlenbeck were able to prove [15, 16] that the spectrum is discrete and can be conveniently described by attaching two indexes to the eigenvalues λ_{nl} which are given by:

$$\lambda_{nl} = 2\pi\rho_0 m^{-1}\int_0^{\pi/2}\Big\{P_l(\sin\theta)\sin^{2n+l}\theta\Big[\beta(\theta) + \beta\Big(\frac{\pi}{2} - \theta\Big)\Big] - \\ - (\delta_{n0}\,\delta_{l0} + 1)\beta(\theta)\Big\}\,d\theta \qquad (n, l = 0, 1, 2, \ldots) \quad (6.10)$$

where $\beta(\theta)$ is given by Eq. (II.5.22), with $\gamma = 0$ (as appropriate for Maxwell's molecules) and P_l denotes the l-th Legendre polynomial. For each eigenvalue λ_{nl} there are $2l + 1$ eigenfunctions which are conveniently labelled with three indexes and are given by

$$g_{nlm} = \left[\frac{2n!}{\Gamma(n + l + \frac{3}{2})}\right]^{\frac{1}{2}} \left(\frac{\xi}{2RT_0}\right)^l L_n^{(l+\frac{1}{2})} \left(\frac{\xi^2}{2RT_0}\right) Y_l^m(\theta, \varphi)$$

$$(n, l = 0, 1, 2, \ldots ; -l \leqslant m \leqslant l) \quad (6.11)$$

where ξ, θ, φ are polar coordinates in the velocity space, Γ denotes the gamma function, $L^{(l+\frac{1}{2})}$ are the associated Laguerre polynomials and $Y_l^m(\theta, \varphi)$ the spherical harmonics [17, 18].

The above eigenfunctions satisfy the orthogonality relation:

$$(2RT_0)^{-\frac{3}{2}} \int \bar{g}_{nlm}(\boldsymbol{\xi}) g_{n'l'm'}(\boldsymbol{\xi}) \exp[-\xi^2(2RT_0)^{-1}] \, d\boldsymbol{\xi} = \delta_{nn'} \delta_{ll'} \delta_{mm'} \quad (6.12)$$

where the bar denotes a complex conjugate. In particular, for $n = 0$, $l = 0$, $n = 0$, $l = 1$, $n = 1$, $l = 0$, Eq. (6.10) gives $\lambda_{nl} = 0$ and the corresponding eigenfunctions are five linear combinations of the five collision invariants ψ_α. For $n \to \infty$, it can be shown that the eigenvalues λ_{nl} tend to $-\infty$ (they grow asymptotically as the fourth root of n) [16, 19]. Hence in the complex plane of the parameter λ we have the situation illustrated in Fig. 18. A table of the first few eigenvalues is given below (Table 1). More detailed tables are given in Ref. [20].

Hard spheres molecules have also been investigated in some detail. In this case a useful piece of information is obtained from a theorem of Weyl on the perturbation of the spectrum of a self-adjoint operator V, when we add to it a sufficiently regular integral operator K to obtain an operator $W = V + K$. According to Weyl's theorem [2–4], if K is completely continuous and so transforms a bounded sequence of functions $\{g_k\}$ into a converging sequence $\{Kg_k\}$ (convergence being understood in a suitable function space, in our case the Hilbert space $\mathscr{H}$, where the norm is defined by Eq. (1.9)), then the continuous spectra of W and V are the same. That is, the only effect of K is to

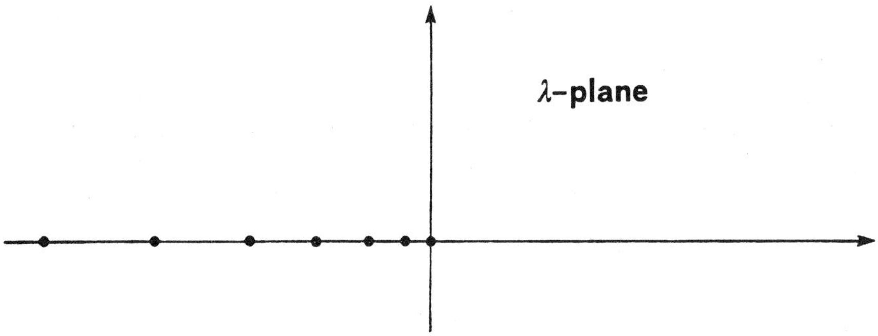

Fig. 18. Spectrum of the collision operator for Maxwell's molecules.

change the discrete spectrum. To be precise, it is the so called essential spectrum (i.e. the set of limit points of the spectrum) which remains unchanged. Now Grad [21] and Dorfman [22] were able to show that the operator K in Eq. (5.20) is completely continuous in $\mathscr{H}$ for power law potentials, when grazing collisions are cutoff or rigid spheres are involved. Hence, the continuous spectrum, if any, of Lh is given by the continuous spectrum of the multiplication operator $-\nu(\xi)$ (where we write $-\nu(\xi)$ in place of $-\nu(\xi)I$ and I is the identity operator). For this multiplication operator the eigenvalue equation reduces to

$$-\nu(\xi)g = \lambda g \tag{6.13}$$

or

$$[\lambda + \nu(\xi)]g = 0 \tag{6.14}$$

We must distinguish two cases: either $\nu(\xi)$ is constant (as in the case of Maxwellian molecules with angular cutoff) and Eq. (6.14) implies $\lambda = -\nu$ so that the spectrum reduces to a single point; or $\nu(\xi)$ is not a constant and $\lambda + \nu(\xi)$ can be zero at one point at most (at least if $\nu(\xi)$ is monotonic, as is the case for power law potentials and rigid spheres, as was shown in Section 5). Eq. (6.14) then says that, if $\nu(\xi) \neq$ const., then g must be zero everywhere except at one point. Since we do not want g to be identically zero, we take g to be, except for a factor, a delta function

$$g = \delta(\lambda + \nu(\xi)) \tag{6.15}$$

In fact $x\,\delta(x) = 0$ the sense of the theory of generalized functions (Chapter I, Section 2) and it can be shown that no other generalized function has this property (an arbitrary constant factor, is, of course, allowed). Eq. (6.15) provides a solution different from zero if and only if $\lambda + \nu(\xi)$ is actually zero for some ξ; that is if λ is one of the values taken by the function $-\nu(\xi)$. These values form a continuum and hence Weyl's theorem informs us that, for rigid spheres and power law interactions with angular cutoff,

TABLE I. Values of $\lambda_{nl}/\lambda_{02}$

n \ l	0	1	2	3	4	5	6
0	0	0	1	1.5	1.8731	2.1828	2.4532
1	0	0.6667	1.6667	1.5704	1.9106	2.2066	2.4703
2	0.6667	1	1.3432	1.6670	1.8633	2.2415	2.4936
3	1	1.2281	1.4915	1.7631	2.0288	2.2824	2.5215
4	1.2281	1.4037	1.6193	1.8533	2.0917	2.3262	2.5525
5	1.4037	1.5475	1.7310	1.9371	2.1533	2.3710	
6	1.5475	1.6698	1.8302	2.0148			
7	1.6698	1.7767					
8	1.7767						

there is a continuous spectrum which can be explicitly described. In particular, for rigid spheres and inverse power laws with exponent larger than five, we saw that $\nu(\xi)$ ranges from a minimum value ν_0 corresponding to $\xi = 0$, to $+\infty$ (Section 5). Hence the collision operator has a continuous spectrum extending from $-\nu_0$ to $-\infty$ (see Fig. 19). For inverse power laws with exponent smaller than five, the continuous spectrum ranges from $-\nu_0$ to 0 (Fig. 20); the case $n = 5$ (Fig. 21) is clearly a separating case. We note

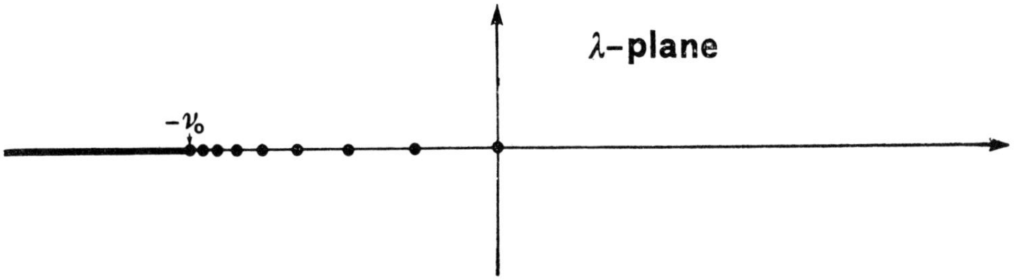

Fig. 19. Spectrum of the collision operator for rigid spheres and inverse power laws with angular cutoff and exponent larger than five.

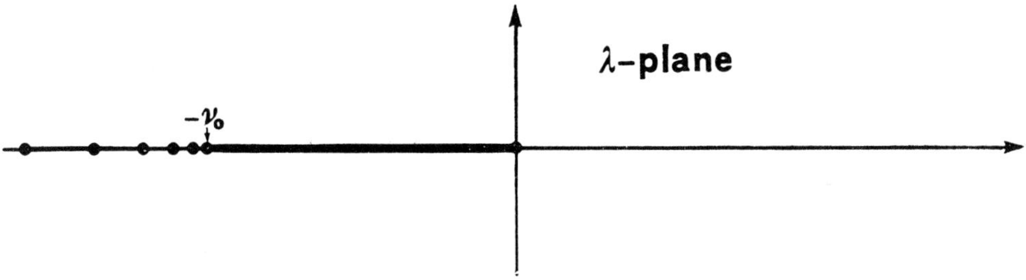

Fig. 20. Spectrum of the collision operator for inverse power laws with exponent smaller than five and angular cutoff.

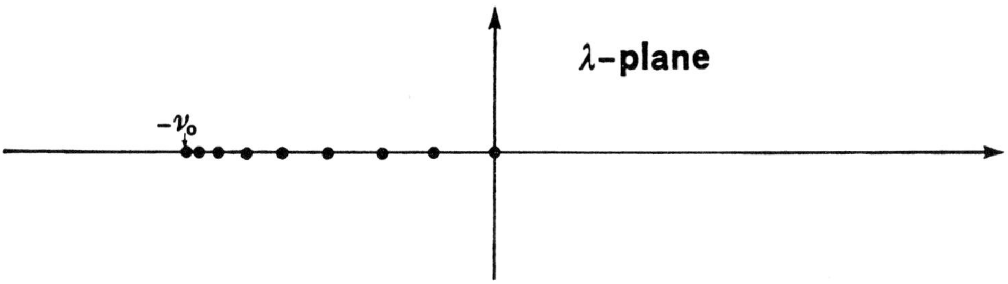

Fig. 21. Spectrum of the collision operator for Maxwell's molecules with angular cutoff.

that these results imply that the decay to equilibrium is exponential for rigid spheres and inverse power laws with exponent larger than, or equal to, five; in fact, the discrete eigenvalues cannot arbitrarily approach the origin ($\lambda = 0$ cannot be a limit point of the spectrum because of Weyl's theorem) and, consequently, $\mu_0 > 0$ in Eqs. (6.5) and (6.9).

The next question is whether there are indeed discrete eigenvalues (other than $\lambda = 0$) besides the continuous spectrum. The question has been answered in the affirmative, for rigid spheres, by Kuščer and Williams [23], who applied an idea of Lehner and Wing [24] which had already been used in neutron transport theory. This idea is to introduce an artificial parameter c as a multiplier of $[\nu(\xi) + \lambda]h$ in the eigenvalue equation, Eq. (6.8) (where Lh is replaced by its expression $Kh - \nu h$) and to study the eigenvalues $c = c_n(\lambda)$ as functions of the continuous parameter λ ranging from 0 to $-\nu(0)$. The eigenvalues of the original problem are then obtained as roots of $c_n(\lambda) = 1$. The result is that L has an infinite number of discrete negative eigenvalues in the interval $(-\nu(0), 0)$ which accumulate at the point $\lambda^* = -\nu(0)$.

We next consider the case of molecules interacting with central forces. As we saw above, for inverse power laws of interaction, a cutoff in the scattering angle leads to results similar to those holding for rigid spheres; Kuščer and Williams' result has not been extended to this case, but it seems reasonable to guess that it can be so extended. If we do not introduce a cutoff in the angle then the matter is very complicated for potentials extending to infinity; the only case which is simply analysed is the case of Maxwellian molecules, discussed above. It is interesting to point out that the spectrum in this case is exactly what one would expect by a naive limiting procedure on the results for angular cutoff; in fact Fig. 18 is simply Fig. 19, except for the fact that $-\nu_0$ has been moved to $-\infty$. It is tempting to argue that an analogous procedure is valid for inverse power laws with exponent $n \geqslant 5$; this would lead to a purely discrete spectrum. Recently, Pao [53] has given a rigorous proof of the truth of this conjecture.

In the case of a potential with a strictly finite range, an angular cutoff is not needed to obtain the splitting shown in Eq. (5.20), but the operator K is much harder to handle; in particular, it is not easy to prove, or disprove, that K is completely continuous in $\mathscr{H}$ [25].

A rather general result which is easily proved [25] is the following

THEOREM I *Inequality* (6.5) *applies to any h such that Eq.* (6.4) *is satisfied and* (h, Lh) *exists, provided the intermolecular potential satisfies*

$$|U(\rho)| \geqslant 0(\rho^{-n+1}) \qquad (\rho \to 0, n \geqslant 5) \tag{6.16}$$

To prove this result, let us consider the general expression for L, Eq. (1.7), and split L as follows

$$L = L^\delta + L_\delta \tag{6.17}$$

where L^δ and L_δ are given by the same expression as L, Eq. (1.7), with $B(\theta, V)$ replaced by B^δ and B_δ, which are defined as follows:

$$B^\delta(\theta, V) = \begin{cases} B(\theta, V) & \text{for } \theta < \delta \\ 0 & \text{for } \theta > \delta \end{cases} \tag{6.18}$$

$$B_\delta(\theta, V) = \begin{cases} 0 & \text{for } \theta < \delta \\ B(\theta, V) & \text{for } \theta > \delta \end{cases} \tag{6.18}$$

where δ is any fixed angle between 0 and $\pi/2$. Now L^δ and L_δ, having the general expression given by Eq. (1.7), have the property (1.12); in particular

$$(h, L_\delta h) \leqslant 0 \tag{6.19}$$

But, if Eq. (6.16) is satisfied, L^δ is an operator with angular cutoff for a potential harder than Maxwell's and hence, according to the results discussed above:

$$(h, L^\delta h) \leqslant -\mu_0 \|h\|^2 \qquad (\mu_0 > 0; (\psi_\alpha, h) = 0) \tag{6.20}$$

By adding Eqs. (6.19) and (6.20), Eq. (6.5) is obtained, as was to be shown.

By extending this kind of argument, it is possible to show that when $\delta \to \pi/2$ all the eigenvalues increase in absolute value, so that they move to the left.

When the splitting shown in Eq. (5.20) applies, the existence of (h, Lh), which is an obvious prerequisite for Eq. (6.5) to be meaningful, follows from the existence of $(h, \nu h)$ as shown by the following

THEOREM II *The operator K appearing in Eq. (5.20) satisfies the inequality*

$$|(h, Kh)| \leqslant \lambda(\nu h, h) \qquad (h \in \mathcal{N}) \tag{6.21}$$

for some positive λ. Here $\mathcal{N}$ denotes the (Hilbert) space of the functions such that $(\nu h, h)$ exists ($\mathcal{N}$ is a subspace of $\mathcal{H}$ for potential harder than Maxwell's).

A preliminary result is the trivial statement

$$(h, Kh) \leqslant (\nu h, h) \tag{6.22}$$

which follows from Eqs. (1.12) and (5.20). We note also that according to Eqs. (5.1) and (5.20), $K = K_2 - K_1$, where K_1 and K_2 are operators, whose kernels, given by Eqs. (5.12) and (5.13), are nonnegative; hence, by means of trivial inequalities and Eq. (6.22)

$$\begin{aligned} |(h, Kh)| &\leq (|h|, K_2 |h|) + (|h|, K_1 |h|) \\ &= (|h|, (K_2 - K_1) |h|) + 2(|h|, K_1 |h|) \\ &= (|h|, K |h|) + 2(|h|, K_1 |h|) \\ &\leqslant 2(|h|, K_1 |h|) + (h, \nu h) \end{aligned} \tag{6.23}$$

where the bars denoting absolute value are not required in the last term. Now, because of Eq. (5.3) and the Schwartz inequality [2]:

$$\begin{aligned}(|h|, K_1 |h|) &= \frac{2\pi}{m}\int d\theta \int d\boldsymbol{\xi}_* \, d\boldsymbol{\xi}\{[f_0(\boldsymbol{\xi})f_0(\boldsymbol{\xi}_*)B(\theta, |\boldsymbol{\xi}_* - \boldsymbol{\xi}|)]^{\frac{1}{2}} \, |h(\boldsymbol{\xi})|\} \times \\ &\quad \times \{[f_0(\boldsymbol{\xi})f_0(\boldsymbol{\xi}_*)B(\theta, |\boldsymbol{\xi}_* - \boldsymbol{\xi}|)]^{\frac{1}{2}} \, |h(\boldsymbol{\xi}_*)|\} \\ &\leqslant \frac{2\pi}{m}\int d\theta \left\{\int d\boldsymbol{\xi} \, d\boldsymbol{\xi}_* f_0(\boldsymbol{\xi})f_0(\boldsymbol{\xi}_*)B(\theta, |\boldsymbol{\xi} - \boldsymbol{\xi}_*|) \, |h(\boldsymbol{\xi})|^2\right\}^{\frac{1}{2}} \times \\ &\quad \times \left\{\int d\boldsymbol{\xi} \, d\boldsymbol{\xi}_* f_0(\boldsymbol{\xi})f_0(\boldsymbol{\xi}_*)B(\theta, |\boldsymbol{\xi}_* - \boldsymbol{\xi}|) \, |h(\boldsymbol{\xi}_*)|^2\right\}^{\frac{1}{2}} \end{aligned} \tag{6.24}$$

Both integrals within brackets have the same value, as can be seen by interchanging $\boldsymbol{\xi}$ and $\boldsymbol{\xi}_*$. Then, if Eq. (5.4) is also used, we obtain:

$$\begin{aligned}(|h|, K_1 |h|) &\leqslant \int d\boldsymbol{\xi}\left\{\frac{2\pi}{m}\int d\theta \, d\boldsymbol{\xi}_* f_0(\boldsymbol{\xi}_*)B(\theta, |\boldsymbol{\xi}_* - \boldsymbol{\xi}|)\right\} f_0(\boldsymbol{\xi}) \, |h(\boldsymbol{\xi})|^2 \\ &= \int d\boldsymbol{\xi}\nu(\boldsymbol{\xi})f_0(\boldsymbol{\xi}) \, |h(\boldsymbol{\xi})|^2 = (\nu h, h) \end{aligned} \tag{6.25}$$

Putting this result into Eq. (6.23) gives Eq. (6.21) with $\lambda = 3$.

The result embodied in Theorem I has a very important

COROLLARY: *The equation*

$$Lh = g \qquad (g \in \mathscr{H}) \tag{6.26}$$

has a solution if and only if the source term g is orthogonal to the five collision invariants:

$$(\psi_\alpha, g) = 0 \tag{6.27}$$

provided the potential satisfies Eq. (6.16). *The solution is determined up to an additive linear combination of the ψ_α's.*

To prove this, let $g \in \mathscr{H}$ satisfy Eq. (6.27) and let us solve Eq. (6.26) in the subspace $W \subset \mathscr{H}$ orthogonal to the subspace spanned by the five collision invariants (note that $g \in W$ by assumption and $Lh \in W$ because of Eq. (1.16)). In W, L is a self-adjoint operator whose spectrum does not contain the origin, by virtue of theorem I; because of the definition of the spectrum, L^{-1} exists in W and a unique solution $h^{(0)} \in W \subset \mathscr{H}$ is found ($h^{(0)} = L^{-1}g$). While $h^{(0)}$ is the only solution in W, if we return to $\mathscr{H}$ we can add to $h^{(0)}$ any linear combination of the five collision invariants and still satisfy Eq. (6.26); this completes the argument.

We also note that the solution $h^{(0)} \in W$ of Eq. (6.26) minimizes the functional

$$J(\tilde{h}) = (2g - L\tilde{h}, \tilde{h}) \qquad (\tilde{h} \in W) \tag{6.28}$$

This follows from the simple fact that if we let $\tilde{h} = h^{(0)} + \delta h$, then

$$\begin{aligned} J(h^{(0)} + \delta h) &= J(h^{(0)}) + 2(g - Lh^{(0)}, \delta h) - (L\,\delta h, \delta h) \\ &\geqslant J(h^{(0)}) + \mu_0 \|\delta h\|^2 \qquad (\delta h \in W) \end{aligned} \tag{6.29}$$

where Eq. (6.5) has been used together with the fact that $h^{(0)}$ satisfies Eq. (6.26). Eq. (6.29) shows that $J(\tilde{h}) \geqslant J(h^{(0)})$, the equality sign being valid if and only if $\delta h = 0$ or $\tilde{h} = h^{(0)}$.

7. Steady one-dimensional problems. Transport coefficients

The discussion given in Section 6 provides us with a fairly complete picture of the behavior of the time evolution of the solution of the linearized Boltzmann equation in the space homogeneous case. The success achieved in that kind of problems stimulates a similar investigation of the formally similar problems, whose solution is independent of time t and two space coordinates, say x_2 and x_3; then we have to solve the equation

$$\xi_1 \frac{\partial h}{\partial x_1} = Lh \tag{7.1}$$

in the unknown $h = h(x_1, \xi_1, \xi_2, \xi_3) = h(x_1, \boldsymbol{\xi})$.

The similarity between Eqs. (6.1) and (7.1) suggests that we look for solutions of the form

$$h = e^{\lambda x_1} g(\boldsymbol{\xi}) \tag{7.2}$$

where g satisfies

$$Lg = \lambda \xi_1 g \tag{7.3}$$

which is the analogue of Eq. (6.8). The first question is whether the solutions of Eq. (7.3) are sufficient to construct the general solution of Eq. (7.1) by superposition, as in Eq. (6.9). Next comes a study of the set of values of λ for which Eq. (7.3) has a solution.

The problem here is more difficult because there is an interplay between L and the multiplicative operator ξ_1. In addition, the existence of collision invariants prevents L from being a strictly negative operator. In order to alleviate the latter difficulty, we decompose the Hilbert space $\mathscr{H}$ into two orthogonal subspaces, $\mathscr{H}^+$, containing all the even functions of ξ_1 and $\mathscr{H}^-$ containing the odd ones. Any function g is accordingly split as follows

$$g = g^+ + g^- \tag{7.4}$$

where

$$g^+ = \tfrac{1}{2}(g + P_1 g) \qquad g^- = \tfrac{1}{2}(g - P_1 g) \tag{7.5}$$

where P_1 is the operator which reflects ξ_1:

$$P_1 g(\xi_1, \xi_2, \xi_3) = g(-\xi_1, \xi_2, \xi_3) \tag{7.6}$$

Clearly:

$$P_1(\xi_1 g) = -\xi_1 P_1 g \tag{7.7}$$

whereas:

$$P_1 L g = L P_1 g \tag{7.8}$$

because, for central interactions, L is invariant with respect to any reflection in velocity space, provided the velocity in the Maxwellian f_0 has no components along the x_1-axis. By applying the operators $\frac{1}{2}(I \pm P_1)$, where I denotes the identity operator, to both sides of Eq. (7.3), we obtain the equivalent system:

$$Lg^+ = \lambda \xi_1 g^- \qquad (g^+ \in \mathscr{H}^+, g^- \in \mathscr{H}^-) \tag{7.9}$$

$$Lg^- = \lambda \xi_1 g^+ \tag{7.10}$$

where Eqs. (7.7) and (7.8) have been used. In Eqs. (7.9) and (7.10), of course, L denotes the appropriate restriction of L to $\mathscr{H}^+$ and $\mathscr{H}^-$, respectively.

Now, four of the collision invariants ($\psi_0, \psi_2, \psi_3, \psi_4$) are in $\mathscr{H}^+$ and only one (ψ_1) is in $\mathscr{H}^-$. Also, any solution of the system of Eqs. (7.9) and (7.10) for $\lambda \neq 0$ is such that g^- is orthogonal to $\psi_1 = \xi_1$, because forming the scalar product in $\mathscr{H}$ of both sides of Eq. (7.9) with $\psi_0 = 1$ gives

$$\lambda(\xi_1, g^-) = 0 \tag{7.11}$$

We take advantage of this fact to replace Eq. (7.10) by the following equation:

$$-Ng^- = \lambda \xi_1 g^+ \tag{7.12}$$

where

$$Ng^- = -Lg^- \quad \text{if} \quad (\xi_1, g^-) = 0$$

$$N\xi_1 = \kappa \xi_1 \tag{7.13}$$

where κ is any fixed positive constant. The system of Eqs. (7.9) and (7.12) is essentially equivalent to the system of Eqs. (7.9) and (7.10) because $-N$ differs from L in a one-dimensional subspace $\mathscr{R}$ which is orthogonal to g^-, due to Eq. (7.11), unless $\lambda = 0$; there we lose only one solution $g^- = \xi_1$, corresponding to $\lambda = 0$, which we can add later, if necessary. Thus $\mathscr{H}^-$ splits into $\mathscr{H}_1^- \oplus \mathscr{R}$ where $\mathscr{H}_1^-$ is orthogonal to ψ_1 and hence to all the collision invariants. N has a unique inverse N^{-1} in $\mathscr{H}_1^-$ (N^{-1} is even bounded according to the corollary proved at the end of the previous equation, provided the interaction potential is sufficiently hard). This implies that N also has an inverse in $\mathscr{H}^- = \mathscr{H}_1^- \oplus \mathscr{R}$, given by

$$N^{-1}g^- = -L^{-1}g^- \quad \text{if} \quad (\xi_1, g^-) = 0$$

$$N^{-1}\xi_1 = \xi_1/\kappa \tag{7.14}$$

Hence we can write:

$$g^- = -\lambda N^{-1}(\xi_1 g^+) \tag{7.15}$$

and substituting into Eq. (7.9) gives

$$Lg^+ = -\lambda^2(\xi_1 N^{-1}\xi_1)g^+ \qquad (g^+ \in \mathscr{H}^+) \tag{7.16}$$

This is an equation for g^+; once g^+ is known, g^- is determined by Eq. (7.15). At first sight, it could appear that we have unnecessarily complicated the problem, because we have replaced Eq. (7.3), which contained the operator L and the multiplication operator ξ_1, by Eq. (7.16), which contains L and the operator $\xi_1 N^{-1}\xi_1$; the latter is much more complicated than a multiplication operator and not available in an explicit form. The advantage which has been gained is, however, the following: the operator $\xi_1 N^{-1}\xi_1$, acting on functions of $\mathscr{H}^+$, is positive definite. This is easily seen because N^{-1} is positive definite, because of Eqs. (7.14) (L^{-1} is negative definite in any subspace orthogonal to the collision invariants and $\kappa > 0$); and as a consequence:

$$(g^+, \xi_1 N^{-1}\xi_1 g^+) = (\xi_1 g^+, N^{-1}(\xi_1 g^+)) > 0 \tag{7.17}$$

Hence [3], a selfadjoint operator C exists such that

$$\xi_1 N^{-1}\xi_1 = C^2 \tag{7.18}$$

We observe that C will be unbounded in general, but will have a uniquely determined inverse C^{-1}, since $\xi_1 N^{-1}\xi_1$ is positive definite; C^{-1} will be also unbounded in general. We work with C formally at first and then consider the restrictions imposed by the unboundedness of C.

By means of Eq. (7.18), Eq. (7.16) can be rewritten as follows:

$$Lg^+ = -\lambda^2 C^2 g^+ \tag{7.19}$$

or

$$M\varphi = -\lambda^2\varphi \tag{7.20}$$

where

$$\varphi = Cg^+ \tag{7.21}$$

$$M = C^{-1}LC^{-1} \tag{7.22}$$

We remark that M is self-adjoint and nonnegative:

$$\begin{aligned}(\varphi, M\psi) &= -(\varphi, C^{-1}LC^{-1}\psi) = -(C^{-1}\varphi, LC^{-1}\psi)\\ &= -(LC^{-1}\varphi, C^{-1}\psi) = -(C^{-1}LC^{-1}\varphi, \psi) = (M\varphi, \psi)\end{aligned} \tag{7.23}$$

$$(\varphi, M\varphi) = -(\varphi, C^{-1}LC^{-1}\varphi) = -([C^{-1}\varphi], L[C^{-1}\varphi]) \geqslant 0 \tag{7.24}$$

the equality sign holding if and only if $C^{-1}\varphi$ is a collision invariant. Hence Eq. (7.20) will admit of a set of eigenfunctions φ_λ with $\lambda^2 > 0$ or $\lambda^2 = 0$; the latter are four in number and are given by $\varphi_\alpha = C\psi_\alpha$ ($\alpha = 0, 2, 3, 4$), while the former, generally speaking, will correspond to a partly continuous and partly discrete spectrum. The set of φ_λ's, according to general theorems [14], is complete for the functions of $\mathscr{H}_1^+$, so that for any $\varphi = \mathscr{H}_1^+$ we can write:

$$\varphi(\boldsymbol{\xi}) = \sum_{\substack{\alpha=0\\ \alpha\neq 1}}^{4} A_\alpha \varphi_\alpha + \int_{\lambda_0}^{\lambda_\infty} \varphi_\lambda(\boldsymbol{\xi}) A_\lambda^+ \, d\lambda \tag{7.25}$$

where $0 \leqslant \lambda_0 < \infty$, $0 < \lambda_\infty \leqslant \infty$ and the A_α are arbitrary constants, A_λ^+ is an arbitrary function of λ and the integral has to be replaced by a sum over discrete sets whenever the latter are part of the spectrum. Eq. (7.25) implies that any function h^+ such that $Ch^+ \in \mathscr{H}_1^+$ or

$$|(h^+, \xi_1 N^{-1} \xi_1 h^+)| < \infty \tag{7.26}$$

can be expanded as follows

$$h^+ = \sum_{\substack{\alpha=0\\ \alpha\neq 1}}^{4} A_\alpha \psi_\alpha + \int_{\lambda_0}^{\lambda_\infty} g_\lambda^+(\boldsymbol{\xi}) A_\lambda^+ \, d\lambda \tag{7.27}$$

where

$$g_\lambda^+ = C^{-1} \varphi_\lambda \tag{7.28}$$

Of course, here one has to extend the meaning of C^{-1} outside $\mathscr{H}_1^+$ when φ_λ^+ belongs to the continuous spectrum; this can easily be done by interpreting Eq. (7.28) in the sense of generalized functions (this interpretation is needed for Eqs. (7.25) and (7.27) as well).

We remark that, given the function h^+, the coefficients A_α, A_λ^+ in Eqs. (7.25) and (7.27) are determined by the orthogonality relations, which follow from Eq. (7.20):

$$(\varphi_\lambda, \varphi_{\lambda'}) = 0 \quad \text{if} \quad \lambda \neq \lambda' \tag{7.29}$$

or

$$(g_\lambda, \xi_1 N^{-1}[\xi_1 g_{\lambda'}]) = 0 \quad \text{if} \quad \lambda \neq \lambda' \tag{7.30}$$

Thus the coefficients A_α in Eq. (7.27) are given by ($h = h^+$):

$$A_\alpha = (\xi_1 N^{-1}[\xi_1 \psi_\alpha], h)/(\xi_1 N^{-1}[\xi_1 \psi_\alpha], \psi_\alpha) \qquad (\alpha = 0, 2, 3, 4) \tag{7.31}$$

provided that ψ_α denote not the four elementary collision invariants 1, ξ_2, ξ_3, ξ^2, but their linear combinations which satisfy

$$(\xi_1 N^{-1}[\xi_1 \psi_\alpha], \psi_\beta) = 0 \qquad (\alpha \neq \beta) \tag{7.32}$$

In particular, Eq. (7.31) gives:

$$A_0 = (\xi_1^2, h)/(\xi_1^2, 1)$$

$$A_\alpha = (\xi_1 L^{-1}[\xi_1 \psi_\alpha], h)/(\xi_1 L^{-1}[\xi_1 \psi_\alpha], \psi_\alpha) \qquad (\alpha = 2, 3, 4) \tag{7.33}$$

Once we have computed g_λ^+, Eq. (7.15) gives us the odd part g_λ^- of the eigensolution of Eq. (7.3); to be precise, each g_λ^+ ($\lambda \neq 0$) produces two different g_λ^- (opposite to each other in sign), because we can take for λ, in Eq. (7.12), either square root of λ^2. Now, from Eq. (7.27) we conclude that any h^- which can be written in the form

$$h^- = N^{-1}\xi_1 h^+ \qquad (\xi_1 h^+ = Nh^-) \tag{7.34}$$

with h^+ satisfying Eq. (7.27) and even, will have an expansion of the following form

$$h^- = \sum_{\substack{\alpha=0\\ \alpha\neq 1}}^{4} A_\alpha N^{-1}(\xi_1\psi_\alpha) + \int_{\lambda_0}^{\lambda_\infty} N^{-1}[\xi_1 g_\lambda^+(\boldsymbol{\xi})]A_\lambda^+\, d\lambda \tag{7.35}$$

or, due to Eqs. (7.14) and (7.15):

$$h^- = A_1\psi_1 + \sum_{\alpha=2}^{4} B_\alpha L^{-1}(\xi_1\psi_\alpha) + \int_{\lambda_0}^{\lambda_\infty} g_\lambda^- A_\lambda^-\, d\lambda \tag{7.36}$$

where we have called A_1 what was previously A_0/κ, B_α what was previously A_α ($\alpha = 2, 3, 4$) and set $A_\lambda^+ = -A_\lambda^-/\lambda$. Eq. (7.36) holds provided

$$|(Nh^-, h^-)| < \infty \tag{7.37}$$

as follows from Eqs. (7.26) and (7.34).

We are now in position to prove the following

THEOREM *Any function $h = h(\boldsymbol{\xi})$ whose even and odd parts h^+ and h^- satisfy Eqs.* (7.26) *and* (7.37) *can be expanded in the form*

$$h = \sum_{\alpha=0}^{4} A_\alpha\psi_\alpha + \sum_{\alpha=2}^{4} B_\alpha L^{-1}(\xi_1\psi_\alpha) + \left[\int_{-\lambda_\infty}^{-\lambda_0} + \int_{\lambda_0}^{\lambda_\infty}\right] g_\lambda(\boldsymbol{\xi})A_\lambda\, d\lambda \tag{7.38}$$

where $g_\lambda = g_\lambda^+ + g_\lambda^-$ are the eigensolutions of Eq. (7.3) *corresponding to $\lambda \neq 0$ and the coefficients are given by Eqs.* (7.33), *and*

$$\begin{aligned} A_\lambda &= (h, \xi_1 g_\lambda)\\ A_1 &= (h, \xi_1)/(\xi_1, \xi_1)\\ B_\alpha &= (h, \xi_1\psi_\alpha)/(\xi_1\psi_\alpha, L^{-1}[\xi_1\psi_\alpha]) \end{aligned} \tag{7.39}$$

provided the ψ_α satisfy Eq. (7.32) *and the g_λ are suitably normalized (we proceed as if $g_\lambda \neq g_{\lambda'}$ for $\lambda \neq \lambda'$, which is not true; the degeneracy is easily disposed of, however, and will not be mentioned at all in the proof).*

To prove the theorem, it is sufficient to decompose h in the form $h^+ + h^-$ and apply Eqs. (7.27) and (7.36) which give Eq. (7.38) with

$$A_\lambda = \frac{A_\lambda^+ + A_\lambda^-}{2} \qquad A_{-\lambda} = \frac{A_\lambda^+ - A_{-\lambda}^-}{2} \qquad (\lambda > 0) \tag{7.40}$$

provided we recall that $g_\lambda^+ = g_{-\lambda}^+$ and $g_\lambda^- = -g_{-\lambda}^-$. The computation of the coefficients follows from the fact that Eq. (7.3) gives, by standard manipulation:

$$(g_\lambda, \xi_1 g_{\lambda'}) = 0 \qquad (\lambda \neq \lambda') \tag{7.41}$$

In particular if $\lambda' = 0$:

$$(g_\lambda, \xi_1 \psi_\alpha) = 0 \qquad (\alpha = 0, 1, 2, 3, 4, \lambda \neq 0) \tag{7.42}$$

and, as a consequence:

$$(\xi_1 L^{-1}[\xi_1\psi_\alpha], g_\lambda) = (L^{-1}[\xi_1\psi_\alpha], \xi_1 g_\lambda) = \frac{1}{\lambda}(\xi_1\psi_\alpha, g_\lambda) = 0$$

$$(\alpha = 2, 3, 4; \lambda \neq 0) \tag{7.43}$$

By suitably normalizing g_λ, Eqs. (7.39) follow by taking the scalar products of Eq. (7.38) with $\xi_1 g_\lambda$, $\xi_1\psi_\alpha$, $\xi_1 L^{-1}(\xi_1\psi_{\alpha'})$.

We remark that the assumptions on h are unnecessarily strong. In fact, it is sufficient that the scalar products appearing in Eqs. (7.33) and (7.39) exist (in the sense of the theory of generalized functions, if necessary) for Eq. (7.38) to be valid (in the sense of generalized functions).

The result which has just been proved is strikingly similar to a completeness theorem for the eigensolutions of Eq. (7.3). In fact, in Eq. (7.38) we find all such eigensolutions corresponding to $\lambda \neq 0$, g_λ, and to $\lambda = 0$, ψ_α. The theorem says, however, that these eigensolutions are not sufficient to represent an arbitrary function h; we have to add three functions $L^{-1}(\xi_1\psi_\alpha)$ ($\alpha = 2, 3, 4$) which are *not* eigensolutions. This fact is extremely remarkable. In fact, we expect to be able to find a solution h of Eq. (7.1) which takes essentially arbitrary values at a point x (say $x = 0$); hence the result just discussed implies that it is *not* possible to represent an arbitrary solution of Eq. (7.1) as a superposition of separated variable solutions having the form shown in Eq. (7.2). The above theorem also says, however, that although the goal is missed one goes very close to it; it is sufficient to find three particular solutions h_α (= 2, 3, 4) of Eq. (7.1) which take at a given position x, say $x = 0$, exactly the values $L^{-1}(\xi_1\psi_\alpha)$, and add these to the set of solutions having the form shown in Eq. (7.2) to obtain the general solution of Eq. (7.1) as a linear superposition of the solutions of the resulting set. Three solutions of the required form, h_α ($\alpha = 2, 3, 4$) are

$$h_\alpha = x_1\psi_\alpha + L^{-1}(\xi_1\psi_\alpha) \tag{7.44}$$

as may be easily checked by inserting Eq. (7.44) into Eq. (7.1).

Hence we can prove the following

THEOREM *Any solution h of Eq.* (7.1) *such that the scalar products in Eqs.* (7.33) *and* (7.39) *exist (in the sense of the theory of generalized functions, if* *necessary) can be written as follows:*

$$h = \sum_{\alpha=0}^{4} A_\alpha \psi_\alpha + \sum_{\alpha=2}^{4} B_\alpha[x_1\psi_\alpha + L^{-1}(\xi_1\psi_\alpha)] + \int_{-\lambda_\infty}^{-\lambda_0} + \int_{\lambda_0}^{\lambda_\infty} g_\lambda(\boldsymbol{\xi}) e^{\lambda x_1} A_\lambda \, d\lambda$$

$$(0 \leqslant \lambda_0 < \lambda_\infty \leqslant \infty) \quad (7.45)$$

where ψ_α $(\alpha = 0, 1, 2, 3, 4)$ *are the five collision invariants,* $g_\lambda(\boldsymbol{\xi})$ *the eigenfunctions of Eq. (7.3) corresponding to* $\lambda = 0$ *and* $A_\alpha, B_\alpha, A_\lambda$ *arbitrary coefficients (as usual a sum must replace the integral if the values of* λ *form a discrete set).*

In fact, let h be a solution of Eq. (7.1), then scalarly multiplying both sides by $g_\lambda(\boldsymbol{\xi})$ we conclude:

$$\frac{d}{dx_1}(\xi_1 g_\lambda, h) = (g_\lambda, Lh) = (Lg_\lambda, h) = \lambda(\xi_1 g_\lambda, h) \quad (7.46)$$

or

$$(\xi_1 g_\lambda, h) = A_\lambda e^{\lambda x_1} \quad (7.47)$$

where A_λ is a constant. Analogously:

$$\frac{d}{dx}(\xi_1\psi_\alpha, h) = (\psi_\alpha, Lh) = 0 \quad (7.48)$$

or

$$(\xi_1\psi_\alpha, h) = \bar{A}_\alpha \quad (7.48')$$

where $\bar{A}_\alpha$ is a constant. Finally:

$$\frac{d}{dx_1}(\xi_1 L^{-1}[\xi_1\psi_\alpha], h) = (L^{-1}[\xi_1\psi_\alpha], Lh) = (\xi_1\psi_\alpha, h) = \bar{A}_\alpha \quad (\alpha = 2, 3, 4) \quad (7.49)$$

or

$$(\xi_1 L^{-1}[\xi_1\psi_\alpha], h) = \bar{A}_\alpha x_1 + \bar{B}_\alpha \quad (7.49')$$

where $\bar{B}_\alpha$ is a constant. Now h can be expanded, for any fixed x_1, according to Eq. (7.35) with coefficients $A_\alpha(x_1)$, $B_\alpha(x_1)$, $A_\lambda(x_1)$ which can be computed by means of Eqs. (7.33) and (7.37). By means of Eqs. (7.47), (7.48′) and (7.49′), we immediately find that:

$$A_\lambda(x_1) = A_\lambda e^{\lambda x_1}; \qquad A_\alpha(x_1) = A_\alpha, \qquad (\alpha = 0, 1);$$

$$A_\alpha(x_1) = A_\alpha + B_\alpha x_1, \qquad (\alpha = 2, 3, 4); \qquad B_\alpha(x_1) = B_\alpha \quad (7.50)$$

where A_λ is the constant appearing in Eq. (7.45) while A_α and B_α are constants which are simply related to $\bar{A}_\alpha$, $\bar{B}_\alpha$. Eq. (7.45) then follows.

The general solution given in Eq. (7.45) is made up of two parts, h_A and h_B, given by

$$h_A = \sum_{\alpha=0}^{4} A_\alpha \psi_\alpha + \sum_{\alpha=2}^{4} B_\alpha [x_1 \psi_\alpha + L^{-1}(\xi_1 \psi_\alpha)] \tag{7.51}$$

$$h_B = \int_{-\lambda_\infty}^{-\lambda_0} + \int_{\lambda_0}^{\lambda_\infty} e^{\lambda x_1} g_\lambda(\boldsymbol{\xi}) A_\lambda \, d\lambda \tag{7.52}$$

where the eigenvalues λ are of the order of the inverse of the mean free path l; in fact according to the discussion concerning the order of magnitude of the eigenvalues of L (Section 6), $Lh/h \simeq \bar{\xi}/l$ and, since $\xi_1 \simeq \bar{\xi}$, $\lambda \simeq l^{-1}$ immediately follows from Eq. (7.3). It is clear that h_B describes space transients which will be of importance in the neighbourhood of boundaries and become negligible a few mean free paths far from them. The circumstance that Eq. (7.52) contains exponentials with both $\lambda > 0$ and $\lambda < 0$ is exactly what is required to describe a decay either for $x_1 > \bar{x}_1$ or $x_1 < \bar{x}_1$, where $\bar{x}_1$ is the location of a boundary.

The general solution given by Eq. (7.45) then shows that, if the region where the gas is contained (either a half space or a slab of thickness d, because of the assumption that h is independent of two space coordinates) is much thicker than the mean free path ($d \gg l$) then h_B will be negligible except in boundary layers a few mean free paths thick. These layers receive the name of "Knudsen layers" or "kinetic boundary layers", to distinguish them from the Prandtl boundary layers, familiar from continuum fluid dynamics. Outside the Knudsen layers, the solution will be accurately described by the asymptotic part h_A, defined by Eq. (7.51), which we rewrite as follows:

$$h_A = a + \mathbf{b} \cdot \boldsymbol{\xi} + c(\xi^2 - 5RT_0) + \sum_{i=2}^{3} [x_1 \xi_i + L^{-1}(\xi_1 \xi_i)] \, d_i +$$
$$+ [x_1(\xi^2 - 5RT_0) + L^{-1}(\xi_1[\xi^2 - 5RT_0])] g \tag{7.53}$$

where a, $\mathbf{b}$, c, d_i, g are eight arbitrary constants and ψ_α has been taken in the form $\psi_4 = \xi^2 - 5RT_0$ in order to comply with Eq. (7.32), which for $\alpha = 0$, $\beta = 4$ gives ($\psi_0 = 1$):

$$(\xi_1^2, \psi_4) = 0 \tag{7.54}$$

It is interesting to compute the density ρ, velocity u_i, temperature T, stresses p_{ij} and heat flux q_i corresponding to h_A. To this end, we first remark that if we linearize the Boltzmann equation about a zero-velocity Maxwellian

f_0, with density ρ_0 and temperature T_0, we have

$$\rho = \int f_0(1 + h)\, d\boldsymbol{\xi} = \rho_0 + \int f_0 h\, d\boldsymbol{\xi} \tag{7.55}$$

$$u_i = \rho^{-1}\int \xi_i f_0(1 + h) = \rho^{-1}\int \xi_i f_0 h\, d\boldsymbol{\xi} = \rho_0^{-1}\int \xi_i f_0 h\, d\boldsymbol{\xi} + O(h^2) \tag{7.56}$$

$$T = (3R\rho)^{-1}\int \xi^2 f_0(1 + h)\, d\boldsymbol{\xi} = \rho_0 T_0 \rho^{-1} + (3R\rho)^{-1}\int \xi^2 f_0 h\, d\boldsymbol{\xi}$$

$$= T_0 - \rho_0^{-1}\int f_0 h\, d\boldsymbol{\xi} + (3R\rho_0)^{-1}\int \xi^2 f_0 h\, d\boldsymbol{\xi} + O(h^2)$$

$$= T_0 + (3R\rho_0)^{-1}\int (\xi^2 - 3RT_0) f_0 h\, d\boldsymbol{\xi} + O(h^2) \tag{7.57}$$

$$p_{ij} = \int c_i c_j f_0(1 + h)\, d\boldsymbol{\xi} = p_0\, \delta_{ij} + \int \xi_i \xi_j f_0 h\, d\boldsymbol{\xi} + O(h^2) \tag{7.58}$$

$$q_i = \frac{1}{2}\int c_i c^2 f_0(1 + h)\, d\boldsymbol{\xi} = \frac{1}{2}\int \xi_i \xi^2 f_0 h\, d\boldsymbol{\xi} - \frac{1}{2}\int u_i \xi^2 f_0\, d\boldsymbol{\xi} -$$

$$- \frac{1}{2}\int \xi_i \xi_k u_k f_0\, d\boldsymbol{\xi} + O(h^2)$$

$$= \frac{1}{2}\int \xi_i \xi^2 f_0 h\, d\boldsymbol{\xi} - \tfrac{5}{2} R\rho_0 T_0 u_i + O(h^2)$$

$$= \frac{1}{2}\int \xi_i(\xi^2 - 5RT_0) f_0 h + O(h^2) \tag{7.59}$$

where $O(h^2)$ denotes terms of second order in the perturbation and hence negligible in a linearized theory. If we insert $h = h_A$ and neglect the terms $O(h^2)$ we find:

$$\rho^{(A)} = \rho_0[1 + a - 2RT_0(c + gx_1)] \tag{7.60}$$

$$u_i^{(A)} = b_i RT_0 + x_1\, d_i RT_0 \qquad (i = 2, 3) \tag{7.61}$$

$$u_1^{(A)} = b_1 RT_0 \tag{7.62}$$

$$T^{(A)} = T_0[1 + 2RT_0(c + gx)] \tag{7.63}$$

$$p_{ij}^{(A)} = RT_0\rho_0(1 + a)\delta_{ij} - \mu(d_j\, \delta_{i1} + d_i\, \delta_{j1})RT_0 \tag{7.64}$$

$$q_j^{(A)} = -\delta_{j1}\kappa 2RT_0{}^2 g \tag{7.65}$$

where

$$\mu = -(RT_0)^{-1}\int \xi_1\xi_2 f_0 L^{-1}(\xi_1\xi_2)\, d\boldsymbol{\xi} = -(RT_0)^{-1}\int \xi_1\xi_3 f_0 L^{-1}(\xi_1\xi_3)\, d\boldsymbol{\xi} \quad (7.66)$$

$$\kappa = (4RT_0^2)^{-1}\int f_0\xi_1\xi^2 L^{-1}[\xi_1(\xi^2 - 5RT_0)]\, d\boldsymbol{\xi} \quad (7.67)$$

and the invariance of L with respect to reflections in velocity space has been used to conclude that some of the integrals involving L^{-1} vanish.

Eqs. (7.64), (7.65), (7.61), (7.63) show that

$$p_{j1}^{(A)} = p_{1j}^{(A)} = -\mu \frac{\partial u_j^{(A)}}{\partial x_1} \qquad (j \neq 1); \quad (7.68)$$

$$p_{23} = 0; \qquad p_{11} = p_{22} = p_{33} = \rho^{(A)}RT^{(A)} + O(h^2) \quad (7.69)$$

$$q_1^{(A)} = -\kappa \frac{\partial T^{(A)}}{\partial x_1}; \qquad q_2^{(A)} = q_3^{(A)} = 0 \quad (7.70)$$

These equations show that a few mean free paths from a solid wall (where the distribution $f = f_0(1 + h_A)$ applies), the stress tensor and heat flux vector are related to the velocity and temperature by the Navier-Stokes-Fourier relations, Eq. (II.8.27), where the viscosity coefficient μ and the heat conduction coefficient κ are given by Eqs. (7.66) and (7.67). The value of the second viscosity coefficient does not follow from the above treatment since $\partial v_k/\partial x_k = \partial v_1/\partial x_1 = 0$. However, Eq. (II.8.27) gives for the trace of p_{ij}:

$$p_{ii} = 3p - (2\mu + 3\lambda)\frac{\partial v_k}{\partial x_k} \quad (7.71)$$

where p, the pressure, is p_{ii} evaluated at equilibrium. Eqs. (II.8.13) and (II.8.16), however, say that (for a monatomic gas) $p_{ii} = 3p$, no matter what f may be; hence $\lambda = -(\frac{2}{3})\mu$ so that the so called Stokes relation is satisfied.

The above results are satisfactory; we found that when the mean free path is negligible compared to macroscopic lengths, the constitutive equations for a Navier-Stokes compressible fluid are satisfied (see Section 11 for an extension to three-dimensional problems). In addition, general formulas, Eq. (7.66) and (7.67) have been obtained for the viscosity and heat conduction coefficients. It is to be remarked that these transport coefficients turn out to depend only upon temperature and molecular constants (density drops out because it is contained in f_0 with exponent 1 and in L^{-1} with exponent -1); this independence of viscosity from density was one of the first successes of kinetic theory, since it was predicted before any experimental measurement was made.

It is also clear that both κ and μ are proportional to the mean free path (since $Lh \sim (\bar{\xi}/l)h$ according to the discussion in Section 6); this explains why the Prandtl number

$$\mathrm{Pr} = c_p(\mu/\kappa) \tag{7.74}$$

where c_p is the specific heat at constant pressure, is always of order unity for a gas. In the case of Maxwellian molecules it is very easy to calculate μ, κ and hence Pr because $\xi_1\xi_2$ and $\xi_1(\xi^2 - 5RT_0)$ are eigenfunctions of the collision operator with eigenvalues λ_{02} and λ_{11}, given by Eq. (6.9). Accordingly, since $c_p = 5R/2$ for a monatomic gas, we obtain

$$\mu = -\rho_0 RT_0/\lambda_{02}, \qquad \kappa = -(\tfrac{5}{2})\rho_0 RT_0{}^2/\lambda_{11} = -c_p\rho_0 RT_0/\lambda_{11}$$
$$\mathrm{Pr} = \lambda_{11}/\lambda_{02} \tag{7.73}$$

and the latter ratio can be shown to be $\frac{2}{3}$ from Eq. (6.9) (see also Table I in Section 6). For non-Maxwellian molecules, the Prandtl number (which for potential laws other than power is slightly temperature dependent) is always close to $\frac{2}{3}$.

These results are in good agreement with experimental data.

The main result of this section is Eq. (7.45), which exhibits the structure of the general solution of the linearized Boltzmann equation for one-dimensional problems. The next step would be to find the values of λ_0 and λ_∞ for any assigned operator L. The main question is whether $\lambda_0 = 0$ or $\lambda_0 \neq 0$; in the second case the decay to the asymptotic solution h_A is exponential, on a length scale independent of the boundary data. A qualitative discussion is possible when the splitting $Lh = Kh - \nu h$, with a completely continuous K, holds. In fact, we can apply Weyl's theorem or a suitable extension of it [3] to conclude that if K is completely continuous, the set of singular points of $(L - \lambda\xi_1)^{-1} = \{K - [\nu(\xi) + \lambda\xi_1]\}^{-1}$ (i.e. the spectrum of our problem) differs from the set of zeros of $\nu(\xi) + \lambda\xi_1$ only at a set of discrete points.

This is the case for rigid spheres and power law interactions with angular cutoff. For these cases we conclude that the continuous spectrum is the set of values taken by the function $-\nu(\xi)/\xi_1$ when the components of $\boldsymbol{\xi}$ range from $-\infty$ to $+\infty$. Accordingly:

$$\lambda_0 = \lim_{\xi_1\to\infty} \frac{\nu(\xi)}{\xi_1} \qquad \lambda_\infty = \lim_{\xi_1\to 0} \frac{\nu(\xi)}{\xi_1} \tag{7.76}$$

In particular, for rigid spheres $\lambda_0 > 0$, $\lambda_\infty = \infty$ and for power law potentials with angular cutoff $\lambda_0 = 0$, $\lambda_\infty = \infty$ (see Fig. 22). The question whether there are eigenvalues in the intervals $0 < |\lambda| < \lambda_0$ for rigid spheres does not seem to have been investigated.

We finally comment on the extension of these results to the case of neutron transport. It is obvious that if fission is absent or is dominated by absorption in such a way that L is a negative operator, then the above machinery is not needed since the spectrum can be obtained from the eigenvalue equation for the operator $(-L)^{-\frac{1}{2}}\xi_1(-L)^{-\frac{1}{2}}$; the eigensolutions form a complete set. In the purely scattering case the above proof applies in an even simpler form because no odd collision invariants exist. The above treatment fails, however, if the spectrum of L contains points of the positive real semi-axis; in this case we can only say that, if the fission term is a completely continuous operator, then (by the extended Weyl theorem [3]) the nonreal eigenvalues form a discrete set.

8. The general case

The fairly general results obtained in Sections 6 and 7 lead us to enquire whether it is possible to write the general solution of the linearized Boltzman

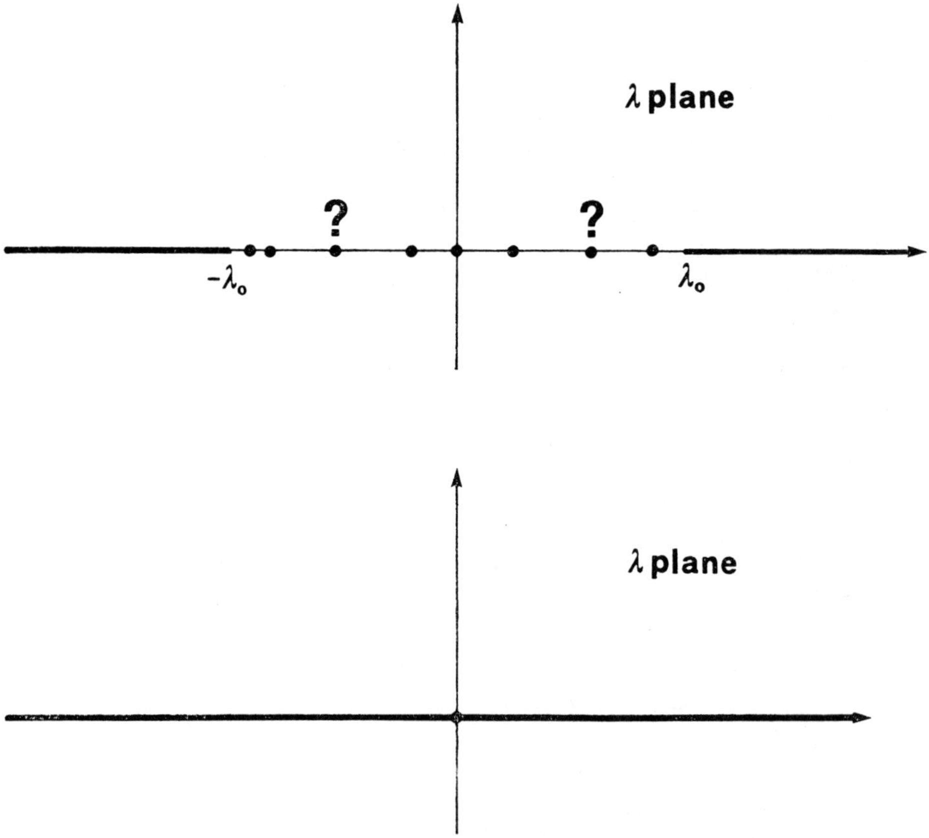

Fig. 22. Spectrum of the eigenvalue problem (7.3). The two figures refer to rigid spheres and power law potentials with angular cutoff.

equation as a superposition of "elementary solutions" with separated variables. When we separate the variables we find, in general, that the space and time dependence is exponential

$$h = g(\boldsymbol{\xi})e^{i\omega t - i\mathbf{k}\cdot\mathbf{x}} \tag{8.1}$$

where the function $g(\boldsymbol{\xi})$ describing the dependence upon molecular velocity satisfies:

$$(i\omega - i\mathbf{k}\cdot\boldsymbol{\xi})g = Lg \tag{8.2}$$

and ω, $\mathbf{k}$ are, in general, a complex scalar and a complex vector ($\omega = -i\lambda$, $\mathbf{k} = 0$ in Section 6; $\omega = 0$, $\mathbf{k} = (i\lambda, 0, 0)$ in Section 7). Eq. (8.2) will have solutions only when particular relations between ω and $\mathbf{k}$ are satisfied; corresponding to the allowed values of ω and $\mathbf{k}$ it is possible to find g's which are either in $\mathscr{H}$ ("eigensolutions") or not ("generalized eigensolutions").

In the former case the relation between ω and $\mathbf{k}$ is usually called the dispersion relation and the solution $g \exp(i\omega t - i\mathbf{k}\cdot\mathbf{x})$ a normal mode. Both the "eigensolutions" and the "generalized eigensolutions" should combine to form the general solution of the linearized Boltzmann equation:

$$h = \int_{D(\omega,\mathbf{k})} A(\omega, \mathbf{k})\exp(i\omega t - i\mathbf{k}\cdot\mathbf{x})g(\boldsymbol{\xi}; \omega, \mathbf{k})\, dD \tag{8.3}$$

where g is a solution of Eq. (8.1) properly normalized and the integral is a set integral over a complete set $D(\omega, \mathbf{k})$ of allowed values of ω, $\mathbf{k}$ (accordingly it will contain summations over discrete values of ω and $\mathbf{k}$ and integrals over continuous sets). In general, Eq. (8.3) will not be the most general solution, as Eq. (7.45) shows, but it is reasonable to expect that a study of the set $D(\omega, \mathbf{k})$ will shed light on this question as well.

In the study of Eq. (8.2), there are several possibilities because we can fix three of the four complex parameters ω, k_1, k_2, k_3 and find out what is the spectrum of the values of the remaining one. In particular, we can fix $\mathbf{k}$ and investigate ω; a case which has been studied in detail is when $\mathbf{k}$ is taken to be real. This case arises in the study of free sound waves [26–28] and is a standard eigenvalue problem for the operator $L + i\mathbf{k}\cdot\boldsymbol{\xi}$:

$$(L + i\mathbf{k}\cdot\boldsymbol{\xi})g = \lambda g \qquad (\lambda = i\omega) \tag{8.4}$$

There is a major difficulty, however, because the operator $L + i\mathbf{k}\cdot\boldsymbol{\xi}$ is not self-adjoint. This implies that no general theorems on the completeness of the eigenfunctions are known. In addition, it is not an easy matter to make a qualitative study of the spectrum by means of Weyl's theorem when $L = K - \nu$ with K compact. In fact, Weyl's theorem states that, if A is closed and self-adjoint and K completely continuous and self-adjoint, then

$$\sigma_e(A + K) = \sigma_e(A) \tag{8.5}$$

where $\sigma_e(A)$ denotes the essential spectrum, that is, the set of limit points of the spectrum, of A. But $A = -\nu + i\mathbf{k}\cdot\boldsymbol{\xi}$ is not self-adjoint, and Weyl's theorem does not hold [29]. In order to avoid this unpleasant circumstance, several modified definitions of σ_e have been proposed; for some of them, Weyl's theorem does not hold in its full generality [29, 30]. As noticed by Nicolaenko [28], this circumstance generated a great deal of confusion in some papers using the invariance of σ_e in connection with the Boltzmann equation; several authors [27, 31, 34] have claimed that

$$\sigma_e([-\nu + i\mathbf{k}\cdot\boldsymbol{\xi}] + K) = \sigma_e([-\nu + i\mathbf{k}\cdot\boldsymbol{\xi}]) \tag{8.6}$$

thus following a definition of essential spectrum for which Weyl's theorem holds, but implicitly assuming σ_e to be the set of limit points of the spectrum; that is, the complement of σ_e to be the set where $(L + i\mathbf{k}\cdot\boldsymbol{\xi})^{-1}$ exists except at isolated points. The basic error, as noticed by Nicolaenko [28], is that σ_e can exclude whole open sets of the point spectrum. Nicolaenko was also able to show that this does not happen for the operator under study.

Hence the continuous spectrum of $L + i\mathbf{k}\cdot\boldsymbol{\xi}$, when $\mathbf{k}$ is real and $L = K - \nu$ with K completely continuous (as in the case of rigid spheres), is made up of the points

$$\lambda = -\nu(\boldsymbol{\xi}) + i\mathbf{k}\cdot\boldsymbol{\xi} \tag{8.7}$$

where $\boldsymbol{\xi}$ ranges over all the possibility values (see Fig. 23).

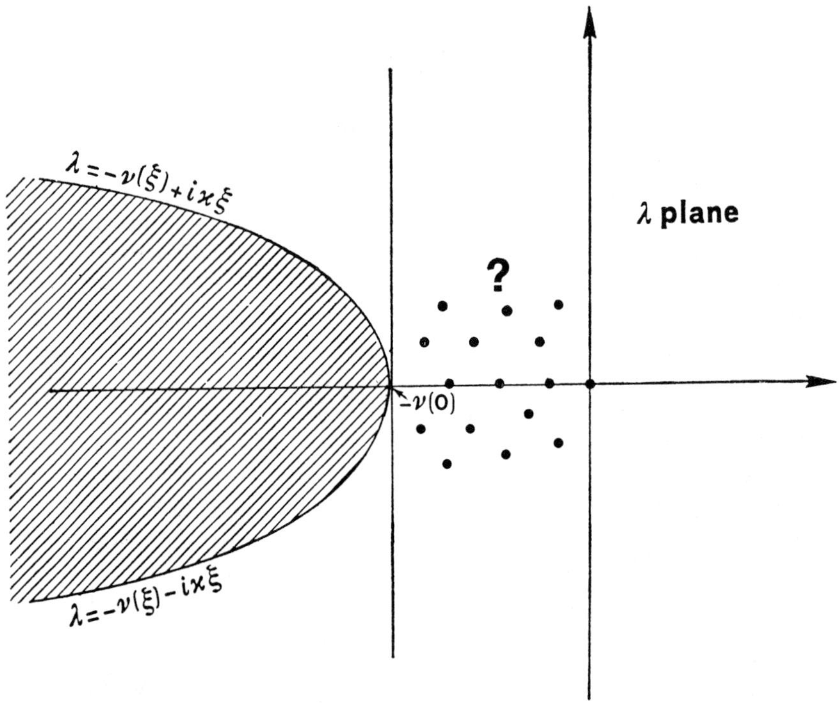

Fig. 23. Spectrum of the eigenvalue problem (8.4) for a fixed real value of **k**.

Nothing seems to have been proved about the discrete spectrum except for a result due to McLennan [32] in the case of rigid spheres. According to this result, $L + i\mathbf{k} \cdot \boldsymbol{\xi}$ has, for a sufficiently small $\mathbf{k}$, five isolated point eigenvalues (not necessarily distinct) $\lambda^{(\alpha)}(\mathbf{k})$ ($\alpha = 0, 1, 2, 3, 4$) which reduce to $\lambda^{(\alpha)} = 0$ for $\mathbf{k} = 0$ and are analytic in $|\mathbf{k}|$. This follows from a theorem on the perturbation of spectra [3], which states that if λ is a multiple point eigenvalue of a self-adjoint operator L with multiplicity m and L' a self-adjoint operator which satisfies

$$\|L'h\| \leqslant M(\|h\| + \|Lh\|) \tag{8.8}$$

(M is independent of h, for any h in the domain of both L and L') then the operator $L + zL'$ has, for sufficiently small values of the complex parameter z, m isolated point eigenvalues which reduce to λ for $z = 0$ and are analytic in z.

In order to apply this result to the fivefold eigenvalue $\lambda = 0$ of L we have only to put $\mathbf{k} = k\mathbf{e}$ ($\mathbf{e}$ a real unit vector), $z = ik$, $L' = \mathbf{e} \cdot \boldsymbol{\xi}$ and prove that Eq. (8.8) applies. But this is a very simple consequence of the following facts:

1. $L = K - \nu$
2. $\nu(\xi)$ grows linearly for large ξ in the case of rigid spheres [hence an M_1 exists such that $|\mathbf{e} \cdot \boldsymbol{\xi}| < M_1\nu(\xi)$].
3. K is completely continuous, and hence bounded, for rigid spheres (thus an M_2 exists such that $\|Kh\| < M_2\|h\|$).
4. The triangle inequality ($\|f + g\| \leqslant \|f\| + \|g\|$).

Using these properties, we have:

$$\begin{aligned} \|L'h\| &= \|(\mathbf{e} \cdot \boldsymbol{\xi})h\| \leqslant M_1 \|\nu h\| = M_1 \|(K - L_1)h\| \leqslant M_1(\|Kh\| + \|Lh\|) \\ &< M_1(M_2 \|h\| + \|Lh\|) \leqslant M(\|h\| + \|Lh\| \qquad (M = \max(M_1, M_1M_2)) \end{aligned} \tag{8.9}$$

as was to be shown. This result is only local, since the radius of convergence of the series for $\lambda^{(\alpha)} = \lambda^{(\alpha)}(k)$ is presumably not larger than λ_0 defined by Eq. (7.67). In fact, for z real, rather than purely imaginary, $L + z(\boldsymbol{\xi} \cdot \mathbf{e})$ has a continuous spectrum which reaches the origin for $z = \pm\lambda_0$. For the same reason it is to be expected that for potentials with angular cutoff ($\lambda_0 = 0$) there are no discrete eigenvalues expandable into a series of powers of k.

This completes the discussion of the case of $\mathbf{k}$ real and fixed. The completeness of the eigenfunctions does not seem to have been discussed in the literature, although it should not be difficult to obtain results by means of semigroup techniques.

The other case which has been considered in detail is when $\lambda = i\omega$ is fixed (with ω real) and $\mathbf{k} = k\mathbf{e}$ where $\mathbf{e}$ is a real unit vector, also fixed. Here k plays the role of an eigenvalue, but occurs as a multiplier of the unbounded operator

$\boldsymbol{\xi} \cdot \mathbf{e}$ and this, of course, constitutes an additional difficulty. The equation to be studied is:

$$(i\omega - L)g = ik(\boldsymbol{\xi} \cdot \mathbf{e})g \tag{8.10}$$

If $L = K - \nu$, with K completely continuous, then Nicolaenko [28] was able to prove that the continuous spectrum for the generalized eigenvalue problem in Eq. (8.10) is given by the values assumed by

$$k = \frac{\omega - i\nu(\xi)}{\boldsymbol{\xi} \cdot \mathbf{e}} \tag{8.11}$$

when $\boldsymbol{\xi}$ ranges over all the possible values (see Fig. 24).

A more complicated situation arises when $\mathbf{k}$ cannot be written in the form $\mathbf{k} = k\mathbf{e}$ with $\mathbf{e}$ real unit vector. Then the eigenvalue problem is decidedly

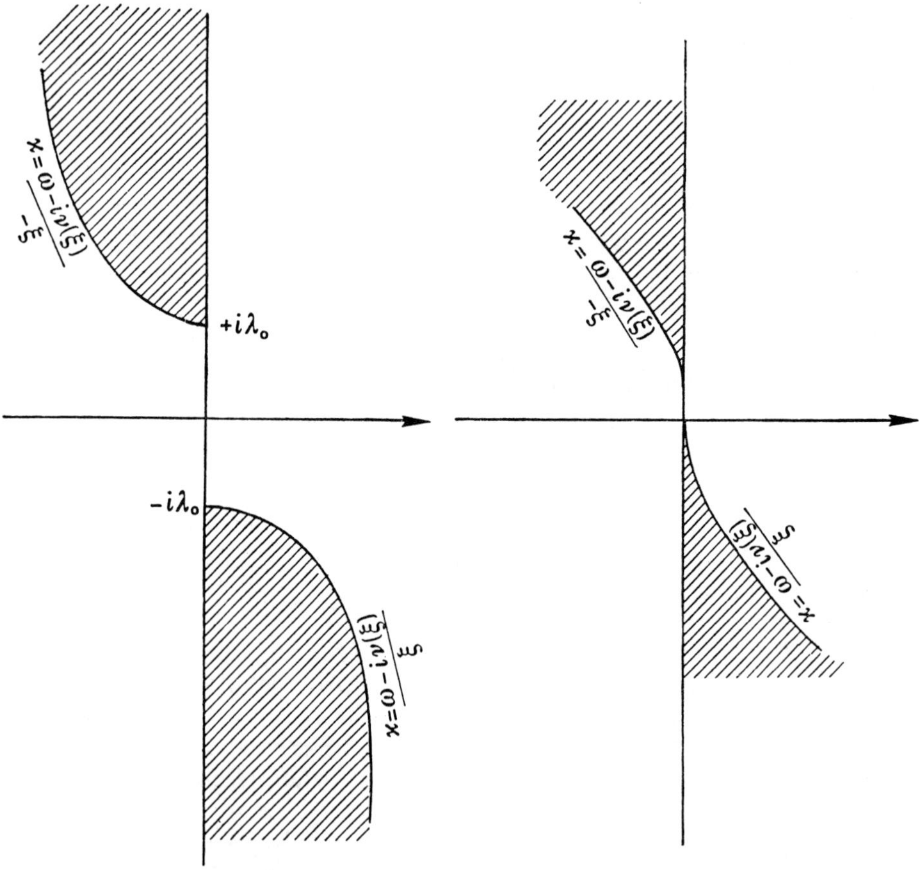

Fig. 24. Spectrum of the eigenvalue problem (8.10) for a fixed real value of ω.

nonstandard. The case $\omega = 0$ corresponding to

$$-i\mathbf{k} \cdot \boldsymbol{\xi} g = Lg \tag{8.12}$$

is particularly interesting because it corresponds to steady state problems. When $\mathbf{k} = k\mathbf{e}$($\mathbf{e} \cdot \mathbf{e} = 1$, $\mathbf{e}$ real), then, taking the first Cartesian axis in velocity space along $\mathbf{e}$ and putting $-ik = \lambda$ we are back to Eq. (7.3) and, as we know, the solutions of Eq. (8.12) do not form a complete set. Nothing is known, in general, when $\mathbf{e}$ is not real.

It is to be remarked, however, that in general the set of all values of ω and $\mathbf{k}$ for which Eq. (8.2) has a solution is too large, in the sense that not all the solutions of the set are linearly independent. This circumstance makes it hard to draw conclusions from a study of "eigenvalue problems" with more than two complex parameters. An alternative is to use the Laplace-Fourier transform

$$h = \int_0^\infty \int_R e^{-i\omega t + i\mathbf{k}\cdot\mathbf{x}} h(\mathbf{x}, t)\, d\mathbf{x}\, dt \tag{8.13}$$

Then

$$(i\omega - i\mathbf{k} \cdot \boldsymbol{\xi})\tilde{h} = L\tilde{h} + \tilde{g}_0 \tag{8.14}$$

where the source term $\tilde{g}_0$ contains boundary and initial values. Eq. (8.14) is the inhomogeneous equation corresponding to Eq. (8.2), with $\mathbf{k}$ real. The study of the spectrum and eigenfunctions of Eq. (8.2) leads, at least in principle, to the construction of the inverse operator $(L - i\omega + i\mathbf{k} \cdot \boldsymbol{\xi})^{-1}$ (see Section 11 for the steady case, $\omega = 0$).

Thus $\tilde{h} = (L - i\omega + i\mathbf{k} \cdot \boldsymbol{\xi})^{-1}\tilde{g}_0$ is known, if $\tilde{g}_0$ is. The difficulty is that $\tilde{g}_0$ is not known, in general, for a boundary value problem; thus a solution can be found only by suitable techniques in particular cases (in particular, some half-space problem can be solved by the so-called Wiener-Hopf technique [35]).

9. Linearized kinetic models

In Chapter II, Section 10, we discussed the possibility of replacing the collision term in the Boltzmann equation by a simpler expression, called a collision model. The idea was that the large amount of detail of the two body interaction contained in the collision term and reflected, for example, in the details of the spectrum of the linearized operator, is not likely to influence significantly the values of many experimentally measured quantities. This led to the BGK model and the variants which were discussed in Chapter II.

The same possibility arises in connection with the linearized and linear Boltzmann equations, for which satisfactory and systematic methods for constructing models have been devised.

The simplest model for the linearized collision operator is obtained by linearizing the BGK model. If we let $f = f_0(1 + h)$ in Eq. (II.10.3) and neglect powers of h higher than first, the linearized BGK model turns out to be given by

$$Lh = \nu\left[\sum_{\alpha=0}^{4} \psi_\alpha(\psi_\alpha, h) - h\right] \tag{9.1}$$

where the collision invariants ψ_α are normalized in such a way that

$$(\psi_\alpha, \psi_\beta) = \delta_{\alpha\beta} \qquad (\alpha, \beta = 0, 1, 2, 3, 4) \tag{9.2}$$

and (,) denotes the scalar product in $\mathscr{H}$, defined by Eq. (1.9). It is obvious that Eq. (9.1) exhibits an operator with a structure definitely simpler than the true linearized operator.

We remark that Eq. (9.1) can also be written as follows:

$$Lh = \nu(\Pi h - h) = -\nu(I - \Pi)h \tag{9.3}$$

where Π is the projection operator onto the five dimensional space $\mathscr{F}$ spanned by the ψ's, and I is the identity operator (accordingly, $I - \Pi$ is the projector onto W, the orthogonal complement of $\mathscr{F}$ in $\mathscr{H}$). Eq. (9.1), or (9.2), implies that the properties expressed by Eqs. (1.10), (1.12), (1.15) and, consequently, (1.16) are satisfied by the linearized BGK operator. This is almost evident for Eqs. (1.10), (1.12) and (1.16) while Eq. (1.15) follows from Eq. (9.2), by noticing that

$$(h, Lh) = -\nu(h, (I - \Pi)h) = -\nu\|(I - \Pi)h\|^2 \leqslant 0$$

where equality obviously holds if and only if $(I - \Pi)h = 0$, or

$$h = \sum_\alpha c_\alpha \psi_\alpha.$$

A systematic procedure for improving the linearized BGK model and characterizing the latter as the first step in a hierarchy of models approximating the collision operator for Maxwell molecules with arbitrary accuracy (in a suitable norm) was proposed by Gross and Jackson [36]. The idea is to start from an expansion of h into a series of eigenfunctions of the collision operator for Maxwell molecules (Section 6), which form a complete set of orthogonal functions:

$$h = \sum_{\alpha=0}^{\infty} \psi_\alpha(\psi_\alpha, h) \tag{9.4}$$

where α is a single label which summarizes the triplet (n, l, m) in such a way that the collision invariants ψ_α correspond to $\alpha = 0, 1, 2, 3, 4$. Then, since, for Maxwell molecules, $L\psi_\alpha = \lambda_\alpha \psi_\alpha$ by definition:

$$Lh = \sum_{\alpha=0}^{\infty} L\psi_\alpha(\psi_\alpha, h) = \sum_{\alpha=0}^{\infty} \lambda_\alpha \psi_\alpha(\psi_\alpha, h) \tag{9.5}$$

A systematic procedure for approximating L consists in partially destroying the fine structure of the spectrum of L by collapsing all the eigenvalues corresponding to $\alpha > N$ into a single eigenvalue, which we shall denote by $-\nu_N$ (remember that $\lambda_\alpha \leqslant 0$).

This amounts to replacing L by an approximate operator L_N defined as follows:

$$L_N h = \sum_{\alpha=0}^{N} \lambda_\alpha \psi_\alpha(\psi_\alpha, h) - \nu_N \sum_{\alpha=N+1}^{\infty} \psi_\alpha(\psi_\alpha, h) \tag{9.6}$$

Eq. (9.4) gives:

$$\sum_{\alpha=N+1}^{\infty} \psi_\alpha(\psi_\alpha, h) = h - \sum_{\alpha=0}^{N} \psi_\alpha(\psi_\alpha, h) \tag{9.7}$$

and Eq. (9.6) becomes

$$L_N h = \sum_{\alpha=0}^{N} (\lambda_\alpha + \nu_N)\psi_\alpha(\psi_\alpha, h) - \nu_N h \tag{9.8}$$

In particular, if $N = 4$, $\lambda_\alpha = 0$ for $0 \leqslant \alpha \leqslant N$ and, consequently, Eq. (9.8) reduces to Eq. (9.1) (with $\nu_4 = \nu$); by taking N larger and larger, we include more and more details of the spectrum of L into the model. If we take $N = 9$ by including the five eigenfunctions corresponding to $n = 0$, $l = 2$ in Eq. (6.10), we obtain the linearized version of the ES model, Eq. (II.10.7).

The above procedure applies only to the case of Maxwell's molecules. However, a slight generalization of the expansion (9.5) is capable of producing collision models in correspondence with any kind of linearized collision operator [37]. In fact, nothing prevents our using Eq. (9.4) even if the ψ_α (eigenfunctions of the Maxwellian collision operator) are not the eigenfunctions of the operator under consideration. Applying L to both sides of Eq. (9.4) gives

$$Lh = \sum_{\beta=0}^{\infty} L\psi_\beta(\psi_\beta, h) \tag{9.9}$$

but we cannot perform the second step in Eq. (9.5). We can, however, expand the right hand side of Eq. (9.9) into a series involving the ψ_α and obtain

$$Lh = \sum_{\alpha,\beta=0}^{\infty} \lambda_{\alpha\beta}(\psi_\beta, h)\psi_\alpha \tag{9.10}$$

where

$$\lambda_{\alpha\beta} = (\psi_\alpha, L\psi_\beta) = \lambda_{\beta\alpha} \tag{9.11}$$

Eq. (9.10) generalizes Eq. (9.5) and reduces to the latter when $\lambda_{\alpha\beta} = \lambda_\alpha\, \delta_{\alpha\beta}$. If we now introduce the approximation $\lambda_{\alpha\beta} = -\nu_N\, \delta_{\alpha\beta}$ for $\alpha, \beta > N$, we obtain the model

$$L_N h = \sum_{\alpha,\beta=0}^{N} (\nu_N\, \delta_{\alpha\beta} + \lambda_{\alpha\beta})\psi_\alpha(\psi_\beta, h) - \nu_N h \tag{9.12}$$

which generalizes Eq. (9.8) to operators other than Maxwell's. Taking $N = 4$ gives the BGK model again.

We remark that L_N can be written in the form $L_N = K_N - \nu_N I$, where ν_N is a constant, I is the identity operator, and K_N maps any function onto the finite dimensional space $\mathscr{R}_N$ spanned by the $\psi_\alpha(\alpha \leqslant N)$. If we write the eigenvalue equation, Eq. (6.7), for L_N and take the projections of this equation onto $\mathscr{R}_N$, and its orthogonal complement, we find that the two equations are uncoupled; it is then easy to show that the spectrum comprises the N eigenvalues between $-\nu_N$ and 0 (with polynomial eigenfunctions; in particular, the ψ_α in the case of the model defined by Eq. (9.8)) and an eigenvalue $-\nu_N$, infinitely many times degenerate.

We recall now that the collision operators for hard spheres and hard potentials with angular cutoff are unbounded and display a continuous spectrum; the operators for hard potentials without cutoff are also unbounded. If these features of the operator have any influence on the solution of particular problems, this influence is lost when we adopt the models described by Eq. (9.12) (or (9.8) as a particular case). It is therefore convenient to introduce and investigate models which retain the abovementioned features of the linearized collision operator; this can be done in several ways.

Conceptually, the simplest procedure is based upon exploiting the fact that we either know (rigid spheres and angular cutoff) or conjecture (potentials with finite range) that $L = K - \nu I$ where the operator $K^* = \nu^{-\frac{1}{2}} K \nu^{-\frac{1}{2}}$ is self-adjoint and completely continuous in $\mathscr{H}$ (see Section 6); accordingly, the kernel of K^* can be expanded into a series of its square summable eigenfunctions $\varphi_\alpha \nu^{\frac{1}{2}}$ (such that $K\varphi_\alpha = \mu_\alpha \nu \varphi_\alpha$). In other words, we can write

$$Kh = \nu(\xi) \sum_{\alpha=0}^{\infty} \mu_\alpha \varphi_\alpha (\nu \varphi_\alpha, h) \tag{9.13}$$

Truncating this series (degenerate kernel approximation), we obtain the model [38]:

$$L_N h = \nu(\xi) \sum_{\alpha=0}^{N} \mu_\alpha \varphi_\alpha (\nu \varphi_\alpha, h) - \nu(\xi) h \tag{9.14}$$

This operator automatically satisfies the basic requirements, expressed by Eqs. (1.10), (1.12), (1.15), (1.16). Since the first five φ_α are the collision invariants ψ_α and correspond to $\mu_\alpha = 1$ $(0 \leqslant \alpha \leqslant 4)$, if we take $N = 4$ in Eq. (9.14), we have

$$L_4 h = \nu(\xi) \sum_{\alpha=0}^{4} \psi_\alpha (\nu \psi_\alpha, h) - \nu(\xi) h \tag{9.15}$$

where the collision invariants are normalized according to

$$(\psi_\alpha, \nu \psi_\beta) = \delta_{\alpha\beta} \tag{9.16}$$

Eq. (9.15) is just the linearized version of the nonlinear model with velocity-dependent collision frequency which was briefly discussed in Chapter II, Section 10.

If we want to obtain models corresponding to $N \geqslant 5$ we are faced with the difficulty that we have no analytical expressions for the φ_α ($\alpha \geqslant 5$); it is true that we can compute them numerically but this is obviously a significant complication. Also, the procedure by which we have derived Eqs. (9.14) and (9.15) shows that $\nu(\xi)$ is fixed by the original molecular model; hence we do not have any parameters to be adjusted in order to reproduce the correct viscosity and heat conduction coefficients (which, as we know from Eqs. (7.68) and (7.69) are determined by the operator L in a unique fashion). We remark that the BGK model, as constructed by the Gross and Jackson method, contains the adjustable parameter ν_N which can be used to reproduce the correct value of one of the two transport coefficients. In order to obtain the same flexibility, we can introduce an adjustable parameter by a slight modification of the above procedure; instead of simply truncating the series in Eq. (9.13) (i.e., putting $\mu_\alpha = 0$ for $\alpha > N$), we can put $\lambda_\alpha = k_N$ for $\alpha > N$ (with the understanding that $k_N \to 0$ for $N \to \infty$). The result is exactly the same model as in Eq. (9.14), except for the fact that $\nu(\xi)$ is replaced by $(1 - k_N)\nu(\xi)$ and μ_α by $(\mu_\alpha - k_N)/(1 - k_N)$; in particular, for $N = 4$, the only change is an adjustable factor in front of $\nu(\xi)$, which gives the model the desired flexibility (but shifts the continuous spectrum).

A greater flexibility can be obtained by using higher order models; here we meet the abovementioned trouble that we do not know the φ_α ($\alpha \geqslant 5$). The difficulty can be avoided by a procedure analogous to the non-Maxwell modeling considered above; that is, we take a complete set of orthogonal functions and expand everything in terms of these functions, as we did above (Eq. (9.10)). Of course, if we want to retain the basic features of the models appearing in Eqs. (9.14) and (9.15), we have to make a proper choice of the set; the simplest one is to take polynomials orthogonalized with respect to the weight $f_0(\xi)\nu(\xi)$, f_0 being the basic Maxwellian. The simplest model again takes the form shown in Eq. (9.15), but now we can extend it to arbitrarily high order without much trouble (the only things to be computed numerically are coefficients, not functions); the procedure is obvious and will not be described in detail.

A useful procedure for constructing models can be based upon the requirement that the model be able, for small mean free paths, to reproduce the behavior not only of some coefficients but also of the distribution function itself. According to the results of Section 7, the solutions of the Boltzmann equation for steady one-dimensional problems take on the form $h = h_A$ a few mean free paths far from solid boundaries, h_A being given by Eq. (7.51). Hence, in order to obtain the correct asymptotic behavior from a collision model, we should have $L_N^{-1}(\xi_1\psi_\alpha) = L^{-1}(\xi_1\psi_\alpha)$, if L_N is the collision operator replacing L. Recalling that L_N^{-1} and L^{-1} are the inverses of L and L_N in the subspace W orthogonal to the collision operators, the above condition becomes:

$$L_N^{-1}(\xi_i\psi_\alpha - \Pi[\xi_i\psi_\alpha]) = L^{-1}(\xi_i\psi_\alpha - \Pi[\xi_i\psi_\alpha]) \tag{9.17}$$

where $i = 1$ and Π is the projector onto $\mathscr{F}$, the subspace spanned by the ψ_α. Since the first axis is not privileged, Eq. (9.17) should also be imposed for $i = 2, 3$. The condition expressed by Eq. (9.17) was pointed out by Ferziger and Loyalka [39]. We note that although α can take five values, the value $\alpha = 0$ in Eq. (9.17) produces the identity $0 = 0$. Hence we are left with:

$$L_N^{-1}(\xi_i\xi_j - \tfrac{1}{3}\xi^2\,\delta_{ij}) = L^{-1}(\xi_i\xi_j - \tfrac{1}{3}\xi^2\,\delta_{ij}) = A(\xi)(\xi_i\xi_j - \tfrac{1}{3}\xi^2\,\delta_{ij}) \tag{9.18}$$

$$L_N^{-1}(\xi_i[\xi^2 - 5RT_0]) = L^{-1}(\xi_i[\xi^2 - 5RT_0]) = \xi_i B(\xi) \tag{9.19}$$

where the dependence of $L^{-1}(\xi_i\xi_j - \frac{1}{3}\xi^2\,\delta_{ij})$ and $L^{-1}(\xi_i[\xi^2 - 5RT_0])$ upon $\boldsymbol{\xi}$ is determined up to a speed dependent factor because of parity considerations. In addition $B(\xi)$ must be such that

$$(\xi^2, B(\xi)) = 0$$

because $\xi_i B(\xi)$ must be in W. Eqs. (9.18) and (9.19) can be satisfied by means of a model of the following kind:

$$L_{13}h = \nu(\xi)[\psi_0(\nu\psi_0, h) + \psi_i(\nu\psi_i, h) + \psi_4(\nu\psi_4, h) + \\ + \psi_{ij}(\nu\psi_{ij}, h) + \psi_{i4}(\nu\psi_{i4}, h) - h] \tag{9.20}$$

where as usual one sums over the repeated indexes i, j from 1 to 3 and ψ_{ij}, ψ_{i4} are given by

$$\psi_{ij} = (\xi_i\xi_j - \tfrac{1}{3}\xi^2\,\delta_{ij})C(\xi) \tag{9.21}$$

$$\psi_{i4} = \xi_i D(\xi)$$

where $C(\xi)$ and $D(\xi)$ are functions of the molecular speed ξ, determined by Eqs. (9.18) in terms of $A(\xi)$, $B(\xi)$ and $\nu(\xi)$. A simple calculation gives:

$$C(\xi) = \{A(\xi) + [\nu(\xi)]^{-1}\}\{\tfrac{1}{15}[(A, \xi^4) + (\nu A, A\xi^4)]\}^{-\frac{1}{2}}$$

$$D(\xi) = \{B(\xi) + [\nu(\xi)]^{-1}(\xi^2 - 5RT_0) - [(\nu, \xi^2)]^{-1}(\nu, \xi^2 B)\} \times \\ \times \{\tfrac{1}{3}(\xi^2 B, \xi^2 - 5RT_0) + (\nu B, \xi^2 B)] - [3(\nu, \xi^2)]^{-1}[(\nu, \xi^2 B)]^2\}^{-\frac{1}{2}} \tag{9.22}$$

Again the difficulty is that except for Maxwell's molecules (A = const., B = const.), we have no analytical tools for computing $A(\xi)$ and $B(\xi)$. Sometimes, however, the procedure of using Eq. (9.20) can be worthwhile because it is capable of giving accurate results for molecules other than Maxwell's; this is particularly true when the final results can be expressed in terms of quadratures to be performed numerically.

Similar considerations apply to the case of neutron transport. Here the situation is simpler because only one conservation law is present, at most. Hence we can delete four of the five terms in Eq. (9.15) and replace the right hand side of Eq. (3.9) by

$$Lf = \gamma\nu(\xi)f_0(\xi)\int \nu(\xi')f(\boldsymbol{\xi}')\,d\boldsymbol{\xi}' - \nu(\xi)f(\boldsymbol{\xi}) \tag{9.23}$$

where

$$\gamma^{-1} = \int \nu(\xi)f_0(\xi)\,d\boldsymbol{\xi} \tag{9.24}$$

This model seems to have been proposed for neutron transport by Corngold et al. [40]. Of course, we can improve Eq. (9.23) by adding more integral terms, exactly as in the case of gases. These models, however, are not sufficient to deal with crystalline moderators where the phenomenon of coherent scattering, due to interference of neutron waves, arises. This phenomenon produces important effects in the spectrum discussed in Section 8 and cannot be neglected [41, 42]. Care must also be exercised in using simplified models when high energy sources are present; in this case a simplified model exists [43], which violates reciprocity (the latter becomes meaningless when the energies involved are, say, 10^6 times the energies of thermal motions of the moderator).

Eq. (9.23) leads to the following linear Boltzmann equation:

$$\frac{\partial f}{\partial t} + \boldsymbol{\xi}\cdot\frac{\partial f}{\partial \mathbf{x}} = \gamma\nu(\xi)f_0(\xi)\int \nu(\xi')f(\boldsymbol{\xi}')\,d\boldsymbol{\xi}' - \nu(\xi)f(\boldsymbol{\xi}) \tag{9.25}$$

Another kind of approximation which is frequently made in neutron transport is to neglect the speed dependence by assuming that all particles have the same speed $\bar{\xi}$ and to write Eq. (3.10) as follows:

$$\frac{1}{\bar{\xi}}\frac{\partial \Psi}{\partial t} + \boldsymbol{\Omega}\cdot\frac{\partial \Psi}{\partial \mathbf{x}} + \sigma(\mathbf{x})\Psi = \int \sigma(\boldsymbol{\Omega}'\cdot\boldsymbol{\Omega}, \mathbf{x})\psi(\boldsymbol{\Omega}')\,d\boldsymbol{\Omega}' + S(\mathbf{x}, \boldsymbol{\Omega}, t) \tag{9.26}$$

where $\Psi = \Psi(\mathbf{x}, \boldsymbol{\Omega})$ is a suitable average of f with respect to speed and $\bar{\xi}$ is a suitable average speed. An additional assumption is isotropic scattering, which neglects the dependence of $\sigma(\boldsymbol{\Omega}'\cdot\boldsymbol{\Omega}, \mathbf{x})$ upon $\boldsymbol{\Omega}\cdot\boldsymbol{\Omega}'$. Then, since the fission source is also approximately isotropic, we can write (if $q(\mathbf{x}, \boldsymbol{\Omega}, t)$ denotes an external source):

$$\frac{1}{\bar{\xi}}\frac{\partial \Psi}{\partial t} + \boldsymbol{\Omega}\cdot\frac{\partial \Psi}{\partial \mathbf{x}} + \sigma\Psi = \frac{c\sigma}{4\pi}\int \psi(\boldsymbol{\Omega}')\,d\boldsymbol{\Omega}' + q(\mathbf{x}, \boldsymbol{\Omega}, t) \tag{9.27}$$

where, if σ_s, σ_a, σ_f denote the cross sections for scattering, absorption and fission respectively ($\sigma = \sigma_s + \sigma_a$) and c denotes the average number of secondary neutrons per collision

$$c = \frac{\sigma_s + \sigma_f}{\sigma_s + \sigma_a} \tag{9.28}$$

Eqs. (9.26) and (9.27) arise also in connection with radiative transfer. In fact, if we assume that σ_a and σ_s in Eq. (3.25) are frequency-independent (the so called "grey" case) and integrate Eq. (3.25) with respect to frequency, we obtain Eq. (9.26) with $\bar{\xi} = c_l$ (speed of light) and

$$\sigma(\boldsymbol{\Omega}' \cdot \boldsymbol{\Omega}) = \sigma_s g(\boldsymbol{\Omega}', \boldsymbol{\Omega}) + \sigma_a/(4\pi)$$

provided Eq. (3.26) is taken into account. The fourth power of absolute temperature plays then the role played by neutron density in neutron transport. In particular if scattering is absent or isotropic, Eq. (9.27) (with $c = 1$ and $\bar{\xi} = c_l$) follows.

10. The variational principle

In this section we shall investigate the possibility of obtaining a variational principle for the linearized Boltzmann equation. We shall consider the steady case, although the unsteady case can be dealt with by introducing the convolution product with respect to the time variable or by taking the Laplace Transform. Hence we shall consider the equation

$$\boldsymbol{\xi} \cdot \frac{\partial h}{\partial \mathbf{x}} = Lh + g_0 \tag{10.1}$$

where g_0 is a source term, which is usually zero but is kept in Eq. (10.1) for the sake of generality. The boundary conditions to be matched to Eq. (10.1) have the form shown in Eq. (2.14) and the basic Maxwellian f_0 has zero mass velocity.

A nontrivial variational principle for a linear equation is usually based on the self-adjointness of the linear operator $\mathscr{L}$ which appears in the equation

$$\mathscr{L}h = S \tag{10.2}$$

where S is the source term. Once $\mathscr{L}$ has been shown to be self-adjoint with respect to a certain scalar product ((,)), then the functional

$$J(\tilde{h}) = ((\tilde{h}, \mathscr{L}\tilde{h})) - 2((S, \tilde{h})) \tag{10.3}$$

is easily shown to satisfy

$$\delta J = 0 \tag{10.4}$$

if and only if $\tilde{h} = h + \delta h$, where h solves Eq. (10.2) and δh is infinitesimal. Although L is self-adjoint in $\mathscr{H}$ and hence also in the Hilbert space $\mathscr{H} \otimes \mathscr{R}$, where the scalar product is given by Eq. (4.3), the operator $D = \boldsymbol{\xi} \cdot \partial/\partial \mathbf{x}$

with the boundary conditions given by Eq. (2.14) with $h_0 = 0$ (h_0 can be, formally at least, lumped in the source term) is not self-adjoint in $\mathscr{H} \otimes \mathscr{R}$ or in any Hilbert space with positive definite norm. It is not hard to see, however, that

$$((g, PDh)) = ((PDg, h)) + ((g^+, Ph^-))_B - ((Pg^-, h^+))_B \tag{10.5}$$

where P is the parity operator in velocity space defined by Eq. (III.5.7). Eqs. (2.14), (2.19) and (4.5), where g and h are replaced by Pg^- and Ph^-, then give

$$((g, PDh)) = ((PDg, h)) \tag{10.6}$$

so that the operator PD is self-adjoint with respect to the scalar product in $\mathscr{H} \otimes \mathscr{R}$. We finally remark that, for any central interaction, L commutes with P provided the velocity u_0 in the basic Maxwellian f_0 is zero (see Eq. (7.8)). Accordingly

$$(g, PLh) = (g, LPh) = (Lg, Ph) = (PLg, h) \tag{10.7}$$

and so, not only L, but also PL, is self-adjoint in $\mathscr{H}$ and, as a consequence, in $\mathscr{H} \otimes \mathscr{R}$. In other words, although the operator $D - L$ is not self-adjoint in $\mathscr{H} \otimes \mathscr{R}$, $P(D - L)$ is, or, if we prefer, $D - L$ is self-adjoint in the pseudo-Hilbert space with a bilinear form $((f, g))_P \equiv ((f, Pg))$ which has all usual properties of the scalar product except that $((f, f))_P$ is not a norm because it can take negative as well as positive values.

The self-adjointness of $P(D - L)$ implies that we can immediately write a variational principle by identifying $P(D - L)$ with $\mathscr{L}$ and Pg_0 with S in Eq. (10.3). In this variational principle $\tilde{h}$ must satisfy the same boundary conditions as h and this would turn out to be inconvenient in many applications. Hence we generalize the form of $J(h)$ as follows [44]

$$J(\tilde{h}) = ((\tilde{h}, P(D - L)\tilde{h} - 2Pg_0)) + ((P\tilde{h}^-, \tilde{h}^+ - AP\tilde{h}^- - 2h_0))_B \tag{10.8}$$

It is easy to compute

$$\delta J = ((\delta\tilde{h}, P(D - L)\tilde{h} - 2Pg_0)) + ((\tilde{h}, P(D - L)\,\delta\tilde{h})) + ((P\tilde{h}^-, \delta\tilde{h}^+))_B - \\ - ((P\tilde{h}^-, AP\,\delta\tilde{h}^-))_B + ((\tilde{h}^+ - AP\tilde{h}^- - 2h_0, P\,\delta\tilde{h}^-))_B \tag{10.9}$$

The symmetry of $P(D - L)$ can now be exploited; we must take into account, however, that in general $\tilde{h}$ will not satisfy the boundary conditions. Eq. (10.5) then gives

$$((\tilde{h}, P(D - L)\,\delta\tilde{h})) = ((P(D - L)\tilde{h}, \delta\tilde{h})) + ((\tilde{h}^+, P\,\delta\tilde{h}^-))_B - ((P\tilde{h}^-, \delta\tilde{h}^+))_B \tag{10.10}$$

Also

$$((P\tilde{h}^-, AP\,\delta\tilde{h}^-))_B = ((AP\tilde{h}^-, P\,\delta\tilde{h}^-)) \tag{10.11}$$

because of Eq. (2.19) with $h = P\,\delta\tilde{h}^-$, $g = P\tilde{h}^-$. Substituting Eqs. (10.10) and (10.11) into Eq. (10.9) gives

$$\delta J = 2((P(D-L)\tilde{h} - Pg_0, \delta\tilde{h})) + 2((\tilde{h}^+ - AP\tilde{h}^- - h_0, P\,\delta\tilde{h}^-))_B \quad (10.12)$$

Accordingly, if $\tilde{h} = h$, where h is the solution of Eq. (10.1) with the boundary conditions given by Eq. (2.14), then $\delta J = 0$. If, *vice versa*, $\delta J = 0$ for arbitrary $\delta\tilde{h}$ in the region occupied by the gas and on the boundary, it follows $\tilde{h} = h$.

We remark that this variational principle is different from the minimum principle given at the end of Section 6 for the equation $Lh = g$.

It is interesting to examine the value attained by J when $\tilde{h} = h$; Eq. (10.8) gives:

$$J(h) = -((h, Pg_0)) - ((h_0, Ph^-))_B \quad (10.13)$$

This result acquires its full meaning only when we examine the expressions for h_0 and g_0. In general $g_0 = 0$ (see, however, Chapter VI, Section 5) and h_0 is given by Eq. (2.15); in this case

$$\begin{aligned} J(h) &= -((\psi_0^+, Ph^-))_B + ((AP\psi_0^-, Ph^-))_B \\ &= -((\psi_0^+, Ph^-))_B + ((P\psi_0^-, APh^-))_B \\ &= -((\psi_0^+, Ph^-))_B + ((P\psi_0^-, h^+ - \psi_0^+ + AP\psi_0^-))_B \\ &= -((\psi_0^+, Ph^-))_B + ((P\psi_0^-, h^+))_B + ((P\psi_0^-, AP\psi_0^- - \psi_0^+))_B \quad (10.14) \end{aligned}$$

The last term is a known quantity, while the first and second terms can be written as follows (we take $\mathbf{u}_0 = 0$ in Eq. (2.16), for Eq. (10.7) to be valid):

$$\begin{aligned} ((P\psi_0^-, h^+))_B - ((\psi_0^+, Ph^-))_B &= \int \boldsymbol{\xi}\cdot\mathbf{n}\psi_0(\mathbf{x}, -\boldsymbol{\xi})h(\mathbf{x}, \boldsymbol{\xi})f_0(\xi)\,d\boldsymbol{\xi}\,dS \\ &= -\int \frac{\mathbf{u}_w}{RT_0}\cdot\left[\int \boldsymbol{\xi}(\boldsymbol{\xi}\cdot\mathbf{n})f_0(\xi)h(\mathbf{x}, \boldsymbol{\xi})\,d\boldsymbol{\xi}\right] dS \\ &\quad + \int \frac{T_w - T_0}{T_0}\left[\int \boldsymbol{\xi}\cdot\mathbf{n}\left(\frac{\xi^2}{2RT_0} - \frac{3}{2}\right) f_0 h(\mathbf{x}, \boldsymbol{\xi})\,d\boldsymbol{\xi}\,dS\right] \\ &= -(RT_0)^{-\frac{1}{2}}\left[\int \mathbf{u}_w\cdot\mathbf{p}_{(n)}\,dS + \int \frac{T_0 - T_w}{T_0} q_n\,dS\right] \quad (10.15) \end{aligned}$$

Here, $\mathbf{p}_{(n)}$ is the normal stress vector, and $q_n = \mathbf{q}\cdot\mathbf{n}$ is the normal component of heat flux. The mass conservation at the wall has been taken into account.

We observe now that it is convenient, without loss of generality, to consider separately the two cases $\mathbf{u}_w = 0$ and $T_w = T_0$ on account of the linearity. It is also frequent, albeit not necessary, that in the two cases T_w and $\mathbf{u}_w$ can be taken to be constant (this is, for example, the situation for the important case of flow past a solid body). When the latter circumstance holds, the functional becomes proportional to a global quantity of great physical interest (e.g., a drag or heat transfer coefficient).

As was remarked by Lang and Loyalka [45, 46] it is possible to generalize the above principle by introducing, along with h, the solution h^* of a linearized Boltzmann equation with a different source term $g_0{}^*$; in this case we then introduce a functional $J(\tilde{h}, \tilde{h}^*)$ which is stationary for $\tilde{h} = h$ and $\tilde{h}^* = h^*$ and can choose the quantity which is related to $J(h, h^*)$ by suitably choosing $g_0{}^*$.

11. Green's function

As pointed out at the end of the previous section, it is sometimes convenient to consider, along with the Boltzmann equation linearized about a Maxwellian with zero mass velocity:

$$Dh = Lh + g_0 \tag{11.1}$$

where D is the operator $\boldsymbol{\xi} \cdot \partial/\partial \mathbf{x}$ considered in Section 10, the same equation with a different source term:

$$Dh^* = Lh^* + g_0{}^* \tag{11.2}$$

We have considered the steady case because we can take care of the time dependence by taking a Laplace transform and replacing L by $L - sI$, where s is the parameter of the Laplace transform (in this case we must treat s as real, even when it is not, so that we must not conjugate s when it appears in the scalar products). The self-adjointness of L and the parity operator P gives:

$$\begin{aligned}((Ph^*, g_0)) - ((h, Pg_0{}^*)) &= ((Ph^*, Dh - Lh)) - ((P(D - L)h^*, h)) \\ &= ((h^*, PDh)) - ((PDh^*, h))\end{aligned} \tag{11.3}$$

Eq. (10.5) then gives

$$((Ph^*, g_0)) - ((h, Pg_0{}^*)) = ((h^{*+}, Ph^-))_B - ((Ph^{*-}, h^+))_B \tag{11.4}$$

This is a general relation between the solution h, h^* and the corresponding sources g_0, $g_0{}^*$. Particular cases are interesting. If, for example, g satisfies Eq. (11.1) in a region R with the usual boundary conditions at ∂R and h^* satisfies Eq. (11.2) with

$$g_0{}^* = \delta(\mathbf{x} - \mathbf{x}')\, \delta(\boldsymbol{\xi} - \boldsymbol{\xi}') \tag{11.5}$$

(we remark that the notation ((,)) has a meaning even for elements not belonging to the Hilbert space $\mathscr{H}$) in all of space with suitable conditions at infinity, then Eq. (11.4) becomes, provided $\mathbf{x} \in R$:

$$\iint_R G(\mathbf{x}, -\boldsymbol{\xi}; \mathbf{x}', \boldsymbol{\xi}') g_0(\mathbf{x}, \boldsymbol{\xi}) f_0(\xi)\, d\boldsymbol{\xi}\, d\mathbf{x} - h(\mathbf{x}', -\boldsymbol{\xi}') f_0(\xi')$$

$$= -\iint_{\partial R} \boldsymbol{\xi} \cdot \mathbf{n} G(\mathbf{x}, -\boldsymbol{\xi}; \mathbf{x}', \boldsymbol{\xi}') h(\mathbf{x}, \boldsymbol{\xi}) f_0(\xi)\, d\boldsymbol{\xi}\, dS \quad (11.6)$$

where $G(\mathbf{x}, \boldsymbol{\xi}; \mathbf{x}', \boldsymbol{\xi}')$ denotes the solution of Eq. (11.2) with the source (22.5) and appropriate boundary conditions at infinity. A suitable rearrangement of Eq. (11.6) gives

$$f_0(\xi) h(\mathbf{x}, \boldsymbol{\xi}) = \iint_R G(\mathbf{x}', -\boldsymbol{\xi}'; \mathbf{x}, -\boldsymbol{\xi}) g_0(\mathbf{x}', \boldsymbol{\xi}') f_0(\xi')\, d\boldsymbol{\xi}'\, d\mathbf{x}' +$$

$$+ \iint_{\partial R} \boldsymbol{\xi}' \cdot \mathbf{n} G(\mathbf{x}', -\boldsymbol{\xi}'; \mathbf{x}, -\boldsymbol{\xi}) h(\mathbf{x}', \boldsymbol{\xi}') f_0(\xi')\, d\boldsymbol{\xi}'\, dS' \qquad (\mathbf{x} \in R) \quad (11.7)$$

This equation shows that the solution h of Eq. (11.1) can be expressed in terms of the source g_0 and the boundary values of h on ∂R, once the Green's function for infinite space is known. The problem of course is that the values of h at the boundary are not known (in the easiest case, $A = 0$ in Eq. (2.14), h is known for $\boldsymbol{\xi}' \cdot \mathbf{n} > 0$, but not for $\boldsymbol{\xi}' \cdot \mathbf{n} < 0$).

If we take $g_0 = g_0^*$ where g_0^* is given by Eq. (11.5) and assume appropriate boundary conditions for h, then $h = h^* = G(\mathbf{x}, \boldsymbol{\xi}; \mathbf{x}', \boldsymbol{\xi}')$ and Eq. (11.7) becomes

$$f_0(\xi) G(\mathbf{x}, \boldsymbol{\xi}; \mathbf{x}', \boldsymbol{\xi}') = G(\mathbf{x}', -\boldsymbol{\xi}'; \mathbf{x}, -\boldsymbol{\xi}) f_0(\boldsymbol{\xi}') \quad (11.8)$$

where the surface integral has been neglected, under the assumption that it vanishes for $\partial R \to \infty$. Eq. (11.8) is a reciprocity relation for the Green's function of the linearized Boltzmann equation. By means of Eq. (11.8), Eq. (11.7) can be rewritten

$$h(\mathbf{x}, \boldsymbol{\xi}) = \iint_R G(\mathbf{x}, \boldsymbol{\xi}; \mathbf{x}', \boldsymbol{\xi}') g_0(\mathbf{x}', \boldsymbol{\xi}')\, d\boldsymbol{\xi}'\, d\mathbf{x}' +$$

$$+ \iint_{\partial R} \boldsymbol{\xi}' \cdot \mathbf{n}' G(\mathbf{x}, \boldsymbol{\xi}; \mathbf{x}', \boldsymbol{\xi}') h(\mathbf{x}', \boldsymbol{\xi}')\, d\boldsymbol{\xi}'\, dS' \quad (11.9)$$

In order to make full use of Eq. (11.9) we can try to use it to determine $h(\mathbf{x}, \boldsymbol{\xi})$ for $\mathbf{x} \in R$. To this end we let $\mathbf{x}$ tend to a point of the boundary in

Eq. (11.7) and obtain

$$h(\mathbf{x}, \boldsymbol{\xi}) = \iint_R G(\mathbf{x}, \boldsymbol{\xi}; \mathbf{x}', \boldsymbol{\xi}')g_0(\mathbf{x}', \boldsymbol{\xi}')\, d\boldsymbol{\xi}'\, d\mathbf{x}' + $$

$$+ \iint_{\partial R} \boldsymbol{\xi}' \cdot \mathbf{n}' G_+(\mathbf{x}, \boldsymbol{\xi}; \mathbf{x}', \boldsymbol{\xi}')h(\mathbf{x}', \boldsymbol{\xi}')\, d\boldsymbol{\xi}'\, dS' \qquad (\mathbf{x} \in \partial R) \quad (11.10)$$

where G_+ denotes the limit of $G(\mathbf{x}'', \boldsymbol{\xi}; \mathbf{x}', \boldsymbol{\xi}')(\mathbf{x}' \in \partial R)$ for $\mathbf{x}'' \to \mathbf{x}$ from within R (G suffers a discontinuity proportional to a delta function when crossing the boundary, as is seen by integrating Eq. (11.2) with $g_0{}^*$ given by Eq. (11.5) over a slice containing a piece of ∂R).

In particular, if we take for h the infinite space Green's function, $G_+(\mathbf{x}, \boldsymbol{\xi}; \mathbf{x}'', \boldsymbol{\xi}'')$, for $\mathbf{x}'' \in \partial R$, then $g_0 = 0$ and Eq. (11.10) becomes

$$G_+(\mathbf{x}, \boldsymbol{\xi}; \mathbf{x}'', \boldsymbol{\xi}'') = \iint_{\partial R} \boldsymbol{\xi}' \cdot \mathbf{n}' G_+(\mathbf{x}, \boldsymbol{\xi}; \mathbf{x}', \boldsymbol{\xi}') \times$$

$$\times\, G_+(\mathbf{x}', \boldsymbol{\xi}'; \mathbf{x}'', \boldsymbol{\xi}'')\, d\boldsymbol{\xi}'\, dS' \qquad (\mathbf{x}, \mathbf{x}'' \in \partial R) \quad (11.11)$$

If, on the other hand, we take for h the Green's function $G(\mathbf{x}, \boldsymbol{\xi}; \mathbf{x}'', \boldsymbol{\xi}'')$ for $\mathbf{x}'' \in R$ then $g_0 = \delta(\mathbf{x} - \mathbf{x}'')\, \delta(\boldsymbol{\xi} - \boldsymbol{\xi}'')$, and Eq. (11.10) gives (the first integral on the right cancels the left hand side):

$$0 = \iint_{\partial R} \boldsymbol{\xi}' \cdot \mathbf{n}' G_+(\mathbf{x}, \boldsymbol{\xi}; \mathbf{x}', \boldsymbol{\xi}')G(\mathbf{x}', \boldsymbol{\xi}'; \mathbf{x}'', \boldsymbol{\xi}'')\, d\boldsymbol{\xi}'\, dS' \qquad (\mathbf{x} \in \partial R; \mathbf{x}'' \in R) \quad (11.12)$$

Eqs. (11.11) and (11.12) are basic identities satisfied by the Green's function for the linearized Boltzmann equation. Some of them were pointed out by Case [47]. These identities are useful in dealing with a paradoxical remark about Eq. (11.10). The latter equation is an integral equation for $h(\mathbf{x}, \boldsymbol{\xi})(\mathbf{x} \in \partial R)$; we might attempt to solve it and find h on ∂R and hence h in R by means of Eq. (11.9). On the other hand, the values of h on the boundary for $\boldsymbol{\xi} \cdot \mathbf{n} > 0$ can be assigned at will or, in any case, must be related to those for $\boldsymbol{\xi} \cdot \mathbf{n}$ by the boundary conditions, Eq. (2.14), where A is an assigned operator and h_0 a given source term. Hence we conclude that Eq. (11.10) cannot be solved, but the system of Eqs. (2.14) and (11.10) must be solvable in the unknowns h^+, h^-. In this connection we note that Eq. (11.11) can be written as follows

$$h = Wg_0 + Bh \qquad (11.13)$$

where W and B are integral operators defined in an obvious manner. Eqs. (11.11) and (11.12) are identities for the kernels of these operators and are equivalent to:

$$B^2 = B \qquad BW = 0 \qquad (11.14)$$

If we apply B to both sides of Eq. (11.13) and use Eq. (11.14), we obtain the identity $Bh = Bh$ and this confirms that Eq. (11.13) is not able to determine h uniquely. Let us introduce h^+, h^- and the operator B^{++}, B^{+-}, B^{-+}, B^{--}, W^+, W^-, defined as the restrictions of W and B from one of the two subspaces of functions defined for $\boldsymbol{\xi}\cdot\mathbf{n} > 0$ and $\boldsymbol{\xi}\cdot\mathbf{n} < 0$, respectively, to another (possibly the same) subspace. Then Eq. (11.13) becomes

$$h^+ = W^+g_0 + B^{++}h^+ + B^{+-}h^- \tag{11.15}$$

$$h^- = W^-g_0 + B^{-+}h^+ + B^{--}h^- \tag{11.16}$$

and Eqs. (11.14) give rise to six identities; in particular, for example,

$$\begin{aligned} B^{++}B^{++} + B^{+-}B^{-+} &= B^{++} \\ B^{++}B^{+-} + B^{+-}B^{--} &= B^{+-} \\ B^{++}W^{+} + B^{+-}W^{-} &= 0 \end{aligned} \tag{11.17}$$

By applying the operator B^{++} to both sides of Eq. (11.15) and taking into account Eq. (11.17), we obtain

$$0 = B^{+-}(h^- - W^-g_0 - B^{-+}h^+ - B^{--}h^-) \tag{11.18}$$

and on the plausible assumption that B^{+-} has an inverse, Eqs. (11.15) and (11.16) turn out to be equivalent and hence we can dismiss one of them and make a system of the other and Eq. (2.14):

$$\begin{aligned} h^+ &= h_0^+ + APh^- \\ h^- &= W^-g_0 + B^{-+}h^+ + B^{--}h^- \end{aligned} \tag{11.19}$$

or

$$h^- = (W^-g_0 + B^{-+}h_0^+) + (B^{-+}AP + B^{--})h^- \tag{11.20}$$

This is an integral equation for h^-, which in principle solves the boundary value problem for the linearized Boltzmann equation. Since it is not a practical method, we shall not enter into the details and the justification of the formal treatment given above.

We consider now the related problem of computing the Green's function; without loss of generality we shall take $\mathbf{x}' = 0$ since $G(\mathbf{x}, \boldsymbol{\xi}; \mathbf{x}', \boldsymbol{\xi}') = G(\mathbf{x} - \mathbf{x}', \boldsymbol{\xi}; 0, \boldsymbol{\xi}')$. We have then to solve the equation

$$\boldsymbol{\xi}\cdot\frac{\partial G}{\partial \mathbf{x}} = Lg + \delta(\mathbf{x})\,\delta(\boldsymbol{\xi} - \boldsymbol{\xi}') \tag{11.21}$$

A suitable method is to take the Fourier transform of G with respect to the space variables

$$\tilde{G}(\mathbf{k}, \boldsymbol{\xi}; \boldsymbol{\xi}') = \int \tilde{G}(\mathbf{x}, \boldsymbol{\xi}; \boldsymbol{\xi}')e^{i\mathbf{k}\cdot\mathbf{x}}\, d\mathbf{x} \tag{11.22}$$

inverted by

$$G(\mathbf{x}, \boldsymbol{\xi}; \boldsymbol{\xi}') = (2\pi)^{-3}\int \tilde{G}(\mathbf{k}, \boldsymbol{\xi}; \boldsymbol{\xi}')e^{-i\mathbf{k}\cdot\mathbf{x}}\,d\mathbf{k} \tag{11.23}$$

Eq. (11.21) then becomes (if G does not tend to zero to $\mathbf{x} \to \infty$ the equations are to be interpreted in the sense of generalized functions):

$$-i\mathbf{k}\cdot\boldsymbol{\xi}\tilde{G} = L\tilde{G} + \delta(\boldsymbol{\xi} - \boldsymbol{\xi}') \tag{11.24}$$

If we take the first axis along $\mathbf{k}$ we obtain

$$-ik\xi_1\tilde{G} = L\tilde{G} + \delta(\boldsymbol{\xi} - \boldsymbol{\xi}') \tag{11.25}$$

where k is the absolute value of $\mathbf{k}$. The same equation would have been found if we had started from the one-dimensional equation corresponding to Eq. (11.21):

$$\xi_1 \frac{\partial G_1}{\partial x_1} = LG_1 + \delta(x_1)\,\delta(\boldsymbol{\xi} - \boldsymbol{\xi}') \tag{11.26}$$

where the source is not a point source but a plane source (in the plane $x_1 = 0$), and had taken the one-dimensional Fourier transform

$$\tilde{G}_1(k, \xi_1, \boldsymbol{\xi}_\perp; \boldsymbol{\xi}', \boldsymbol{\xi}_\perp') = \int G_1(x_1, \xi_1, \boldsymbol{\xi}_\perp; \xi_1', \boldsymbol{\xi}_\perp')e^{ikx_1}\,dx_1 \tag{11.27}$$

where $\boldsymbol{\xi}_\perp$ is the two-dimensional vector with components ξ_2, ξ_3. Hence, if we are able to solve Eq. (11.26), then we can compute $\tilde{G}(\mathbf{k}, \boldsymbol{\xi}; \boldsymbol{\xi}') = \tilde{G}_1(k, \boldsymbol{\xi}; \boldsymbol{\xi}')$ by means of Eq. (11.27) provided $\xi_1 = (\boldsymbol{\xi}\cdot\mathbf{k})/k$ and $\boldsymbol{\xi}_\perp = \boldsymbol{\xi} - [(\boldsymbol{\xi}\cdot\mathbf{k})/k]$. Now it is possible to express the solution of Eq. (11.26) with suitable conditions at infinity by means of the eigensolutions discussed in Section 7. We shall assume that G_1 does not grow exponentially for $x_1 \to \infty$ and $x_1 \to -\infty$. Eq. (11.26) shows that, for $x_1 > 0$ and $x_1 < 0$, G_1 satisfies Eq. (7.1) and hence is given by an expression of the form shown in Eq. (7.43):

$$\begin{aligned} G_1 &= \sum_{\alpha=0}^{4} A_\alpha^+\psi_\alpha + \sum_{\alpha=2}^{4} B_\alpha^+[x_1\psi_\alpha + L^{-1}(\xi_1\psi_\alpha)] + \\ &\qquad + \int_{-\lambda_\infty}^{-\lambda_0} g_\lambda(\boldsymbol{\xi})e^{\lambda x_1}A_\lambda^+\,d\lambda \qquad (x_1 > 0) \\ G_1 &= \sum_{\alpha=0}^{4} A_\alpha^-\psi_\alpha + \sum_{\alpha=2}^{4} B_\alpha^-[x_1\psi_\alpha + L^{-1}(\xi_1\psi_\alpha)] + \\ &\qquad + \int_{\lambda_0}^{\lambda_\infty} g_\lambda(\boldsymbol{\xi})e^{\lambda x_1}A_\lambda^-\,d\lambda \qquad (x_1 < 0) \end{aligned} \tag{11.28}$$

where the conditions at $+\infty$ and $-\infty$ have been taken into account. On the other hand if we integrate Eq. (4.26) from $x_1 = -\varepsilon$ to $x_1 = \varepsilon$ and then let

$\varepsilon \to 0$, we find

$$\xi_1\left[\lim_{x_1\to 0+} G_1 - \lim_{x_1\to 0-} G_1\right] = \delta(\boldsymbol{\xi} - \boldsymbol{\xi}') \tag{11.29}$$

If we determine A_α^+, A_α^-, B_α^+, B_α^-, A_λ^+, A_λ^- by means of Eq. (11.29), then Eq. (11.28) gives the solution of Eq. (11.26) with the stated conditions at infinity. Hence we must have

$$\sum_{\alpha=0}^{4}(A_\alpha^+ - A_\alpha^-)\psi_\alpha + \sum_{\alpha=0}^{4}(B_\alpha^+ - B_\alpha^-)L^{-1}(\xi_1\psi_\alpha) + \int_{-\lambda_\infty}^{-\lambda_0} g_\lambda(\boldsymbol{\xi})A_\lambda^+\,d\lambda + \\ + \int_{\lambda_0}^{\lambda_\infty} g_\lambda(\boldsymbol{\xi})A_\lambda^-\,d\lambda = \frac{1}{\xi_1}\delta(\boldsymbol{\xi} - \boldsymbol{\xi}') = \frac{1}{\xi_1'}\delta(\boldsymbol{\xi} - \boldsymbol{\xi}') \tag{11.30}$$

According to the theorem proved in Section 7, for any function h we have:

$$h(\boldsymbol{\xi}) = \int\left[\frac{\xi_1'^2 + \xi_1\xi_1'}{C_1} + \sum_{\alpha=2}^{4}\frac{\xi_1'\psi_\alpha[L^{-1}(\xi_1\psi_\alpha)]' + \xi_1'[L^{-1}(\xi_1\psi_\alpha)]\psi_\alpha'}{C_\alpha} + \\ + \int_{-\lambda_\infty}^{-\lambda_0} + \int_{\lambda_0}^{\lambda_\infty}\xi_1' g_\lambda(\boldsymbol{\xi}')g_\lambda(\boldsymbol{\xi})\,d\lambda\right]h(\boldsymbol{\xi}')\,d\boldsymbol{\xi}' \tag{11.31}$$

$$C_1 = \int \xi_1^2 f_0\,d\boldsymbol{\xi}, \qquad C_\alpha = \int \xi_1\psi_\alpha L^{-1}(\xi_1\psi_\alpha) f_0\,d\boldsymbol{\xi} \qquad (\alpha = 2, 3, 4) \tag{11.32}$$

Eq. (11.31) implies that

$$\frac{\xi_1' + \xi_1}{C_1} + \sum_{\alpha=2}^{4}\frac{\psi_\alpha[L^{-1}(\xi_1\psi_\alpha)]' + [L^{-1}(\xi_1\psi_\alpha)]\psi_\alpha'}{C_\alpha} + \\ + \int_{-\lambda_\infty}^{-\lambda_0} + \int_{\lambda_0}^{\lambda_\infty} g_\lambda(\boldsymbol{\xi}')g_\lambda(\boldsymbol{\xi})\,d\lambda = \frac{1}{\xi_1'}\delta(\boldsymbol{\xi} - \boldsymbol{\xi}') \tag{11.33}$$

where the integral over the continuous spectrum has to be interpreted in the sense of the theory of generalized functions (it is such, in other words, that we can multiply Eq. (11.33) by $h(\boldsymbol{\xi}')$, integrate with respect to $\boldsymbol{\xi}'$ and interchange the order of integrations in order to obtain Eq. (7.38) with Eqs. (7.39) and (7.33)). The values of $A_\alpha^+ - A_\alpha^-$, $B_\alpha^+ - B_\alpha^-$, $A_\lambda^\pm$ may be picked out by comparison of Eq. (11.33) with Eq. (11.30). We determine A_α^+, A_α^-, B_α^+, B_α^- uniquely by assuming $A_\alpha^- = A_\alpha^+$, $B_\alpha^- = -B_\alpha^+$.

Then Eq. (11.28) gives

$$G_1 = \left\{\frac{\xi_1' + \xi_1}{2C_1} + \sum_{\alpha=2}^{4}\frac{\psi_\alpha[L^{-1}(\xi_1\psi_\alpha)]' + \psi_\alpha'[L^{-1}(\xi_1\psi_\alpha)]}{2C_\alpha}\right\}\operatorname{sgn} x_1 + \\ + \sum_{\alpha=2}^{4}|x_1|\frac{\psi_\alpha\psi_\alpha'}{2C_\alpha} + \int_{\lambda_0}^{\lambda_\infty} e^{-\lambda|x_1|}g_\lambda(\boldsymbol{\xi}')g_\lambda(\boldsymbol{\xi})\,d\lambda \tag{11.34}$$

where $\bar{\lambda} = \lambda \operatorname{sgn} x_1$. From G_1 we can compute G by means of Eq. (11.27); Eq. (11.23) then gives G.

It is clear that G splits into an asymptotic part G_A and a remainder G_B arising from the corresponding splitting of h into h_A and h_B, discussed in Section 7 (Eqs. (7.51) and (7.52)). G_B is only important close to the origin and negligible a few mean free paths far from it (the decay is exponential if $\lambda_0 > 0$). It is interesting to investigate the asymptotic part which arises from

$$G_{1A} = \left\{\frac{\xi_1' + \xi_1}{2C_1} + \sum_{\alpha=2}^{4} \frac{[L^{-1}(\xi_1\psi_\alpha)]'\psi_\alpha + [L^{-1}(\xi_1\psi_\alpha)]\psi_\alpha'}{2C_\alpha}\right\} \operatorname{sgn} x_1 + \\ + \sum_{\alpha=2}^{4} |x_1| \frac{\psi_\alpha\psi_\alpha'}{2C_\alpha} \quad (11.35)$$

Instead of computing the Fourier transforms of $\operatorname{sgn} x_1$ and $|x_1|$ which requires the use of the theory of generalized functions, we remark that Eqs. (7.70)–(7.72) hold for G_{1A}, which is a particular solution of Eq. (7.1) (for $x_1 \neq 0$). Hence the Fourier transforms of the pressure, velocity, temperature, stress tensor and heat flux satisfy (remember that $u_1^{(A)} = \text{const.}$):

$$\tilde{\mathbf{q}}^{(A)} = i\kappa k T \mathbf{e}_1; \qquad \tilde{p}_{jl}^{(A)} = \tilde{p}^{(A)}\,\delta_{jl} + i\mu k u_j^{(A)} e_{1l} + i\mu k u_l^{(A)} e_{ij} \quad (11.36)$$

where $\mathbf{e}_1 = (e_{11}, e_{12}, e_{13})$ is the unit vector along the x_1-axis (hence $e_{1l} = \delta_{1l}$). Now, if we replace $\mathbf{e}_1$ by $\mathbf{e} = \mathbf{k}/k$, Eq. (11.36) gives the corresponding relations for the Fourier transform of the stress tensor and heat flux vector in the three-dimensional case. We obtain

$$\tilde{q}_r^{(A)} = i\kappa k_r \tilde{T}^{(A)} \qquad \tilde{p}_{jl}^{(A)} = \tilde{p}^{(A)}\,\delta_{jl} + i\mu(k_l \tilde{u}_j^{(A)} + k_j u_l^{(A)}) \quad (11.37)$$

If we now take the inverse Fourier transform in three dimensions we obtain that $q_r^{(A)}$ and $p_{jl}^{(A)}$ are related to $T^{(A)}$ and $u_l^{(A)}$ by the Navier-Stokes-Fourier relations, given by Eq. (II.8.27) ($\partial v_k/\partial x_k = 0$ for $\mathbf{x} \neq 0$, as follows by integrating Eq. (11.21) after multiplication by f_0).

Since Eq. (11.9) shows that, if $g_0 = 0$, the solution of a boundary value problem can be expressed in terms of an integral involving $G(\mathbf{x} - \mathbf{x}', \boldsymbol{\xi}, \boldsymbol{\xi}')$ for $\mathbf{x}' \in \partial R$, we conclude that when $|\mathbf{x} - \mathbf{x}'|$ is large, for any $\mathbf{x}' \in \partial R$, compared with the mean free path (i.e. $\mathbf{x}$ is several mean free paths from the boundary), then the solution depends upon the space coordinates and velocity in the same way as G_A. In particular we conclude that all the solutions will satisfy the Navier-Stokes-Fourier relations a few mean free paths from the boundaries, with viscosity and heat conduction coefficients given by Eqs. (7.68) and (7.69).

12. The integral equation approach

The results of Section 11 show that it is possible to reduce the solution of a boundary value problem for the linearized Boltzmann equation to the solution of an integral equation, Eq. (11.20), for the values of the distribution function of the molecules impinging upon the boundary. Even for the simplest boundary conditions ($A = 0$), Eq. (11.20) is not easy to solve because the kernel of B^{--} is expressed in terms of the Green's function, which in turn is expressed in terms of the eigensolutions of Eq. (7.3), which are not explicitly known, in general. For some problems, however, and in connection with questions about existence and uniqueness of the solutions, it is useful to reduce a boundary value problem to the solution of an integral equation. This is particularly true, as we shall see, for the model equations described in Section 9.

It is possible to obtain a useful integral equation when the collision operator can be written in the form $Lh = Kh - \nu h$, where K is an integral operator. This, as we know, can be achieved for rigid spheres, cutoff potentials and model equations.

In these cases we can write Eq. (11.1) in the form

$$Dh + \nu h = g_1 \qquad (g_1 = g_0 + Kh) \tag{12.1}$$

and regard g_1 as a source term. Then we can repeat the considerations of Section 11, with the Green's function $U(\mathbf{x}, \boldsymbol{\xi}; \mathbf{x}', \boldsymbol{\xi}')$ of Eq. (12.1) ($g_1 = \delta(\mathbf{x} - \mathbf{x}')\,\delta(\boldsymbol{\xi} - \boldsymbol{\xi}')$) in place of the Green's function of Eq. (11.1) ($g_0 = \delta(\mathbf{x} - \mathbf{x}')\,\delta(\boldsymbol{\xi} - \boldsymbol{\xi}')$). The result is analogous to Eq. (11.10) ($G \mapsto U$, $g_0 \mapsto g_1$):

$$h(\mathbf{x}, \boldsymbol{\xi}) = \iint_R U(\mathbf{x}, \boldsymbol{\xi}; \mathbf{x}', \boldsymbol{\xi}')g_1(\mathbf{x}', \boldsymbol{\xi}')\, d\boldsymbol{\xi}'\, d\mathbf{x}' + \iint_{\partial R} \boldsymbol{\xi}' \cdot \mathbf{n} U_+(\mathbf{x}, \boldsymbol{\xi}; \mathbf{x}', \boldsymbol{\xi}')h(\mathbf{x}', \boldsymbol{\xi}')\, d\boldsymbol{\xi}'\, dS' \qquad (\mathbf{x} \in \partial R) \tag{12.2}$$

or, if U and U_0 denote the operators with kernels $U(\mathbf{x}, \boldsymbol{\xi}; \mathbf{x}', \boldsymbol{\xi}')$ and $\boldsymbol{\xi}' \cdot \mathbf{n}U_+(\mathbf{x}, \boldsymbol{\xi}; \mathbf{x}', \boldsymbol{\xi}')(\mathbf{x} \in R; \mathbf{x}' \in \partial R)$ and we let $g_1 = g_0 + Kh$ according to Eq. (12.1):

$$h = Ug_0 + UKh + U_0 h \tag{12.3}$$

The main advantage is that $U(\mathbf{x}, \boldsymbol{\xi}; \mathbf{x}', \boldsymbol{\xi}')$ can be computed explicitly

$$U(\mathbf{x}, \boldsymbol{\xi}; \mathbf{x}', \boldsymbol{\xi}') = \xi^{-1}H(\boldsymbol{\xi} \cdot [\mathbf{x} - \mathbf{x}'])\exp\left[-\frac{\boldsymbol{\xi} \cdot (\mathbf{x} - \mathbf{x}')\nu(\xi)}{\xi^2}\right] \times \delta_2\left(\frac{\boldsymbol{\xi}}{\xi} \wedge [\mathbf{x} - \mathbf{x}']\right)\delta(\boldsymbol{\xi} - \boldsymbol{\xi}') \tag{12.4}$$

where H denotes the Heaviside step function, δ_2 is the two-dimensional delta function in the plane orthogonal to $\boldsymbol{\xi}$ and $\wedge$ denotes the vector product. Eq. (12.4) is easily obtained by observing that, since the operator $D + \nu$ does not mix different values of $\boldsymbol{\xi}$, $\boldsymbol{\xi}$ can be treated as a parameter and $D + \nu$ is a first order partial differential operator with constant coefficients. In any case, Eq. (12.4) is easily checked by direct substitution. Eq. (12.3) is then equivalent to:

$$h(\mathbf{x}, \boldsymbol{\xi}) = \int_R \xi^{-1} H(\boldsymbol{\xi} \cdot [\mathbf{x} - \mathbf{x}']) \exp\left[-\frac{\boldsymbol{\xi} \cdot (\mathbf{x} - \mathbf{x}')\nu(\xi)}{\xi^2}\right] \delta_2\left(\frac{\boldsymbol{\xi}}{\xi} \wedge [\mathbf{x} - \mathbf{x}']\right) \times$$
$$\times \left[g_0(\mathbf{x}', \boldsymbol{\xi}) + \int K(\boldsymbol{\xi}', \boldsymbol{\xi}) h(\mathbf{x}', \boldsymbol{\xi}')\, d\boldsymbol{\xi}'\right] d\mathbf{x}' +$$
$$+ \int_{\partial R} \boldsymbol{\xi} \cdot \mathbf{n}' \xi^{-1} H(\boldsymbol{\xi} \cdot [\mathbf{x} - \mathbf{x}'])\, \delta_2\left(\frac{\boldsymbol{\xi}}{\xi} \wedge [\mathbf{x} - \mathbf{x}']\right) \times$$
$$\times \exp\left[-\frac{\boldsymbol{\xi} \cdot (\mathbf{x} - \mathbf{x}')\nu(\xi)}{\xi^2}\right] h(\mathbf{x}', \boldsymbol{\xi})\, dS \tag{12.5}$$

We observe that Eq. (12.5) holds for any shape of the region R, which could also extend to infinity. However, for nonconvex regions (see Fig. 25), it is sufficient to extend the volume and surface integrals to the parts $R_{\mathbf{x}}$ and $\partial R_{\mathbf{x}}$ of R and ∂R, respectively, which are "seen" from $\mathbf{x}$ (i.e. to those $\mathbf{x}'$ such that the straight line joining $\mathbf{x}'$ with $\mathbf{x}$ does not intersect the boundary ∂R).

This is physically obvious because the Green's function describes the influence of $\mathbf{x}'$ upon $\mathbf{x}$, and U is the Green's function for a process in which particles travel along straight lines. Mathematically, this follows by applying Eq. (11.9) with g_1 in place of g_0, U in place of G to the "unseen" region $R - R_{\mathbf{x}}$; then, since $\mathbf{x} \in R_{\mathbf{x}}$, the left hand side is zero rather than $h(\mathbf{x}, \boldsymbol{\xi})$ and we obtain

$$\iint_{R - R_{\mathbf{x}}} U(\mathbf{x}, \boldsymbol{\xi}; \mathbf{x}', \boldsymbol{\xi}') g_1(\mathbf{x}', \boldsymbol{\xi}')\, d\mathbf{x}'\, d\boldsymbol{\xi}' +$$
$$+ \int_{\partial R - \partial R_{\mathbf{x}}} U(\mathbf{x}, \boldsymbol{\xi}; \mathbf{x}', \boldsymbol{\xi}') \boldsymbol{\xi}' \cdot \mathbf{n} h(\mathbf{x}', \boldsymbol{\xi}')\, d\boldsymbol{\xi}'\, dS' +$$
$$+ \int_{\Sigma} U(\mathbf{x}, \boldsymbol{\xi}; \mathbf{x}', \boldsymbol{\xi}') \boldsymbol{\xi}' \cdot \mathbf{n} h(\mathbf{x}', \boldsymbol{\xi}')\, d\boldsymbol{\xi}'\, dS' = 0 \tag{12.6}$$

where Σ is the surface separating $R_{\mathbf{x}}$ from $R - R_{\mathbf{x}}$. But Eq. (12.4) gives

$$\int_{\Sigma} \boldsymbol{\xi}' \cdot \mathbf{n}' U(\mathbf{x}, \boldsymbol{\xi}; \mathbf{x}', \boldsymbol{\xi}') h(\mathbf{x}', \boldsymbol{\xi}')\, d\boldsymbol{\xi}'\, dS'$$
$$= \int_{\Sigma} \boldsymbol{\xi} \cdot \mathbf{n}' \xi^{-1} H(\boldsymbol{\xi} \cdot [\mathbf{x} - \mathbf{x}']) \times$$
$$\times \delta_2\left(\frac{\boldsymbol{\xi}}{\xi} \wedge [\mathbf{x} - \mathbf{x}']\right) \exp\left[-\frac{\boldsymbol{\xi} \cdot (\mathbf{x} - \mathbf{x}')\nu(\xi)}{\xi^2}\right] dS' = 0 \tag{12.7}$$

because $\boldsymbol{\xi} \cdot \mathbf{n}' = 0$ along Σ. Hence, according to Eq. (12.6), the contributions from $R - R_{\mathbf{x}}$ and $\partial R - \partial R_{\mathbf{x}}$ cancel each other in Eq. (12.2), or (12.5), as was to be shown. Due to this circumstance, we can write Eq. (12.5) as follows:

$$\begin{aligned} h(\mathbf{x}, \boldsymbol{\xi}) = \int_{R_x} & \xi^{-1} H(\boldsymbol{\xi} \cdot [\mathbf{x} - \mathbf{x}']) \times \\ & \times \exp\left[-\frac{\boldsymbol{\xi} \cdot (\mathbf{x} - \mathbf{x}')\nu(\xi)}{\xi^2}\right] \delta_2\left(\frac{\boldsymbol{\xi}}{\xi} \wedge [\mathbf{x} - \mathbf{x}']\right) \times \\ & \times \left[g_0(\mathbf{x}', \boldsymbol{\xi}) + \int K(\boldsymbol{\xi}', \boldsymbol{\xi}) h(\mathbf{x}', \boldsymbol{\xi}')\, d\boldsymbol{\xi}'\right] d\mathbf{x}' + \\ & + \int_{\partial R_x} \boldsymbol{\xi} \cdot \mathbf{n}' \xi^{-1} H(\boldsymbol{\xi} \cdot \mathbf{n}')\, \delta_2\left(\frac{\boldsymbol{\xi}}{\xi} \wedge [\mathbf{x} - \mathbf{x}']\right) \times \\ & \times \exp\left[-\frac{\boldsymbol{\xi} \cdot (\mathbf{x} - \mathbf{x}')\nu(\xi)}{\xi^2}\right] h(\mathbf{x}', \boldsymbol{\xi})\, dS' \end{aligned} \tag{12.8}$$

where we have replaced $H(\boldsymbol{\xi} \cdot [\mathbf{x} - \mathbf{x}'])$ by $H(\boldsymbol{\xi} \cdot \mathbf{n}')$, since for $\mathbf{x} \in \partial R_{\mathbf{x}}$ and $\mathbf{x} - \mathbf{x}'$ aligned with $\boldsymbol{\xi}$ (as required by the delta function) the signs of $\boldsymbol{\xi} \cdot (\mathbf{x} - \mathbf{x}')$ and $\boldsymbol{\xi} \cdot \mathbf{n}'$ are equal. Hence the surface term in Eq. (12.8) involves only values of h corresponding to emerging molecules, which is physically satisfactory. If the boundary conditions give h^+ explicitly at the boundary, Eq. (12.8) is an integral equation solving the original boundary value problem; otherwise we must consider the system of Eqs. (12.8) and (2.14). In particular, for example, we can use Eq. (12.8) to express $h^-(\mathbf{x} \in \partial R)$ in terms of $h(\mathbf{x} \in R)$ and $h^+(\mathbf{x} \in \partial R)$ and by inserting the result into Eq. (2.14) find an equation which together with Eq. (12.8) yields a system of two coupled integral equations for h and h^+.

The integral equation approach is particularly convenient when one of the collision models described in Section 9 is used. In this case, in fact, Kh is an explicitly known function of $\boldsymbol{\xi}$ depending upon several unknown functions of $\mathbf{x}$, $M_\alpha = (\nu\varphi_\alpha, h)$, in a linear fashion. Hence Eq. (12.8) expresses h in terms of the M_α's (and the boundary values); using this expression to compute $M_\alpha = (\nu\varphi_\alpha, h)$ $(\alpha = 1, \ldots, N)$ gives each M_α in terms of all M's, which is a system of N equations for the N unknowns M_α $(\alpha = 1, \ldots, N)$. The advantage is that we deal now with N functions of three variables (the space coordinates) rather than one function of 6 variables (components of $\mathbf{x}$, $\boldsymbol{\xi}$). This is a great advantage, especially if N is small, and there are symmetries which further reduce the number of variables.

Let us consider in more detail the case of models of the form shown in Eq. (9.14) when the boundary condition is that of complete diffusion (Eq. (2.14) with A given by Eqs. (III.4.14) and (III.5.1) with $\alpha = 1$):

$$h^+ = h_0 + N(\mathbf{x}) \qquad (\mathbf{x} \in \partial R, \boldsymbol{\xi} \cdot \mathbf{n} > 0) \tag{12.9}$$

where, if T_0 is the temperature of the basic Maxwellian f_0:

$$N(\mathbf{x}) = [2\pi(RT_0)^2]^{-1} \int_{\xi \cdot n < 0} \exp[-\xi^2(2RT_0)^{-1}] h^-(\boldsymbol{\xi}) \, |\boldsymbol{\xi} \cdot \mathbf{n}| \, d\boldsymbol{\xi} \quad (12.10)$$

We shall also assume $g_0 = 0$ in Eq. (12.8) to make the equations a little shorter (the case $g_0 \neq 0$ is very easily dealt with). Eq. (12.8) takes on the

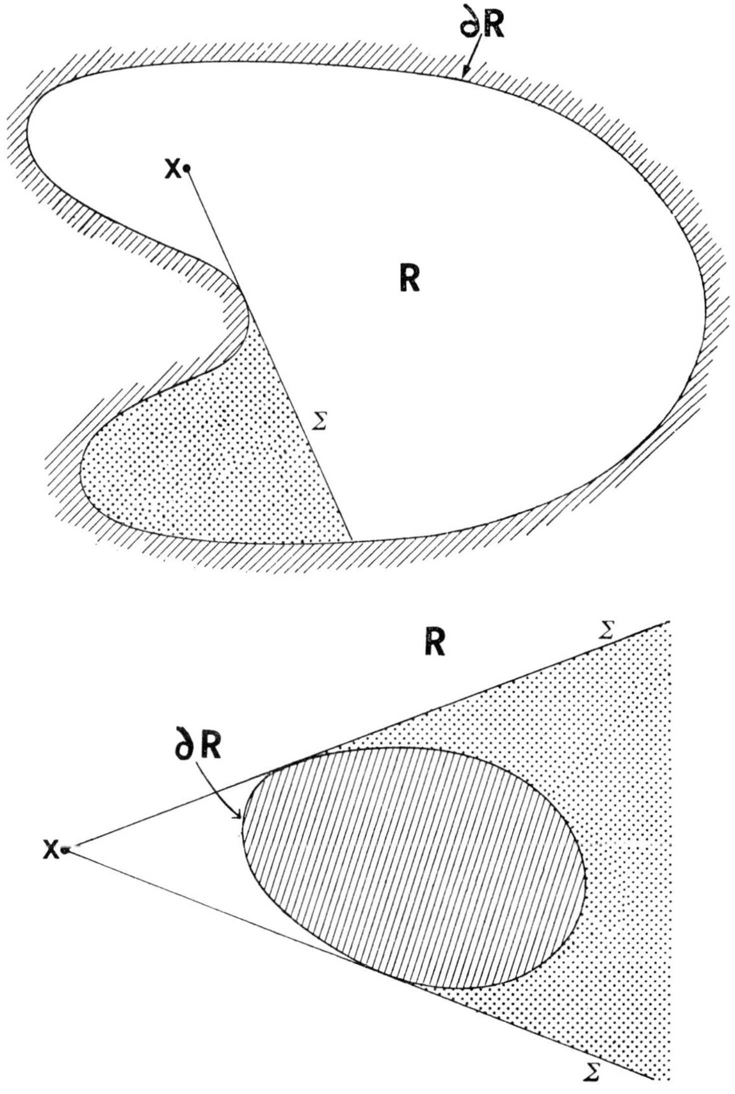

Fig. 25. Flows in nonconvex regions. The region R in Eq. (12.5) can be taken to be the region "seen" from $\mathbf{x}$.

form ($(\mathbf{x}-\mathbf{x}')\cdot\boldsymbol{\xi} = |\mathbf{x}-\mathbf{x}'|\,\xi$ because of the delta and step functions):

$$h(\mathbf{x},\boldsymbol{\xi}) = \sum_{\beta=0}^{N}\mu_\beta\nu(\xi)\xi^{-1}\int_{R_x} H(\boldsymbol{\Omega}\cdot[\mathbf{x}-\mathbf{x}'])\exp\left[-\frac{|\mathbf{x}-\mathbf{x}'|\,\nu(\xi)}{\xi}\right]\times$$
$$\times\,\delta_2(\boldsymbol{\Omega}\wedge[\mathbf{x}-\mathbf{x}'])\varphi_\beta(\xi\boldsymbol{\Omega})M_\beta(\mathbf{x}')\,d\mathbf{x}' + S(\mathbf{x},\boldsymbol{\xi}) +$$
$$+\int_{\partial R_x}\boldsymbol{\Omega}\cdot\mathbf{n}'H(\boldsymbol{\Omega}\cdot\mathbf{n}')\,\delta_2(\boldsymbol{\Omega}\wedge[\mathbf{x}-\mathbf{x}'])\times$$
$$\times\exp\left[-\frac{|\mathbf{x}-\mathbf{x}'|\,\nu(\xi)}{\xi}\right]N(\mathbf{x}')\,dS' \qquad (12.11)$$

where $\boldsymbol{\Omega} = \boldsymbol{\xi}/\xi$, $S(\mathbf{x},\boldsymbol{\xi})$ is given by the last integral in Eq. (12.8) with h_0 in place of h, and

$$M_\alpha(\mathbf{x}) = (\nu\varphi_\alpha, h) \qquad (\alpha = 0, 1, \ldots, N) \qquad (12.12)$$

We now compute M_α by means of Eqs. (12.12) and (12.11). We obtain

$$M_\alpha(\mathbf{x}) = \sum_{\beta=0}^{N}\int_{R_x} K_{\alpha\beta}(\mathbf{x}-\mathbf{x}')M_\beta(\mathbf{x}')\,d\mathbf{x}' +$$
$$+\int_{\partial R_x} K_\alpha(\mathbf{x}-\mathbf{x}')N(\mathbf{x}')\,dS' + S_\alpha(\mathbf{x}) \qquad (\alpha = 0, 1, \ldots, N) \quad (12.13)$$

where

$$K_{\alpha\beta}(\mathbf{x}-\mathbf{x}') = \mu_\beta\iint[\nu(\xi)]^2\xi H(\boldsymbol{\Omega}\cdot[\mathbf{x}-\mathbf{x}'])\,\delta_2(\boldsymbol{\Omega}\wedge[\mathbf{x}-\mathbf{x}'])\times$$
$$\times\exp\left[-\frac{|\mathbf{x}-\mathbf{x}'|\,\nu(\xi)}{\xi}\right]\varphi_\alpha(\xi\boldsymbol{\Omega})\varphi_\beta(\xi\boldsymbol{\Omega})f_0(\xi)\,d\xi\,d\boldsymbol{\Omega} \quad (12.14)$$

$$K_\alpha(\mathbf{x}-\mathbf{x}') = \int\nu(\xi)\xi^2\boldsymbol{\Omega}\cdot\mathbf{n}'H(\boldsymbol{\Omega}\cdot[\mathbf{x}-\mathbf{x}'])\,\delta_2(\boldsymbol{\Omega}\wedge[\mathbf{x}-\mathbf{x}'])\times$$
$$\times\exp\left[-\frac{|\mathbf{x}-\mathbf{x}'|\,\nu(\xi)}{\xi}\right]\varphi_\alpha(\xi\boldsymbol{\Omega})f_0(\xi)\,d\xi\,d\boldsymbol{\Omega} \quad (12.15)$$

$$S_\alpha(\mathbf{x}) = (\nu\varphi_\alpha, S) \qquad (12.16)$$

Eqs. (12.13) form a system of $N+1$ integral equations for the $N+2$ unknowns $M_\beta(\mathbf{x})$ ($\beta = 0, 1, \ldots, N$) and $N(\mathbf{x})$. A further equation is obtained by using Eq. (12.11) for $\mathbf{x}\in\partial R$ to compute $N(\mathbf{x})$:

$$N(\mathbf{x}) = \sum_{\beta=0}^{N}\int_{R_\mathbf{x}} K_\beta'(\mathbf{x}-\mathbf{x}')M_\beta(\mathbf{x}')\,d\mathbf{x}' +$$
$$+\int_{\partial R_\mathbf{x}} K(\mathbf{x}-\mathbf{x}')N(\mathbf{x}')\,d\mathbf{x}' + S(\mathbf{x}) \quad (12.17)$$

where

$$K_\beta'(\mathbf{x}-\mathbf{x}') = (2\pi/RT_0)^{\frac{1}{2}}\rho_0^{-1}\mu_\beta K_\beta(\mathbf{x}-\mathbf{x}');$$

$$K(\mathbf{x}-\mathbf{x}') = (2\pi/RT_0)^{\frac{1}{2}}\rho_0^{-1}\int \xi^3\,|\boldsymbol{\Omega}\cdot\mathbf{n}|\,|\boldsymbol{\Omega}\cdot\mathbf{n}'|\,\delta_2(\boldsymbol{\Omega}\wedge[\mathbf{x}-\mathbf{x}'])\times$$

$$\times\exp\left[-\frac{\nu(\xi)\,|\mathbf{x}-\mathbf{x}'|}{\xi}\right]f_0(\xi)\,d\xi\,d\boldsymbol{\Omega} \quad (12.18)$$

$$S(\mathbf{x}) = (2\pi/RT_0)^{\frac{1}{2}}\rho_0^{-1}(1, S(\mathbf{x}, \boldsymbol{\xi}))_B \quad (12.19)$$

Eqs. (12.13) and (12.17) form a system of integral equations whose solution yields $M_\alpha(\mathbf{x})$ ($\alpha = 0, 1, \ldots, N$) and $N(\mathbf{x})$, and hence $h(\mathbf{x}, \boldsymbol{\xi})$ through Eq. (12.11).

Due to the presence of the delta function δ_2, the integrations with respect to the unit vector $\boldsymbol{\Omega}$ in Eqs. (12.14), (12.15) and (12.18) can be performed explicitly with following results:

$$K_{\alpha\beta} = \mu_\beta\int_0^\infty [\nu(\xi)]^2\xi\,|\mathbf{x}-\mathbf{x}'|^{-2}\times$$

$$\times\exp[-|\mathbf{x}-\mathbf{x}'|\,\nu(\xi)/\xi]\varphi_\alpha\left(\xi\frac{\mathbf{x}-\mathbf{x}'}{|\mathbf{x}-\mathbf{x}'|}\right)\varphi_\beta\left(\xi\frac{\mathbf{x}-\mathbf{x}'}{|\mathbf{x}-\mathbf{x}'|}\right)f_0(\xi)\,d\xi \quad (12.20)$$

$$K_\alpha = \int_0^\infty \nu(\xi)\xi^2(\mathbf{x}-\mathbf{x}')\cdot\mathbf{n}'\,|\mathbf{x}-\mathbf{x}'|^{-3}\times$$

$$\times\exp[-|\mathbf{x}-\mathbf{x}'|\,\nu(\xi)/\xi]\varphi_\alpha\left(\xi\frac{\mathbf{x}-\mathbf{x}'}{|\mathbf{x}-\mathbf{x}'|}\right)f_0(\xi)\,d\boldsymbol{\xi} \quad (12.21)$$

$$K = (2\pi/RT_0)^{\frac{1}{2}}\rho_0^{-1}\int \xi^3\,|(\mathbf{x}-\mathbf{x}')\cdot\mathbf{n}|\,|(\mathbf{x}-\mathbf{x}')\cdot\mathbf{n}'|\,|\mathbf{x}-\mathbf{x}'|^{-4}\times$$

$$\times\exp[-|\mathbf{x}-\mathbf{x}'|\,\nu(\xi)/\xi]f_0(\xi)\,d\xi \quad (12.22)$$

The additional integration with respect to ξ cannot be performed unless $\nu(\xi)$ is specified; even if $\nu(\xi)$ has a simple expression, however, the kernels $K_{\alpha\beta}$, K_α and K are nonelementary functions of $\mathbf{x}-\mathbf{x}'$. The simplest case is when ν is constant and the φ_α are polynomials (Eqs. (9.8) and (9.12)); then the kernels can be expressed in terms of a family of transcendental functions $T_n(x)$, defined by:

$$T_n(x) = \int_0^\infty t^n\exp[-t^2-(x/t)]\,dt \quad (12.23)$$

The T_n functions have some notable properties which are easily derived from their definition:

$$\frac{dT_n}{dx} = -T_{n-1} \tag{12.24}$$

$$T_n(x) = \frac{n-1}{2} T_{n-2}(x) + \frac{x}{2} T_{n-3} \tag{12.25}$$

$$T_n(0) = \tfrac{1}{2}\Gamma\left(\frac{n+1}{2}\right) = \begin{cases} \dfrac{1}{2}\left(\dfrac{n-1}{2}\right)! & (n \text{ odd}) \\ \dfrac{n-1}{2}\,\dfrac{n-3}{2}\cdots\dfrac{1}{2}\,\dfrac{\sqrt{\pi}}{2} & (n \text{ even}) \end{cases} \tag{12.26}$$

where Γ denotes the gamma-function. Expansions valid for small x and large x are available (see [18], p. 1001–1003, where the T_n functions are denoted by f_n) and can be used to compute $T_n(x)$ to any desired accuracy. Note that many authors use the notation J_n in place of T_n.

A particularly simple case arises in neutron transport when the constant speed approximation Eq. (9.26) is employed. In this case, if the cross section is not space dependent and we adopt a degenerate kernel approximation, we can repeat the above discussion without the Maxwellian and the integration over the speed variable in Eqs. (12.14)–(12.16), (12.18)–(12.22). This means that the kernels $K_{\alpha\beta}$, K_β, K are now elementary functions. A particularly simple case arises when the scattering is assumed to be isotropic (Eq. (9.27)). Then, with the usual boundary condition $\psi = 0$ for $\mathbf{\Omega} \cdot \mathbf{n} > 0$, there is just one integral equation

$$\rho(\mathbf{x}) = \int \frac{\exp[-\sigma\,|\mathbf{x}-\mathbf{x}'|]}{|\mathbf{x}-\mathbf{x}'|^2}\left[\frac{c\sigma}{4\pi}\,\rho(\mathbf{x}') + q(\mathbf{x}')\right] d\mathbf{x}' \tag{12.27}$$

Eq. (12.27) is known as the Pejerl's integral equation [48]. A slightly more complicated case arises when although the scattering is isotropic, the cross section σ varies with $\mathbf{x}$; in this case the exponent in Eq. (12.27) is replaced by $-\int_0^1 \sigma(\mathbf{x}' + \lambda[\mathbf{x}-\mathbf{x}'])\,d\lambda$ and the factor σ in front of $\rho(\mathbf{x}')$ by $\sigma(\mathbf{x}')$. The integral in the exponent is frequently called the "optical depth", a term obviously arising from problems of radiative transfer.

If we denote by $\mathbf{u}$ the column vector formed with the $N+2$ quantities M_α, N, Eqs. (12.3) and (12.17) can be written as follows:

$$\mathbf{u} = \mathbf{u}^{(0)} + \mathsf{A}\mathbf{u} \tag{12.28}$$

where A is a matrix integral operator and $\mathbf{u}^{(0)}$ a source term.

The kernels of the elements of A are $K_{\alpha\beta}$, K_α, K_β', K. It is easy to check that the variational principle of Section 10 can be applied to Eq. (12.28) as well

[49–52] provided we write the functional $J_1(\tilde{u})$ such that $\delta J_1 = 0$ at $\mathbf{u} = \tilde{\mathbf{u}}$ in the following form:

$$J_1(\tilde{u}) = (\mathsf{PK}, \tilde{\mathbf{u}} - \mathsf{A}\tilde{\mathbf{u}} - 2\mathbf{u}^{(0)}) \tag{12.29}$$

where P is the diagonal matrix whose elements are $(-1)^P \delta_{\alpha\beta}$, $(-1)^P = \pm 1$ according to the parity of φ_α and K is a diagonal matrix whose elements are μ_α, $(2\pi/RT_0)^{-\frac{1}{2}}\rho_0$. Eq. (12.29) holds because

$$\widetilde{\mathsf{PKA}} = \mathsf{PKA} \tag{12.30}$$

where $\tilde{\mathsf{A}}$ denotes the adjoint operator of A, which is obtained by interchanging rows and columns, $\mathbf{x}$ and $\mathbf{x}'$. Eq. (12.30) is easily obtained from the expressions of P, K, A.

The advantage of using the variational principle $\delta J_1 = 0$ with the functional given by Eq. (12.29) is that J_1 involves only functions of the space variables about which guesses are easier than about h; in addition, for some problems with particular symmetries, the principle becomes a true maximum or minimum principle [49, 50, 52]. Finally, we can frequently relate the value attained by J_1 for $\tilde{u} = u$ to the global quantities of basic interest (drag, flow rate, heat flux, torque, etc.). This follows from the fact that $\mathsf{PK}\mathbf{u} = PKh$, $\mathsf{A}\mathbf{u} = UKh$, $\mathbf{u}^{(0)} = \Pi U g_0$ where g_0 is the source in Eq. (10.1) and Π is the projector upon the subspace spanned by the φ_α ($\alpha = 0, 1, \ldots, N$) (we assume $h_0 = 0$ to make things simpler). Then, for $\tilde{h} = h$:

$$\begin{aligned} J(h) - J_1(\mathbf{u}) &= -((h, Pg_0) + ((PKh, \Pi U g_0)) = -((P[h - UKh], g_0)) \\ &= -((P[Ug_0], g_0)) \end{aligned} \tag{12.31}$$

that is, the value attained by the functionals J, given by Eq. (10.8), and J_1, given by Eq. (12.29), at the point where they are stationary differ from each other by a known quantity. The connection between $J_1(\mathbf{u})$ and a suitable global quantity then follows from the analogous result about $J(h)$.

References

[1] N. I. Akhiezer and I. M. Glazman, "Theory of Linear Operators in Hilbert Space" (translated by M. Nestell), Frederick Ungar Publishing Co., New York (1963).

[2] F. Riesz and B. Sz. Nagy, "Functional Analysis" (translated by L. Boron), Frederick Ungar Publishing Co., New York (1955).

[3] T. Kato, "Perturbation Theory for Linear Operators", Springer-Verlag, N.Y. (1966).

[4] M. Schechter, "Principles of Functional Analysis", Academic Press, N.Y. (1971).

[5] N. Dunford and J. Schwartz, "Linear Operators", Interscience, vol. I (1958) and II (1963).

[6] C. CERCIGNANI, "High Mach Number Flow on an Almost Specularly Reflecting Plate with Sharp Leading Edge", Presented at the VIII Symposium on Rarefied Gas Dynamics, Stanford (1972).

[7] A. M. WEINBERG and E. P. WIGNER, "The Physical Theory of Neutron Chain Reactors", University of Chicago Press, Chicago (1958).

[8] H. HURWITZ, M. S. NELKIN and G. J. HABETLER, "Nucl. Sci. Eng." **1,** 280 (1956).

[9] I. KUŠČER and G. C. SUMMERFIELD, "Phys. Rev.", **188,** 1445 (1969).

[10] S. CHANDRASEKHAR, "Radiative Transfer", Oxford University Press, Oxford (1950).

[11] I. MAREK, "Appl. Mat." **8,** 442 (1963).

[12] J. MIKA, "J. Quant. Spectrosc. Radiat. Transfer" **1,** 869 (1971).

[13] G. N. WATSON, "Theory of Bessel Functions", Cambridge University Press, Cambridge (1958).

[14] I. M. GEL'FAND and co-authors, "Generalized Functions", English translation from the Russian, Academic Press, New York (1964).

[15] C. S. WANG CHANG and G. E. UHLENBECK, "On the Propagation of Sound in Monatomic Gases", University of Michigan Press, Project M 999, Ann. Arbor, Michigan (1952). Reprinted in "The Kinetic Theory of Gases", Studies in Statistical Mechanics, vol. V, North-Holland, Amsterdam (1970).

[16] C. CERCIGNANI, "Mathematical Methods in Kinetic Theory", Plenum Press, New York (1969).

[17] M. ABRAMOWITZ and I. A. STEGUN, "Handbook of Mathematical Functions", Dover, New York (1965).

[18] Bateman Manuscript Project, "Higher Transcendental Functions", McGraw-Hill Book Co., New York (1953).

[19] H. M. MOTT SMITH, "A New Approach in the Kinetic Theory of Gases", MIT Lincoln Laboratory Group Report V, 2 (1954).

[20] Z. ALTERMAN, K. FRANKOWSKI and C. L. PEKERIS, "Astrophys. J. Suppl.", **7,** 291 (1962).

[21] G. GRAD, in "Rarefied Gas Dynamics", J. A. Laurmann, Ed., Vol. 1, 26, Academic Press, New York (1963).

[22] J. R. DORFMAN, "Proc. of the Nat. Acad. of Sciences", U.S.A., **50,** 804 (1963).

[23] I. KUŠČER and M. M. R. WILLIAMS, "Phys. Fluids", **10,** 1922 (1967).

[24] G. M. WING, "An Introduction to Transport Theory", John Wiley & Sons, Inc., New York (1962).

[25] C. CERCIGNANI, "Phys. Fluids", **10,** 2097 (1967).

[26] L. SIROVICH and J. K. THURBER, in "Rarefied Gas Dynamics" J. H. de Leeuw, Ed., Vol. I, 21 (1965).

[27] L. SIROVICH and J. K. THURBER, "J. Math. Phys.", **10,** 239 (1969).

[28] B. NICOLAENKO, "Dispersion Laws for Plane Wave Propagation", in

"The Boltzmann Equation", Edited by F. A. Grunbaum, Courant Institute, New York University, New York (1972).
[29] M. SCHECHTER, "J. Math. Anal. and Appl.", **13,** 205 (1966).
[30] K. GUSTAFSON and T. WEISMANN, "J. Math. Anal. and Appl." **25,** 121 (1969).
[31] M. BIXON, J. R. DORFMAN amd K. C. MO, "Phys. Fluids", **14,** 1049 (1971).
[32] J. A. MCLENNAN, "Phys. Fluids", **8,** 1580 (1965).
[33] J. A. MCLENNAN, "Phys. Fluids", **9,** 1581 (1966).
[34] G. SCHARF, "Helvetica Physica Acta", **40,** 929 (1967).
[35] K. M. CASE, "The Soluble Boundary Value Problems of Transport Theory", in "The Boltzmann Equation", Edited by F. A. Grunbaum, Courant Institute, New York University, New York (1972).
[36] E. P. GROSS and E. A. JACKSON, "Phys. Fluids", **2,** 432 (1959).
[37] L. SIROVICH, "Phys. Fluids", **5,** 908 (1962).
[38] C. CERCIGNANI, "Ann. Phys. (N.Y.)", **40,** 469 (1966).
[39] S. K. LOYALKA and J. H. FERZIGER, Phys. Fluids, **10,** 1833 (1967).
[40] N. CORNGOLD et al., "Nucl. Sci. Engng.", **15,** 13 (1963).
[41] I. K. KUŠČER and N. CORNGOLD, "Phys. Rev.", **139,** A981 (1965).
[42] N. CORNGOLD, in "Transport Theory", R. Bellman et al., Eds., p. 79, American Mathematical Society, Providence, R.I. (1969).
[43] M. M. R. WILLIAMS, "The Slowing Down and Thermalisation of Neutrons", North-Holland, Amsterdam (1966).
[44] C. CERCIGNANI, "J. Stat. Phys.", **1,** 297 (1969).
[45] H. LANG and S. K. LOYALKA, "On Variational Principles in the Kinetic Theory", Presented at the VII Symposium on Rarefied Gas Dynamics, Pisa (1970).
[46] H. LANG, "Acta Mechanica", **5,** 163 (1968).
[47] K. M. CASE, in "Transport Theory", R. Bellman et al., Eds., p. 17, American Mathematical Society, Providence, R.I. (1969).
[48] R. E. PEJERLS, "Proc. Camb. Phil. Soc. Math. Phys. Sci.", **35,** 610 (1939).
[49] C. CERCIGNANI and C. D. PAGANI, "Phys. Fluids", **9,** 1167 (1966).
[50] C. CERCIGNANI and C. D. PAGANI, in "Rarefied Gas Dynamics". C. L. Brundin, Ed., Vol. I, **55,** Academic Press, New York (1967).
[51] C. CERCIGNANI and C. D. PAGANI, "Phys. Fluids", **11,** 1399 (1968).
[52] P. BASSANINI, C. CERCIGNANI and C. D. PAGANI, "Int. J. Heat and Mass Transfer", **11,** 1359 (1968).
[53] Y. P. PAO, in "Rarefied Gas Dynamics", M. Becker and M. Fiebig, Eds., Vol. I, A.6-1 DFVLR-Press, Porz-Wahn (1974).

V SMALL AND LARGE MEAN FREE PATHS

1. The Knudsen number

It was pointed out in Chapter IV, Section 1, that if we want to solve the Boltzmann equation for realistic nonequilibrium situations, we must rely upon approximation methods, in particular, perturbation procedures. In order to do this, we have to look for a parameter ε which can be considered to be small in some situations. In Chapter IV, Section 2, ε was assumed not to appear directly in the Boltzmann equation. This led us to considering the linearized Boltzmann equation, which turns out to be useful for describing situations in which deviations of velocity and temperature from their average values are small. If we look for different expansions, a first step consists in investigating the order of magnitude of the various terms appearing in the Boltzmann equation. If we denote by τ a typical time scale, by d a typical length scale and by $\bar{\xi}$ a typical molecular velocity, then [see, for example, Eq. (II.3.15)]:

$$\frac{\partial f}{\partial t} \sim \tau^{-1} f; \qquad \boldsymbol{\xi} \cdot \frac{\partial f}{\partial \mathbf{x}} \sim \bar{\xi}\, d^{-1} f, \qquad Q(f, f) \sim n \bar{\xi} \sigma^2 f \tag{1.1}$$

where $\sim$ denotes that two quantities are of the same order of magnitude, $n = \rho/m$ is the number density of molecules and σ the molecular diameter (or range of the interaction potential).

According to Eq. (I.4.17), then

$$Q(f, f) \sim \bar{\xi} l^{-1} f \tag{1.2}$$

where l is the mean free path. In this connection, it is to be remarked that the mean free path concept can be rigorously defined only for rigid-sphere molecules or cutoff potentials ($\sigma < \infty$), for long-range potentials the molecules are always interacting ($\sigma = \infty$) and accordingly, the mean free path is, strictly speaking, zero. In spite of this, the mean free path concept can be retained as a tool for expressing the order of magnitude of the right hand side of the Boltzmann equation. In such a way the mean free path is defined in a qualitative fashion. The definition can be made quantitative, if desired, by recalling [Eq. (IV.7.51)] that the viscosity coefficient μ turns out to be proportional to the mean free path; this suggests a definition of the latter in terms of the former. A definition frequently used is

$$l = \mu(\pi RT/2)^{\frac{1}{2}}/p = \mu[\pi/(2RT)]^{\frac{1}{2}}/\rho \tag{1.3}$$

where, as usual, $p = \rho RT$ denotes pressure. This definition associates a definite number with the concept of mean free path; it has the advantage that we can easily compare results corresponding to different microscopic models because the above definition is in terms of macroscopically measurable quantities and, accordingly, does not depend upon any assumption concerning the molecular interactions. The combination $\bar{\xi}\rho^{-1}$ appearing in Eq. (1.2) can be considered as a measure of the collision frequency and its inverse $\bar{\xi}^{-1}l = \theta$ as defining a mean free time, as remarked in Section 6 of Chapter IV.

Eqs. (1.1) and (1.2) show the existence of two basic nondimensional numbers in the Boltzmann equation: θ/τ and l/d. The nondimensional number l/d is called the Knudsen number and is denoted by Kn:

$$Kn = l/d \tag{1.4}$$

It is clear that Kn ranges from 0 to ∞; $Kn \to 0$ corresponding to a fairly dense gas and $Kn \to \infty$ to a free molecular flow (i.e. a flow where molecules have negligible interactions with each other). The other nondimensional number θ/τ is related to Kn as follows:

$$\theta/\tau = \bar{\xi}^{-1}l/\tau = (d/\bar{\xi}\tau)Kn = ShKn \tag{1.5}$$

where

$$Sh = d/(\bar{\xi}\tau) \tag{1.6}$$

is the analog of the Strouhal number, used in continuum theory (more exactly, since $\bar{\xi}$ is usually of the order of the sound speed, we should write ShM, where M is the Mach number, in place of Sh). Frequently θ/τ is also called Knudsen number, with the implicit assumption that $Sh \simeq 1$. There are, however, important phenomena, such as viscous decay to equilibrium for which $\theta d^2/\tau l^2 \simeq 1$ or $Sh \simeq Kn$ with $Kn \ll 1$ and hence also $Sh \ll 1$. In steady problems, of course, only one parameter, namely the Knudsen number, appears. This discussion, however, applies to problems with only one characteristic length beside the mean free path: otherwise, several Knudsen numbers are to be considered.

The Knudsen number can be related to the Reynolds number Re. According to Eq. (1.3)

$$l \sim \mu/\rho c \tag{1.7}$$

where $c = (\gamma RT)^{\frac{1}{2}}$ ($\gamma = \frac{5}{3}$ for a monatomic gas) is the sound speed; accordingly:

$$Kn = \mu/(\rho c d) = (U/c)(\mu/\rho U d) = M/\mathrm{Re} \tag{1.8}$$

where U denotes a typical mass velocity (hence, in general, $c \sim \bar{\xi}$, but $U \neq \bar{\xi}$).

Two kinds of perturbation methods suggest themselves in connection with the Knudsen number, one for $\mathrm{Kn} \to 0$, the other for $\mathrm{Kn} \to \infty$. The two expansions will describe situations corresponding to extremely small and extremely large mean free paths, respectively. These situations will be examined in this Chapter.

2. The Hilbert expansion

The first instance of a perturbation expansion for solving the Boltzmann equation was considered by Hilbert in 1912 [1, 2]. It is based on the assumption that the Knudsen number Kn is small and the Strouhal number Sh of order unity. Then the order of magnitude of the ratio of a typical term of the left hand side of the Boltzmann equation to a typical term of the right-hand side is Kn and we can formalize Hilbert's method by putting an artificial parameter ε (to be treated as small) in front of the left-hand side (or, equivalently, by introducing nondimensional quantities and denoting the Knudsen number by ε):

$$\varepsilon\left(\frac{\partial f}{\partial t} + \boldsymbol{\xi}\cdot\frac{\partial f}{\partial \mathbf{x}}\right) = Q(f, f) \tag{2.1}$$

The singular nature of a perturbation procedure in the limit $\varepsilon \to 0$ is made clear by the fact that ε multiplies all the derivatives which appear in the Boltzmann equation. *In spite of this*, we try a series expansion in powers of ε (Hilbert's expansion):

$$f = \sum_{n=0}^{\infty} \varepsilon^n f_n \tag{2.2}$$

Substituting into Eq. (2.1) gives:

$$\sum_{n=1}^{\infty} \varepsilon^n \left(\frac{\partial f_{n-1}}{\partial t} + \boldsymbol{\xi}\cdot\frac{\partial f_{n-1}}{\partial \mathbf{x}}\right) = \sum_{n=0}^{\infty} \varepsilon^n Q_n \tag{2.3}$$

where Q_n is given by Eq. (IV.1.5). Accordingly:

$$Q_0 = 0 \tag{2.4}$$

$$\frac{\partial f_{n-1}}{\partial t} + \boldsymbol{\xi}\cdot\frac{\partial f_{n-1}}{\partial \mathbf{x}} = Q_n \qquad (n \geqslant 1) \tag{2.5}$$

Eq. (2.4) ensures that f_0 is Maxwellian; as a consequence, Eq. (1.6) of Chapter III gives

$$\left(\frac{\partial}{\partial t} + \boldsymbol{\xi}\cdot\frac{\partial}{\partial \mathbf{x}}\right)(f_0 h_{n-1}) = f_0 L h_n + S_n \qquad (n = 1, 2, \ldots) \tag{2.6}$$

$$S_1 = 0; \qquad S_n = \sum_{k=1}^{n-1} Q(f_0 h_k, f_0 h_{n-k}) \qquad (n \geqslant 2) \tag{2.7}$$

$$f_n = f_0 h_n; \qquad h_0 = 1 \tag{2.8}$$

Accordingly, we have a sequence of equations for the unknowns h_n; we can solve these equations step by step, by noting that they have the form shown in Eq. (IV.6.26). According to the Corollary at the end of Section 6 of Chapter IV, at each step we can find h_n provided the five conditions expressed by Eq. (IV.6.27) (orthogonality to the five collision invariants) are satisfied by the source term, but h_n is determined up to a linear combination of the five collision invariants ψ_α, involving five parameters c_n^{α} (which can depend upon the time and space variables). The source term in the n-th step

$$g_n = \left[\left(\frac{\partial}{\partial t} + \boldsymbol{\xi} \cdot \frac{\partial}{\partial \mathbf{x}}\right)(f_0 h_{n-1}) - S_n\right] \Big/ f_0 \tag{2.9}$$

is constructed by means of the previous approximations. The two circumstances mentioned above combine cyclically in such a way that the five orthogonality conditions on the n-th source term "determine" the five parameters left unspecified by the $(n - 1)$-th step; the start of the cycle is made possible by the fact that the zero-th order approximation already contains five disposable parameters (the density, the mass velocity components and the temperature of the Maxwellian f_0).

Let us now see in what sense the orthogonality conditions, Eq. (IV.6.27) with $g = g_n$ determine the five disposable parameters c_{n-1}^{α}.

Because of Eqs. (2.7) and (II.6.40), S_n/f_0 automatically satisfies the orthogonality conditions and we are left with

$$\left(f_0^{-1}\psi_\alpha, \left[\frac{\partial}{\partial t} + \boldsymbol{\xi} \cdot \frac{\partial}{\partial \mathbf{x}}\right] f_0 h_{n-1}\right) = 0 \qquad (n \geqslant 1) \tag{2.10}$$

or

$$\int \psi_\alpha \left(\frac{\partial}{\partial t} + \boldsymbol{\xi} \cdot \frac{\partial}{\partial \mathbf{x}}\right) f_n \, d\boldsymbol{\xi} = 0 \qquad (n \geqslant 0) \tag{2.11}$$

Since the collision invariants depend only upon $\boldsymbol{\xi}$, we can exchange the order of the differentiations and integrations under suitable assumptions and write

$$\frac{\partial \rho_n^{\alpha}}{\partial t} + \operatorname{div} \mathbf{j}_n^{\alpha} = 0 \qquad (n \geqslant 0; \alpha = 0, 1, 2, 3, 4) \tag{2.12}$$

where

$$\rho_n^{\alpha} = \int \psi_\alpha f_n \, d\boldsymbol{\xi}; \qquad \mathbf{j}_n^{\alpha} = \int \boldsymbol{\xi} \psi_\alpha f_n \, d\boldsymbol{\xi} \tag{2.13}$$

constitute the n-th order terms of the expansion of

$$\rho^{\alpha} = \int \psi_\alpha f \, d\boldsymbol{\xi}; \qquad \mathbf{j}^{\alpha} = \int \boldsymbol{\xi} \psi_\alpha f \, d\boldsymbol{\xi} \tag{2.14}$$

into a series of powers of ε (any expansion of f into a series implies a related expansion of its moments $\int \xi_1^{n_1} \xi_2^{n_2} \xi_3^{n_3} f \, d\xi$).

The ρ^α can be described as a five-dimensional (abstract) vector and $\mathbf{j}^\alpha$ as a 3×5 matrix; their components can be expressed in terms of the physical quantities defined in Chapter II, Section 8;

$$((\rho^\alpha)) = \left(\left(\begin{matrix} \rho \\ \rho v_i \\ \rho(\frac{1}{2}v^2 + e) \end{matrix}\right)\right); \qquad ((\mathbf{j}^\alpha)) = \left(\left(\begin{matrix} \rho v_i \\ \rho v_i v_j + p_{ij} \\ \rho v_i(\frac{1}{2}v^2 + e) + p_{ij}v_j + q_i \end{matrix}\right)\right)$$

$$(i, j = 1, 2, 3; \qquad \alpha = 0, 1, 2, 3, 4) \quad (2.15)$$

In terms of ρ^α and $\mathbf{j}^\alpha$ the conservation equations, Eq. (II.8.24) with $X_i = 0$, can be written as follows:

$$\frac{\partial \rho^\alpha}{\partial t} + \operatorname{div} \mathbf{j}^\alpha = 0 \qquad (\alpha = 0, 1, 2, 3, 4) \quad (2.16)$$

Accordingly, the compatibility equations, Eq. (2.12), are simply the n-th terms of the expansion of Eqs. (2.16). The expanded equations, however, contain more than Eqs. (2.16). In order to show this, let us remark that the only unknown quantities in the expression for f_n are the five coefficients $c_n{}^\alpha$ of the linear combination of ψ_α's to be added to $f_0 h_n^{(0)}$, where $h_n^{(0)} \in W$ (the subspace orthogonal to the collision invariants) is uniquely determined in terms of the previous approximations; hence

$$\rho_n{}^\alpha = \int \psi_\alpha f_0 h_n^{(0)} \, d\boldsymbol{\xi} + \sum_{\beta=0}^{4} c_n{}^\beta \int \psi_\alpha f_0 \psi_\beta \, d\boldsymbol{\xi} \quad (2.17)$$

This equation can be used to express the $c_n{}^\beta$ in terms of the $\rho_n{}^\alpha$ (the determinant of the matrix $((\int \psi_\alpha f_0 \psi_\beta \, d\boldsymbol{\xi}))$ is easily shown to be different from zero); since this is true for each n, the coefficients $c_n{}^\alpha$ are expressed in terms of the $\rho_n{}^\alpha$; that is, the only unknown quantities in f_n can be considered to be the five $\rho_n{}^\alpha$. But this implies that the $\mathbf{j}_n{}^\alpha$ are also expressed in terms of the $\rho_n{}^\alpha$, so that we have found (in the form of a series expansion) a relation between the stress tensor and the heat flux, on the one hand, and the basic macroscopic unknowns (density, velocity, temperature, or internal energy) on the other. This means that the present procedure accomplishes (at least formally) the closure of the system of conservation equations, and extracts a macroscopic model based upon the concepts of density, velocity and temperature from the microscopic description based upon a distribution function!

In order to obtain a better appreciation of the situation, let us consider the conservation equations written as follows:

$$E^\alpha(\rho^\beta) = S^\alpha \quad (2.18)$$

where E^α is the (nonlinear) Euler operator (such that $E^\alpha(\rho^\beta) = 0$ gives the inviscid fluid equations) and

$$S^\alpha = \left(\left(\begin{array}{c} 0 \\ -\dfrac{\partial}{\partial x_j}(p_{ij} - p\delta_{ij}) \\ -\dfrac{\partial}{\partial x_i}(p_{ij}v_j - pv_i + q_i) \end{array}\right)\right) \tag{2.19}$$

When ρ^β is expanded into a series of powers of ε

$$E^\alpha(\rho^\beta) = E^\alpha(\rho_0{}^\beta) + \sum_{n=1}^{\infty} \varepsilon^n E^{\alpha\beta}\rho_n{}^\beta + \sum_{n=2}^{\infty} \varepsilon^n E_n{}^\alpha(\rho_k{}^\beta) \tag{2.20}$$

where a sum from 0 to 4 over repeated greek indices is understood.

Here $E^{\alpha\beta}$ is an operator which depends upon the $\rho_0{}^\beta$ and acts linearly, if the latter are regarded as known; the nonlinear expressions $E_n{}^\alpha(\rho_k{}^\beta)$ contain only the $\rho_k{}^\alpha$ with $k \leqslant n - 1$. When we expand the ρ^β, Eq. (2.18) gives

$$E^\alpha(\rho_0{}^\beta) = 0 \tag{2.21}$$

$$E^{\alpha\beta}\rho_n{}^\beta = S_n{}^\alpha - E_n{}^\alpha \qquad (n \geqslant 1;\, E_1{}^\alpha = 0) \tag{2.22}$$

where

$$S_n{}^\alpha = \left(\left(\begin{array}{c} 0 \\ -\dfrac{\partial}{\partial x_j}(p_{ij}^{(n)} - p^{(n)}\,\delta_{ij}) \\ -\dfrac{\partial}{\partial x_j}\left[\sum_{k=1}^{n}(p_{ij}^{(k)} - p^{(k)}\,\delta_{ij})v_i^{(n-k)} + q_j^{(n)}\right] \end{array}\right)\right) \tag{2.23}$$

Here $v_i^{(n)}$, $p_{ij}^{(n)}$, $p^{(n)}$ and $q_i^{(n)}$ denote, respectively, the n-th terms of the power expansions of v_i, p_{ij}, p and q_i. The fact that the zero-th order equation, Eq. (2.22), has zero source term comes from the fact that f_0 is Maxwellian and hence has an isotropic stress tensor and a zero heat flux vector.

We must now explain why we have treated the terms in $S_n{}^\alpha$ as source terms. We observe that

$$\mathbf{j}_n{}^\alpha = \int \psi_\alpha \boldsymbol{\xi} f_n \, d\boldsymbol{\xi} = \int \psi_\alpha \boldsymbol{\xi} f_0 h_n^{(0)} \, d\boldsymbol{\xi} + \sum_{\beta=0}^{4} c_n{}^\beta \int \boldsymbol{\xi}\psi_\alpha\psi_\beta f_0 \, d\boldsymbol{\xi} \tag{2.24}$$

It can easily be seen [by explicitly performing the integrations not involving $h_n^{(0)}$ in Eq. (2.17) and (2.24)] that $\partial\rho_n{}^\alpha/\partial t + \operatorname{div} \mathbf{j}_n{}^\alpha (n \geqslant 1)$ can be written as $E^{\alpha\beta}\rho_n{}^\beta$ plus something depending only on $h_n{}^0$, and hence only on the $\rho_k{}^\alpha$ $(k \leqslant n - 1)$; accordingly, the n-th order contributions to S^α are expressed in terms of the $\rho_k{}^\alpha$ $(k \leqslant n - 1)$ and can be regarded as known at the n-th step.

In conclusion, the situation is the following: we have to solve the ordinary inviscid fluid equations at the zero-th level of approximation, and inhomogeneous linearized inviscid equations at the next steps.

As pointed out above, the main result of the Hilbert expansion is that, if we grant the possibility of expanding the distribution function into a power series in the Knudsen number, then we can extract a macroscopic description of the gas in terms of density, mass, velocity and temperature. This description is essentially in terms of the inviscid fluid equations, but contains corrections which can be computed by solving linearized equations. It must be noted, however, that we have proceeded formally, and an obvious question is: when are we allowed to make the Hilbert expansion?

In order to answer this question, let us remark that the Hilbert expansion cannot provide uniformly valid solutions. This is suggested by the singular manner in which the parameter ε enters into the Boltzmann equation (compare with the nonanalytic character of the solutions of $\varepsilon\, \partial f/\partial t + f = 0$ at $\varepsilon = 0$). Also, we know that the inviscid fluid equations are unrealistic and incapable of dealing with certain situations; what is worse, we know that regular perturbation methods are not capable of correcting the unsatisfactory features of an inviscid fluid description at subsequent steps. The latter difficulty shows up, in particular, in studying (viscous) boundary layers and the last stage of evolution in time dependent problems.

These circumstances, however, do not prevent a truncated Hilbert expansion from representing solutions of the Boltzmann equation with arbitrary accuracy in suitably chosen space-time regions (which will be called the normal regions), provided that we stay at a finite distance from certain singular surfaces and ε is sufficiently small. In fact, if we substitute a truncated Hilbert expansion into the Boltzmann equation, the latter is satisfied except for an error term of order ε^n; therefore the Hilbert expansion can be used to approximate certain solutions (normal solutions) of the Boltzmann equation, the error being arbitrarily small for a sufficiently small ε [a rigorous proof with estimates is available in the case of the linearized Boltzmann equation [3]]. In order to see what kind of solutions these normal solutions are, we can observe that the above mentioned remainder of order ε^n contains space derivatives of order n; accordingly, a high degree of smoothness is required for the normal solutions to exist. This suggests that the normal solutions cease to be valid in space-time regions where the density, velocity and temperature profiles tend to be very steep; such regions are immediately identified in the neighborhood of boundaries (boundary layers), the initial stage (initial layer) and shock waves (shock layers). The first two kinds of layer show up also in the linearized Boltzmann equation, as appears from Eqs. (IV.6.9), where $\mu \sim \varepsilon^{-1}$, and Eqs. (II.7.45), where $\lambda_0 \sim \varepsilon^{-1}$; a shock layer is a region of large gradients produced inside the region where the gas flows and endowed with a structure intimately tied to the nonlinearity of the Boltzmann equation. In the abovementioned layers f sensibly changes on the scale of a mean free path and, accordingly, $\partial f/\partial t$ or $\partial f/\partial \mathbf{x}$ (or both) are of order f/ε, while the Hilbert expansion treats the space and time derivatives of f as if they were of the same order as f. An additional region, where the Hilbert expansion fails, is given by the "final layer", which is the evolution at times of order $1/\varepsilon$; on such a scale $\partial f/\partial t \sim \varepsilon f$ is negligible with respect to f, but a nonuniform expansion arises because $\partial f/\partial \mathbf{x} \sim f$ is forced to be of the same order of magnitude as $\partial f/\partial t$ and, as a consequence, secular terms are introduced by higher approximations [4, 5].

According to this discussion, the normal solutions are capable of approximating (for sufficiently small ε) the solutions of arbitrary problems, provided the abovementioned layers are excluded. However, in order to solve the differential equations which, according to the Hilbert method, regulate the fluid variables ρ^α, it is necessary to complete them with suitable initial data, boundary conditions, or matching conditions across a shock. That is, it is necessary to pass through those regions where the theory does not hold. It is evident that to complete the theory it is necessary to solve the three connection problems across the layers within which the Hilbert expansion fails:

(1) To relate a given initial distribution function to the Hilbert solution which takes over after an initial transient.

(2) To relate a given boundary condition on the distribution function to the Hilbert solution which holds outside the boundary layer.

(3) To find the correct matching conditions for the two Hilbert solutions prevailing on each side of a shock layer.

The present state of these connection problems will be reviewed in Section 5.

3. The Chapman-Enskog expansion

As was remarked above, the solutions of both the continuum conservation equations and the Boltzmann equation are, in general, nonanalytic in a certain parameter ε describing the deviation from the inviscid fluid equations (the viscosity and heat conduction coefficients in continuum theory and the mean free path in kinetic theory). Accordingly, series expansions in powers of ε fail to give uniformly valid solutions for specific initial and boundary value problems; some of the troubles can be avoided, however, if we do not expand the solutions, and expand the equations instead, as is done by the so-called Chapman-Enskog expansion. To understand what we mean, let us remark that if we multiply Eqs. (2.22) by ε^n, sum from 1 to ∞ and add Eq. (2.21) to the result, we have

$$E^\alpha(\rho^\beta) = \sum_{n=1}^{\infty} \varepsilon^n S_n{}^\alpha = S^\alpha(\rho^\beta) \tag{3.1}$$

where $S^\alpha(\rho^\beta)$ is the same source term as in Eq. (2.18), but is now expressed as a nonlinear operator acting on the ρ^β, while E^α is the nonlinear Euler operator [Eq. (2.20) has been used]. The operator S^α approaches zero with ε, and at present is known only when the ρ^β are expanded into a power series in ε (this is the basic result of the Hilbert expansion). We can, however, assume that S^α exists (at least in an asymptotic sense for $\varepsilon \to 0$) even when the ρ^β are not expanded. In fact, although the Hilbert expansion does not give uniformly valid solutions, it does give valid solutions if we restrict ourselves to the normal regions, as discussed in Section 2; hence the existence of S^α almost everywhere follows (at least in an asymptotic sense for $\varepsilon \to 0$). We can think of $S_n{}^\alpha$ as containing only space derivatives and not time derivatives, because the latter can always be eliminated from Eq. (2.6) by observing that $f_{n-1} = f_0 h_{n-1}$ varies with time and space variables through its dependence upon the $\rho_k{}^\alpha(k \leqslant n-1)$, whose time derivatives are known in terms of space derivatives according to Eq. (2.12). As a consequence, the operator S^α appearing in Eq. (3.1) can be thought of as acting only upon the space dependence of the ρ^β.

The idea of the Chapman-Enskog method [6–8] is to expand S^α while leaving the ρ^β unexpanded; this assumes that although the dependence of the ρ^β upon ε is nonanalytic at $\varepsilon = 0$ in many cases, the operator S^α is analytic in ε at $\varepsilon = 0$ (or, at least, possesses an asymptotic expansion in powers of ε). This assumption is far from being contradictory, since, for example, the Navier-Stokes equations depend analytically (actually linearly) upon the viscosity and heat conduction coefficients, but have, in general, solutions which cannot be expanded into a power series in these parameters. In order to formalize this notion into an algorithm, we remark that Eq. (3.1) can be written as follows:

$$\partial \rho^\beta / \partial t = D_\beta(\rho^\alpha) \tag{3.2}$$

where D_β is a nonlinear operator acting upon the space dependence of the ρ^α and is obtained by subtracting the space derivatives in E^α from $S^\alpha(\rho^\beta)$. Expanding S^α clearly means expanding D_β as follows:

$$D_\beta = \sum_{n=0}^{\infty} \varepsilon^n D_\beta^{(n)} \tag{3.3}$$

where $D_\beta^{(n)}$ are nonlinear operators to be found by the present method; in particular, $D_\beta^{(0)}$ will consist of the space derivatives contained in $E^{\alpha\beta}$ (with a sign change). The expansion of D_β means that we regard the time evolution of ρ^β as influenced by processes of different orders of magnitude in ε.

Since our present expansion essentially amounts to a reordering of the Hilbert expansion according to a new criterion, we must also maintain a basic result of the latter expansion which is also a prerequisite for extracting a self-contained macroscopic theory from the Boltzmann equation; that is, that the distribution function depends upon the space and time variables only through a functional dependence on the ρ^α. In other words, we have

$$\frac{\partial f}{\partial t} = \sum_{k=0}^{\infty} \frac{\partial f}{\partial(\nabla^k \rho^\alpha)} \frac{\partial(\nabla^k \rho^\alpha)}{\partial t} \tag{3.4}$$

where ∇^k formally denotes the n-th order space derivatives. Since Eq. (3.2) gives:

$$\frac{\partial}{\partial t} \nabla^k \rho^\beta = \nabla^k D_\beta(\rho^\alpha) \tag{3.5}$$

the expansion of D_β, Eq. (3.2), implies an expansion of the operator giving the time evolution of f as well. We can write this formally as follows:

$$\frac{\partial f}{\partial t} = \sum_{n=0}^{\infty} \varepsilon^n \frac{\partial^{(n)} f}{\partial t} \tag{3.6}$$

where $\partial^{(n)} f/\partial t$ simply denotes the contribution to $\partial f/\partial t$ coming from $D_\beta^{(n)}$ through Eqs. (3.3)–(3.5). The expressions of the operators $\partial^{(n)}/\partial t$ (or equivalently, $D_\beta^{(n)}$) and the dependence of f upon the ρ^α are the unknowns of the Chapman-Enskog method.

As noted above, the ρ^α are left unexpanded; we must, however, expand f if we want to avoid trivial results. To use these two requirements consistently, we simply have to write

$$f = \sum_{n=0}^{\infty} \varepsilon^n f_n \tag{3.7}$$

and

$$\int \psi_\alpha f_n \, d\boldsymbol{\xi} = 0 \qquad (n = 0, 1, \ldots; \alpha = 0, 1, 2, 3, 4) \tag{3.8}$$

This implies that f as a function of $\mathbf{x}$, $\boldsymbol{\xi}$, t *is not* expanded into a series of functions of ε; in fact, the coefficients f_n depend on ε in a complex way. However, the dependence of f_n upon ε is only through the ρ^α; this defines the algorithm uniquely, since it implies that f as a functional of ρ^α is expanded into a series of powers of ε. If we now substitute Eqs. (3.6) and (3.7) into Eq. (2.1), we find

$$Q_0 = 0 \tag{3.9}$$

$$\sum_{k=0}^{n-1} \frac{\partial^{(k)}(f_0 h_{n-k-1})}{\partial t} + \boldsymbol{\xi} \cdot \frac{\partial(f_0 h_{n-1})}{\partial \mathbf{x}} = f_0 L h_n + S_n \qquad (n = 1, 2, \ldots) \tag{3.10}$$

where the notation is the same as in Section 2, Eqs. (2.6)–(2.8).

Eq. (3.9) shows that f_0 is again Maxwellian. There is, however, a basic difference from the Hilbert method; the fluid variables (density, velocity, temperature) which appear in the Maxwellian are now exact (unexpanded), while only the zero-th order approximation to the fluid variables appeared in the zero-th-order distribution function of the Hilbert method.

Eq. (3.9) is again of the general form, Eq. (IV.6.26), discussed in Chapter IV, Section 6. The orthogonality conditions, Eq. (IV.6.27), now take the following form

$$\sum_{k=0}^{n} \frac{\partial^{(k)}}{\partial t} \int \psi_\alpha f_{n-k}\, d\boldsymbol{\xi} + \operatorname{div}\left(\int \boldsymbol{\xi} \psi_\alpha f_n \, d\boldsymbol{\xi} \right) = 0 \qquad (n = 0, 1, \ldots) \quad (3.11)$$

Now, however, no disposable parameters are available, since Eq. (3.8) implies that $h_n \in W(n \geqslant 1)$, and so $h_n(n \geqslant 1)$ is uniquely determined as a function of $\boldsymbol{\xi}$ and as a functional of the ρ^α.

Eq. (3.8) also implies that all the integrals appearing in the sum in Eq. (3.11) except one are zero, and the nonzero one is simply equal to ρ^α; accordingly:

$$\frac{\partial^{(n)} \rho^\alpha}{\partial t} + \operatorname{div}\left(\int \boldsymbol{\xi} \psi_\alpha f_n \, d\boldsymbol{\xi} \right) = 0 \qquad (n = 0, 1, \ldots) \qquad (3.12)$$

Satisfying these compatibility conditions means solving our problem and finding $\partial^{(n)} \rho^\alpha / \partial t$ (or $D_\alpha^{(n)}$); as a matter of fact, Eq. (3.12) gives $D_\alpha^{(n)}$ explicitly in terms of space derivatives of the ρ^α once we have solved the n-th equation of the hierarchy, Eq. (3.10).

The net result of the Chapman-Enskog expansion is that we can write the N-th approximation to the macroscopic equations in the following way:

$$\frac{\partial \rho^\alpha}{\partial t} + \operatorname{div}\left[\int \boldsymbol{\xi} \psi_\alpha \left(\sum_{n=0}^{N} \varepsilon^n f_n \right) d\boldsymbol{\xi} \right] = 0 \qquad (3.13)$$

where the f_n are functionals of the ρ^α and functions of $\boldsymbol{\xi}$, explicitly known from the step by step solution. The main difference from the Hilbert procedure, which was based on a nonlinear zeroth-order equation plus linear equations to evaluate higher-order corrections, is that now we choose a fixed N and evaluate the solution at that level; if we want a higher-order solution, we have to solve a completely new, more complicated set of nonlinear equations.

At the zeroth level we have the inviscid fluid equations again, because f_0 is still a Maxwellian. In order to see what happens at the next level, we write Eq. (3.10) for $n = 1$

$$\frac{1}{\rho}\left(\frac{\partial^{(0)}}{\partial t} + \xi_i \frac{\partial}{\partial x_i} \right) \rho + \frac{1}{T}\left(\frac{c^2}{2RT} - \frac{3}{2} \right)\left(\frac{\partial^{(0)}}{\partial t} + \xi_i \frac{\partial}{\partial x_i} \right) T +$$
$$+ \frac{1}{RT} c_k \left(\frac{\partial^{(0)}}{\partial t} + \xi_i \frac{\partial}{\partial x_i} \right) v_k = Lh \quad (3.14)$$

where the explicit form of f_0 has been used and $\mathbf{c} = \boldsymbol{\xi} - \mathbf{v}$, as usual. Eq.

(3.12) now gives ($n = 0$):

$$\frac{\partial^{(0)}\rho}{\partial t} = -\frac{\partial}{\partial x_i}(\rho v_i)$$

$$\frac{\partial^{(0)}v_k}{\partial t} = -v_i\frac{\partial v_k}{\partial x_i} - \frac{1}{\rho}\frac{\partial p}{\partial x_i} \qquad (p = \rho RT) \tag{3.15}$$

$$\frac{\partial^{(0)}T}{\partial t} = -v_i\frac{\partial T}{\partial x_i} - \tfrac{2}{3}T\frac{\partial v_j}{\partial x_j}$$

These relations, substituted back into Eq. (3.14), give

$$\left(\frac{c^2}{2RT} - \frac{5}{2}\right)c_i\frac{1}{T}\frac{\partial T}{\partial x_i} + \frac{1}{RT}(c_ic_k - \tfrac{1}{3}c^2\,\delta_{ik})\frac{\partial v_i}{\partial x_k} = Lh_1 \tag{3.16}$$

This equation can be used to evaluate h_1 or, equivalently, $f_1 = f_0h_1$ which is required in order to write Eq. (3.13) for $N = 1$ in an explicit form. Eq. (3.16) must be solved in W, since both the left hand side and h_1 belong to W; since L has an inverse L^{-1} in W, we can write

$$h_1 = L^{-1}(c_ic_k - \tfrac{1}{3}c^2\,\delta_{ik})\frac{1}{RT}\frac{\partial v_i}{\partial x_k} + L^{-1}[c_i(c^2 - 5RT)]\frac{1}{2RT^2}\frac{\partial T}{\partial x_i}$$

$$= (c_ic_k - \tfrac{1}{3}c^2\,\delta_{ik})A(c;\,T)(\rho RT)^{-1}\frac{\partial v_i}{\partial x_k} + c_i(2RT^2)^{-1}B(c;\,T)\frac{\partial T}{\partial x_i} \tag{3.17}$$

where the functions $A(c;\,T)$, and $B(c;\,T)$ defined by

$$\rho^{-1}A(c;\,T)(c_ic_j - \tfrac{1}{3}c^2\,\delta_{ij}) = L^{-1}(c_ic_j - \tfrac{1}{3}c^2\,\delta_{ij})$$

$$c_i\rho^{-1}B(c;\,T) = L^{-1}[c_i(c^2 - 5RT)] \tag{3.18}$$

were already introduced in Section 9 of Chapter IV, where the dependence upon $T(= T_0)$ was not explicitly indicated. f_0h_1 gives the following contributions to the stress tensor and heat flux:

$$p_{ij}^{(1)} = -\mu\left(\frac{\partial v_i}{\partial x_j} + \frac{\partial v_j}{\partial x_i}\right) + \tfrac{2}{3}\mu\frac{\partial v_k}{\partial x_k}\delta_{ij}$$

$$q_i^{(1)} = -\kappa\frac{\partial T}{\partial x_i} \tag{3.19}$$

where μ and κ are given by Eqs. (IV.7.51) and (IV.7.54) with T in place of T_0. Accordingly the first two terms ($n = 0, 1$) of the Chapman-Enskog expansion provide us with a macroscopic model of the Navier-Stokes type, the transport coefficients depending only upon temperature and molecular constants (see remarks in Section 7 of Chapter IV).

The actual computation of μ and κ [via $A(c; T)$ and $B(c; T)$ which according to Eqs. (3.18) appear in Eqs. (IV.7.51) and (IV.7.52)] from specific molecular models is a thoroughly investigated subject [8–10] and will not be considered in detail here. We only remark that, according to Eqs. (3.18), the problem of computing $A(c; T)$ reduces to the solution of Eq. (IV.6.26) for a given polynomial $g \in W$ (and $h \in W$). A method for solving such equation is to expand h into a series of the eigenfunctions of the Maxwell molecules, g_{nlm} ($l = 1, 2$; $0 \leqslant n \leqslant \infty$); this method is trivially exact if we are dealing with Maxwell molecules, for which the result is $A(c; T) = A(T)$, $B(c; T) = B(T)(c^2 - 5RT)$ with $A(T)$ and $B(T)$ independent of c. Another method is to use the variational principle discussed at the end of Section 6 of Chapter IV. One takes a trial function $\tilde{h}$ containing a certain number of constants and minimizes the functional $J(\tilde{h})$ defined by Eq. (IV.6.28) by a suitable choice of the constants; the resulting $\tilde{h}$ gives an approximation to h. The usefulness of the method is further enhanced by the circumstance that one can relate the value assumed by $J(\tilde{h})$ for $\tilde{h} = h$ to the values of the transport coefficients; in fact, one finds from Eq. (IV.6.28) with $\tilde{h} = h$ (and hence $Lh = g$):

$$J[\rho^{-1}A(c; T)c_1c_2] = (c_1c_2, \rho^{-1}A(c; T)c_1c_2) = \int f_0 c_1 c_2 L^{-1}(c_1c_2)\, d\mathbf{c} = \mu \tag{3.20}$$

$$\begin{aligned} J[c_1\rho^{-1}B(c; T)] &= (c_1(c^2 - 5RT), \rho^{-1}c_1 B(c; T]) \\ &= -\int f_0 c_1 c^2 L^{-1}[c_1(c^2 - 5RT)]\, d\mathbf{c} \\ &= 4RT^2\kappa \end{aligned} \tag{3.21}$$

where Eqs. (IV.7.51) and (IV.7.52) (with $\boldsymbol{\xi} = \mathbf{c}$ and $T_0 = T$) have been used. Due to this fact a relative error, of order, say, δ, in approximating h by $\tilde{h}$ implies a relative error of order δ^2 in evaluating μ and κ.

Thus, the basic result of the Chapman-Enskog procedure is that we can recover the Navier-Stokes-Fourier macroscopic description of a gas by a suitable expansion of certain solutions of the Boltzmann equation. In this way, one of the many nonuniformities of the Hilbert expansion is avoided; the viscous boundary layers (with thickness of order $\varepsilon^{\frac{1}{2}}$) and the final layer (of order ε^{-1}) are incorporated into a single description together with the normal regions, while the initial and Knudsen layers with thickness of order ε^{-1} are still left out. The Chapman-Enskog theory simply takes into account the existence of regimes with $d^2(\varepsilon\tau)^{-1} \simeq 1$ (where τ and d are a typical time and a typical length; τ may be replaced by another length, different from d). Thus we may expect the Chapman-Enskog theory to be much more accurate than the Hilbert theory. If we consider, however, higher order approximations of the Chapman-Enskog method, we obtain differential equations of higher

The presence of the second term in Eq. (4.6) requires a restriction on the expansion of f. If we take into account that the splitting in Eq. (4.6) is based on the interplay between terms of the n-th and $(n + 1)$-th order in ε, the simplest choice is to assume that there is no contribution to the fluid variables from the odd order terms:

$$\int \psi_\alpha f_n \, d\boldsymbol{\xi} = 0 \qquad (n = 1, 3, 5, \ldots) \tag{4.8}$$

It is clear that the proposed expansion is only a particular instance of the infinitely many possible expansions of this kind. The most general is based upon a truncated expansion of the time derivatives [Eq. (3.6) truncated at $n = N$] plus certain conditions regulating the contribution of different orders to the fluid variables [Eq. (3.8) for $n \neq (N + 1)k$; $k = 1, 2, 3, \ldots$]. The Hilbert expansion corresponds to $N = 0$ and the Chapman-Enskog expansion to $N = \infty$; the particular procedure which has just been described corresponds to $N = 1$.

This particular choice is suggested by the information available from the macroscopic description, but is also dictated by a comparison with the results of Chapter IV. There (Sections 7 and 11) we found that the solution of a steady problem, linearized about a Maxwellian with zero mass velocity, is made of two parts, one h_A, essentially equivalent to a Chapman-Enskog solution with variables satisfying the Navier-Stokes-Fourier equations and the other h_B, which contains exponentials such as $\exp(-x/\varepsilon)$ (remember that $\lambda_0 \simeq \varepsilon$) and hence cannot be expanded into a power series in ε. By taking $N = 1$, that is, by choosing Eqs. (4.6)–(4.8) to compute f we are sure that, at least for steady linearized flows, we keep all the necessary terms and reject all the spurious solutions. It is interesting to remark that the *Hilbert* expansion gives the same result when applied to the particular case of steady flows [14] linearized about a Maxwellian with zero mass velocity.

All the above expansions when truncated satisfy the Boltzmann equation with an error term $\varepsilon^n R_n(\mathbf{x}, \boldsymbol{\xi}, t; \varepsilon)$ which is formally of order ε^n. For the Hilbert expansion R_n does not depend upon ε but grows algebraically as t^n in time dependent problems (because of secular terms); hence the Hilbert expansion is asymptotic for only a limited time $t_0 < t < t_1$. No estimates are of course available for the Chapman-Enskog remainder beyond the Navier-Stokes level. The procedure embodied in Eqs. (4.6)–(4.8) leads to a remainder which decays for large t at any order $n > 1$; the corresponding expansion is therefore superior to the Hilbert series in range of validity and to the Chapman-Enskog series in avoiding spurious solutions and leading to a known set of partial differential equations.

We end this section with some remarks about a question which is frequently asked: does the Chapman-Enskog expansion converge? This is a difficult question in general, but some results can be obtained in the case of the linearized Boltzmann equation. In this connection, we remark that both the linearization of the Boltzmann equation and the Chapman-Enskog expansion are the result of suitable perturbation procedures applied to the Boltzmann equation. The two procedures are basically different because the expansion parameters are completely different: the deviation of the initial and boundary distributions from a uniform Maxwellian distribution in the case of linearization and the ratio of the mean free path or mean free time with respect to other typical lengths or times in the case of the Chapman-Enskog expansion. Thus local gradients in the latter case and global differences in the former are to be small. It is clear that when both circumstances are realized we can apply the Chapman-Enskog method to the linearized Boltzmann equation. One of the main reasons for doing this comes from the possibility of answering certain questions concerning the linearized case in order to gain insight into the more complicated nonlinear case.

As we know, the Chapman-Enskog series essentially consists in the expansion of an operator; accordingly, its convergence has a meaning only if referred to a certain class of functions upon which the operator acts. In the case of the linearized Boltzmann equation, it is easy to exhibit normal solutions whose expansions are trivially convergent because they contain just a finite number of terms; such as, for example, the asymptotic part, h_A, of the general solution for one-dimensional problems, Eq. (IV.7.44). In order to investigate less trivial cases, we consider the separated variable solutions, discussed in Chapter IV, Section 8. If such a solution is written in the form shown in Eq. (IV.8.1), it is clear that the Chapman-Enskog series for such a solution will converge or not, depending on whether the expansion of $\omega = \omega(\mathbf{k})[\omega(0) = 0]$ into a series of powers of $\mathbf{k}$ converges or not. According to the discussion in Chapter IV, Section 8 we can conclude that there is convergence for rigid spheres when $|\mathbf{k}|$ is sufficiently small and conjecture that convergence fails for potentials with angular cutoff. An analogous conjecture is that the radius of convergence in $|\mathbf{k}|$ for rigid spheres is not infinity. It is clear that a convergence for $|\mathbf{k}| < \varepsilon_0$ implies convergence of the Chapman-Enskog expansion for a very restricted type of space dependence; all the derivatives of the fluid variables must be bounded uniformly with respect to the order, and this means that they are not only analytic, but also entire functions.

5. Initial, boundary and shock layers

As mentioned in Section 2, the Hilbert theory is not complete. In order to complete it, we have to solve the three connection problems concerning the initial, boundary and shock layers; the same connection problems arise for the Chapman-Enskog expansion as well as for the modified expansion proposed in Section 4.

We first consider the problem of the initial layer, following a paper by Grad [3]. A complete theory should deal with the connection of the above-mentioned expansions with an arbitrary initial datum; such a theory, however, would imply the solution of the nonlinear Boltzmann equation, and would be completely impractical. If we take the spirit of the Hilbert series into account, we can, however, restrict ourselves to matching an initial datum of the same type as the solution and adopt a datum which reduces to a Maxwellian when $\varepsilon \to 0$; accordingly the initial datum $\bar{f}$ will be assumed to be arbitrary within the condition that it can be written in the form $f_M + \varepsilon f_N$, where f_M is Maxwellian.

Let us rescale the time variable as follows:

$$\tau = t/\varepsilon \tag{5.1}$$

in order to give the initial layer a finite duration. We now look for solutions of the Boltzmann equation, Eq. (2.1), in the form

$$f = f_H(\mathbf{x}, \boldsymbol{\xi}, \varepsilon\tau; \varepsilon) + \varepsilon f_R(\mathbf{x}, \boldsymbol{\xi}, \tau; \varepsilon) \tag{5.2}$$

where the leading term $f_H = f_H(\mathbf{x}, \boldsymbol{\xi}, t; \varepsilon)$ is given by a Hilbert expansion and the "remainder" f_R is expanded as follows:

$$\varepsilon f_R = \sum_{n=0}^{\infty} \varepsilon^{n+1} f_{R(n+1)}(\mathbf{x}, \boldsymbol{\xi}, \tau) \tag{5.3}$$

The additional factor ε comes from the above assumption on the allowed initial states. We remark that, because we let $t = \varepsilon\tau$, the Hilbert expansion for f_H is no longer a power series in ε; we have then to re-expand as follows:

$$f_H = \sum_{n=0}^{\infty} \varepsilon^n f_{H(n)}(\mathbf{x}, \boldsymbol{\xi}, \tau) \tag{5.4}$$

in which the leading term $f_{H(0)}$ is time independent and locally Maxwellian. We also remark that the expansion in Eq. (5.4) coincides with the original Hilbert expansion at $t = \tau = 0$. The Hilbert expansion formally satisfies the Boltzmann equation; as a consequence, substituting Eq. (5.2) into Eq. (2.1) gives:

$$\frac{\partial f_R}{\partial t} + \varepsilon\boldsymbol{\xi} \cdot \frac{\partial f_R}{\partial \mathbf{x}} = 2Q(f_H, f_R) + \varepsilon Q(f_R, f_R) \tag{5.5}$$

Substituting Eqs. (5.3) and (5.4) into Eq. (5.5) yields:

$$\partial g_n/\partial\tau = Lg_n + G_n \tag{5.6}$$

where

$$f_{R(n)} = f_{H(0)} g_n \qquad (n \geqslant 1)$$

$$G_n = \left\{ -\boldsymbol{\xi} \cdot \frac{\partial_{R(n-1)}}{\partial \mathbf{x}} + \sum_{k=1}^{n-1} [2Q(f_{H(n-k)}, f_{R(k+1)}) + \right.$$

$$\left. + Q(f_{R(k+1)}, f_{R(n-k)})] \right\} f_{H(0)}^{-1} \tag{5.7}$$

The linearized collision operator L is based on the Maxwellian $f_{(H0)}$; accordingly, its eigenvalues and eigenfunctions are space- but not time-dependent. Since f, as given by Eq. (5.2) must satisfy the initial conditions, $f_{H(0)}$ is nothing other than the Maxwellian f_H appearing in the initial data: this defines L completely. The next step is to expand the initial datum $\bar{f}$ into a power series in ε:

$$\bar{f} = \sum_{n=0}^{\infty} \varepsilon^n \bar{f}_n \tag{5.8}$$

and satisfy the initial conditions term by term:

$$f_{H(n)} + f_{R(n)} = \bar{f}_n \qquad (n \geqslant 1) \tag{5.9}$$

Eq. (5.6) can be projected onto the space $\mathscr{F}$ spanned by the collision invariants to yield

$$\frac{\partial \rho_{(n)}^{\alpha}}{\partial \tau} = \int \psi_\alpha f_{H(0)} G_n \, d\boldsymbol{\xi} = -\int \psi_\alpha \boldsymbol{\xi} \cdot \frac{\partial f_{R(n-1)}}{\partial \mathbf{x}} \, d\boldsymbol{\xi} \tag{5.10}$$

where $\rho_{(n)}^{\alpha}$ is the contribution to ρ^α from the n-th term of the expansion of the *remainder* $f_{R(n)}$. Eq. (5.10) gives

$$\rho_{(n)}^{\alpha}(\mathbf{x}, \tau) = \rho_{(n)}^{\alpha}(\mathbf{x}, 0) - \int_0^\tau d\tau \int \psi_\alpha \boldsymbol{\xi} \cdot \frac{\partial f_{R(n-1)}}{\partial \mathbf{x}} \, d\boldsymbol{\xi} \tag{5.11}$$

Let $\bar{\rho}_n{}^\alpha$ be the n-th order contribution to the initial value $\bar{\rho}_\alpha$ of ρ_α; $\bar{\rho}_n{}^\alpha$ can be evaluated, of course, from the initial datum $\bar{f}$. Then

$$\rho_n^\alpha(\mathbf{x}, 0) - \rho_{(n)}^{\alpha}(\mathbf{x}, 0) = \bar{\rho}_n^\alpha(\mathbf{x}) \tag{5.12}$$

It would not be correct to take $\rho_{(n)}^{\alpha}(\mathbf{x}, 0) = 0$, and so to assign to $\rho_n{}^\alpha$ the true initial value $\bar{\rho}_n{}^\alpha(\mathbf{x})$. In fact, the Hilbert expansion takes over after several mean free times and it would be meaningless to pretend that the correct solution and the asymptotic solution coincide at $t = 0$; we have to require instead that for long times the complete solution f differs from f_H by a negligible quantity (see Fig. 26, where the dependence upon $\mathbf{x}$ and $\boldsymbol{\xi}$ have been suppressed). In particular, $\rho_{(n)}^{\alpha}(\mathbf{x}, t)$, the contribution of the remainder to the fluid variables must go to zero when $\tau \to \infty$; this is easily accomplished if we use Eq. (5.11) to obtain:

$$\rho^{\alpha}_{(n)}(\mathbf{x}, 0) = \int_0^{\infty} d\tau \int \psi_{\alpha} \boldsymbol{\xi} \cdot \frac{\partial f_{R(n-1)}}{\partial \mathbf{x}} \, d\boldsymbol{\xi} \tag{5.13}$$

Inserting this into Eq. (5.12), we obtain the initial conditions for the Hilbert equations:

$$\rho_n{}^{\alpha}(\mathbf{x}, 0) = \bar{\rho}_n{}^{\alpha} - \int_0^{\infty} d\tau \int \psi_{\alpha} \boldsymbol{\xi} \cdot \frac{\partial f_{R(n-1)}}{\partial \mathbf{x}} \, d\boldsymbol{\xi} \tag{5.14}$$

These conditions involve the physical data $\bar{\rho}_n{}^{\alpha}$ as well as a contribution from the solution $f_{R(n-1)}$ at the previous step. It can be verified that the integral term gives no contribution to first-order initial data, so that the first correction is of order ε^2; to this order, there is no correction to the initial density, and the corrected initial conditions for velocity and temperature take on the form [3]

$$\begin{aligned} \mathbf{v}(\mathbf{x}, 0) &= \bar{\mathbf{v}} - a[\operatorname{div}(a' \operatorname{grad} \bar{\mathbf{v}}) + \tfrac{1}{3} \operatorname{grad}(a' \operatorname{div} \bar{\mathbf{v}})] \\ T(\mathbf{x}, 0) &= \bar{T} - b \operatorname{div}(b' \operatorname{grad} \bar{T}) \end{aligned} \tag{5.15}$$

where $\bar{\mathbf{v}}$ and $\bar{T}$ denote the physical initial data for velocity and temperature respectively, and a, a', b and b' are four coefficients of the order of the mean

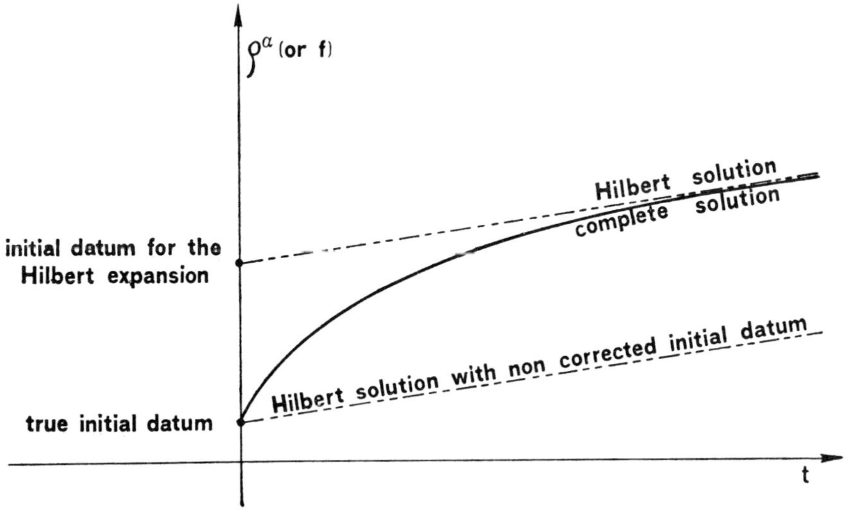

Fig. 26. Illustration of the correct initial datum for the Hilbert solution.

free path. These coefficients can be computed exactly for Maxwell molecules with the following result [3]:

$$aa' = l^2/\pi; \qquad bb' = 15l^2/(4\pi) \tag{5.16}$$

where l is the mean free path defined by Eq. (1.3).

The above results show that the naive approach based upon letting $\rho_n^{\alpha}(\mathbf{x}, 0) = \bar{\rho}_n^{\alpha}(\mathbf{x})$ is essentially correct at the Euler and Navier-Stokes level; it is insufficient at the level of the Burnett equations, whose practical importance is, however, negligible. This means that an expansion of the Hilbert type treats the initial layer in a correct way except for (usually negligible) terms of order ε^2 (initial slip).

The situation is similar in principle but essentially different in practice for boundary layers. We already know that the Hilbert expansion misses completely not only the kinetic boundary layers but also the viscous boundary layers; the latter are recovered by the Chapman-Enskog method and the method described briefly in Section 4. The kinetic layers of order ε, however, are missed by all the expansions in powers of ε described so far; in order to recover them, we have to use a magnified variable $X = x/\varepsilon$, analogous to the variable τ used before for the initial layer. We shall consider the case of boundaries whose radius of curvature is large with respect to the mean free path, and boundary conditions which do not change appreciably along the boundary (on the scale of the mean free path) and in time (on the scale of the mean free time); if these conditions are not satisfied the analysis is complicated and becomes two- or three-dimensional instead of one-dimensional in space variables. We also assume that the deviation of the distribution from a Maxwellian remains of order ε in the vicinity of the boundary in complete analogy with the case of the initial layer.

Under our assumptions we can take, in the neighbourhood of the boundary, a non-Cartesian reference frame made up as follows (see Fig. 27): we take a pair of coordinates $\alpha_i (i = 1, 2)$ on the surface Σ bounding the flow region; then, through each point $\mathbf{x}$ we draw the straight line joining $\mathbf{x}$ to the point $\mathbf{x}_0 \in \Sigma$ which is the closest to $\mathbf{x}$ (this straight line is obviously normal to Σ); finally, we take as coordinates of $\mathbf{x}$ the distance x along the normal at $\mathbf{x}_0$ and the coordinates α_i $(i = 1, 2)$ of $\mathbf{x}_0$. If $\mathbf{n}$ is the normal unit vector at $\mathbf{x}$, we have:

$$x_j = x_{0j} + xn_j \qquad (j = 1, 2, 3) \tag{5.17}$$

From the parametric equations of the surface $[x_{0j} = x_{0j}(\alpha_1, \alpha_2), j = 1, 2, 3]$, we can obtain $\partial\alpha_i/\partial x_k$ $(i = 1, 2; k = 1, 2, 3)$ and n_k as functions of the α's, and hence $x_j = x_j(x, \alpha_i)$ through Eq. (5.17). Thus

$$f = f_c(\varepsilon X, \alpha_i, \boldsymbol{\xi}, t; \varepsilon) + \varepsilon f_R(X, \alpha_i, \boldsymbol{\xi}, t; \varepsilon) \qquad (X = x/\varepsilon) \tag{5.18}$$

The leading term f_c is given by the Chapman-Enskog expansion (or any expansion capable of describing the viscous layer), while f_R satisfies:

$$\boldsymbol{\xi}\cdot\mathbf{n}\frac{\partial f_R}{\partial X}+\varepsilon\left(\frac{\partial f_R}{\partial t}+\sum_{i=1}^{2}\boldsymbol{\xi}\cdot\frac{\partial\alpha_i}{\partial\mathbf{x}}\frac{\partial f_R}{\partial\alpha_i}\right)=2Q(f_c,f_R)+\varepsilon Q(f_R,f_R) \quad (5.19)$$

If we now expand as above (interchanging the roles of X and τ), the basic equation turns out to be

$$\boldsymbol{\xi}\cdot\mathbf{n}\frac{\partial g_m}{\partial X}=Lg_m+G_m \qquad (m\geqslant 1) \quad (5.20)$$

where

$$f_{R(m)}=f_{c(0)}g_m \qquad (m\geqslant 1)$$

$$G_m=-\left[\frac{\partial f_{R(m-1)}}{\partial t}+\sum_{i=1}^{2}\boldsymbol{\xi}\cdot\frac{\partial\alpha_i}{\partial\mathbf{x}}\frac{\partial f_{R(m-1)}}{\partial\alpha_i}+\sum_{k=1}^{m-1}2Q(f_{c(m-k)},f_{R(k+1)})+\right.$$
$$\left.+\sum_{k=1}^{m-1}Q(f_{R(k+1)},f_{R(m-k)})\right]\Big/f_{H(0)} \quad (5.21)$$

The problem of solving Eq. (5.20) is much more difficult than the analogous problem of solving Eq. (5.6) (compare Sections 6 and 7 of Chapter IV); in particular, we cannot project Eq. (5.20) onto $\mathscr{F}$ in order to obtain an equation for the fluid variables $\rho^{\alpha}_{(m)}$, because the factor $\boldsymbol{\xi}\cdot\mathbf{n}$ couples the fluid variables to the whole distribution function. This does not allow us to write equations similar to Eq. (5.14) until we have constructed a theory for solving

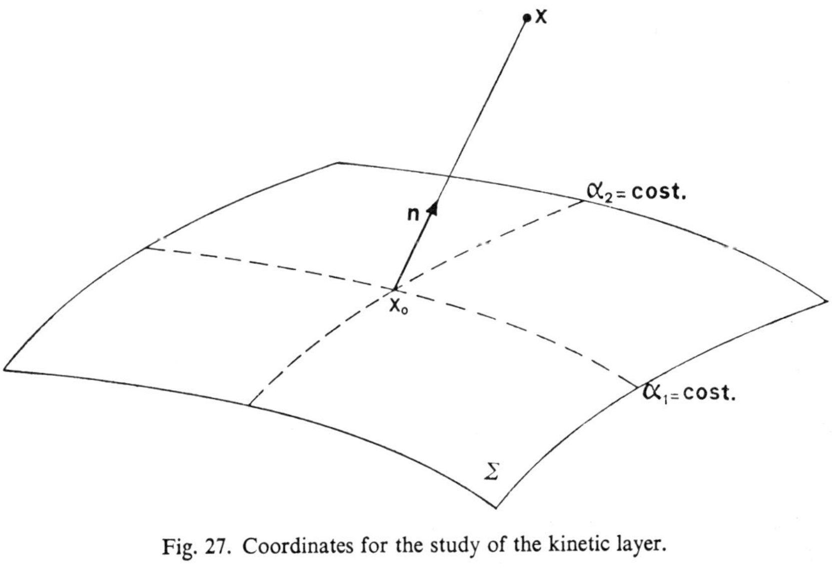

Fig. 27. Coordinates for the study of the kinetic layer.

Eq. (5.20); this was qualitatively done in Chapter IV, Section 7, but quantitative results are not easy to obtain even for Maxwell molecules. The theory can be partly constructed for model equations (see Chapter VI). Furthermore, this solution depends upon the boundary conditions, which, according to the discussion in Chapter III, are much more complicated than the initial conditions.

Let us consider the first order terms in more detail. For $n = 1$, Eq. (5.20) becomes [$f_{R(0)} = 0$ according to the analogous of Eq. (5.3)]:

$$\boldsymbol{\xi} \cdot \mathbf{n} \frac{\partial g_1}{\partial X} = L g_1 \tag{5.22}$$

that is, the one-dimensional steady linearized Boltzmann equation, studied in Chapter IV, Section 7 (Eq. (7.1) with $\xi_1 = \boldsymbol{\xi} \cdot \mathbf{n}$, $x_1 = X$). Because of our assumptions L is the collision operator linearized about the Maxwellian f_w of the wall and a suitable density. The analogue of Eq. (5.9) is (for $n = 1$):

$$|\boldsymbol{\xi} \cdot \mathbf{n}| \, (f_{c(1)} + f_w g_1) = \int_{\boldsymbol{\xi}' \cdot \mathbf{n} > 0} R(\boldsymbol{\xi}' \to \boldsymbol{\xi}; \mathbf{x}_0)(f'_{c(1)} + f_w' g_1') \, |\boldsymbol{\xi}' \cdot \mathbf{n}| \, d\boldsymbol{\xi}'$$

$$(\boldsymbol{\xi} \cdot \mathbf{n} > 0; X = 0) \tag{5.23}$$

where $f_{c(1)}$ is given by:

$$f_{c(1)} = f_w\{\mathbf{A} \cdot (\boldsymbol{\xi} \wedge \mathbf{n}) + B_{ij} L^{-1}(\xi_i \xi_j - \tfrac{1}{3}\xi^2 \, \delta_{ij}) + C(\xi^2 - 5RT_w) +$$

$$+ \mathbf{D} \cdot L^{-1}(\boldsymbol{\xi}[\xi^2 - 5RT_w])\} \tag{5.24}$$

where the velocity $\mathbf{u}_w$ of the wall was assumed to be zero (otherwise $\boldsymbol{\xi}$ must be replaced by $\boldsymbol{\xi} - \mathbf{u}_w$), T_w is the temperature of the wall and

$$\begin{aligned} \mathbf{A} &= \mathbf{v}/(RT_w) \\ B_{ij} &= \frac{(\partial v_i/\partial x_j)}{(RT_w)} \\ C &= \frac{T - T_w}{2RT_w^{\ 2}} \\ \mathbf{D} &= \frac{\partial T}{\partial \mathbf{x}} \Big/ (RT_w) \end{aligned} \tag{5.25}$$

where $\mathbf{v}$, T, $\partial \mathbf{v}/\partial n$, $\partial T/\partial n$ denote the velocity, temperature and their derivatives evaluated at the wall. The constants $\mathbf{A}$ and C arise from the re-expansion of the zero-th order Chapman-Enskog solution about the wall Maxwellian, the remaining part of $f_{c(1)}$ being the first-order Chapman-Enskog terms evaluated at the wall.

By the same argument as for the initial layer we conclude that g_1 must go to zero for $X \to \infty$; this means that in the general solution, Eq. (IV7.4), we must put $A_\alpha = 0$, $B_\alpha = 0$; $A_\lambda = 0$ for $\lambda > 0$. As we see from Eq. (5.23), only the combination $f_{c(1)} + f_w g_1$ enters into Eq. (5.23). Thus, to satisfy the boundary conditions at $X = 0$, the function A_λ and the constants $\mathbf{A}$, B_{ij}, C, $\mathbf{D}$ are available. From what is known from particular cases (see Chapter VI, Section 3) and general theorems (see Chapter VIII, Section 4, Refs. 38–40) the constants $\mathbf{A}$, C and the function A_λ should be sufficient; because of this, rather than determining the constants completely, Eq. (5.23) leads to three scalar relations between $\mathbf{v}$, T, $\partial\mathbf{v}/\partial\mathbf{x}$, $\partial T/\partial\mathbf{x}$. Because of the invariance of the equations and the boundary conditions with respect to rotations in the tangential plane, the relations must have the following form:

$$\begin{aligned} \mathbf{v} - \zeta \frac{\partial \mathbf{v}_t}{\partial n} - \omega(2R/T)^{\frac{1}{2}}\left(\frac{\partial T}{\partial \mathbf{x}} - \mathbf{n}\frac{\partial T}{\partial n}\right) &= \bar{\mathbf{v}} \\ T - \tau\frac{\partial T}{\partial n} - \chi(2RT)^{-\frac{1}{2}}\left(\operatorname{div} \mathbf{v}_t - \mathbf{n}\frac{\partial \mathbf{v}_t}{\partial n}\right) &= \bar{T} \end{aligned} \tag{5.26}$$

where $\bar{\mathbf{v}}$, $\bar{T}$ are, respectively, the velocity and temperature of the boundary ζ, ω, τ, χ are coefficients of the order of the mean free path, $\mathbf{v}_t = \mathbf{v} - \mathbf{n}(\mathbf{n} \cdot \mathbf{v})$ and $\partial/\partial n = \mathbf{n} \cdot \partial/\partial\mathbf{x}$. In particular ζ measures the tendency of the gas to slip over a solid wall in the presence of velocity gradients, and is called the slip coefficient; τ measures the tendency of the gas to have a temperature different from the wall temperature, and is called the temperature jump coefficient. The coefficients ζ, ω, τ have been evaluated by means of kinetic models for the case of a completely diffuse re-emission at the wall. In this case Eq. (5.23) reduces to

$$f_{c(1)} + f_w g_1 = 0 \qquad (\boldsymbol{\xi} \cdot \mathbf{n} > 0;\ X = 0) \tag{5.27}$$

so that, if the so-called half-range orthogonality relations (see Chapter VI, Section 3) are known, the problem can be explicitly solved. Alternatively, we may use the variational method described in Sections 10 and 12 of Chapter IV or solve numerically the system of integral equations associated with the problem (see Chapter IV, Section 12). All these methods have been used [17–30]. The slip coefficient ζ has been computed with particular accuracy for several models; for the BGK model the computation of ζ can be even reduced to quadratures (Chapter VI, Section 4) and thus a very accurate numerical value can be obtained [19]:

$$\zeta = 1.1466 l \tag{5.28}$$

where l is the mean free path defined by Eq. (1.3). An analogous result is obtained for the simplest model with velocity dependent collision frequency

[27]. The numerical values for realistic choices of the collision frequency [28] are somewhat lower (3%) than for the model with constant collision frequency. The latter seems essentially accurate for Maxwell's molecules [24, 26]. τ has also been computed with good accuracy [17, 23] for the BGK model. In this case, due to the fact that the BGK model has a Prandtl number equal to 1 (Section 6) rather than $\frac{2}{3}$ as is appropriate for a monatomic gas, we must be careful in the choice of the value of the collision frequency; for the computation of τ, the appropriate choice is the value which leads to the correct heat conduction coefficient. If the correct value of $\frac{2}{3}$ for the Prandtl number is used, the following value of τ is found [23]:

$$\tau = \tfrac{15}{8}(1.1682)l = 2.1904l \tag{5.29}$$

where l is again given by Eq. (1.3). ω (the thermal creep coefficient) has been computed by Sone [29], Williams [30] and Loyalka [22] with the following result

$$\omega = \frac{3}{2\sqrt{\pi}}(0.7662)l = 0.648l \tag{5.30}$$

provided the collision frequency in the BGK model is taken to have the value appropriate for heat conduction. The influence of the boundary conditions on the values of ζ, τ, ω has been studied by Cercignani [31], Cercignani and Pagani [32], Loyalka and Cipolla [33], Klinc and Kuščer [34].

We note that when the mean free path is not only small, but completely negligible, Eqs. (5.26) reduce to $\mathbf{v} = \bar{\mathbf{v}}$, $T = \bar{T}$, so that the gas does not slip and completely accommodates to the wall temperature.

The second order boundary conditions are much more complicated; in the case in which the boundary is flat, calculations were performed by the author [12]. The effects of the wall curvature were studied by Grad [14] and Sone [35, 36]. In general, terms involving second derivatives of velocity and temperature are to be added to Eqs. (5.26) together with terms involving the products of first derivatives and the principal curvatures of the boundary; the coefficients are of course of order l^2. Some of these coefficients have been computed for the BGK model [12, 35–37].

A different approach is due to Darrozès [38] who considered expansions of the Hilbert type but in powers of $\sqrt{\varepsilon}$ rather than ε. As a consequence, he finds two boundary layers: the outer one, of thickness $O(\varepsilon^{\frac{1}{2}})$ to be identified with the Prandtl viscous boundary layer and the inner one of thickness $O(\varepsilon)$ to be identified with the Knudsen or kinetic boundary layer. Within the Prandtl layer, the distribution function is not of the Hilbert class, but keeps the property of being functionally related to the fluid variables (as is known from the success of the Chapman-Enskog method at the Navier-Stokes level). As a result of the adopted expansion, however, the Navier-Stokes equations never appear; they are replaced by the Prandtl boundary layer equations.

Another nonuniformity and hence another kind of boundary layer arises from molecules travelling almost parallel to the boundary. In fact, if $|\boldsymbol{\xi} \cdot \mathbf{n}|/\xi \simeq \varepsilon$, the first term on the right hand side of Eq. (5.19) is of order $\varepsilon\, \partial f_R/\partial X$; on the other hand, this term must be of finite order at the boundary in order to give rise to a regular boundary value problem. Hence, near the boundary, f_R must vary on a scale $X \simeq \varepsilon$ or $x \simeq \varepsilon^2$ in a significant fashion. Of course, since ε is of the order of magnitude of the mean free path, a relation of the form $x \simeq \varepsilon^2$ is correct only if it is understood to mean $x \simeq l^2/L'$ where L' is some characteristic length (analogously $|\boldsymbol{\xi} \cdot \mathbf{n}|/\xi \simeq \varepsilon$ means $|\boldsymbol{\xi} \cdot \mathbf{n}|/\xi \simeq l/L'$). This length must be defined by the local geometry of the boundary and is to be identified with the radius of curvature R_c of the boundary itself. Hence the new boundary layer is due to the curvature of the boundary and will not arise for a flat wall ($R_c = \infty$).

Let us understand how the effect arises. We remark that the influence of the boundary on the value of the distribution function $f(\mathbf{x}, \boldsymbol{\xi})$ primarily depends upon the ratio of the distance $|\mathbf{x} - \mathbf{x}_0|$ to the mean free path l, provided $\mathbf{x}$ is sufficiently close to the boundary such that $\mathbf{x} - \mathbf{x}_0 = |\mathbf{x} - \mathbf{x}_0|\, \boldsymbol{\xi}/\xi$. If the origin of the position vector is at the center of curvature relative to $\mathbf{x}_0$, and if R_c is the corresponding radius of curvature (we consider the plane case for simplicity), we obtain (see Fig. 28):

$$|\mathbf{x} - \mathbf{x}_0| = \sqrt{R_c^{\,2} - \frac{(\mathbf{x} \wedge \boldsymbol{\xi})^2}{\xi^2}} + \frac{\mathbf{x} \cdot \boldsymbol{\xi}}{\xi} = \sqrt{R_c^{\,2} - |\mathbf{x}|^2 \sin^2 \varphi} + |\mathbf{x}| \cos \varphi \quad (5.31)$$

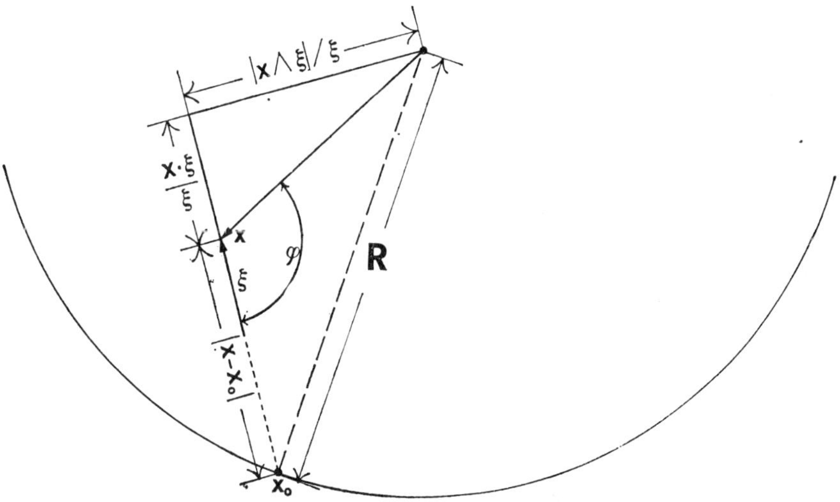

Fig. 28. Illustration of the computation of $|\mathbf{x} - \mathbf{x}_0|$.

where φ is the angle between $\mathbf{x}$ and $\boldsymbol{\xi}$. Let $|\mathbf{x}| = R_c \pm \delta$ where the minus sign holds for a concave boundary and the plus sign for a convex one. Eq. (5.31) then gives

$$\frac{|\mathbf{x} - \mathbf{x}_0|}{l} = \frac{R_c\,|\cos\varphi|}{l}\sqrt{1 \pm \frac{2\delta}{R_c}\operatorname{tg}^2\varphi + \frac{\delta^2}{R_c^{\,2}}\operatorname{tg}^2\varphi} - \frac{(R_c \pm \delta)}{l}|\cos\varphi| \qquad (5.32)$$

If $\operatorname{tg}\varphi \simeq O(1)$, then for $\delta \ll R_c$ we have:

$$\frac{|\mathbf{x} - \mathbf{x}_0|}{l} = \pm\frac{\delta}{l}\frac{1}{|\cos\varphi|} + O\left(\frac{\delta^2}{R_c l}\right) \qquad (5.33)$$

and there is an appreciable change when δ varies on the scale of l; on this scale the remainder is $O(l/R_c)$ and hence behaves regularly for $l \to 0$. If $\operatorname{tg}\varphi \simeq O(R_c/l)$, however,

$$\frac{2\delta}{R_c}\operatorname{tg}^2\varphi = O(\delta R_c/l^2)$$

becomes comparable with unity for $\delta = O(l^2/R_c)$. Hence $|\mathbf{x} - \mathbf{x}_0|/l$ as given by Eq. (5.33) varies significantly on the scale $\delta = O(l^2/R_c)$ when $\cos\varphi \simeq \operatorname{cotg}\varphi = O(l/R_c)$.

When we proceed to compute density, velocity, etc. or, in general, "moments" of the distribution function:

$$M_{ijk\cdots l}(\mathbf{x}, t) = \int \xi_i\xi_j\xi_k \cdots \xi_l f\, d\boldsymbol{\xi} \qquad (5.34)$$

we can integrate with respect to the cylindrical variables $\rho = \sqrt{\xi_x^{\,2} + \xi_y^{\,2}}$, φ, ξ_z, if the plane considered before is taken to be $z = 0$. If f is uniformly continuous with respect to φ, then the curvature boundary layer of thickness $\delta^2/R_c l$ disappears, because the contribution from the grazing molecules is smoothed out by integration. This is easily seen by a change of angular variable. Instead of φ, the angle between $\boldsymbol{\xi}$ and $\mathbf{x}$, we take θ, the angle between $\mathbf{x}$ and $\mathbf{x}_0$, as our integration variable. Then:

$$\begin{aligned}\frac{|\mathbf{x} - \mathbf{x}_0|}{l} &= \frac{1}{l}\sqrt{R_c^{\,2} + |\mathbf{x}| \;- 2R_c\,|\mathbf{x}|\cos\theta} \\ &= \sqrt{\frac{\delta^2}{l^2} + \frac{2R_c(R_c \pm \delta)}{l^2}(1 - \cos\theta)} \qquad (5.35)\end{aligned}$$

If φ ranges from 0 to 2π, θ ranges from $-\pi$ to π and the factor multiplying $\cos\theta$ varies very little ($\delta \ll R_c$); hence only the dependence on δ/l appears. In other words, moments of the distribution function do not exhibit the boundary layer due to curvature but only the ordinary Knudsen layer, unless integration with respect to φ is suitably restricted. This restriction automatically occurs if the boundary surrounds a (locally) convex solid body. In fact, in this case, molecules arriving at P come from just a part of the boundary, $\sin\varphi \leqslant R_c/|\mathbf{x}|$ or $|\cos\theta| > \sqrt{|\mathbf{x}|^2 - R_c^2}/|\mathbf{x}|$. Accordingly, when computing a moment, the integral has limits of integration depending upon the ratio $|\mathbf{x}|/R_c$. The angles corresponding to the integration limits are such that $\operatorname{tg}\varphi = R_c(|\mathbf{x}|^2 - R_c^2)^{-\frac{1}{2}} = R_c[\delta(2R_c + \delta)]^{-\frac{1}{2}}$ which is of order R_c/l if δ is of order l^2/R_c. Hence for a flow past a (locally) convex body, the boundary layer due to curvature is present in the moments as well. This boundary layer is formed by points which cannot be reached from the boundary along straight lines much longer than the mean free path (see Fig. 29 where $PA = PB \simeq l$). This fact was discovered by Sone [39], who worked with a particular model at the level of the integral equation for one of the

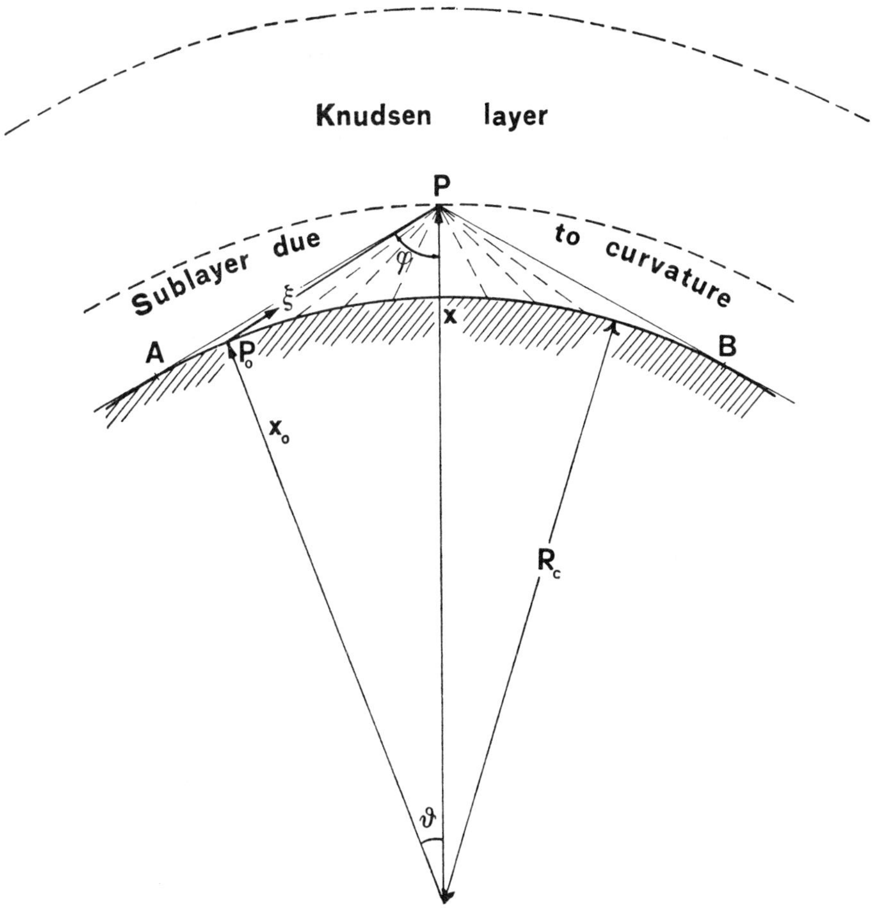

Fig. 29. The kinetic sublayer due to curvature in the case of a convex body.

moments and thus missed the fact that the new boundary layer is present even for concave walls, though only at the level of the distribution function.

The third connection problem (shock layer) should lead to the evaluation of the correction to the classical Rankine-Hugoniot relations which is required for a calculation made at the continuum level to yield the same results as the solution of the Boltzmann equation far from the shock layer. The same need arises in the Navier-Stokes theory [40] in order to take into account the interaction between a shock and a boundary layer. Although the Navier-Stokes equations give a smooth shock structure, they must be allowed to have discontinuities in order to incorporate the kinetic effects correctly. The kinetic theory solution of the zero-th order connection problem for the Hilbert expansion is already a difficult one (the shock structure problem; see Chapter VII, Section 6) but the matching relations are trivial (the Rankine-Hugoniot relations); the setting up of the analogous problem for the Chapman-Enskog theory (or the modified expansion discussed in Section 4) has never been attempted.

6. Further remarks on the Chapman-Enskog method and the computation of transport coefficients

The procedures expounded in Sections 2 to 5 apply not only to the nonlinear Boltzmann equation for a monatomic gas, but also to the linearized Boltzmann equation, to the Boltzmann-like equations which are obtained when the quadratic collision term is replaced by a model term $J(f)$ (Chapter II, Section 10, Chapter IV, Section 9) and to the generalized Boltzmann equations describing mixtures or polyatomic gases. We shall make a few remarks on these topics.

The use of the Chapman-Enskog method for the linearized Boltzmann equation has already been considered at the end of Section 4. Here we just remark that the procedure is simpler than for the full equation, because the term S_n is now missing in Eqs. (2.6) and (3.10).

When nonlinear collision models such as those described in Section 10 of Chapter II are used, the only changes arise in connection with the expansion of the nonlinear terms in powers of ε, because the nonlinearity of the model is, in general, more complicated than quadratic. This circumstance, however, does not come in before the second order approximation (terms in ε^2). As a consequence, the models reproduce correctly the Euler and Navier-Stokes equations, and even the viscosity and heat conduction coefficients can be adjusted to agree with the correct ones, provided that the models contain at least two adjustable parameters. This is not true for the simplest models as, for example, the BGK model and we have to decide whether to adjust viscosity or heat conduction.

Since the BGK model is frequently used, we briefly describe the first steps of the Chapman-Enskog theory for this model. As noted above, the zero-th and first order equations are formally the same as for the Boltzmann equation; f_0 is Maxwellian and $f_1 = f_0 h_1$ is to be found by solving Eq. (3.16), where now, however, L is the linearized BGK operator. The procedure used in Chapter IV, Section 9 to construct the Gross and Jackson models shows that the linearized BGK collision operator has the same eigenfunctions as the Maxwell collision operator; the distinct eigenvalues are now only two, $\lambda = 0$ (corresponding to the five collision invariants) and $\lambda = -\nu$ (corresponding to the remaining eigenfunctions). Accordingly, the solution of Eq. (3.16) for the BGK model is again given by Eq. (3.17) with $A(c, T) = A(T)$; $B(c, T) = B(T)(c^2 - 5RT)$ as in the case of Maxwell's molecules and μ and κ are given by Eq. (IV.7.60) with $\lambda_{02} = \lambda_{11} = -\nu$. The main consequence of the infinite-fold degeneracy of the eigenvalue $\lambda = -\nu$ is that the Prandtl number [(still given by Eq. (IV.7.60)] is now 1. This result implies the above-mentioned result that we cannot adjust μ and κ at the same time if the BGK model is being used; the adjustment can be achieved, however, if the *ES* model or more complicated nonlinear models are used.

The general problem of computing transport coefficients for a gas mixture can be solved in a manner analogous to that employed for a simple gas [8–10]. Two new transport phenomena arise in addition to viscosity and thermal conduction, namely, diffusion and thermal diffusion: the average velocity of a particular species is different, in general, from the mass velocity of the mixture and the difference, comprising the diffusion velocity, turns out to contain terms proportional to the concentration gradient, the pressure gradient, the difference between the external forces acting on the various molecular species and the temperature gradient. The first three terms correspond to ordinary diffusion and the fourth to thermal diffusion. Thermal diffusion was first predicted by Enskog [41] and Chapman [6] on purely theoretical grounds and confirmed experimentally by Chapman and Dootson [42]. The reason why it had escaped the attention of previous workers in the field is that the thermal diffusion coefficient turns out to be precisely zero for Maxwell's molecules.

As remarked by Grad [43], variations on the theme of power series expansions in the mean free path are possible in the case of a mixture. Instead of placing a factor ε^{-1} in front of each collision integral $Q_{\alpha\beta}$, we can distinguish between two different kinds of collisions by placing different factors such as ε^{-3}, ε^{-2}, ε^{-1}, 1 in front of different integrals. The new expansions should be useful for mixtures with disparate mass. The method has been recently reconsidered by E. A. Johnson [44].

Even for polyatomic gases, the problem of carrying out a Chapman-Enskog expansion is almost entirely a matter of lengthy, but relatively straightforward, computation. We refer to the book of Ferziger and Kaper [10] for some details and references.

7. Free molecule flow past a convex body

In Section 1, free molecule flow was defined as the flow obtained in the limit when the Knudsen number Kn tends to infinity. In that case, the Boltzmann equation (in the absence of body forces) takes on the form:

$$\frac{\partial f}{\partial t} + \boldsymbol{\xi} \cdot \frac{\partial f}{\partial \mathbf{x}} = 0 \tag{7.1}$$

Since the collisions of the molecules among themselves are neglected, the interaction of the molecules with solid walls plays a major role. This situation is typical for satellites (the mean free path is about 50 m at 200 Km of altitude). The general solution of Eq. (7.1) has the form

$$f(t, \mathbf{x}, \boldsymbol{\xi}) = g(\mathbf{x} - \boldsymbol{\xi} t, \boldsymbol{\xi}) \tag{7.2}$$

where g is an arbitrary function of two vectors (formally g is the value of f for $t = 0$). Eq. (7.2) simply states that f is constant along the straight-line trajectories of the molecules ($\mathbf{x} - \boldsymbol{\xi} t = \mathbf{x}_0$), as is appropriate in the absence of collisions. In the steady case Eq. (6.1) reduces to

$$\boldsymbol{\xi} \cdot \frac{\partial f}{\partial \mathbf{x}} = 0 \tag{7.3}$$

and the general solution takes the form

$$f(\mathbf{x}, \boldsymbol{\xi}) = g(\mathbf{x} \wedge \boldsymbol{\xi}, \boldsymbol{\xi}) \tag{7.4}$$

which can be derived either directly or from Eq. (7.2) (note that a function of $\mathbf{x} - \boldsymbol{\xi} t$, $\boldsymbol{\xi}$ is also a function of $\mathbf{x} \cdot \boldsymbol{\xi} - \xi^2 t$, $\mathbf{x} \wedge \boldsymbol{\xi}$, $\boldsymbol{\xi}$).

Frequently, it is easier to work with the implicit statement that f is constant along the trajectories than with the analytic solutions given by Eqs. (7.2) and (7.4).

Let us examine, as an example, the steady flow past a convex body of a uniform equilibrium stream of density ρ_∞, velocity $\mathbf{V}$ and temperature T_∞. The molecules impinging upon the body come directly from space at infinity (see Fig. 30) and hence their distribution function f^- is equal to

$$f_\infty = \rho_\infty (2\pi R T_\infty)^{-\frac{3}{2}} \exp[-(\boldsymbol{\xi} - \mathbf{V})^2 (2RT_0)^{-1}] \tag{7.5}$$

At a point P outside the body (see Fig. 30) the velocity vectors fall into two classes. The first class contains the vectors $\boldsymbol{\xi}$ such that the straight line along

$\boldsymbol{\xi}$ through P intersects the body at some point P_0 and $\boldsymbol{\xi}$ points from the body to infinity; the vectors $\boldsymbol{\xi}$ of this class (applied at P) form a cone with apex at P. The distribution function for $\boldsymbol{\xi}$ outside this cone (and hence belonging to the second class) equals f_∞, whereas the distribution function for $\boldsymbol{\xi}$ inside the cone equals the distribution function f^+ of the molecules re-emitted at P_0. Hence the solution is completely determined once the distribution function of the molecules re-emitted at the body surface is known. But this is what is done by the boundary conditions, Eq. (III.I.6): in fact $f^- = f_\infty$ and hence

$$f^+ = \int_{\boldsymbol{\xi}\cdot\mathbf{n}<0} R(\boldsymbol{\xi}' \to \boldsymbol{\xi}; \mathbf{x})\, |\boldsymbol{\xi}' \cdot \mathbf{n}|\, |\boldsymbol{\xi} \cdot \mathbf{n}|^{-1} f_\infty(\boldsymbol{\xi})\, d\boldsymbol{\xi} \qquad (\mathbf{x} \in \partial R) \qquad (7.6)$$

Hence the solution depends in a crucial way on the scattering kernel $R(\boldsymbol{\xi}' \to \boldsymbol{\xi}; \mathbf{x})$ whose properties were studied in Chapter III.

Interest is usually confined to the total momentum and energy exchanged between the molecules and the solid body. In terms of them the aerodynamic forces exerted on the body as well as the heat transfer between the body and the gas are easily computed. The momentum and energy carried to unit

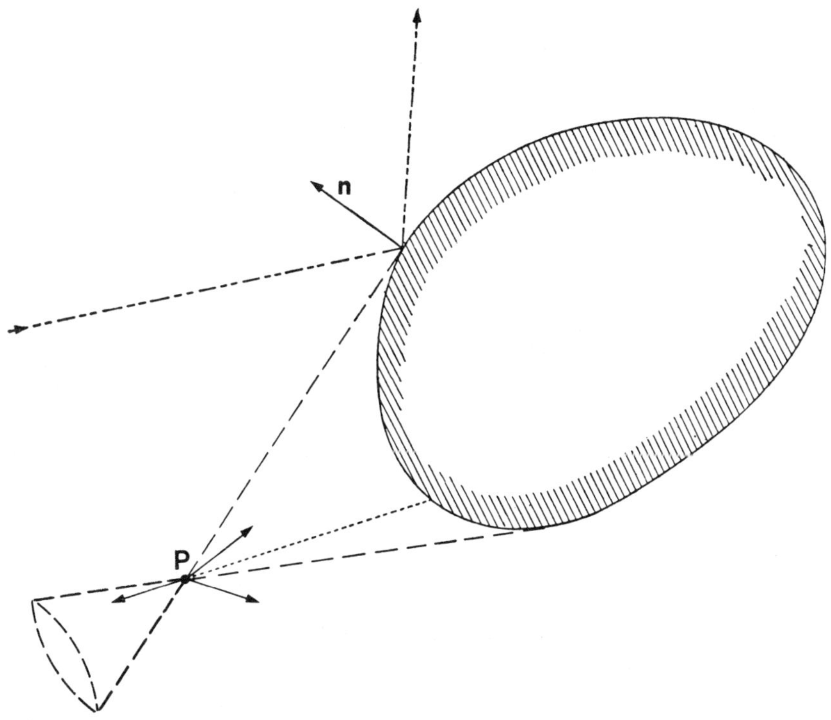

Fig. 30. Classes of velocity vectors and other details in the flow past a convex body.

surface area in unit time are:

$$\mathbf{p}^- = \int_{\boldsymbol{\xi}\cdot\mathbf{n}<0} \boldsymbol{\xi}\,|\boldsymbol{\xi}\cdot\mathbf{n}|\,f^-\,d\boldsymbol{\xi} \tag{7.7}$$

$$q^- = \int_{\boldsymbol{\xi}\cdot\mathbf{n}<0} \frac{\xi^2}{2}\,|\boldsymbol{\xi}\cdot\mathbf{n}|\,f^-\,d\boldsymbol{\xi} \tag{7.8}$$

whereas the momentum and energy carried away from unit surface area in unit time are

$$\mathbf{p}^+ = \int_{\boldsymbol{\xi}\cdot\mathbf{n}>0} \boldsymbol{\xi}\,|\boldsymbol{\xi}\cdot\mathbf{n}|\,f^+\,d\boldsymbol{\xi} \tag{7.9}$$

$$q^+ = \int_{\boldsymbol{\xi}\cdot\mathbf{n}>0} \frac{\xi^2}{2}\,|\boldsymbol{\xi}\cdot\mathbf{n}|\,f^-\,d\boldsymbol{\xi} \tag{7.10}$$

The computation of $\mathbf{p}^-$ and q^- is easy. Let us consider a surface element inclined at an angle of attack θ to the stream; in other words, the free stream velocity $\mathbf{V}$ forms an angle $(\pi/2) + \theta$ with the normal pointing into the gas (see Fig. 31). Then, inserting $f^- = f_\infty$ into Eqs. (7.7) and (7.8) gives:

$$\mathbf{p}^- = p_D^- \frac{\mathbf{V}}{V} - p_\Lambda^- \left(\mathbf{n} + \frac{\mathbf{V}}{V}\sin\theta\right) \tag{7.11}$$

$$p_D^- = \frac{\rho_\infty V^2}{2S^2}\left\{\sin\theta(\tfrac{1}{2} + S^2)[1 + \operatorname{erf}(S\sin\theta)] + \frac{S}{\sqrt{\pi}}\exp(-S^2\sin^2\theta)\right\} \tag{7.12}$$

$$p_\Lambda^- = \frac{\rho_\infty V^2}{4S^2}\,[1 + \operatorname{erf}(S\sin\theta)] \tag{7.13}$$

$$q^- = \tfrac{1}{4}\rho_\infty(2RT_\infty)^{\frac{3}{2}}\left[(S^2 + \tfrac{5}{2})S\sin\theta[1 + \operatorname{erf}(S\sin\theta)] + \right.$$
$$\left. + \frac{S^2 + 2}{\sqrt{\pi}}\exp(-S^2\sin^2\theta)\right] \tag{7.14}$$

Here, S is the so-called speed ratio:

$$S = \frac{V}{\sqrt{2RT_\infty}} = \sqrt{\frac{\gamma}{2}}\,M_\infty \tag{7.15}$$

where M_∞ is the free stream Mach number, γ is the ratio of the specific heats, and

$$\operatorname{erf} x = \frac{2}{\sqrt{\pi}}\int_0^x e^{-t^2}\,dt \tag{7.16}$$

is the error function [45].

$\mathbf{p}^-$ has been written in such a way that it is easy to project it along $\mathbf{V}$ (the result being p_D^-) and any direction perpendicular to $\mathbf{V}$ (the result being $p_L^- = -p_\Lambda^-\cos\beta$, if β is the angle between this direction and $\mathbf{n}$). This form of the result is useful for the purpose of computing the drag and lift exerted upon the body. Frequently, however, the results are presented in terms of the components along **n**:

$$p_n^- = -p_D^- \sin\theta - p_\Lambda^- \cos^2\theta = \frac{-\rho_\infty V^2}{4S^2}\left\{(\tfrac{1}{2} + S^2 \sin^2\theta) \times \right.$$

$$\left. \times [1 + \operatorname{erf}(S\sin\theta)] + \frac{S\sin\theta}{\sqrt{\pi}} \exp(-S^2\sin^2\theta)\right\} \quad (7.17)$$

and $\mathbf{t} \equiv \mathbf{V}/(V\cos\theta) + \mathbf{n}\operatorname{tg}\theta$:

$$p_t^- = p_D^- \cos\theta - p_\Lambda^- \sin\theta\cos\theta = \frac{\rho_\infty V^2}{2S}\cos\theta \times$$

$$\times \left\{S\sin\theta[1 + \operatorname{erf}(S\sin\theta)] + \frac{1}{\sqrt{\pi}}\exp(-S^2\sin^2\theta)\right\} \quad (7.18)$$

The computation of $\mathbf{p}^+$ and q^+ is crucially dependent upon the form of the scattering kernel. The classical way of treating the problem [46–48] is to by-pass the computation by introducing suitable "accommodation coefficients" for the normal momentum (α_N), tangential momentum (α_T) and energy (α_E); these coefficients are formally given by Eq. (III.5.3) with $\varphi = |\boldsymbol{\xi} \cdot \mathbf{n}|$, $\varphi = \boldsymbol{\xi} - \mathbf{n}(\boldsymbol{\xi} \cdot \mathbf{n})$, $\varphi = \xi^2/2$ and are assumed to depend on V, $\mathbf{V} \cdot \mathbf{n}$, T_∞,

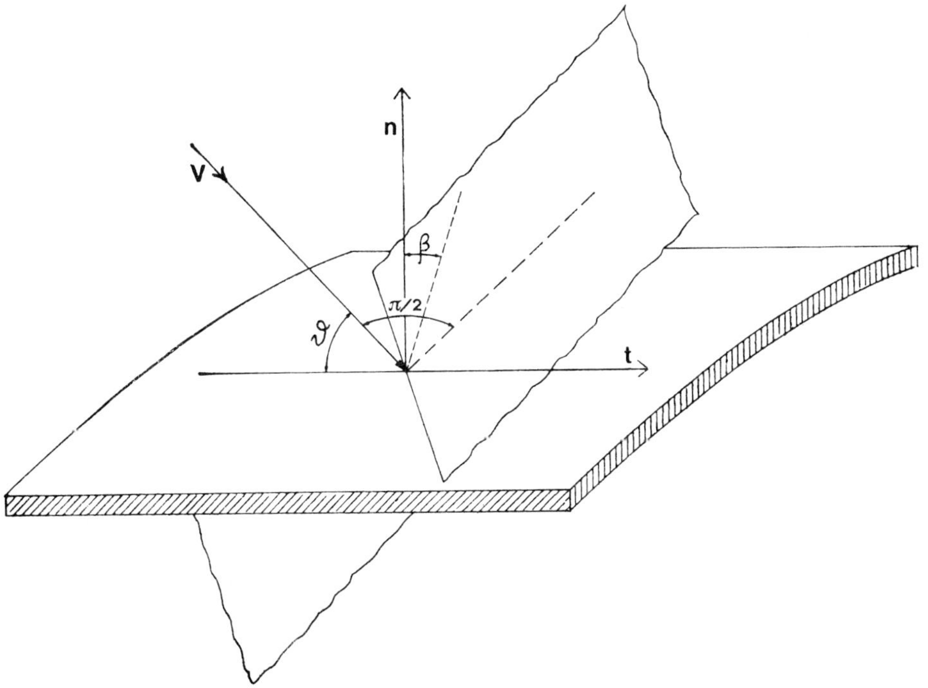

Fig. 31. Nomenclature for the computation of momentum and energy exchanges at a surface element.

as well as on the wall temperature T_w. Thus the approach is largely phenomenological; it gives, however, the correct results for scattering kernels with constant α_N, α_T, α_E. Eq. (III.5.3) then becomes (for $\varphi = |\xi_n|, \boldsymbol{\xi}_t, \xi^2/2$):

$$\alpha_N = \frac{p_n^- + p_n^+}{p_n^- + p_n^w} \qquad \alpha_T = \frac{p_t^- - p_t^+}{p_t^- - p_t^w} \qquad \alpha_E = \frac{q^- - q^+}{q^- - q^w} \tag{7.19}$$

where p_n^w, p_t^w, q^w are equal to p_n^+, p_t^+, q^+ when $f^+ = J_w f_w$, f_w is the wall Maxwellian and J_w a normalizing factor chosen in such a way that $J_w f_w$ gives the same mass flow as f. Then $p_t^w = 0$, while

$$p_n^w = \frac{\int |\boldsymbol{\xi}\cdot\mathbf{n}| f_\infty \, d\boldsymbol{\xi}}{\int |\boldsymbol{\xi}\cdot\mathbf{n}| f_w \, d\boldsymbol{\xi}} \int |\boldsymbol{\xi}\cdot\mathbf{n}|^2 f_w \, d\boldsymbol{\xi} = \frac{\sqrt{2\pi R T_w}}{2} \int |\boldsymbol{\xi}\cdot\mathbf{n}| f_\infty \, d\boldsymbol{\xi}$$

$$= \frac{\rho_\infty V^2}{4S^2} \sqrt{\frac{T_w}{T_\infty}} \{\exp(-S^2 \sin^2\theta) + \sqrt{\pi} S \sin\theta[1 + \operatorname{erf}(S\sin\theta)]\} \tag{7.20}$$

$$q^w = \frac{\int |\boldsymbol{\xi}\cdot\mathbf{n}| f_\infty \, d\boldsymbol{\xi}}{\int |\boldsymbol{\xi}\cdot\mathbf{n}| f_w \, d\boldsymbol{\xi}} \int \tfrac{1}{2} |\boldsymbol{\xi}\cdot\mathbf{n}| \, \xi^2 f_w \, d\boldsymbol{\xi} = 2RT_w \int |\boldsymbol{\xi}\cdot\mathbf{n}| f_\infty \, d\boldsymbol{\xi}$$

$$= \rho_\infty \frac{T_w}{2T_\infty} (2RT_\infty)^{\frac{3}{2}} \left\{ \frac{1}{\sqrt{\pi}} \exp(-S^2 \sin^2\theta) + S\sin\theta[1 + \operatorname{erf}(S\sin\theta)] \right\} \tag{7.21}$$

Then

$$p_n = p_n^- - p_n^+ = (2 - \alpha_N) p_n^- - \alpha_N p_n^w$$

$$= \frac{-\rho_\infty V^2}{2S^2} \left\{ \left(\frac{2 - \alpha_N}{\sqrt{\pi}} S \sin\theta + \frac{\alpha_N}{2} \sqrt{\frac{T_w}{T_\infty}} \right) \exp(-S^2 \sin^2\theta) + \right.$$

$$+ [1 + \operatorname{erf}(S\sin\theta)] \left[(2 - \alpha_N)(S^2 \sin^2\theta + \tfrac{1}{2}) + \right.$$

$$\left. \left. + \frac{\alpha_N}{2} \sqrt{\frac{\pi T_w}{T_\infty}} S \sin\theta \right] \right\} \tag{7.22}$$

$$p_t = p_t^- - p_t^+ = \alpha_T p_t^- = \alpha_T \frac{\rho_\infty V^2}{2S} \cos\theta \times$$

$$\times \{ S\sin\theta[1 + \operatorname{erf}(S\sin\theta)] + \pi^{-\frac{1}{2}} \exp(-S^2 \sin^2\theta) \} \tag{7.23}$$

$$q = q^- + q^+ = \alpha_E (q^- + q^w) = \tfrac{1}{4} \alpha_E \rho_\infty (2RT_\infty)^{\frac{3}{2}} \times$$

$$\times \left\{ \left(S^2 + \tfrac{5}{2} - 2\frac{T_w}{T_\infty} \right) S \sin\theta[1 + \operatorname{erf}(S\sin\theta)] + \right.$$

$$\left. + \pi^{-\frac{1}{2}} \left[S^2 + 2\left(1 - \frac{T_w}{T_\infty}\right) \right] \exp(-S^2 \sin^2\theta) \right\} \tag{7.24}$$

It is to be noted that Eq. (7.24) determines the energy transmitted to the wall by a monatomic gas. If the gas is polyatomic, Eq. (7.24) gives the contribution from the translational degrees of freedom, to which the contribution from the rotational degrees of freedom must be added [47].

The determination of the global aerodynamic coefficients is thus reduced to quadratures for scattering kernels with constant accommodation coefficients (as, for example, Maxwell's kernel, Eq. (III.6.13), for which $\alpha_N = \alpha_T = \alpha_E = \alpha$ is independent of V, $\mathbf{V} \cdot \mathbf{n}$, T_∞). In fact one has to integrate the above expressions over the entire surface of the body past which the flow takes place; of course, the quadratures will be very complicated unless the shape is particularly simple. The calculations have been carried out for several bodies [46–48], among which are flat plates, spheres, cones and cylinders.

An interesting fact follows from Eq. (7.24), if we look for the temperature $T_{\text{eq.}}$ attained by a thermally insulated element ($q = 0$).

We obtain

$$\frac{T_w}{T_\infty} = 1 + \frac{S^2}{2} + \frac{1}{4}\left\{1 - \frac{1}{1 + \pi^{\frac{1}{2}} S \sin\theta \exp(S^2 \sin^2\theta)[1 + \operatorname{erf}(S \sin\theta)]}\right\} \tag{7.25}$$

Now the function in the denominator is obviously never smaller than unity and hence the factor multiplying $\frac{1}{4}$ is always nonnegative. Accordingly:

$$T_{\text{eq.}} \geqslant T_\infty\left(1 + \frac{S^2}{2}\right) \tag{7.26}$$

The adiabatic stagnation temperature T_0 of the gas is equal to

$$\begin{aligned} T_0 &= T_\infty\left(1 + \frac{\gamma - 1}{2} M_\infty{}^2\right) \\ &= T_\infty\left(1 + \frac{\gamma - 1}{\gamma} S^2\right) = T_\infty(1 + \tfrac{2}{5}S^2) < T_{\text{eq.}} \end{aligned} \tag{7.27}$$

where $\gamma = \frac{5}{3}$ as is appropriate for a monatomic gas. Accordingly, the recovery factor, which characterizes the difference between $T_{\text{eq.}}$ and T_0 and is defined by

$$r = \frac{T_{\text{eq.}} - T_\infty}{T_0 - T_\infty} \tag{7.28}$$

turns out to be larger than $\frac{5}{4}$, so that dissipation of mechanical energy predominates over heat conduction. This result is in sharp contrast with the known result of laminar boundary layer theory, according to which r is always less than unity [49]. It is to be remarked that for a finite body, $T_{\text{eq.}}$ is defined to be the *uniform* temperature at which the body is to be kept in order to have a zero *global* heat transfer. Hence the above result for $T_{\text{eq.}}$ is not strictly valid for a finite body; it is to be expected, however, that $r > 1$ for finite bodies of sufficiently symmetrical shape. This result is confirmed by the computations of heat transfer for finite bodies as well as by the experiments performed by Stalder, Goodwin and Creager [50].

In practice, of course, the temperature of the body is determined by a balance of all the forms of heat transfer at the body surface. For an artificial satellite, a considerable part of the heat is lost by radiation and this process must duly be taken into account in the balance.

The above results take on a particularly simple form if we formally let $S \to \infty$ in Eqs. (7.22), (7.23), (7.24). Then, if T_w/T_∞ is of order unity (and $\sin\theta > 0$):

$$p_n = -\rho_\infty V^2(2-\alpha_N)\sin^2\theta \quad p_+ = \alpha_T \rho_\infty V^2 \sin\theta\cos\theta \quad q = \tfrac{1}{2}\alpha_E \rho_\infty V^3 \tag{7.29}$$

Accordingly

$$\mathbf{p} = \alpha_T \rho_\infty V \mathbf{V} \sin\theta - (2-\alpha_T-\alpha_N)\rho_\infty V^2 \sin^2\theta \mathbf{n} \tag{7.30}$$

Then each element's contribution to drag and lift is

$$\begin{aligned} p_D &= \alpha_T \rho_\infty V^2 \sin\theta + (2-\alpha_N-\alpha_T)\rho_\infty V^2 \sin^3\theta \\ p_L &= -(2-\alpha_T-\alpha_N)\rho_\infty V^2 \sin^2\theta\cos\beta \end{aligned} \tag{7.31}$$

In particular if $\alpha_T = \alpha_N = 1$ the lift vanishes and

$$p_D = \rho_\infty V^2 \sin\theta = \rho_\infty V^2 \frac{dA_*}{dA} \tag{7.32}$$

where dA is the area of the element and dA_* the area of the projection of the element upon a plane perpendicular to $\mathbf{V}$. Accordingly, if we refer the drag coefficient to the frontal area $A_* = \int dA_*$, we obtain (no contribution arises from the back part of the body surface for $S \to \infty$):

$$C_D = \frac{\int p_D \, dA}{\frac{1}{2}\rho_\infty V^2 A_*} = 2 \tag{7.33}$$

a particularly simple result.

Eqs. (7.29) are not uniformly valid in θ; so that for any fixed speed ratio, there is a range of values of θ (those for which $S\sin\theta \lessdot 1$) for which Eqs. (7.29) cannot be applied. In practice, one can use Eqs. (6.29) if $S \gg 1$ and the area of the regions in which $S\sin\theta \lessdot 1$ is small. More complicated results are obtained if terms of first order in S^{-1} are retained [51–52].

The above treatment finds its strength and weakness in the use of the accommodation coefficients α_T, α_N and α_E. If the latter cannot be regarded as constant, the above procedure collapses. In particular, the lift coefficient depends in a crucial way on the departure from complete accommodation, and thus on the mechanism of gas-surface interaction. Since the lift to drag ratio is of critical importance for widening the depth of the corridor of re-entry of manned spacecraft, we cannot be content with the above results.

More sophisticated ways of treating the boundary conditions have thus been considered [53–59]. In particular, the scattering model described in Section 6 of Chapter III, Eq. (III.6.20), has been applied to the computation of aerodynamic forces [57–58] and heat transfer in free molecular flow. Eqs. (7.11)–(7.13) are, of course, still valid for $\mathbf{p}^-$ and q^- whereas $\mathbf{p}^+$ and q^+ are given by:

$$\mathbf{p}^+ = -p_D{}^+\frac{\mathbf{V}}{V} + p_\Lambda{}^+\left(\mathbf{n} + \frac{\mathbf{V}}{V}\sin\theta\right) \tag{7.34}$$

$$p_D{}^+ = p_D{}^1(\alpha_t) + p_D{}^2(\alpha_n) \qquad p_\Lambda{}^+ = p_\Lambda{}^1(\alpha_t) + p_\Lambda{}^2(\alpha_n) \tag{7.35}$$

$$\begin{aligned} p_D{}^1 = -(1-\alpha_t)\frac{V^2}{2S}\rho_\infty\cos^2\theta \times \\ \times \{\pi^{-\frac{1}{2}}\exp(-S^2\sin^2\theta) + S\sin\theta[1+\operatorname{erf}(S\sin\theta)]\} \end{aligned} \tag{7.36}$$

$$\begin{aligned} p_D{}^2 = \frac{\rho_\infty\alpha_n{}^{\frac{3}{2}}V^2\sin\theta}{\pi S^2} \times \\ \times \left\{\exp(-S^2\sin^2\theta)\int_0^{\pi/2} d\varphi \frac{\left[\frac{1}{2} + \frac{\tau(1-\alpha_n)\cos^2\varphi}{\alpha_n+\tau(1-\alpha_n)\sin^2\varphi}(1+U^2)\right]}{\alpha_n+\tau(1-\alpha_n)\sin^2\varphi} + \right. \\ + \sqrt{\pi}\int_0^{\pi/2} d\varphi \frac{\exp\left[-S^2\frac{\sin^2\theta\tau(1-\alpha_n)\sin^2\varphi}{\alpha_n+\tau(1-\alpha_n)\sin^2\varphi}\right]}{\alpha_n+\tau(1-\alpha_n)\sin^2\varphi} U(1+\operatorname{erf} U) \times \\ \left. \times \left[\tfrac{1}{2} + (\tfrac{3}{2}+U^2)\frac{\tau(1-\alpha_n)\cos^2\varphi}{\alpha_n+\tau(1-\alpha_n)\sin^2\varphi}\right]\right\} \end{aligned} \tag{7.37}$$

$$p_\Lambda{}^1 = -\frac{\sin\theta}{\cos^2\theta}p_D{}^1 \qquad p_\Lambda{}^2 = p_D{}^2/\sin\theta \tag{7.38}$$

$$\begin{aligned} q^+ = \frac{\rho_\infty}{4}(2RT_\infty)^{\frac{3}{2}}\{\pi^{-\frac{1}{2}}\exp(-S^2\sin^2\theta) + S\sin\theta[1+\operatorname{erf}(S\sin\theta)] \times \\ \times [(1-\alpha_n)(1+S^2\sin^2\theta) + (1-\alpha_t)^2(1+S^2\cos^2\theta) + \\ + \tau^{-1}(\alpha_n + \alpha_t[2-\alpha_t])] + \tfrac{1}{2}(1-\alpha_n)S\sin\theta[1+\operatorname{erf}(S\sin\theta)]\} \end{aligned} \tag{7.39}$$

where α_t and α_n are the two parameters appearing in Eq. (III.6.20) and

$$\tau = T_\infty/T_w \tag{7.40}$$

$$U = \alpha_n{}^{\frac{1}{2}}S\sin\theta[\alpha_n + \tau(1-\alpha_n)\sin^2\varphi]^{-\frac{1}{2}} \tag{7.41}$$

The calculations leading to the above results are somewhat simplified if we recall that α_t and α_n have the meaning of accommodation coefficients for tangential momentum (hence $\alpha_T = \alpha_t$) and the part of the kinetic energy corresponding to normal motion (hence $\alpha_N \neq \alpha_n$). α_N and α_E are *not* constant for the adopted model.

If we let $S \to \infty$ for τ of order unity (and $\sin\theta > 0$), we obtain for p_D, p_L, q:

$$p_D = \rho_\infty V^2 \sin\theta\{\alpha_t + [(1-\alpha_n)^{\frac{1}{2}} + (1-\alpha_t)]\sin\theta\}$$

$$p_L = -\rho_\infty V^2 \sin^2\theta \cos\beta[(1-\alpha_t) + (1-\alpha_n)^{\frac{1}{2}}]$$

$$q = \frac{\rho_\infty V^3}{2}[\alpha_n \sin^2\theta + \alpha_t(2-\alpha_t)\cos^2\theta] \tag{7.42}$$

If we compare these equations with Eq. (7.31), we see that the results for drag and lift are equivalent (for $S \to \infty$) to those of the standard theory based on the accommodation coefficients, provided an effective accommodation coefficient for normal momentum, $\alpha_N{}^*$, is defined as follows:

$$\alpha_N{}^* = 1 - (1-\alpha_n)^{\frac{1}{2}} \tag{7.43}$$

It is to be stressed, however, that this identification is possible because we consider a very special distribution function for the impinging molecules which is the limiting case of a Maxwellian distribution for $S \to \infty$ (with fixed V); that is, a delta function. As for q, identification with Eq. (7.29) gives

$$\begin{aligned}\alpha_E{}^*(\theta) &= \alpha_t(2-\alpha_t)\cos^2\theta + \alpha_n \sin^2\theta \\ &= \alpha_t(2-\alpha_t)\cos^2\theta + \alpha_N{}^*(2-\alpha_N{}^*)\sin^2\theta\end{aligned} \tag{7.44}$$

In other words it is not possible to define a constant accommodation coefficient for energy even for such a special distribution of the impinging molecules as a delta function. If $\alpha_N{}^* = \alpha_T$, that is if $1 - \alpha_n = (1-\alpha_t)^2$, however, then $\alpha_E{}^* = \alpha_T(2-\alpha_T) = \alpha_t(2-\alpha_t)$ is constant.

The calculation of the global aerodynamic coefficients according to the two-parameter model based on Eq. (III.6.20) has been carried out for several bodies [57, 58], such as flat plates, spheres, cylinders, cones and ellipsoids. Of course, due to the complexity of the results for a surface element, most calculations have to be carried out numerically. However, the calculation of heat transfer is easier and can be carried out analytically for flat plates, cylinders, spheres [59]. In the case of a sphere, the result is the same as for the classical treatment, provided α_E is replaced by

$$\bar{\alpha}_E{}^* = \tfrac{1}{2}[\alpha_t(2-\alpha_t) + \alpha_n] \tag{7.45}$$

In particular, $T_{\text{eq.}}$ is the same as in the classical case; for the flat plate and the cylinder, however, $T_{\text{eq.}}$ depends on the ratio $\alpha_n/\alpha_E{}^*$. This can lead to recovery factors smaller than unity for suitable values of the latter ratio.

8. Free molecule flow in presence of nonconvex boundaries

When free molecule flow is studied in the presence of boundaries with concavities, we must take account of the shading of some parts of the body by other parts, as well as of the incidence on some parts of the body of molecules re-emitted from other parts.

Let P be a point on the surface ∂R of the body (see Fig. 32). Let $\omega(\mathbf{x})$ denote the solid angle subtended at the point P (of position vector $\mathbf{x}$) by the rest of the body (which might be actually made of several disconnected pieces). Then the distribution function of the molecules impinging upon the boundary may be represented in the form

$$f^-(x, \boldsymbol{\xi}) = f_\infty(\boldsymbol{\xi}) \qquad (\boldsymbol{\xi} \cdot \mathbf{n} > 0; \mathbf{x} \in \partial R, \boldsymbol{\xi} \notin \omega(\mathbf{x})) \tag{8.1}$$

$$f^-(\mathbf{x}, \boldsymbol{\xi}) = |\boldsymbol{\xi} \cdot \mathbf{n}'|^{-1} \int_{\boldsymbol{\xi}' \cdot \mathbf{n}' < 0} R(\boldsymbol{\xi}' \to \boldsymbol{\xi}; \mathbf{x}') f^-(\mathbf{x}', \boldsymbol{\xi}') \, |\boldsymbol{\xi}' \cdot \mathbf{n}'| \, d\boldsymbol{\xi}'$$
$$(\boldsymbol{\xi} \cdot \mathbf{n} > 0; \mathbf{x} \in \partial R, \boldsymbol{\xi} \in \omega(\mathbf{x})) \tag{8.2}$$

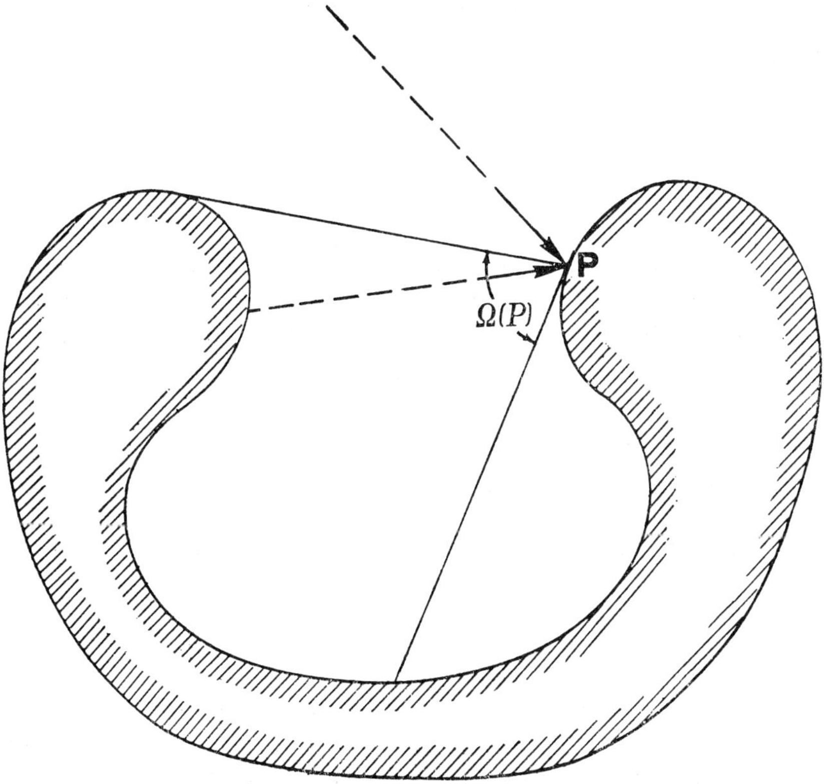

Fig. 32. Shading effect in the flow past a nonconvex body.

where $\mathbf{x}'$ is the point of the boundary which can be reached from the point $\mathbf{x}$ along the vector $-\boldsymbol{\xi}$ (without intersecting the boundary itself) and $\mathbf{n}'$ is the normal unit vector at $\mathbf{x}'$. Eqs. (8.1) and (8.2) form a system which determines f^-, at least in principle; the solution can be found by a converging iteration (if certain singular cases are excluded), but no analytical solutions can be achieved in general.

In practice, diffuse reflection is frequently assumed using Eq. (III.6.13) with $\alpha = 1$. Eq. (8.2) then reduces to

$$f^- = J(\mathbf{x}')[2\pi(RT_w')^2]^{-1}\exp[-\xi^2/(2RT_w')]$$
$$(\mathbf{x} \in \partial R,\ \boldsymbol{\xi}\cdot\mathbf{n} > 0,\ \boldsymbol{\xi} \in \omega(\mathbf{x})) \quad (8.3)$$

where $J(\mathbf{x}')$ is the mass flow of impinging molecules at $\mathbf{x}'$, T_w' the wall temperature at $\mathbf{x}'$. Thus the problem is reduced to finding $J(\mathbf{x})$ for $\mathbf{x} \in \partial R$.

The mass flow at $\mathbf{x}$ is equal to

$$J(\mathbf{x}) = J_\infty(\mathbf{x}) + \int_{\substack{\boldsymbol{\xi}\cdot\mathbf{n}<0\\ \boldsymbol{\xi}\in\omega}} J(\mathbf{x}')[2\pi(RT_w')^2]^{-1}\exp[-\xi^2/(2RT_w')]\,|\boldsymbol{\xi}\cdot\mathbf{n}|\,d\boldsymbol{\xi} \quad (8.4)$$

where $x' \in \partial R$ is given by $\mathbf{x}' = \mathbf{x} - \boldsymbol{\xi}\,|\mathbf{x}' - \mathbf{x}|$ and $J_\infty(\mathbf{x}')$ is the mass flow arriving directly from infinity

$$J_\infty(\mathbf{x}) = \int_{\substack{\boldsymbol{\xi}\cdot\mathbf{n}<0\\ \boldsymbol{\xi}\in\omega'}} f_\infty(\boldsymbol{\xi})\,|\boldsymbol{\xi}\cdot\mathbf{n}|\,d\boldsymbol{\xi} \quad (8.5)$$

where ω' is the complement of ω. Introducing the speed $\boldsymbol{\xi}$ and the unit vector $\boldsymbol{\Omega} = \boldsymbol{\xi}/\xi$, the integration with respect to ξ can be performed in Eq. (8.4) to yield

$$J(\mathbf{x}) = J_\infty(\mathbf{x}) + \frac{1}{\pi}\int_{\substack{\boldsymbol{\Omega}\cdot\mathbf{n}<0\\ \boldsymbol{\Omega}\in\omega}} J(\mathbf{x}')\,|\boldsymbol{\Omega}\cdot\mathbf{n}|\,d\boldsymbol{\Omega} \quad (8.6)$$

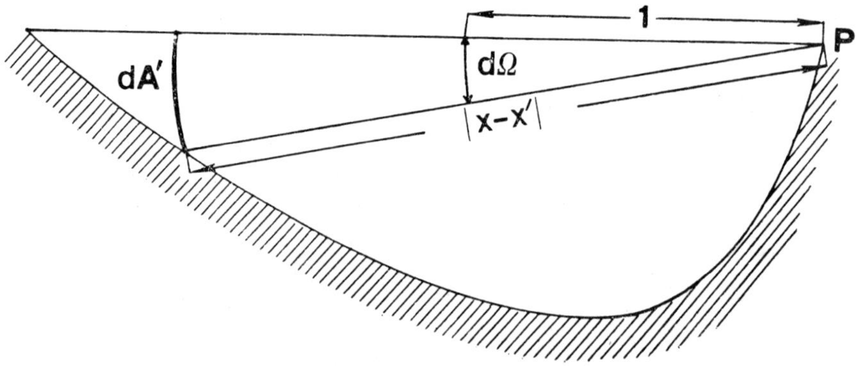

Fig. 33. Illustration of the computation of the surface element $dA' = d\Omega\,|\mathbf{x} - \mathbf{x}'|^2\,|\mathbf{n}'\cdot\boldsymbol{\Omega}|$.

Since (see Fig. 33):

$$\mathbf{\Omega} = \frac{\mathbf{x} - \mathbf{x}'}{|\mathbf{x} - \mathbf{x}'|} \qquad d\mathbf{\Omega} = \frac{dA'\,|\mathbf{n}' \cdot \mathbf{\Omega}|}{|\mathbf{x} - \mathbf{x}'|^2} \tag{8.7}$$

where dA' is the surface element at $\mathbf{x}'$, we can transform Eq. (8.7) as follows:

$$J(\mathbf{x}) = J_\infty(\mathbf{x}') + \frac{1}{\pi} \int_{\partial R(\mathbf{x})} J(\mathbf{x}') \frac{|\mathbf{n} \cdot (\mathbf{x} - \mathbf{x}')|\,|\mathbf{n}' \cdot (\mathbf{x} - \mathbf{x}')|}{|\mathbf{x} - \mathbf{x}'|^4}\, dA' \tag{8.8}$$

where $\partial R(\mathbf{x})$ is the part of the body "seen" from $\mathbf{x}$. Eq. (8.8) is an integral equation with symmetrical kernel (if $\mathbf{x}' \in \partial R(\mathbf{x})$, $\mathbf{x} \in \partial R(\mathbf{x}')$). The kernel is bounded provided, for example, the curvature of the boundary changes in a continuous fashion.

The great advantage of Eq. (8.8) is that it can be used to compute $J(\mathbf{x})$ from the body geometry. In particular, Eq. (8.8) can be used to study free molecule flows in tubes of arbitrary (constant) cross section with diameter much smaller than the mean free path (capillaries). These flows arise when we connect two reservoirs containing gas at equilibrium at different pressures p_1, p_2 and temperatures T_1, T_2 by a thin tube (Fig. 34). Since the molecules are assumed not to collide with each other, the molecule flux arises from the superposition of two independent fluxes comprising the flux of molecules which entered the tube from the left and the flux of molecules which entered the tube from the right. These fluxes can be computed independently; hence we may assume, without loss of generality, that $J_\infty(\mathbf{x})$ is now the mass flow of molecules reaching $\mathbf{x}$ directly from reservoir 1 without collisions.

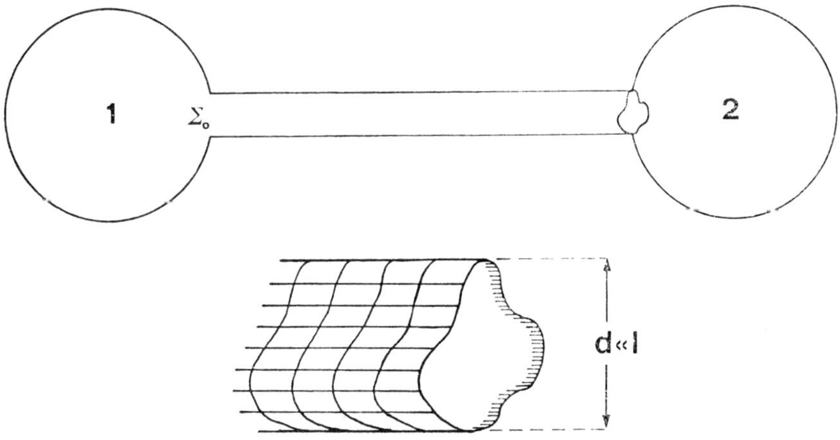

Fig. 34. Two reservoirs connected by a tube much thinner than a mean free path.

Then

$$J_\infty = \frac{J_1}{\pi}\int_{\Sigma_1} \frac{|\mathbf{n}\cdot(\mathbf{x}-\mathbf{x}')|\,|\mathbf{n}'\cdot(\mathbf{x}-\mathbf{x}')|}{|\mathbf{x}-\mathbf{x}'|^4}\,dA' \tag{8.9}$$

where integration extends to the entrance cross section Σ_1, and J_1 is the mass flow through Σ_1:

$$J_1 = \int_{\boldsymbol{\xi}\cdot\mathbf{n}'>0} |\boldsymbol{\xi}\cdot\mathbf{n}'|\, f_{01}\, d\xi_1 = \rho_1\left(\frac{RT_1}{2\pi}\right)^{\frac{1}{2}} = \frac{p_1}{\sqrt{2\pi RT_1}} \tag{8.10}$$

where f_{01} is the Maxwellian corresponding to the density and temperature in reservoir 1.

In particular, if the tube is a circular cylinder of radius r we have, if z is the coordinate of $\mathbf{x}$ along the tube axis and x', y' are the coordinates of $\mathbf{x}'$ with respect to a suitable Cartesian frame:

$$\begin{aligned} J_\infty &= \frac{J_1}{\pi}\, 2\int_0^r dy' \int_{-\sqrt{r^2-y'^2}}^{\sqrt{r^2-y'^2}} dx'\, \frac{z(r-x')}{[(x'-r)^2+y'^2+z^2]^2} \\ &= \frac{J_1 z}{\pi}\int_0^r \left[\frac{1}{z^2+2r^2-2r\sqrt{r^2-y'^2}} - \frac{1}{z^2+2r^2+2r\sqrt{r^2-y'^2}}\right] dy' \\ &= \frac{J_1 z}{2\pi}\int_0^{2\pi} \frac{r\cos\theta\, d\theta}{z^2+2r^2-2r^2\cos\theta} = \frac{J_1}{2}\left(\frac{Z^2+2}{\sqrt{Z^2+4}} - Z\right) \end{aligned} \tag{8.11}$$

where $Z = z/r$ and the last integration is easily effected by contour integration in the complex plane or, less straightforwardly, by standard methods.

Analogously, since $J(\mathbf{x}') = J(z')$ by symmetry:

$$\begin{aligned} &\frac{1}{\pi}\int_{\partial R(\mathbf{x})} J(\mathbf{x}')\,\frac{|\mathbf{n}\cdot(\mathbf{x}-\mathbf{x}')|\,|\mathbf{n}'\cdot(\mathbf{x}-\mathbf{x}')|}{|\mathbf{x}-\mathbf{x}'|^4}\,dA' \\ &\qquad = \frac{1}{\pi}\int_0^L\int_0^{2\pi} J(z')\,\frac{4r^2\sin^4\left(\dfrac{\varphi'}{2}\right) r\,d\varphi'\,dz'}{\left[4r^2\sin^2\left(\dfrac{\varphi'}{2}\right) + (z-z')^2\right]^2} \end{aligned} \tag{8.12}$$

Integration with respect to φ' (the angle between the projections of $\mathbf{x}'$ and $\mathbf{x}$ in a plane normal to the tube axis) gives:

$$\begin{aligned} &\frac{1}{\pi}\int_0^{2\pi} \frac{4r^3\sin^4(\varphi'/2)\,d\varphi'}{[4r^2\sin^2(\varphi'/2)+(z-z')^2]^2} \\ &\qquad = \frac{1}{2r}\left\{1 - |Z-Z'|\,\frac{(Z-Z')^2+6}{[4+(Z-Z')^2]^{\frac{3}{2}}}\right\} \end{aligned} \tag{8.13}$$

where $Z' = z'/r$. Hence Eq. (8.8) becomes:

$$\chi(Z) = \chi_0(Z) + \int_0^{\bar{L}} K(|Z - Z'|)\chi(Z')\, dZ' \tag{8.14}$$

where

$$\chi = J/J_1 \qquad \bar{L} = L/r$$

$$\chi_0(Z) = \frac{1}{2}\left(\frac{Z^2 + 2}{\sqrt{Z^2 + 4}} - Z\right) \tag{8.15}$$

$$K(|Z - Z'|) = \frac{1}{2}\left\{1 - |Z - Z'|\frac{6 + (Z - Z')^2}{[4 + (Z - Z')^2]^{\frac{3}{2}}}\right\}$$

Eq. (8.14) is Clausing's equation [60].

If we consider a very long tube and neglect end effects, it is convenient to put $Z = Y + \bar{L}/2$, $Z' = Y' + \bar{L}/2$ and let $\bar{L} \to \infty$. Then the source term vanishes; hence, for a long tube:

$$\chi(Y) = \int_{-\infty}^{+\infty} K(|Y - Y'|)\chi(Y')\, dY' \tag{8.16}$$

It is easy to check that

$$\int_{-\infty}^{+\infty} K(|Y - Y'|)\, dY' = 1 \tag{8.17}$$

This is a limiting case ($Z = Y + \bar{L}/2$, $Z' = Y' + \bar{L}/2$, $\bar{L} \to \infty$) of the relation

$$\chi_0(\bar{L} - Z) + \chi_0(Z) + \int_0^{\bar{L}} K(|Z - Z'|)\, dZ' = 1 \tag{8.18}$$

which corresponds to the fact that $\chi = 1$, if the flow is the same at both ends of the tube. Eq. (8.18) can also be used to compute $K(|Z - Z'|)$, once $\chi_0(Z)$ has been computed. In fact, differentiation with respect to $\bar{L}$ in Eq. (8.18) yields

$$K(|Z - Z'|) = -\chi_0'(|Z - Z'|) \tag{8.19}$$

where χ_0' is the derivative of χ_0 with respect to its argument.

Hence $\chi(Y) = \text{const.}$ is a solution of Eq. (8.16) as was to be expected (equilibrium solution). In addition, there is a solution with χ proportional to Y. In fact, if we write $\chi(Y') = Y' = Y + (Y' - Y)$, we immediately see that $Y' - Y$ gives a zero contribution to the integral in Eq. (8.16) since $(Y' - Y)K(|Y' - Y|)$ is odd in the variable $t = Y' - Y$. Then, due to Eq. (8.17), the integral reduces to $Y = \chi(Y)$. We conclude that in a very long tube the density, and hence the pressure, change in a linear way along the tube. Numerical results show that this is approximately true for tubes of moderate length [61]. The end effects in a long tube have been investigated by Pao and Tchao [62].

As a consequence of Eq. (8.18), $\chi(Z)$ satisfies

$$\chi(\bar{L} - Z) = 1 - \chi(Z) \tag{8.20}$$

In fact, Eq. (8.14) gives

$$\begin{aligned}\chi(\bar{L} - Z) + \chi(Z) - 1 &= \chi_0(\bar{L} - Z) + \chi_0(Z) - 1 + \\ &\quad + \int_0^{\bar{L}} K(|Z - Z'|)[\chi(\bar{L} - Z') + \chi(Z')]\, dZ' \\ &= \int_0^{\bar{L}} K(|Z - Z'|)[\chi(\bar{L} - Z') + \chi(Z') - 1]\, dZ' \end{aligned} \tag{8.21}$$

where the last step is based on Eq. (8.18). Hence $\chi(\bar{L} - Z) + \chi(Z) - 1$ satisfies a homogeneous Fredholm equation. Since $K > 0$, and according to Eq. (8.18)

$$\int_0^{\bar{L}} K(|Z - Z'|)\, dZ' \leqslant 1 - c < 1 \tag{8.22}$$

where

$$c = \operatorname*{Inf}_{0<Z<\bar{L}} [\chi_0(\bar{L} - Z) + \chi_0(Z)] \tag{8.23}$$

Eq. (8.21) possesses only the trivial solution

$$\chi(\bar{L} - Z) + \chi(Z) - 1 = 0 \tag{8.24}$$

and Eq. (8.20) follows. In particular, Eq. (8.24) implies $\chi(\bar{L}/2) = \frac{1}{2}$.

Once $J(\mathbf{x})$ (in particular, $\chi(Z)$ in the cylindrical case) is computed (by iteration, variational techniques or the discrete ordinate method), any other quantity can be computed. The flow rate is particularly interesting. To compute it, we remark that $[J_\infty(\mathbf{x})/J_1]\, dA$ gives the probability for a molecule to reach the area element dA at $\mathbf{x}$ directly from reservoir 1 without collisions; since the equations of motion are time reversible, this is also the probability for a molecule leaving the area element dA at $\mathbf{x}$ to reach reservoir 1 without collisions. Multiplication by $J(\mathbf{x})$ and integration over the side surface Σ of the tube yields the reverse mass flow through the entrance. Hence the mass flow rate through the exit (difference between J_1A_1, where A_1 is the area of the entrance cross section, and the reverse flow) is:

$$J_1A_1 - \frac{1}{J_1}\int_\Sigma J_\infty(\mathbf{x})J(\mathbf{x})\, dA \tag{8.25}$$

In particular, for a cylindrical tube $J_\infty(\mathbf{x})$ is given by $J_1\chi_0(Z)$, according to Eqs. (8.11), (8.15), dA may be replaced by $2\pi r\, dz = 2\pi r^2\, dZ$, $J(\mathbf{x}) = J_1\chi(Z)$, $A_1 = \pi r^2$; hence:

$$Q = \pi r^2 J_1\left[1 - 2\int_0^{\bar{L}} \chi_0(Z)\chi(Z)\, dZ\right] \tag{8.26}$$

In order to use this equation, it is useful to transform it by means of a relation following from Eqs. (8.14) and (8.19). Let us integrate Eq. (8.14) from 0 to Z'':

$$\int_0^{Z''} \chi(Z)\, dZ = \int_0^{Z''} \chi_0(Z)\, dZ - \int_0^{L}\left[\int_0^{Z''} \chi_0'(|Z - Z'|)\, dZ\right]\chi(Z')\, dZ' \quad (8.27)$$

However

$$\int_0^{Z''} \chi_0'(|Z - Z'|)\, dZ$$
$$= \operatorname{sgn}(Z'' - Z')\chi_0(|Z'' - Z'|) + \chi_0(Z') - H(Z'' - Z') \quad (8.28)$$

where $H(Z - Z') = [1 + \operatorname{sgn}(Z - Z')]/2$ denotes the Heaviside step function (use has been made of $\chi_0(0) = \frac{1}{2}$, according to Eq. (8.15)). Inserting Eq. (8.28) into Eq. (8.27) gives

$$\int_0^{Z''} \chi_0(Z)\, dZ = \int_0^{L} \operatorname{sgn}(Z'' - Z')\chi_0(|Z'' - Z'|)\chi(Z')\, dZ' +$$
$$+ \int_0^{L} \chi_0(Z')\chi(Z')\, dZ' \quad (8.29)$$

If we now multiply by $\chi(Z'')$ and integrate with respect to Z'' from 0 to L we obtain:

$$\int_0^{L}\left[\int_0^{Z''} \chi_0(Z)\, dZ\right]\chi(Z'')\, dZ'' = \frac{L}{2}\int_0^{L} \chi_0(Z')\chi(Z')\, dZ' \quad (8.30)$$

where use has been made of the fact that one of the integrals arising from the right hand side of Eq. (8.29) is zero (it changes its sign when the exchange $Z' \leftrightarrow Z''$ is made) and

$$\int_0^{L} \chi(Z)\, dZ = \frac{L}{2} \quad (8.31)$$

as follows from Eq. (8.24). Finally by direct integration:

$$\int_0^{Z''} \chi_0(Z)\, dZ = \tfrac{1}{4}(Z''\sqrt{Z''^2 + 4} - Z''^2) \quad (8.32)$$

and Eq. (8.30) can be rewritten as follows:

$$\int_0^{L} \chi_0(Z)\chi(Z)\, dZ = (2L)^{-1}\int_0^{L} (Z\sqrt{Z^2 + 4} - Z^2)\chi(Z)\, dZ \quad (8.33)$$

Use of Eqs. (8.31) and (8.33) in Eq. (8.26) yields:

$$Q/Q_0 = L^{-1}\int_0^{L} (Z^2 + 2 - Z\sqrt{Z^2 + 4})\chi(Z) f Z \quad (8.34)$$

where $Q_0 = \pi r^2 J_1$ is the flow rate for $L = 0$ (orifice), according to Eq. (8.26).

In particular, if the tube is very long, $\chi(Z)$ is linear and $\chi(0) = 1$, $\chi(\bar{L}/2) = \frac{1}{2}$; hence

$$\chi(Z) = 1 - Z/\bar{L} \tag{8.35}$$

and Eq. (8.34) gives:

$$\begin{aligned} Q/Q_0 &= \bar{L}^{-1} \int_0^{\bar{L}} (1 - Z/\bar{L})[Z^2 + 2 - Z(Z^2 + 4)^{\frac{1}{2}}]\, dZ \\ &= \frac{1}{12}\bar{L}^2 + 1 + \frac{8}{3\bar{L}} - \frac{1}{12\bar{L}}(\bar{L}^2 + 4)^{\frac{3}{2}} - \frac{1}{2\bar{L}}(\bar{L}^2 + 4)^{\frac{1}{2}} + \\ &\quad + \frac{2}{\bar{L}^2}\log\left(\frac{\bar{L} + \sqrt{\bar{L}^2 + 4}}{2}\right) \\ &= \frac{8}{3\bar{L}} + \frac{2}{\bar{L}^2}\log \bar{L} - \frac{3}{2}\frac{1}{\bar{L}^2} + O(\bar{L}^{-3}) \end{aligned} \tag{8.36}$$

Hence for long tubes

$$Q/Q_0 = 8/(3\bar{L}) \tag{8.37}$$

a formula which goes back to Knudsen [63] and Smoluchowski [64].

9. Nearly free-molecule flows

Let us examine flows which occur when the Knudsen number is relatively large, but the effect of intermolecular collisions is not completely negligible. For steady flows, the Boltzmann equation can be written in the form

$$\boldsymbol{\xi} \cdot \frac{\partial f}{\partial \mathbf{x}} = \varepsilon Q(f, f) \tag{9.1}$$

where ε, the *inverse* Knudsen number, is small.

It seems natural to expand f into a power series in ε as in Eq. (2.2) (with a different meaning of ε, of course). Eq. (9.1) then gives:

$$\boldsymbol{\xi} \cdot \frac{\partial f_0}{\partial \mathbf{x}} = 0 \qquad \boldsymbol{\xi} \cdot \frac{\partial f_n}{\partial \mathbf{x}} = Q_{n-1} \tag{9.2}$$

where Q_n is given by Eq. (IV.1.5). The first of these equations is the equation for free-molecule flow, as was to be expected. In the subsequent equations, the source Q_{n-1} is known in terms of the previous approximations. Thus the solution of the Boltzmann equation is reduced to the solution of a sequence of linear differential equations. These equations are of the form

$$\boldsymbol{\xi} \cdot \frac{\partial f}{\partial \mathbf{x}} = S(\mathbf{x}, \boldsymbol{\xi}) \tag{9.3}$$

where S is the source term. The solution of a given boundary value problem can be obtained by the method of characteristics or by means of the Green's function. The latter is obtained from Eq. (IV.12.4) by letting $\nu = 0$:

$$U_0(\mathbf{x}, \boldsymbol{\xi}; \mathbf{x}', \boldsymbol{\xi}') = \xi^{-1} H(\boldsymbol{\xi} \cdot [\mathbf{x} - \mathbf{x}'])\, \delta_2\left(\frac{\boldsymbol{\xi}}{\xi} \wedge [\mathbf{x} - \mathbf{x}']\right) \delta(\boldsymbol{\xi} - \boldsymbol{\xi}') \quad (9.4)$$

A singularity is present at $\boldsymbol{\xi} = 0$ because of the factor ξ^{-1}. Hence it is to be expected that the solutions obtained by this technique of expanding in powers of ε (the so called Knudsen iteration, since the same techniques may be obtained by iteration with the free molecular flow solution as zero-th iterate) are not valid for low molecular speeds. When moments are computed (see Eq. (5.34)) by integrating the distribution function, however, the contribution of slow molecules is usually negligible, at least in a bounded domain, because $d\boldsymbol{\xi} = \xi^2\, d\xi\, d\boldsymbol{\Omega}$ and the factor ξ^2 cancels the singularity. This remains true for problems with plane two-dimensional symmetry, for which Eq. (9.4) is true with δ_2 replaced by a one-dimensional delta function and ξ replaced by the magnitude of the projection of $\boldsymbol{\xi}$ onto the relevant plane. If the problem has one-dimensional plane symmetry, then δ_2 disappears and U_0 reduces to

$$U_0^{(1)}(x_1, \boldsymbol{\xi}, x_1', \boldsymbol{\xi}') = \xi_1^{-1} H(\xi_1[x_1 - x_1'])\, \delta(\boldsymbol{\xi} - \boldsymbol{\xi}') \quad (9.5)$$

but the singularity at $\xi_1 = 0$ does not disappear, in general, even when computing moments. In addition, it is clear that even the mild singularities in two and three dimensions, though negligible at the first step, can build up to a worse singularity when computing subsequent steps. The difficulties are enhanced in unbounded domains where the subsequent terms diverge at the space infinity. The reason for the latter fact is that the ratio between the mean free path l and the distance d from the body is a local Knudsen number which tends to zero when d tends to infinity; hence collisions certainly arise in an unbounded domain and tend to dominate at large distances. On this basis we are led to expect that a continuum behavior takes place at infinity, even when the dimensions of a solid body moving in the gas are much smaller than the mean free path; this is confirmed by the general solution of the steady linearized Boltzmann equation (Chapter IV, Section 11).

Both difficulties can be removed, as remarked by Willis [65–69] by retaining the collision frequency in the Greens' function given by Eq. (IV.12.4). This requires, of course, that the collision term can be split as follows:

$$Q(f, f) = Q_1(f, f) - \nu f \quad (9.6)$$

where ν, the collision frequency, given by

$$\nu = \nu(f) = \frac{1}{m} \iiint f_* B(\theta, |\boldsymbol{\xi} - \boldsymbol{\xi}_*|)\, d\boldsymbol{\xi}_*\, d\theta\, d\varepsilon \quad (9.7)$$

is finite. ν depends upon $\boldsymbol{\xi}$ and $\mathbf{x}$ through f in an unknown way. This does not prevent the use of an iteration method of the form

$$f_{-1} = 0$$

$$\boldsymbol{\xi} \cdot \frac{\partial f_n}{\partial \mathbf{x}} + \varepsilon \nu_{n-1} f_n = \varepsilon Q_1(f_{n-1}, f_{n-1}) \qquad (n = 0, 1, \ldots) \tag{9.8}$$

where $\nu_{n-1} = \nu(f_{n-1})$. Hence f_n is the n-th order approximation to f and thus corresponds to $\sum_{k=0}^{n} \varepsilon^k f_k$ of the previous method; now, however, the solution is not given by a power series in ε. The Green's function for the differential operator appearing on the left hand side of Eq. (9.8) (ν_{n-1} is assumed to be known from the previous step) is again given by Eq. (IV.12.4) provided the exponent $\nu(\xi)\boldsymbol{\xi} \cdot (\mathbf{x} - \mathbf{x}')/\xi^2$ is replaced by

$$\varepsilon \int U_0(\mathbf{x}, \boldsymbol{\xi}; \mathbf{x}', \boldsymbol{\xi}')\nu_{n-1}(\mathbf{x}', \boldsymbol{\xi}')\, d\boldsymbol{\xi}'\, d\mathbf{x}'$$

[U_0 is given by Eq. (9.4)]. Hence f_n is obtained in terms of quadratures, once f_{n-1} is known. Usually only the first correction to free-molecule flow is carried out. Since we can obtain the same result by first transforming the Boltzmann equation into an integral equation by means of the Green's function (IV.12.4) [with $\varepsilon \int U_0(\mathbf{x}, \boldsymbol{\xi}; \mathbf{x}', \boldsymbol{\xi}')\nu(\mathbf{x}', \boldsymbol{\xi}')\, d\boldsymbol{\xi}'\, d\mathbf{x}'$ in place of $\nu(\xi)\boldsymbol{\xi} \cdot (\mathbf{x} - \mathbf{x}')/\xi^2$ in the exponent] and then solving the integral equation by iteration, with the free-molecule solution as the zero-th iterate; the method is usually called the method of integral iteration. Typical integrals to be evaluated when computing moments (see Eq. (5.34)) have the form:

$$J_n(\varepsilon x) = \int_0^\infty \exp[-\varepsilon x \nu(\xi)/\xi_1] e^{-\xi_1^2} \xi_1^n \, d\xi_1 \qquad (n = -1, 0, 1, \ldots) \tag{9.9}$$

for one-dimensional problems; for two-dimensional problems n starts from 0 and for three-dimensional problems n starts from 1 (in the latter case, $\xi_1 = \xi$).

We can expand the exponent in powers of ε except for small values of ξ_1. In order to take into account the contribution from small values of ξ_1, we must consider integrals of the form

$$I_n(\eta) = \int_0^1 \exp(-\eta/\xi_1)\xi_1^n \, d\xi_1 \qquad (\eta = \varepsilon x \nu_0;\ n = -1, 0, 1 \ldots) \tag{9.10}$$

where ν_0 is the value of $\nu(\xi)$ for $\xi_1 = 0$ and, analogously, $e^{-\xi_1^2}$ has been replaced by 1; we obtain integrals of this form by separating the contributions from the intervals (0, 1) and (1, ∞) in Eq. (9.9) and expanding in powers of ξ_1 in the first integral. If we let $\xi_1 = \eta/t$ in Eq. (9.10), we obtain:

$$I_n(\eta) = \eta^{n+1}\int_\eta^\infty e^{-t}t^{-(n+2)}\,dt$$
$$= \eta^{n+1}\left\{\int_\eta^\infty \sum_{k=0}^{n}\frac{(-t)^k}{k!}\,t^{-(n+2)}\,dt + \int_\eta^1\left[e^{-t} - \sum_{k=0}^{n+1}\frac{(-t)^k}{k!}\right]t^{-(n+2)}\,dt + \right.$$
$$\left. + \int_\eta^1 \frac{(-t)^{n+1}}{(n+1)!}\,t^{-(n+2)}\,dt + \int_1^\infty\left[e^{-t} - \sum_{k=0}^{n}\frac{(-t)^k}{k!}\right]t^{-(n+2)}\,dt\right\}$$
$$(n = -1, 0, 1\ldots) \quad (9.11)$$

where the first term in the last expression is of course missing if $n = -1$. The second and fourth integrals are of order unity for $\varepsilon \to 0$, the former being finite for $\varepsilon \to 0$, the latter independent of ε. The first and third integrals are easily evaluated; hence

$$I_n(\eta) = \sum_{k=0}^{n}\frac{(-\eta)^k}{k!\,(n-k+1)} - \frac{(-\eta)^{n+1}}{(n+1)!}\log\eta + O(\eta^{n+1})$$
$$\eta = \varepsilon v_0 x;\ n = -1, 0, 1, \ldots) \quad (9.12)$$

Again the first term is missing for $n = -1$. In the latter case:

$$I_{-1}(\eta) = -\log\eta + O(1) \qquad (\eta = \varepsilon v_0 x) \quad (9.13)$$

The first correction to a typical moment in a one-dimensional problem is given by ε times an integral of the form (9.9). Hence the first correction to a typical moment is of order $\varepsilon\log\varepsilon = \log(\mathrm{Kn})/\mathrm{Kn}$. In two dimensions, the same correction will be of order ε followed by $\varepsilon^2\log\varepsilon$; in three dimensions we must reach terms of third order to find a logarithm $\varepsilon^3\log\varepsilon$. In any case, these results show the impossibility of expanding f into a uniformly valid power series in ε.

If we look at the dependence upon x then we must consider the effect of a further integration with respect to x (say, from 0 to x, where $x = 0$ is the location of the boundary).

Hence for small values of x (and any fixed ε, since only $\eta = \varepsilon v_0 x$ is important in Eq. (9.13)) a typical moment, in a one dimensional problem, will show a behavior of the form $x\log x$ for $x \to 0$; this implies a logarithmically divergent derivative at a boundary.

In an external domain we have, in addition to the low speed effects, the effect of particles coming from infinity. The contribution from these particles is given by the same integrals as above, integrated from infinity to x; in this case the integration with respect to velocity variables plays a very small role and so we consider:

$$L(\varepsilon) = \int_x^\infty \exp(-\varepsilon v_1 x)Q_0(x)\,dx \quad (9.14)$$

where $v_1 = v/\xi_1$, x is a variable measuring, say, the distance from the boundary, and $Q_0(x)$ is the collision term $Q_1(f, f)$ appearing in Eq. (9.6), evaluated for the deviation f_d of the free-molecule solution from the Maxwellian at infinity. f_d is different from zero at x within the solid angle subtended by the body (or bodies) limiting the gas flow; hence Q_0 (which is obtained from f_d by integration over the velocity space) is proportional to such a solid angle. Accordingly $Q_0 =$ const. in one dimension (half space problems), $Q_1 = O(x^{-1})$ in two dimensions, $Q_2 = O(x^{-2})$ in three dimensions. Hence the integrals to be evaluated have the form

$$L_n(\varepsilon) = \int_1^\infty \exp(-\varepsilon v_1 x)\frac{dx}{x^n} \qquad (n = 0, 1, 2 \ldots) \tag{9.15}$$

where $n = 0$ is typical in one dimension, $n = 1$ in two dimensions, etc. We set the lower limit equal to 1 for definiteness (contributions from finite regions can be handled directly). If $n = 0$, we obtain

$$L_0(\varepsilon) = \int_1^\infty \exp(-\varepsilon v_1 x)\, dx = 1/(\varepsilon v_1) \tag{9.16}$$

and the iteration procedure fails completely, because this term multiplied by ε, as appropriate, gives a contribution of the same order as the zero-th order approximation! Accordingly, let $n \geqslant 1$. If we set $x = t/\eta$, $\eta = \varepsilon v_1$, we obtain

$$\begin{aligned} L_n(\eta) &= \eta^{n-1}\int_\eta^\infty e^{-t} t^{-n}\, dt = I_{n-2}(\eta) \\ &= \sum_{k=0}^{n-2} \frac{(-\eta)^k}{k!\,(n-k-1)} - \frac{(-\eta)^{n-1}}{(n-1)!}\log\eta + O(\eta^{n-1}) \end{aligned}$$

$$(\eta = \varepsilon v_1;\ n = 1, 2 \ldots) \tag{9.17}$$

Hence in two dimensions ($n = 1$) the contribution of distant particles produces a term of order $\log \eta$ in $L_1(\eta)$ and a correction of order $\varepsilon \log \varepsilon$ or log(Kn)/Kn to the moments; in three dimensions, a correction of order log(Kn)/(Kn)^2 is preceded by a correction of order Kn^{-1}. Note that the logarithm enters at a lower order because of distant molecules than it does because of slow molecules. Accordingly, in an exterior problem, the distant particle effect dominates over the slow particle nonuniformity.

Care must be exercised when applying the above results to a concrete numerical evaluation. In fact, for large but not extremely large Knudsen numbers (say $10 \leqslant \text{Kn} \leqslant 100$), log Kn is a small number, although $\log(\text{Kn}) \to \infty$ for $\text{Kn} \to \infty$. Hence terms of order log(Kn)/Kn, though mathematically dominating over terms of order 1/Kn are of the same order as the latter for practical purposes. As a consequence, terms of order 1/Kn must be computed together with terms of order log(Kn)/Kn if numerical accuracy is desired in the abovementioned range of Knudsen numbers.

Related to this remark is the fact that any factor appearing in front of Kn in the argument of the logarithm is meaningless unless the term of order Kn^{-1} is also computed; this is particularly important when the above-mentioned factor depends upon a parameter which can take very large (or small) values (typically, a speed ratio). Thus Hamel and Cooper [70–71] have shown that the first iterate of integral iteration is incapable of describing the correct dependence upon the speed ratio and have applied the method of matched asymptotic expansions to regions near the body and far from the body. In particular, for the hypersonic flow of a gas of hard spheres past a two-dimensional strip, they find for the drag coefficient

$$C_D = C_{Df.m.}[1 + (\varepsilon \log \varepsilon)/(2\pi)] \tag{9.18}$$

where the inverse Knudsen number is based on the mean free path $\lambda = \pi^{\frac{3}{2}}\sigma^2 n_\infty S_w$ (σ is the molecular diameter, n_∞ the number density at infinity and $S_w = S(T_w/T_\infty)^{\frac{1}{2}}$ with notations of Section 7).

If the collision term is such that the splitting shown in Eq. (9.6) is not possible, the expansion of the solution for $\mathrm{Kn} \to \infty$ involves fractional powers of Kn; for molecules interacting with an inverse law force ($X \simeq kr^{-s}$), the first correction in one dimensional problems is of order $\mathrm{Kn}^{-(s-1)/(s+1)}$, where s is the exponent of the force law [72, 69].

These results have the important consequence that approximate methods of solution which are not able to allow for a nonanalytic behavior for $\mathrm{Kn} \to \infty$ produce poor results for large Knudsen numbers.

References

[1] D. Hilbert, "Math. Ann." **72,** 562 (1912).

[2] D. Hilbert, "Grundzüge einer Allgemeinen Theorie der Linearen Integralgleichungen", Chelsea Publishing Co., New York (1953).

[3] H. Grad, "Phys. Fluids", **6,** 147 (1963).

[4] S. Boguslawski, "Math. Ann." **76,** 431 (1915).

[5] T. F. Morse, "Phys. Fluids", **7,** 1691 (1964).

[6] S. Chapman, "Phil. Trans. Roy. Soc." **A216,** 279 (1916); **217,** 118 (1917).

[7] D. Enskog, Dissertation, Uppsala (1917): "Arkiv Mat., Ast. och. Fys" **16,** 1 (1921)

[8] S. Chapman and T. G. Cowling, "The Mathematical Theory of Non-Uniform Gases", Cambridge University Press (Cambridge, 1958).

[9] J. O. Hirschfelder, C. F. Curtiss and R. D. Bird, "Molecular Theory of Gases and Liquids", John Wiley and Sons (New York, 1854).

[10] J. H. Ferziger and H. G. Kaper, "Mathematical Theory of Transport Processes in Gases", North Holland (Amsterdam, 1972).

[11] R. Schamberg, Ph.D. Thesis, Calif. Inst. Technology (1947).

[12] C. CERCIGNANI, University of California, Report No. AS-64-18 (1964).
[13] L. TRILLING, "Phys. Fluids", **7,** 1681 (1964).
[14] H. GRAD, in "Transport Theory", Bellman et al., eds. SIAM-AMS Proceedings, vol. I, p. 269 AMS (Providence, 1968).
[15] C. CERCIGNANI, in "Transport Theory", Bellman et al., eds. SIAM-AMS Proceedings, vol. I, p. 249, AMS (Providence, 1968).
[16] C. CERCIGNANI, "Mathematical Methods in Kinetic Theory", Plenum Press (New York, 1969).
[17] P. WELANDER, "Arkiv Fysik", **7,** 507 (1954).
[18] C. CERCIGNANI, "Ann. Phys." (N.Y.) **20,** 219 (1962).
[19] S. ALBERTONI, C. CERCIGNANI and L. GOTUSSO, "Phys. Fluids", **6,** 993 (1963).
[20] D. R. WILLIS, "Phys. Fluids", **5,** 127 (1962).
[21] S. K. LOYALKA, "Phys. Fluids", **14,** 21 (1971).
[22] S. K. LOYALKA, "Phys. Fluids", **12,** 2301 (1969).
[23] P. BASSANINI, C. CERCIGNANI and C. D. PAGANI, "Int. J. Heat and Mass Transfer", **10,** 447 (1967).
[24] S. K. LOYALKA and J. H. FERZIGER, "Phys. Fluids", **10,** 1833 (1967).
[25] S. K. LOYALKA and J. H. FERZIGER, "Phys. Fluids", **11,** 1668 (1968).
[26] C. CERCIGNANI and G. TIRONI, "Nuovo Cimento", **43,** 64 (1966).
[27] C. CERCIGNANI, "Ann. Phys.", **40,** 469 (1966).
[28] C. CERCIGNANI, P. FORESTI and F. SERNAGIOTTO, "Nuovo Cimento", **X, 57***B*, 297 (1968).
[29] Y. SONE, "J. Phys. Soc. Japan", **21,** 1836 (1966).
[30] M. M. R. WILLIAMS, "J. Fluid Mech.", **45,** 759 (1971).
[31] C. CERCIGNANI, "J. Math. Anal. and Appli.", **10,** 568 (1965).
[32] C. CERCIGNANI and C. D. PAGANI, in "Rarefied Gas Dynamics", L. Trilling and H. Wachman, eds., vol. I, p. 269, Academic Press (New York, 1969).
[33] S. K. LOYALKA and J. W. CIPOLLA, "Phys. Fluids", **14,** 1656 (1971).
[34] T. KLINÇ and I. KUŠČER, "Phys. Fluids", **15,** 1018 (1972).
[35] Y. SONE, in "Rarefied Gas Dynamics", L. Trilling and H. Wachman, eds., vol. I, p. 243, Academic Press, (New York, 1969).
[36] Y. SONE, Presented at the 7th Symposium on Rarefied Gas Dynamics (Pisa, 1970).
[37] A. GANZ and L. SIROVICH, "Phys. Fluids", **16,** 50 (1973).
[38] J. S. DARROZÈS, in "Rarefied Gas Dynamics", L. Trilling and H. Wachman, eds. vol. I, p. 111, Academic Press (New York, 1969).
[39] Y. SONE, Kyoto University Research Report N. 24 (1972).
[40] Y. S. PAN and R. F. PROBSTEIN, in "Rarefied Gas Dynamics", J. A. Laurmann, ed., vol. II, p. 194, Academic Press, New York (1963).
[41] D. ENSKOG, "Physik Zeitschr.", **12,** 533 (1911).
[42] S. CHAPMAN and F. W. DOOTSON, "Phil. Mag.", **33,** 248 (1917).

[43] H. GRAD, in "Rarefied Gas Dynamics", M. Devienne, ed., p. 100, Pergamon Press (New York, 1960).
[44] E. A. JOHNSON, "Phys. Fluids", **16,** 45 (1973).
[45] M. ABRAMOWITZ and I. A. STEGUN, "Handbook of Mathematical Functions", National Bureau of Standards (Washington, 1964).
[46] S. A. SCHAAF, "Handbuch der Physik", vol. VIII/2, p. 591, Springer (Berlin, 1963).
[47] M. N. KOGAN, "Rarefied Gas Dynamics", Plenum Press (New York, 1969).
[48] V. P. SHIDLOVSKI, "Introduction to Dynamics of Rarefied Gases", Elsevier (New York, 1967).
[49] H. SCHLICHTING, "Boundary Layer Theory", McGraw-Hill (New York, 1958).
[50] J. STALDER, C. GOODWIN and M. CREAGER, NASA Rept. Nos. 1032 (1951) and 1093 (1952).
[51] W. A. GUSTAFSON, "A.R.S. Jour.", **29,** 301 (1959).
[52] B. M. SCHRELLO, "A.R.S. Jour.", **30,** 8 (1960).
[53] G. E. COOK, "Planet. Space Sci.", **13,** 929 (1965).
[54] O. K. MOE, Ph.D. Thesis, Univ. of Calif., Los Angeles (1966).
[55] F. C. HURLBUT and F. S. SHERMAN, "Phys. Fluids", **11,** 486 (1968).
[56] R. RIGANTI and M. G. CHIADÒ PIAT, "Meccanica", **6,** 132 (1971).
[57] C. CERCIGNANI and M. LAMPIS, "Entropie", **44,** 40 (1972).
[58] C. CERCIGNANI and M. LAMPIS, "ZAMP", **23,** 713 (1972).
[59] M. LAMPIS, in "Rarefied Gas Dynamics", K. Karamcheti, Ed., p. 369, Academic Press, New York (1974).
[60] D. CLAUSING, "Ann. Physik", **12,** 961 (1932).
[61] E. M. SPARROW and V. K. JOHNSON, "J. of Heat Transfer", **6,** 841 (1963).
[62] Y. P. PAO and J. TCHAO, "Phys. Fluids", **13,** 527 (1970).
[63] M. KNUDSEN, "Ann. Physik", **28,** 75 (1909).
[64] M. SMOLUCHOWSKI, "Ann. Physik", **33,** 1559 (1910).
[65] D. R. WILLIS, Princeton University Aero. Engineering Lab. Report No. 440 (1950).
[66] D. R. WILLIS, in "Rarefied Gas Dynamics", M. Devienne, ed., p. 246, Pergamon Press (New York, 1960).
[67] D. R. WILLIS and P. TAUB, Princeton University Gas Dynamics Laboratory Report No. 726 (1965).
[68] D. R. WILLIS, General Electric Co. TIS 60SD399 (1960).
[69] D. R. WILLIS, Rand Corporation Memorandum TM, 4638, PR (1965).
[70] A. L. COOPER and B. B. HAMEL, "Phys. Fluids", **16,** 35 (1973).
[71] B. B. HAMEL and A. L. COOPER, "Phys. Fluids", **16,** 43 (1973).
[72] J. SMOLDEREN, in "Rarefied Gas Dynamics", J. H. de Leeuw, ed., vol. I, p. 277, Academic Press, New York (1965).

VI ANALYTICAL SOLUTIONS OF MODELS

1. The method of elementary solutions

The theory developed in Chapter IV shows that the study of the linearized Boltzmann equation is worthwhile undertaking and that many of the features of its solutions can be retained by using model equations (Chapter IV, Section 9). We can say more, namely, that practically all the features are retained by a properly chosen model. The advantages offered by the models consist essentially in simplifying both the analytical and numerical procedures for solving boundary value problems of special interest.

In particular, the use of models is invaluable in those cases when the solution of the latter is explicit (in terms of quadratures of functions, whose qualitative behaviour can be studied by analytical means). Accordingly we shall devote this Chapter to the analytical manipulations which can be used to obtain interesting information from the model equations. The method used throughout (with the exception of Sections 13 and 14) is the method of separation of variables already sketched in Sections 6–8 of Chapter IV. The first step is to construct a complete set of separated variable solutions ("elementary solutions") and then to represent the general solution as a superposition of the elementary solutions; the second step is to use the boundary and initial conditions to determine the coefficients of the superposition. While the first problem can be solved for the model equations discussed in Chapter IV, Section 9, the second problem can be solved exactly in only a few cases. The method retains its usefulness, however, even when the second problem is not solvable, or is only approximately solvable, because it is capable of providing an analytical representation of the solution and hence a picture of its qualitative behavior (see Section 5).

It must be stated that the method of separation of variables is not the only one capable of solving these problems; transform techniques of the Wiener-Hopf type [1–3] are completely equivalent to the method of elementary solutions. This point will be further discussed in Section 11.

2. Splitting of a one-dimensional model equation

We begin by considering the simplest kind of problems; that is, the steady problems in one-dimensional geometry, and the simplest collision model comprising the linearized Krook model with velocity dependent frequency, given by Eq. (IV.9.15).

Accordingly we consider the equation

$$\xi_1 \frac{\partial h}{\partial x} = Lh \tag{2.1}$$

where x is the Cartesian coordinate upon which h is assumed to depend, ξ_1 is the x component of the molecular velocity $\boldsymbol{\xi}$ and

$$Lh = \nu(\boldsymbol{\xi})\left[\sum_{\alpha=0}^{4}(\nu\psi_\alpha, h) - h\right]; \qquad (\psi_\alpha, \nu\psi_\beta) = \delta_{\alpha\beta} \tag{2.2}$$

The unknown h can be split as follows

$$h = h_1 + h_2 + h_3 \tag{2.3}$$

where

$$\begin{aligned} h_1 &= \Pi_1 h \equiv \tfrac{1}{2}(I + P_2P_3)h \\ h_2 &= \Pi_2 h \equiv \tfrac{1}{4}(I + P_2)(I - P_3)h \\ h_3 &= \Pi_3 h \equiv \tfrac{1}{4}(I + P_3)(I - P_2)h \end{aligned} \tag{2.4}$$

and P_k denotes, as usual, the operator which reflects the k-th component of $\boldsymbol{\xi}$ $[P_3 f(\xi_1, \xi_2, \xi_3) \equiv f(\xi_1, \xi_2, -\xi_3)]$. In such a way the Hilbert space $\mathscr{H}$ where h can be located is split into three mutually orthogonal subspaces $\mathscr{H}_1$, $\mathscr{H}_2$, $\mathscr{H}_3$ ($P_k^2 = I$ and $P_kP_h = P_hP_k$ imply that the operators Π_k satisfy $\Pi_k\Pi_h = \delta_{kh}\Pi_k$, $\sum_{k=1}^{3}\Pi_k = I$). The collision frequency ν will be assumed to be even in all the components of $\boldsymbol{\xi}$ (usually, it depends only upon the speed ξ). ψ_0, ψ_1, ψ_4 being linear combinations of 1, ξ_1, ξ^2 belong to $\mathscr{H}_1$, ψ_2 to $\mathscr{H}_2$, ψ_3 to $\mathscr{H}_3$. Hence if we apply Π_1, Π_2, Π_3 to Eq. (2.1), we obtain

$$\xi_1 \frac{\partial h_1}{\partial x} = \nu[(\nu\psi_0, h_1)\psi_0 + (\nu\psi_1, h_1)\psi_1 + (\nu\psi_4, h_1)\psi_4 - h_1] \tag{2.5}$$

$$\xi_1 \frac{\partial h_2}{\partial x} = \nu[\psi_2(\nu\psi_2, h_2) - h_2] \tag{2.6}$$

$$\xi_1 \frac{\partial h_3}{\partial x} = \nu[\psi_3(\nu\psi_3, h_3) - h_3] \tag{2.7}$$

The remarkable fact is that Eqs. (2.5), (2.6), (2.7) are uncoupled; in addition, Eqs. (2.6) and (2.7) contain just one "moment", $(\nu\psi_k, h_k)$ $(k = 2, 3)$, and Eq. (2.5) contains three of such terms, instead of the five "moments" in Eq. (2.1).

Eq. (2.5) describes the heat transfer processes taking place along the x-axis, with Eq. (2.6) and (2.7) describing the shear effects due to motions in the y- and z- direction, respectively. We shall begin by considering the simplest case involving Eq. (2.6) (or Eq. (2.5), which differs from Eq. (2.6) by the name of the axes). By letting $h_3 = \psi_3 Y$, we obtain:

$$\xi_1 \frac{\partial Y}{\partial x} = \nu[(\nu\psi_3^2, Y) - Y] \tag{2.7}$$

We remark that in the case of the simplest models for neutron transport, Eqs. (IV.9.23) and (IV.9.27), the above splitting is not required, since the collision term contains one "moment" from the very beginning. By letting $f = f_0 Y$ in Eq. (IV.9.23) we find that Y satisfies Eq. (2.7) with γ in place of ψ_3^2.

3. Elementary solutions of the simplest transport equation

According to the results of Section 2, the simplest problems of neutron transport and rarefied gas dynamics lead to the following equation

$$\xi_1 \frac{\partial Y}{\partial x} + \nu(\xi)Y(x, \boldsymbol{\xi}) = \nu(\xi)\int g_0(\boldsymbol{\xi})Y(x, \boldsymbol{\xi})\, d\boldsymbol{\xi} \tag{3.1}$$

where

$$g_0(\boldsymbol{\xi}) = \nu f_0 \psi_3^2 \qquad \text{(shear flow of a gas)} \tag{3.2}$$

$$g_0(\xi) = \gamma\nu f_0 \qquad \text{(neutron thermalization)} \tag{3.3}$$

Let us use the variable w defined as follows:

$$w = \frac{\xi_1}{\nu(\xi_1, \xi_2, \xi_3)} \tag{3.4}$$

provided this relation is uniquely invertable by

$$\xi_1 = \xi_1(w, \xi_2, \xi_3) \tag{3.5}$$

which is clearly possible if $\partial[\xi_1/\nu(\xi_1, \xi_2, \xi_3)]/\partial\xi_1$ is different from zero for any ξ_2, ξ_3. In such a case, Eq. (3.1) becomes:

$$w \frac{\partial Y}{\partial x} + Y = \int_{-k}^{k} Z(x, w_1)\, dw_1 \tag{3.6}$$

where

$$Z(x, w) = \int g_0[\xi_1(w, \xi_2, \xi_3), \xi_2, \xi_3]Y(x, w, \xi_2, \xi_3) \frac{\partial \xi_1}{\partial w}\, d\xi_2\, d\xi_3 \tag{3.7}$$

and

$$k = \lim_{\xi_1 \to \infty} \frac{\xi_1}{\nu(\xi_1, \xi_2, \xi_3)} \tag{3.8}$$

is assumed to be independent of ξ_2, ξ_3 (which is the case if $\nu = \nu(\xi)$).

Let us put

$$Z_0(w) = \int g_0[\xi_1(w, \xi_2, \xi_3), \xi_2, \xi_3] \frac{\partial \xi_1}{\partial w}\, d\xi_2\, d\xi_3 \tag{3.9}$$

multiply Eq. (3.6) by $g_0(\partial\xi_1/\partial w)$ and integrate with respect to ξ_2, ξ_3. We obtain:

$$w\frac{\partial Z}{\partial x} + Z(x, w) = Z_0(w)\int_{-k}^{k} Z(x, w_1)\, dw_1 \tag{3.10}$$

We arrive at a similar equation if we start from Eq. (IV.9.27), which, in the steady case and absence of external sources, can be written as follows:

$$\mu\frac{\partial\Psi}{\partial x} + \sigma\Psi = \frac{c\sigma}{4\pi}\int_{-1}^{+1}\int_{0}^{2\pi}\Psi(x, \mu', \varphi')\, d\mu'\, d\varphi' \tag{3.11}$$

where $\mu = \cos\theta$ is the cosine of the angle between $\mathbf{x}$ and $\mathbf{\Omega}$, φ is the azimuthal angle of $\mathbf{\Omega}$ (c and σ are constants). If we let

$$w = \frac{\mu}{\sigma}, \qquad Z(x, w) = \frac{1}{2\pi}\int_0^{2\pi}\psi(x, \mu, \varphi')\, d\varphi', \qquad Z_0 = \frac{c}{2} \tag{3.12}$$

and then multiply Eq. (3.11) by $(2\pi)^{-1}$ and integrate with respect to φ we obtain Eq. (3.9), with $k = 1/\sigma$. We remark that

$$\begin{aligned}
&\int Z_0\, dw = \int v f_0 \psi_3^{\,2}\, d\boldsymbol{\xi} = 1 && \text{(shear flow of a gas)}\\
&\int Z_0\, dw = \gamma\int v f_0\, d\boldsymbol{\xi} = 1 && \text{(neutron thermalization)}\\
&\int Z_0\, dw = \int_{-1}^{+1}\frac{c}{2}\, d\mu = c && \text{(one speed neutron transport)}
\end{aligned} \tag{3.13}$$

In the third case the integral is unity only if $c = 1$; that is, in the presence of a purely scattering medium. This circumstance is related to the fact that there is no conservation law if $c \neq 1$. The general solution of Eq. (3.10) was studied by Case [4] for one speed neutron transport, by Cercignani for shear flows of a gas, both with constant [5] and variable [6] collision frequency; the method used in these papers is the method of elementary solutions, to be described in this and the following sections.

Let us begin by separating the variables. Putting

$$Z(x, w) = X(x)g(w)Z_0(w) \tag{3.14}$$

it is easily seen that, either $Z = A_0$ (A_0 arbitrary constant) or

$$Z_u(x, w) = e^{-x/u}\, g_u(w)Z_0(w) \tag{3.15}$$

where $g_u(\xi)$ satisfies

$$(-w/u + 1)g_u(w) = \int_{-k}^{k} g_u(w_1)Z_0(w_1)\, dw_1 \tag{3.16}$$

and u, the separation parameter, has been used to label the elementary solutions.

Though, *a priori*, u may assume any complex value, it is easily seen that if $\int Z_0\,dw \leqslant 1$, then u is a real number. This follows from a direct argument [4–6] or from the general results of Chapter IV, Section 7. If $\int Z_0\,dw > 1$ (one speed neutron transport with $c > 1$) then there will be a discrete set of complex eigenvalues (see Chapter IV, end of Section 7); if $\int_{-k}^{k} Z_0\,dw = 1$ and $k < \infty$ it is also easy to show that there are no real eigenvalues outside the interval $(-k, k)$. These results can be proved [4–6] by the same technique to be employed in Section 6 for discussing the discrete spectrum in the time-dependent case; in particular, if $\int Z_0\,dw > 1$, there are exactly two complex eigenvalues which turn out to be purely imaginary [4]. If we disregard the case $\int Z_0\,dw \neq 1$ which is easily dealt with (see Section 6), the values of u must be real and lie between $-k$ and k. This requires some care, because we cannot divide by $u - w$ in Eq. (3.11). This difficulty is overcome by letting $g_u(w)$ be a generalized function (see Chapter I, Section 2). If we disregard a multiplicative constant (i.e. normalize g_u in such a way that the right hand side of Eq. (3.16) is equal to 1) then $g_u(w)$ will satisfy:

$$\left(\frac{u - w}{u}\right) g_u(w) = 1 \tag{3.17}$$

For $w \neq u$ we find $g_u(w) = u/(u - w)$ but this has no meaning at $w = u$ (in particular, the integral appearing in Eq. (3.16) does not exist in the ordinary sense). It is possible, however, to define a generalized function, for example, as the limit of the sequence

$$g_m(w) = m^2 u(u - w)/[m^2(u - w)^2 + 1] \tag{3.18}$$

This limit, to be understood in the sense explained in Section 2 of Chapter I, exists and satisfies Eq. (3.17) [7]. Such a generalized function is denoted by

$$P\frac{u}{u - w} = \lim_{m\to\infty} \frac{m^2 u(u - w)}{1 + m^2(u - w)^2} \tag{3.19}$$

where P can be read "principal part of". The integrals involving the generalized function, which has been just defined, are to be interpreted as Cauchy principal value integrals [8]:

$$\int P\frac{u}{u - w}\,\varphi(w)\,dw = P\int \frac{u\varphi(w)}{u - w}\,dw = \lim_{\varepsilon\to 0} \int_{|w-u|>\varepsilon} \frac{u\varphi(w)}{u - w}\,du \tag{3.20}$$

We can ask now whether or not $P[u/(u - w)]$ is the only solution of Eq. (3.17). The answer is no. As a matter of fact, the most general solution will be the sum of $P[u/(u-w)]$ and the general solution of the homogeneous equation

$$(u-w)T(w) = 0 \tag{3.21}$$

Now, the most general solution of Eq. (3.21) is a multiple of $\delta(u-w)$ where δ denotes the Dirac delta function defined in Chapter I, Section 2. Therefore the general solution of Eq. (3.17) reads as follows:

$$g_u(w) = P\frac{u}{u-w} + p(u)\,\delta(u-w) \tag{3.22}$$

where the factor in front of the delta function can depend upon u and has been called $p(u)$. In order that Eq. (3.16) be satisfied by Eq. (3.22) the normalization condition for $g_u(w)$ must be satisfied and so the right hand side of Eq. (3.16) must be equal to 1. This condition can be satisfied for any real $u(-k < u < k)$ and serves for determining $p(u)$

$$p(u) = [Z_0(u)]^{-1}\left[1 - P\int_{-k}^{k}\frac{uZ_0(w)}{u-w}\,dw\right] = [Z_0(u)]^{-1}\int_{-k}^{k}\frac{wZ_0(w)}{w-u}\,dw \tag{3.23}$$

where the fact that $\int_{-k}^{k} Z_0(w)\,dw = 1$ has been used.

The generalized eigenfunctions $g_u(w)$ have many properties of orthogonality and completeness. The orthogonality and completeness properties in the full range $(-k < w < k)$ are particular cases of the results of Chapter IV, Section 7. Other properties of orthogonality and completeness in partial ranges (notably $0 < w < k$) are far from trivial to prove, since they require solving singular integral equations. However, standard techniques are available for treating such problems (see [8] and the Appendix) and the following results can be obtained [4–6]:

THEOREM I *The generalized functions* $Z_0(w)g_u(w)$ $(-k < u < k)$ *and* $g_\infty = Z_0(w)$, *complemented by* $g_* = wZ_0(w)$, *form a complete set for the functions* $Z(w)$ *defined on the real axis, satisfying a Hölder condition in any open interval contained in* $(-k, k)$, *and such that*

$$\int_{-k}^{k} w^2 Z(w)\,dw < \infty \tag{3.24}$$

Also, the coefficients of the generalized expansion:

$$Z(w) = \left[A_0 + A_1 w + \int_{-k}^{k} A(u)g_u(w)\,du\right]Z_0(w) \tag{3.25}$$

are uniquely and explicitly determined by

$$A_0 = \left[\int_{-k}^{k} w^2 Z_0(w)\, dw\right]^{-1} \int_{-k}^{k} w^2 Z(w)\, dw \tag{3.26}$$

$$A_1 = \left[\int_{-k}^{k} w^2 Z_0(w)\, dw\right]^{-1} \int_{-k}^{k} w Z(w)\, dw \tag{3.27}$$

$$A(u) = [C(u)]^{-1} \int_{-k}^{k} w Z(w) g_u(w)\, dw \tag{3.28}$$

where

$$C(u) = u Z_0(u)\{[p(u)]^2 + \pi^2 u^2\} \tag{3.29}$$

THEOREM II *The generalized eigenfunctions* $Z_0(w)g_u(w)$ $(0 < u < k)$ *and* $g_\infty = Z_0(w)$ *form a complete set for the functions* $Z(w)$ *defined on* $0 < w < k$, *satisfying a Hölder condition in any open interval contained in* $(0, k)$, *and integrable with respect to the weight* w^2. *Also, the coefficients of the generalized expansion*

$$Z(w) = \left[A_0 + \int_0^k A(u) g_u(w)\, dw\right] Z_0(w) \tag{3.30}$$

are uniquely and explicitly determined by:

$$A_0 = \left[\int_{-k}^{k} w^2 Z_0(w)\, dw\right]^{-1} \int_0^k w P(w) Z(w)\, dw \tag{3.31}$$

$$A(u) = [C(u)P(u)]^{-1} \int_0^k w P(w) g_u(w) Z(w)\, dw \tag{3.32}$$

Here we have put

$$P(w) = w \exp\left\{-\frac{1}{\pi}\int_0^k \tan^{-1}[\pi t/p(t)] \frac{dt}{t + w}\right\} \qquad (w > 0) \tag{3.33}$$

where the inverse tangent varies from $-\pi$ *to* 0 *when* t *varies from* 0 *to* k.

According to the general results of Chapter IV, Section 7, Theorem I ensures that the general solution of Eq. (3.10) is given by

$$Z(x, w) = \left[A_0 + A_1(x - w) + \int_{-k}^{k} A(u) e^{-x/u} g_u(w)\, du\right] Z_0(w) \tag{3.34}$$

Theorem II is equally or, perhaps, more important, because it allows us to solve boundary value problems. This theorem shows that the generalized eigenfunctions are orthogonal on $(0, k)$ with respect to the weight $wZ_0(w)P(w)$. This orthogonality property is more standard than the full range orthogonality, because the weight function is positive. The only trouble now is the complicated expression of $P(w)$; it is to be noted, however, that $P(w)$, though far from being an elementary function, satisfies two important identities which make the manipulation of integrals involving $P(w)$ much easier than would be expected. These identities are (see Refs. [4–6]):

$$\left[\int_{-k}^{k} w^2 Z_0(w)\, dw\right]^{-1} \int_0^k \frac{tZ_0(t)P(t)}{t+u}\, dt = [P(u)]^{-1} \tag{3.35}$$

$$u - \frac{1}{\pi}\int_0^k \tan^{-1}[\pi t/p(t)]\, dt - \left[\int_{-k}^{k} w^2 Z_0(w)\, dw\right] \int_0^k \frac{t[Z_0(t)P(t)]^{-1}}{[p(t)]^2 + \pi^2 t^2} \frac{dt}{t+u} = P(u) \tag{3.36}$$

Also

$$P(0) = \left[\int_{-k}^{k} u^2 Z_0(u)\, du\right]^{\frac{1}{2}} \tag{3.37}$$

Once Eq. (3.10) is solved, the general solution of Eq. (3.6), and hence (3.1), is easily written down, since it is a matter of solving an ordinary differential equation with a given source term. We obtain:

$$Y(x, \boldsymbol{\xi}) = A_0 + A_1\{x - [\xi_1/\nu(\xi)]\} + \\ + \int_{-k}^{k} A(u) e^{-x/u} g_u\left(\frac{\xi_1}{\nu(\xi)}\right) du + B(\boldsymbol{\xi}) e^{-x\nu(\xi)/\xi_1} \tag{3.38}$$

where $B(\boldsymbol{\xi})$ is an arbitrary function, provided only that it satisfies:

$$\int \left[g_0(\boldsymbol{\xi}) B(\boldsymbol{\xi}) \left\{\frac{\partial}{\partial \xi_1}\left(\frac{\xi_1}{\nu(\xi)}\right)\right\}^{-1}\right]_{\xi_1=\xi_1(w,\xi_2,\xi_3)} d\xi_2\, d\xi_3 = 0 \tag{3.39}$$

We end this section by noticing the form taken by the previous results in particular cases. If $\nu(\xi) = \nu$ is constant (BGK model) and we take $(2RT_0)^{\frac{1}{2}}$ as the speed unit and $(2RT_0)^{\frac{1}{2}}\nu^{-1} = 2\pi^{-\frac{1}{2}}l$ as the length unit (T_0 being the unperturbed temperature, l the mean free path given by Eq. (V.I.3)), then $\psi_3 = \sqrt{2}\, \xi_3$, $g_0 = 2\xi_3^2 f_0$, $w = \xi_1$, $k = \infty$ and

$$Z_0(w) = \pi^{-\frac{1}{2}} e^{-w^2} \tag{3.40}$$

Accordingly

$$p(u) = e^{u^2} P\int_{-\infty}^{\infty} \frac{we^{-w^2}}{w-u}\, dw = \pi^{\frac{1}{2}}\left(e^{u^2} - 2u\int_0^u e^{t^2}\, dt\right) \tag{3.41}$$

Thus $p(w)$ can be expressed in terms of tabulated functions [9–10]. Eqs. (3.26), (3.27), (3.31), (3.34), (3.36), (3.37) simplify slightly since

$$\int_{-k}^{k} w^2 Z_0(w)\, dw = \tfrac{1}{2} \tag{3.42}$$

If $\nu(\xi) = \sigma\xi$ (constant mean free path), $\psi_3 = [3/(4\sigma)]^{\frac{1}{2}} \cdot \pi^{\frac{1}{4}}\xi_3$, $w = \xi_1/(\xi\sigma)$, $k = 1/\sigma$ and, if we take $(2RT_0)^{\frac{1}{2}}$ as the speed unit, $1/\sigma$ as length unit then:

$$Z_0(w) = \tfrac{3}{4}(1 - w^2) \tag{3.43}$$

$$p(u) = \frac{1}{1-u^2} P \int_{-1}^{1} \frac{w(1-w^2)}{w-u}\, du = \frac{2(3u^2-1)}{3(1-u^2)} + u \log\left(\frac{1-u}{1+u}\right) \tag{3.44}$$

Finally, in the case of Eq. (3.11) with $c = 1$, if we take σ^{-1} as length unit then:

$$p(u) = P \int_{-1}^{1} \frac{w\, dw}{w-u} = 2 + u \log\left(\frac{1-u}{1+u}\right) \tag{3.45}$$

4. Application of the general method to the Kramers and Milne problems

In this section we shall apply the above results to two typical boundary value problems of transport theory, the Kramers' and Milne problems.

The Kramers' problem consists in finding the molecular distribution function of a gas in the following situation (see Fig. 35); the gas fills the half space $x > 0$ bounded by a physical wall in the plane $x = 0$ and is nonuniform because of a gradient along the x-axis of the z-component of the mass velocity; this gradient tends to a constant a as x tends to infinity. It is seen that this problem can be considered as the limiting case of plane Couette flow (shear flow between two parallel plates), when one of the plates is removed to infinity, while keeping a fixed ratio between the relative speed of the plates and their mutual distance. More generally, the Kramers' problem can be interpreted as a connecting problem through the kinetic boundary layer (see Chapter V, Section 5); in this case "infinity" represents the region where the Hilbert solution holds and the velocity gradient "at infinity" can be regarded as constant because it does not vary sensibly on the scale of the mean free path.

Both of these interpretations of the Kramers' problem suggest that a convenient linearization is about a Maxwellian endowed with a mass velocity ax in the z-direction. Because of the nonuniformity of this Maxwellian, linearization gives an inhomogeneous Boltzmann equation:

$$2ac_1c_3 + c_1 \frac{\partial h}{\partial x} = Lh \tag{4.1}$$

where $\mathbf{c} = (c_1, c_2, c_3) = (\xi_1, \xi_2, \xi_3 - ax)$. Eq. (4.1) can be reduced to the homogeneous Boltzmann equation by subtracting a particular solution. One particular solution, independent of x, is suggested by the Hilbert theory; this solution, $L^{-1}(2ac_1c_3)$ is given by $-2ac_1c_3/\nu(c)$ for the collision model given by Eq. (2.2). Therefore we have:

$$h = -2ac_1c_3/\nu(c) + 2c_3 Y(x, \mathbf{c}) \tag{4.2}$$

where $Y(x, \boldsymbol{\xi})$ satisfies Eq. (3.1). The mass velocity is given by

$$v_3 = ax + 2\rho_0^{-1} \int \xi_3^{\,2} Y(x, \boldsymbol{\xi}) f_0(\boldsymbol{\xi})\, d\boldsymbol{\xi} \tag{4.3}$$

the first term being the contribution from the Maxwellian $f_0(\mathbf{c})$.

Concerning the boundary conditions, we shall assume that the molecules are re-emitted from the wall according to a Maxwellian distribution which is completely accommodated to the state of the wall (for more general assumptions, see, e.g., [11]). Therefore, the boundary condition for h reads as

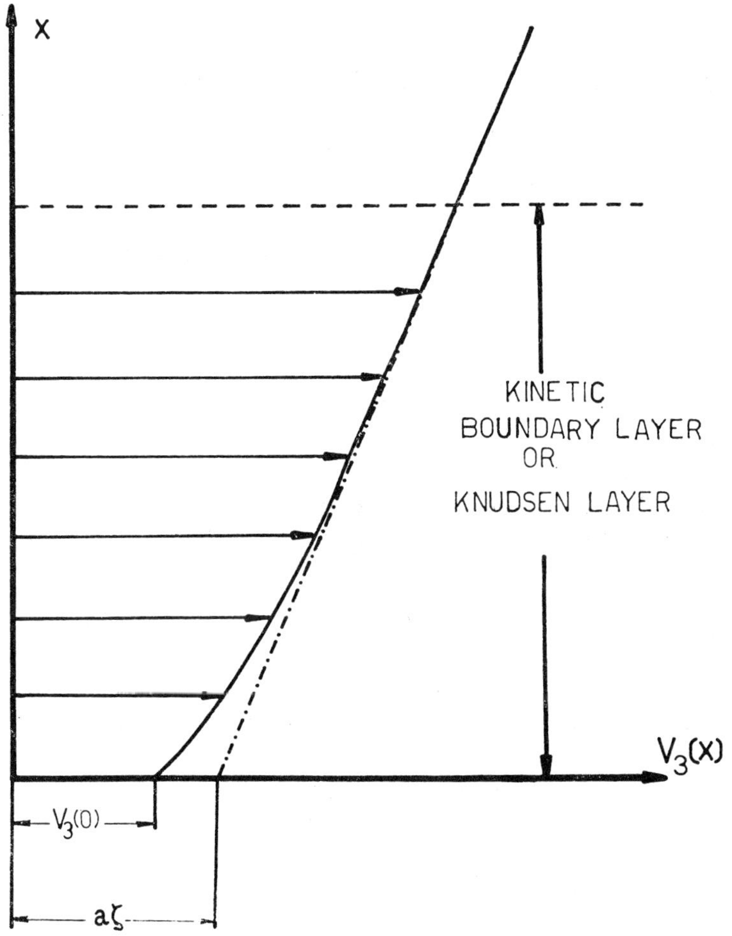

Fig. 35. Sketch of the velocity profile in the neighborhood of a wall in the presence of a velocity gradient in the main body of the flow.

follows:

$$h(0, \mathbf{c}) = 0 \qquad (c_1 > 0) \tag{4.4}$$

and this, in terms of Y, becomes:

$$Y(0, \xi) = a\xi_1/\nu(\xi) \tag{4.5}$$

In addition, Y must satisfy the condition of boundedness at infinity.

According to the discussion in Section 3, the general solution of Eq. (3.1) which also satisfies the condition of boundedness at infinity is given by

$$Y(x, \xi) = A_0 + \int_0^k A(u)e^{-x/u} g_u\left(\frac{\xi_1}{\nu(\xi)}\right) du + B(\boldsymbol{\xi})e^{-x\nu(\xi)/\xi_1} \tag{4.6}$$

where $B(\boldsymbol{\xi})$ satisfies Eq. (3.39). The condition to be satisfied at the plate gives:

$$\alpha\xi_1/\nu(\xi) = A_0 + \int_0^k A(u)g_u\left(\frac{\xi_1}{\nu(\xi)}\right) du + B(\boldsymbol{\xi}) \tag{4.7}$$

Eqs. (4.7) and (3.39) easily give $B(\boldsymbol{\xi}) = 0$. Thus, solving Eq. (4.7), means expanding $Z(w) = awZ_0(w)$ according to Theorem II of Section 3; therefore, A_0 and $A(u)$ are immediately obtained through Eqs. (3.31) and (3.32). The result is as follows:

$$A_0 = -a\pi^{-1}\int_0^k \tan^{-1}[\pi w/p(w)]\, dw \tag{4.8}$$

$$A(u) = -a[Z_0(u)P(u)]^{-1}\{[p(u)]^2 + \pi^2u^2\}^{-1}\int_{-k}^k w^2Z(w)\, dw \tag{4.9}$$

where use has been made of Eq. (3.35), which yields Eq. (4.9) directly, and the following identity, by asymptotically expanding for large values of u and comparing with Eq. (3.36) or Eq. (3.33):

$$\left[\int_{-k}^k w^2Z_0(w)\, dw\right]^{-1}\int_0^k w^2Z_0(w)P(w)\, dw = -\frac{1}{\pi}\int_0^k \tan^{-1}[\pi t/p(t)]\, dt \tag{4.10}$$

Eq. (4.10) yields Eq. (4.8).

Substituting Eqs. (4.8) and (4.9) into Eq. (4.6) gives the solution of the Kramers' problem. The mass velocity is readily obtained from Eqs. (4.3) and (4.6):

$$v_3(x) = ax + A_0 + \int_0^k A(u)n(u)e^{-x/u}\, du \tag{4.11}$$

where

$$n(u) = 2\rho_0^{-1}\int f_0\xi_3^{\,2}g_u\left(\frac{\xi_1}{\nu(\xi)}\right) d\boldsymbol{\xi} \tag{4.12}$$

and A_0 and $A(u)$ are given by Eq. (4.8) and (4.9). From Eq. (4.11) we recognize that A_0 is the macroscopic slip of the gas on the plate (see Chapter V, Section 5); it has the form ζa, where ζ is the slip coefficient:

$$\zeta = -\pi^{-1}\int_0^k \tan^{-1}[\pi w/p(w)]\,dw \tag{4.13}$$

Thus the evaluation of the slip coefficient has been reduced to quadratures for an arbitrary $\nu(\xi)$. In particular, if $\nu(\xi) = \text{const.}$ (BGK model), use of Eq. (3.40) and partial integration yields

$$\zeta = \theta\pi^{\frac{1}{2}}\int_0^\infty \frac{we^w\,dw}{[p(w)]^2 + \pi^2 w^2} = 2l\int_0^\infty \frac{we^{w^2}\,dw}{[p(w)]^2 + \pi^2 w^2} \tag{4.13a}$$

Here $p(w)$ is given by Eq. (3.41) and l is the mean free path defined by Eq. (V.I.3) and hence is related to $\theta = \nu^{-1}$ by $\theta = 2\pi^{-\frac{1}{2}}l$ $(2RT_0 = 1)$. The integral appearing in Eq. (4.13) has been evaluated numerically [12] with the following result

$$\zeta = (1.01615)\theta = (1.1466)l \tag{4.14}$$

If $\nu(\xi)$ is not constant we must evaluate ζ through Eq. (4.13). Some specific cases were considered in [13]. It has been found that for $\nu(\xi)$ increasing linearly when $\xi \to \infty$ the value of ζ is somewhat lower (3%) than for the BGK model (for a fixed value of the viscosity coefficient).

Henceforth in this section we shall restrict our considerations to the case of the BGK model (constant collision frequency). In this case, Eq. (4.11) can be written as follows:

$$v_3(x) = a[x + \zeta - (\pi^{\frac{1}{2}}\theta/2)I(x/\theta)] \tag{4.15}$$

where $I(x/\theta)$ is virtually zero outside the kinetic layer; the velocity profile is sketched in Fig. 35. $I(x/\theta)$ can be easily evaluated [14]; a plot is given in Fig. 36.

A direct evaluation of the microscopic slip, that is the velocity of the gas at the wall, results without any numerical integration. As a matter of fact, we have:

$$v_3(0) = a\left(\zeta - l\int_0^\infty \frac{e^{w^2}[P(w)]^{-1}}{[p(w)]^2 + \pi^2 w^2}\,dw\right) = (2/\pi)^{\frac{1}{2}}al \tag{4.16}$$

where the last result is obtained by letting $u = 0$ in Eq. (3.36) and taking into account Eqs. (3.37), (4.13) and (3.42).

Analogously we can evaluate the distribution function of the molecules arriving at the plate. We obtain:

$$Y(0, \boldsymbol{\xi}) = Y(0, \xi_1) = 2\pi^{-\frac{1}{2}}al\xi_1 + 2\pi^{-\frac{1}{2}}alP(-\xi_1) \qquad (\xi_1 < 0) \tag{4.17}$$

where Eq. (3.36) has been used. Then $h(0, \mathbf{c})$ (the perturbation of the Maxwellian distribution at the plate) is given by

$$h(0, \mathbf{c}) = 4\pi^{-\frac{1}{2}} alc_3 P(|c_1|) \qquad (c_1 < 0) \tag{4.18}$$

and the function $P(w)$ $(w > 0)$ can be given a physical interpretation in terms of the distribution function of the molecules arriving at the wall. From Eqs. (3.36) it is easily inferred that

$$|c_1| + 0.7071 < P(|c_1|) < |c_1| + 1.01615 \tag{4.19}$$

Hence the distribution function of the arriving molecules is rather close to a Hilbert distribution; in fact a Hilbert expansion would predict Eq. (4.16)

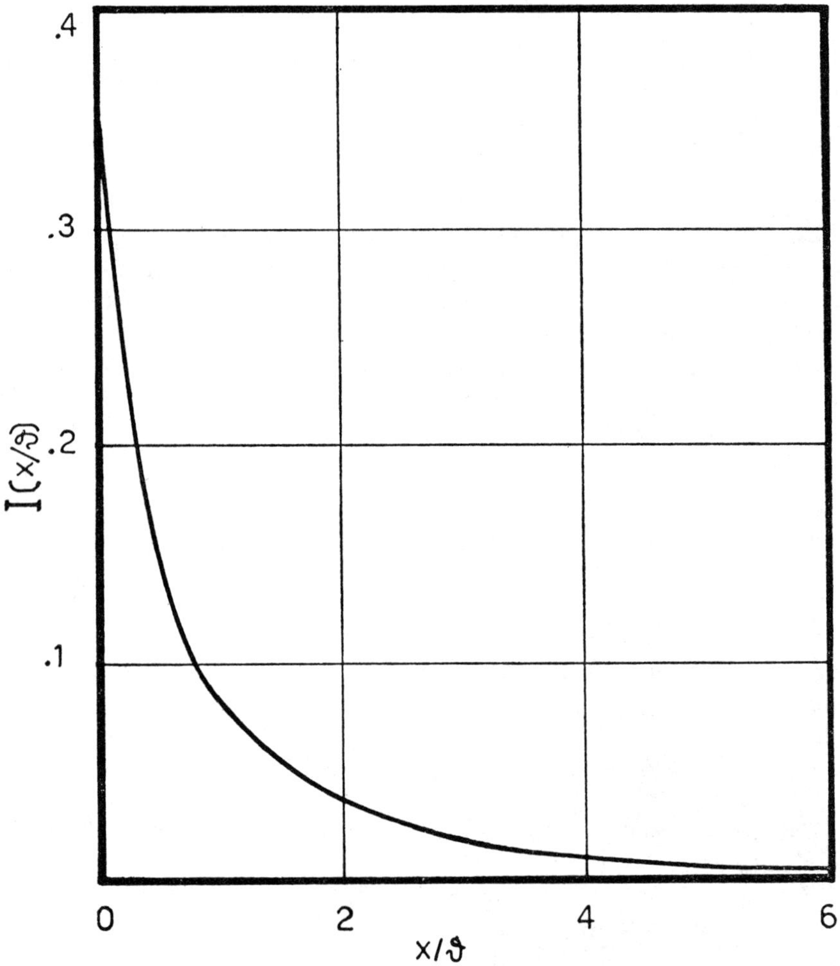

Fig. 36. Plot of the function $I(x/\theta)$ occurring in Eq. (4.15).

with $P(|c_1|)$ linear in $|c_1|$ (such is the distribution holding outside the kinetic layer, see Eqs. (4.2) and (4.6)). The fact that the distribution of the molecules arriving at the plate is close to the one prevailing outside the kinetic boundary layer is not surprising; in fact each molecule has the velocity acquired after its last collision, which, on the average, happened a mean free path from the wall in a region where the distribution function is of the Hilbert type. It is interesting to note that Maxwell [15] assumed that the distribution function of the arriving molecules was exactly the one prevailing far from the wall; by using this assumption together with conservation of momentum, he was able to evaluate the slip coefficient, without solving the Kramers' problem. He found $\zeta = l$ (with an error of 15%) and

$$h(0, \mathbf{c}) = 4\pi^{-\frac{1}{2}} a l c_3 (|c_1| + 0.8863) \qquad (c_1 < 0) \tag{4.20}$$

which is a good approximation to the correct result given by Eqs. (4.18) and (4.19).

The problem of neutron transport corresponding to Kramers' problem is Milne's problem. Mass velocity in the z-direction is replaced by neutron density; thereafter everything proceeds practically unchanged and Eq. (4.11) applies to neutron density. The slip coefficient is replaced by the so called extrapolation length z_0. In particular, in the one speed approximation, the extrapolation length ($c = 1$) is given by Eq. (4.13) with $p(w)$ given by Eq. (3.45)

$$z_0/l = -\pi^{-1} \int_0^1 \tan^{-1} \left[\frac{\pi}{\dfrac{2}{u} + \log\left(\dfrac{1-u}{1+u}\right)} \right] du$$

$$= 2 \int_0^1 \frac{u}{[p(u)]^2 + \pi^2 u^2} \frac{du}{1-u^2} \tag{4.21}$$

where $l = 1/\sigma$ is the mean free path. A numerical evaluation gives:

$$z_0/l = 0.710446 \tag{4.22}$$

Further details are given by Case and Zweifel [16].

5. Application to the flow between parallel plates and the critical problem of a slab

We have just seen that half-space problems connected with Eq. (3.1), or equivalently Eq. (3.10), can be solved by analytical means. This is not true for gas flows between parallel plates, like Couette and Poiseuille flows, or neutron transport in a slab. The method of elementary solutions, however, can be used to obtain series solution and gain insight into the qualitative behaviour of the solution.

Let us consider the flow problems first and restrict our attention to the case of constant collision frequency $\nu = \theta^{-1}$.

The general solution, Eq. (3.38), of Eq. (3.1) can be rewritten as follows

$$Y(x, \xi) = A_0 + A_1(x - \xi) + \int_{-\infty}^{+\infty} A(u)\exp\left(-\frac{x}{u} - \frac{\delta}{2\,|u|}\right) g_u(\xi)\, du \quad (5.1)$$

where the last term in Eq. (3.38) has been omitted because it is usually absent, ξ_1 has been denoted by ξ since no confusion arises and δ is the distance between the plates ($\theta = 1$ and $2RT_0 = 1$, as usual). $A(u)$ has been redefined by inserting a factor $\exp[-\delta/2|u|]$ for convenience, and the plates are assumed to be located at $x = \pm\delta/2$.

According to the discussion in Section 7 of Chapter IV, Eq. (5.1) shows that for sufficiently large δ the picture is the following: a core, where a continuum description (based on the Navier-Stokes equations) prevails, surrounded by kinetic boundary layers, produced by the interaction of the molecules with the walls and described by the integral term in Eq. (5.1).

As δ becomes smaller, however, the exponentials in the latter term are never negligible so that the kinetic layers merge with the core to form a flow field which cannot be described in simple terms. Finally, when δ is negligibly small, $Y(x, \xi)$ does not depend sensibly on x, and the molecules retain the distribution they had just after their last interaction with a boundary. In the case of Couette flow (i.e. when there are two plates at $x = \pm\delta/2$ moving with velocities $\pm V/2$ in the z-direction) the situation is well described by the above short discussion, although it is possible to obtain [17] a more detailed picture by finding approximate expressions for A_1 and $A(u)$ (A_0 is zero and $A(u)$ is odd because of the antisymmetry inherent in the problem). Accordingly we shall consider in more detail the case of Poiseuille flow between parallel plates, which lends itself to more interesting considerations.

Plane Poiseuille flow (Fig. 37) is the flow of a fluid between two parallel plates induced by a pressure gradient parallel to the plates. In the continuum case no distinction is made between a pressure gradient arising from a density gradient and one arising from a temperature gradient. This distinction, on the contrary, is to be taken into account when a kinetic theory description is considered. We shall restrict ourselves to the former case, following Ref. [18]; for the case of a temperature gradient, see Ref. [19].

The basic linearized Boltzmann equation for Poiseuille flow in a channel of arbitrary cross-section (including the slab as a particular case) will now be derived.

We assume that the walls re-emit the molecules with a Maxwellian distribution f_0 with constant temperature and an unknown density $\rho = \rho(z)$ (z being the coordinate parallel to the flow). If the length of the channel is much larger than any other typical length (mean free path, distance between the walls), then we can linearize about the abovementioned Maxwellian f_0; in fact $\rho(z)$ is slowly varying and f_0 would be the solution in the case of a rigorously constant ρ. Accordingly we have

$$\xi_1 \frac{\partial h}{\partial x} + \xi_2 \frac{\partial h}{\partial y} + \xi_3 \frac{\partial h}{\partial z} + \frac{1}{\rho}\frac{d\rho}{dz}\xi_3 = Lh \tag{5.2}$$

Because of the assumption of a slowly varying ρ (long tube), we can regard $(1/\rho)(d\rho/dz)$ as constant (i.e., we disregard higher order derivatives of ρ as well as powers of first order derivatives). If $(1/\rho)(d\rho/dz)$ is constant, it follows that $\partial h/\partial z = 0$, since z does not appear explicitly in the equation nor in the boundary conditions. The latter can be written

$$h(x, y, z, \boldsymbol{\xi}) = 0 \qquad ((x, y) \in \partial\Sigma;\ xn_1 + yn_2 > 0) \tag{5.3}$$

where $\partial\Sigma$ is the boundary of the cross section and $\mathbf{n} = (n_1, n_2)$ is the normal unit vector pointing into the channel. Therefore, we can write

$$\xi_1 \frac{\partial h}{\partial x} + \xi_2 \frac{\partial h}{\partial y} + k\xi_3 = Lh \tag{5.4}$$

where $k = (1/\rho)(d\rho/dz)$. Eq. (5.4) governs linearized Poiseuille flows in a very long tube of arbitrary cross section. If we specialize to the case of a slab and use the BGK model, we have:

$$h = 2\xi_3 W(x, \xi_1) \tag{5.5}$$

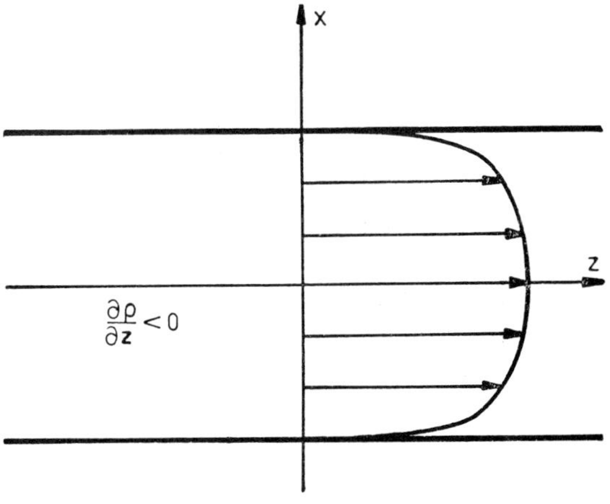

Fig. 37. Sketch of the velocity profile in plane Poiseuille flow.

where $W(x, \xi)$ satisfies

$$\xi \frac{\partial W}{\partial x} + \frac{k}{2} = \pi^{-\frac{1}{2}} \int_{-\infty}^{+\infty} e^{-\xi_1^2} W(x, \xi_1)\, d\xi_1 - W(x, \xi_1) \tag{5.6}$$

$$W\left(-\frac{\delta}{2} \operatorname{sgn} \xi, \xi\right) = 0 \tag{5.7}$$

provided x is measured in θ units. The above equations follow from a splitting analogous to the one considered in Section 2.

Since $W(x, \xi)$ does not depend on the y and $z-$ components of $\boldsymbol{\xi}$ and $\nu = 1$, Eq. (5.6) differs from Eq. (3.1) because of the inhomogeneous term $k/2$. If we find a particular solution of Eq. (5.6), then we can add it to the general solution (5.1) of the homogeneous equation in order to have the general solution of Eq. (5.6). By differentiation of the latter equation, we deduce that $\partial W/\partial x$ satisfies Eq. (3.1) with $\nu = 1$; since the general solution of Eq. (3.1) contains exponentials (which reproduce themselves by integration and differentiation) and a linear function of x, we try a particular solution of Eq. (5.6) in the form of a quadratic function of x (with coefficients depending upon ξ). It is easily verified that solutions of this form exist, and one of them is

$$W_0(x, \xi) = \frac{k}{2}\left[x^2 - \frac{\delta^2}{4} - 2x\xi - (1 - 2\xi^2)\right] \tag{5.8}$$

Therefore

$$W(x, \xi) = W_0(x, \xi) + Y(x, \xi) \tag{5.9}$$

where $Y(x, \xi)$ is given by Eq. (5.1). Eq. (5.7) gives the following boundary condition for $Y(x, \xi)$:

$$Y\left(-\frac{\delta}{2} \operatorname{sgn} \xi, \xi\right) = -\left[|\xi| - (1 - 2\xi^2)\frac{1}{\delta}\right]\frac{k\delta}{2} \tag{5.10}$$

Since the symmetry inherent in our problem implies that

$$Y(x, \xi) = Y(-x, -\xi) \tag{5.11}$$

$A_1 = 0$ and $A(u) = A(-u)$. If we take this into account, Eqs. (5.1) and (5.10) give:

$$A_0 + \int_0^\infty A(u) g_u(\xi)\, du$$

$$= -\frac{k\delta}{2}\left[\xi - (1 - 2\xi^2)\frac{1}{\delta}\right] - \int_0^\infty \frac{uA(u)}{u + \xi} e^{-\delta/u}\, du \qquad (\xi > 0) \tag{5.12}$$

and the equation for $\xi < 0$ is not required because $A(u) = A(-u)$. If we call the right hand side of this equation $\pi^{\frac{1}{2}} e^{\xi^2} Z(\xi)$ and use Eq. (3.40), Eq.

(5.12) becomes Eq. (3.30) ($k = \infty$) and we can apply Eqs. (3.31) and (3.32), thus obtaining:

$$A_0 = -\left\{\sigma + \frac{1}{\delta}\left(\tfrac{1}{2} + \sigma^2\right) - \int_0^\infty u e^{-\delta/u}[P(u)]^{-1}A(u)\,du\right\}\frac{k\delta}{2} \tag{5.13}$$

$$\begin{aligned} A(u) = {} & \frac{k}{2}\pi^{\frac{1}{2}}\left(u + \frac{\delta}{2} + \sigma\right)e^{u^2}[P(u)]^{-1}\{[p(u)]^2 + \pi^2u^2\}^{-1} + \\ & + \frac{\pi^{\frac{1}{2}}}{2}e^{u^2}[P(u)]^{-1}\{[p(u)^2 + \pi^2u^2\}^{-1} \times \\ & \times \int_0^\infty \frac{\xi[P(\xi)]^{-1}}{u+\xi}e^{-\delta/\xi}A(\xi)\,d\xi \qquad (\xi > 0) \end{aligned} \tag{5.14}$$

where permissible inversions of the order of integrations have been performed and Eq. (3.35) used. Here $\sigma = \zeta/\theta$ (see Eqs. (4.13) and (4.13a)). Thus the problem has been reduced to the task of solving an integral equation in the unknown $A(u)$ Eq. (5.14). This equation is a classical Fredholm equation of the second kind with symmetrizable kernel. The corresponding Neumann-Liouville series can be shown to converge for any given positive value of δ [18]. It is also obvious that the larger δ, the more rapid is the convergence. This allows the determination of some results in the near-continuum regime. In particular, if terms of order $\exp[-3(\delta/2)^{\frac{2}{3}}]$ are negligible, only the zero order term of the series need be retained:

$$A(u) = \frac{k}{2}\pi^{\frac{1}{2}}\left\{\left(\frac{\delta}{2} + \sigma\right) + u\right\}e^{u^2}[P(u)]^{-1}\{[p(u)]^2 + \pi^2u^2\}^{-1} \tag{5.15}$$

Within the same limits of accuracy, A_0 is given by:

$$A_0 = -\frac{k\delta}{2}\left[\sigma + \frac{1}{\delta}\left(\tfrac{1}{2} + \sigma^2\right)\right] \tag{5.16}$$

We note that this zero order approximation is far more accurate than a continuum treatment (even if slip boundary conditions are used in the latter). In fact, even in the zero-order approximation:

(1) kinetic boundary layers are present near the walls.

(2) In the main body of the flow the mass velocity satisfies the Navier-Stokes momentum equation; however, the corresponding extrapolated boundary conditions show the presence not only of first order slip but also of a second order slip:

$$v_3\left(\pm\frac{\delta}{2}\right) = \mp\sigma\left(\frac{\partial v_3}{\partial x}\right)_{x=\pm\delta/2} - \tfrac{1}{2}\left(\tfrac{1}{2} + \sigma^2\right)\left(\frac{\partial^2 v_3}{\partial x^2}\right)_{x=\pm\delta/2} \tag{5.17}$$

In order to obtain these results, we observe that the mass velocity is given by:

$$v_3(x) = \pi^{-\frac{1}{2}} \int_{-\infty}^{\infty} W(x, \xi) e^{-\xi^2}\, d\xi$$

$$= \frac{k}{2}(x^2 - \delta^2/4) + A_0 + \int_{-\infty}^{\infty} A(u) \exp\left[-\frac{x}{u} - \frac{\delta}{2\,|u|}\right] du \qquad (5.18)$$

where Eqs. (5.8) and (5.9) have been taken into account. Eq. (5.18) is exact; if terms of order $\exp[-3(\delta/2)^{\frac{2}{3}}]$ can be neglected, A_0 and $A(u)$ are given by Eqs. (5.16) and (5.15). In particular, the integral term in Eq. (5.18) describes the space transients in the kinetic boundary layers; in the main body of the flow the integral term is negligible and we have

$$v_3(x) = \frac{k}{2}[x^2 - \delta^2/4 - \sigma\delta - (\tfrac{1}{2} + \sigma^2)] \qquad (\delta \gg 1) \qquad (5.19)$$

It is easily checked that this expression solves the Navier-Stokes momentum equation for plane Poiseuille flow and satisfies the boundary conditions (5.17).

We can also easily write down the distribution function in the main body of the flow. As a matter of fact, $Y(x, \xi)$ reduces here to A_0, so that Eqs. (5.8) and (5.9) give

$$W(x, \xi) = \frac{k}{2}\left[x^2 - \frac{\delta^2}{4} - 2x\xi - (1 - 2\xi^2) - \sigma\delta - (\tfrac{1}{2} + \sigma^2)\right] \qquad (\delta \gg 1) \quad (5.20)$$

By taking Eq. (5.19) into account, Eq. (5.20) can be rewritten as follows

$$W(x, \xi) = v_3(x) - \theta \frac{\partial v_3}{\partial x} \xi + (\xi^2 - \tfrac{1}{2})\theta^2 \frac{\partial^2 v_3}{\partial x^2} \qquad (\delta \gg 1) \qquad (5.21)$$

where general units for x have been restored (i.e. we have written x/θ in place of x). Eq. (5.21) clearly shows that in the main body of the flow the distribution function is of the Hilbert-Chapman-Enskog type (power series in θ), as was to be expected. It is a truncated series, but the truncation does not occur at the Navier-Stokes level of description. As a matter of fact, Eq. (5.21) gives a Burnett distribution function and this explains the appearance of a second order slip from a formal point of view. From an intuitive standpoint, the second order slip can be attributed to the fact that molecules with nonzero velocity in the z-direction move into a region with different density before having any collision and there is a net transport of mass because of the density gradient; molecules move preferentially toward smaller densities even before suffering any collision, and, therefore, at a mean free path from the wall an effect of additional macroscopic slip appears.

The presence of an additional slip means that, for a given pressure gradient and plate distance, more molecules pass through a cross section than predicted by Navier-Stokes equations with first order slip. This is easily checked, for sufficiently large δ, by using Eqs. (5.18), (5.15), (5.16), which give for the flow rate:

$$F = \int_{-d/2}^{d/2} \rho v_3(x)\, dx \simeq -\frac{1}{2}\frac{dp}{dx} d^2\left\{\frac{1}{6}\,\delta + \sigma + \frac{2\sigma^2 - 1}{\delta}\right\} \qquad (\delta \gg 1) \quad (5.22)$$

Here x and z are in general units and $d = \delta\theta$ is the distance between the plates in the same units. Therefore, for given geometry and pressure gradient, the nondimensional flow rate is

$$Q(\delta) = \tfrac{1}{6}\delta + \sigma + \frac{2\sigma^2 - 1}{\delta} \qquad (\delta \gg 1) \qquad (5.23)$$

The last term is the correction to the first order slip theory; it arises in part from the second order slip and in part from the kinetic boundary layers. In fact the gas near the walls moves more slowly than predicted by an extrapolation of Eq. (5.19); this brings in a contribution to $Q(\delta)$ of the same order as the second order slip, thus reducing the effect of the latter (without eliminating it completely, however). It is also clear that, although Eq. (5.23) is valid for large values of δ, the increase in $Q(\delta)$ with respect to the prediction of the first order slip theory persists for small values of δ because the molecules with velocity almost parallel to the wall give a sensible contribution to the motion by travelling downstream for a mean free path. In particular, in the limiting case of free molecular flow, Eq. (5.8) formally reduces to

$$\xi \frac{\partial W}{\partial x} + \frac{k}{2} = 0 \qquad (5.24)$$

or

$$W = -\frac{k}{2}(x/\xi + d/2\,|\xi|) \qquad (5.25)$$

where general length units are used. Eq. (5.25) clearly shows that molecules travelling almost parallel to the wall ($\xi \simeq 0$) cannot be in free molecular flow. Eq. (5.25) can be assumed to hold for $|\xi| < \delta$ (recall that $|\xi|$ is nondimensional). Hence for $\delta \to 0$

$$v_3 \simeq -\frac{k}{2}\frac{d}{2}\pi^{-\frac{1}{2}} \int_{|\xi|>\delta} \frac{1}{|\xi|} e^{-\xi^2}\, d\xi \simeq -\frac{k\pi^{-\frac{1}{2}}}{2}\, d \log \delta \qquad (\delta \ll 1) \quad (5.26)$$

and

$$Q(\delta) \simeq -\pi^{-\frac{1}{2}} \log \delta \qquad (\delta \ll 1) \qquad (5.27)$$

This approximate argument is confirmed by a study of the nearly-free molecular regime ($\delta \to 0$). This study can be based either on the iteration procedures described in Section 9 of Chapter V [20] or on a different use of the method of elementary solutions [18]. In both cases the conclusion is that Eq. (5.27) is correct and this means that, for $\delta \to 0$, higher order contributions from kinetic layers destroy the $1/\delta$ term in Eq. (5.23) but leave a weaker divergence for $\delta \to 0$ (essentially related to the molecules travelling parallel to the plates). The behavior for large values of δ, Eq. (5.23), and for small values of δ, Eq. (5.27), imply the existence of at least one minimum in the flow rate. This minimum was experimentally found a long time ago by Knudsen [21] and then by different authors for long tubes of various cross section. The above discussion gives a qualitative explanation of the presence of the minimum, although its precise location for slabs and more complicated geometries must be found by appropriate techniques of solution (see Chapter VII, Section 5).

A similar treatment can be applied to problems of neutron transport in a slab in the presence of a plane source (either inside the slab or on the boundary). A new feature appears if we let $c > 1$ in Eq. (3.11). In this case there is a critical value of d (the width of the slab), for which there is a nonzero solution in absence of the source; this corresponds to the fact that the nuclear reactions are self-sustaining without destroying the system, and the slab is said to have reached criticality. Methods similar to those discussed above have been applied to the criticality problem by Mitsis [22]; the two elementary solutions corresponding to complex eigenvalues, which were mentioned in Section 3, play a basic role in the solution of this problem. Details concerning the result by Mitsis are given by Case and Zweifel [16].

6. Unsteady solutions of kinetic models with constant collision frequency

If we consider the time dependent BGK equation in one-dimensional plane geometry, shear effects can be separated from effects related to normal stresses and heat transfer in the same way as for steady situations. The relevant equation for shear flow problems is as follows:

$$\frac{\partial Y}{\partial t} + \xi \frac{\partial Y}{\partial x} + Y(x, \xi) = \pi^{-\frac{1}{2}} \int_{-\infty}^{\infty} e^{-\xi_1^2} Y(x, \xi_1)\, d\xi_1 \tag{6.1}$$

which is the time-dependent analogue of Eq. (3.1) (when we let $\nu = 1$ and assume that Y does not depend upon ξ_2 and ξ_3, which is usually the case). Here both x and t are expressed in θ units, since we have let $\nu = \theta^{-1} = 1$.

The one speed approximation in neutron transport leads to a similar equation whose elementary solutions were studied by Bowden and Williams [23] by a procedure very similar to the method to be employed in this section to deal with Eq. (6.1). This method is taken from Ref. [24] and can be described as follows. A Laplace transform is taken with respect to time and accordingly the time-dependent problem is reduced to a steady one. The solution of the problem depends now on a complex parameter s. After separating the space and velocity variables, the spectrum of values of the separation parameter u must be studied in its dependence on s. This study is essential in order to treat the problem of inversion.

Let us take the Laplace transform of Eq. (6.1). Without any loss of generality, a zero initial value for Y will be assumed. In fact, a particular solution of the inhomogeneous transformed equation, which would result from a nonzero initial condition, can be constructed by using the Greens' function which can easily be obtained when the general solution of the homogeneous equation is known. Accordingly, we shall restrict ourselves to the homogeneous transformed equation:

$$(s+1)\tilde{Y} + \xi \frac{\partial \tilde{Y}}{\partial x} = \pi^{-\frac{1}{2}} \int_{-\infty}^{\infty} e^{-\xi_1^2} \tilde{Y}(x, \xi_1)\, d\xi_1 \tag{6.2}$$

where $\tilde{Y}$ is the Laplace transform of Y. The same equation (with $s = i\omega$) governs the state of a gas forced to undergo steady transverse oscillations with frequency ω.

Separating the variables in Eq. (6.2) gives

$$\tilde{Y}_u(x_1, \xi; s) = g_u(\xi; s)\exp[-(s+1)x/u] \tag{6.3}$$

where u is the separation parameter and $g_u(\xi; s)$ satisfies

$$(s+1)\left(1 - \frac{\xi}{u}\right) g_u(\xi; s) = \pi^{-\frac{1}{2}} \int_{-\infty}^{\infty} g_u(\xi_1; s) e^{-\xi_1^2}\, d\xi_1 \tag{6.4}$$

The right hand side does not depend on ξ and can be normalized to unity. Accordingly, we are led to a typical division problem in complete analogy with the steady case. If the factor $(u - \xi)$ cannot be zero, so that u is not a real number, then $g_u(\xi; s)$ is an ordinary function given by

$$g_u(\xi; s) = \frac{u}{u - \xi} \tag{6.5}$$

with the normalization condition:

$$\pi^{-\frac{1}{2}} \int_{-\infty}^{\infty} \frac{u e^{-\xi^2}}{u - \xi}\, d\xi = s + 1 \tag{6.6}$$

If on the contrary, u is a real number, then $g_u(\xi)$ must be treated as a generalized function and Eq. (6.4) gives:

$$g_u(\xi; s) = P\frac{u}{u - \xi} + p(u; s)\,\delta(u - \xi) \tag{6.7}$$

where $p(u; s)$ which is fixed by Eq. (6.6), is given by

$$p(u; s) = \pi^{\frac{1}{2}} e^{u^2} s + p(u) \tag{6.8}$$

$p(u)$ being given by Eq. (3.41). Eq. (6.7) gives the generalized eigensolutions corresponding to the continuous spectrum ($-\infty < u < \infty$). The essential point, now, is to study the possible values of u which satisfy Eq. (6.6), and therefore form the discrete spectrum. Such values coincide, according to Eq. (6.6), with the zeroes of the following function of the complex variable z:

$$M(z; s) = 1 - \pi^{-\frac{1}{2}}(s + 1)^{-1} \int_{-\infty}^{\infty} \frac{z e^{-t^2}}{z - t}\, dt \tag{6.9}$$

This function is analytic in the complex z-plane with a cut along the real axis where $M(z; s)$ suffers a discontinuity. In fact, the Plemelj formulas (see Appendix) give the following result for the limiting values $M^{\pm}(u; s) = \lim_{\varepsilon\to 0} M(u \pm i\varepsilon; s)$ (u real, $\varepsilon > 0$):

$$M^{\pm}(u; s) = 1 - \pi^{-\frac{1}{2}}(s + 1)^{-1}\left[P\int_{-\infty}^{\infty} \frac{u e^{-t^2}}{u - t}\, dt \pm \pi i u e^{-u^2}\right] \tag{6.10}$$

Eq. (6.10) can also be written as follows:

$$M^{\pm}(u; s) = e^{-u^2}(s + 1)^{-1}\pi^{-\frac{1}{2}}[p(u; s) \pm \pi i u] \tag{6.11}$$

In the limiting case of s such that:

$$p(u; s) \pm \pi i u = 0 \qquad (\text{real } u) \tag{6.12}$$

the discrete spectrum merges into the continuous one. Eqs. (6.12) are satisfied on a closed heart-shaped curve of the complex s-plane (Fig. 38). We have the following parametric representation for such a curve (to be called γ):

$$\begin{aligned} \operatorname{Re} s &= -\pi^{-\frac{1}{2}} e^{-u^2} p(u) \\ \operatorname{Im} s &= -\pi^{\frac{1}{2}} u\, e^{-u^2} \qquad (-\infty < u < \infty) \end{aligned} \tag{6.13}$$

These equations are obtained from Eq. (6.12) and (6.8), taking into account that $p(u)$ is real. The equation $M(z; s) = 0$ defines a mapping from the z-plane to the s-plane: in fact this equation gives unambiguously a point in the s-plane once a point z off the real axis has been fixed. When z tends to a real value u, then $M(z; s) = 0$ becomes Eq. (6.12), because of Eq. (6.11); the double sign of course, is connected with the approach from above or below. Therefore, when u ranges over the real axis, s describes the curve γ counterclockwise if we think of the real axis as the boundary of the upper half plane and clockwise if we think of the real axis as the boundary of the lower half plane. In both cases Eq. (6.13) establishes a one-to-one correspondence between the curve γ of the s-plane and the real axis of the z-plane. From this fact and the argument principle it follows that both the lower and the upper half plane are conformally mapped into the region inside γ by the mapping $M(z; s) = 0$, and for each half plane the mapping is one-to-one. It follows that for any s in the region inside γ there are two complex values of u which satisfy Eq. (6.6) while there are none outside. It is easily seen, from Eq. (6.6), that these values are the negative of each other. We shall denote them by $\pm u_0(s)$.

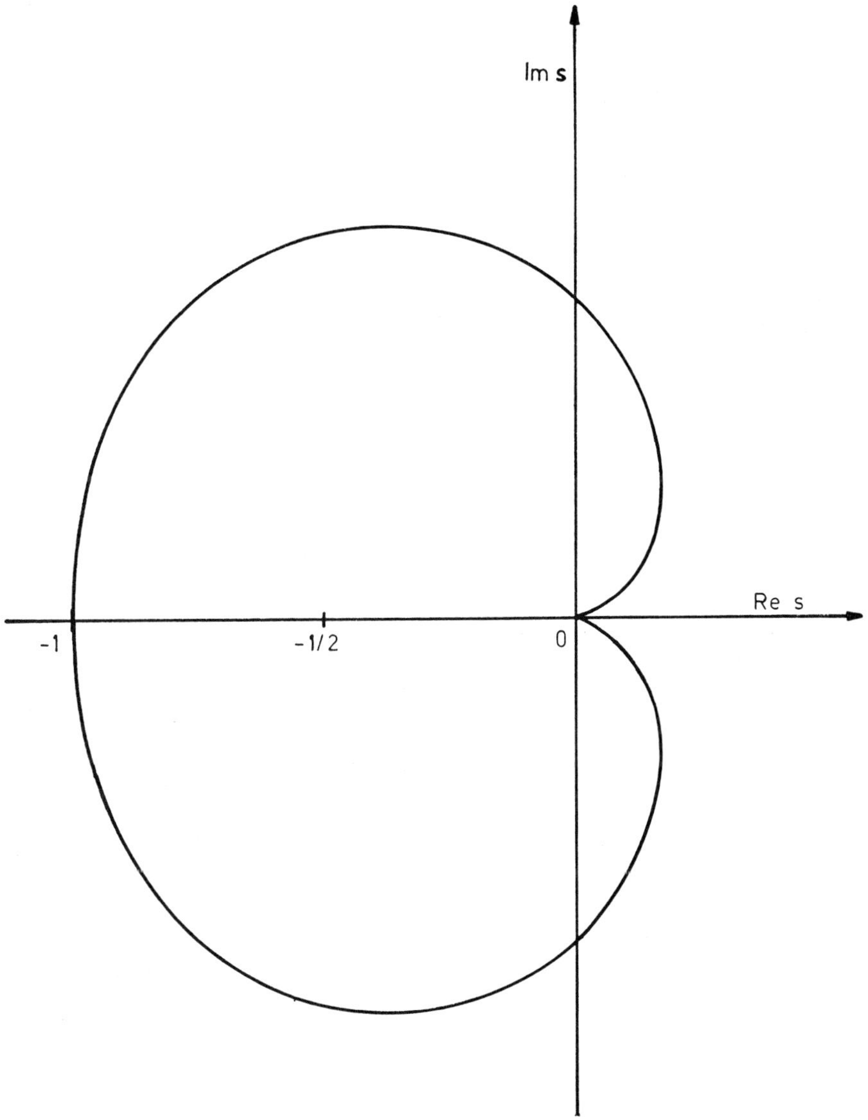

Fig. 38. The curve bounding the region where two discrete eigenvalues exist.

It is now possible to extend the results of the steady case $s = 0$ and, in particular, Theorems I and II of Section 3. The completeness results remain true and there are only slight changes in the equation. Thus, in the case of the full range $(-\infty < \xi < \infty)$ we can expand any function $g(\xi)$ such that $Z(\xi) = e^{-\xi^2}g(\xi)$ satisfies the assumptions of Theorem I as follows:

$$g(\xi) = A_+g_+(\xi; s) + A_-g_-(\xi; s) + \int_{-\infty}^{\infty} A(u)g_u(\xi; s)\, du \tag{6.14}$$

where $g_u(\xi; s)$ is given by Eq. (6.7) and $g_\pm$ by Eq. (6.5), with $u = \pm u_0$. The coefficients A_+ and A_- are zero outside γ (g_+ and g_- do not exist there) while they are given by

$$A_\pm = \pi^{-\frac{1}{2}}[s(1 - 2u_0^2) + 1]^{-1}\int_{-\infty}^{\infty} \frac{\xi e^{-\xi^2}}{\xi \mp u_0}\, g(\xi)\, d\xi \tag{6.15}$$

for s inside γ. For any s, $A(u)$ is given by Eq. (3.28), provided that $p(u)$ is replaced by $p(u; s)$ throughout, $Z(w) = Z_0(w)g(w)$ and $Z_0(w)$ is given by Eq. (3.40). It is obvious that $A_\pm$ and $A(u)$ and possibly $g(\xi)$, depend upon s, though this dependence has not been exhibited in the equations. In the case of the half range $0 < \xi < \infty$, a function $g(\xi)$ can be expanded as follows

$$g(\xi) = A_+g_+(\xi; s) + \int_0^{\infty} A(u)g_u(\xi; s)\, du \tag{6.16}$$

where A_+ is zero for s outside γ and is given by

$$A_+ = [2u_0^2 s\pi^{\frac{1}{2}}P(u_0)]^{-1}\int_0^{\infty} \xi g(\xi)g_+(\xi; s)P(\xi; s)e^{-\xi^2}\, d\xi \tag{6.17}$$

for s inside γ. Here

$$P(u; s) = \frac{u}{u_0 + u}\exp\left\{-\frac{1}{\pi}\int_0^{\infty} \tan^{-1}[\pi t/p(t; s)]\frac{dt}{t + u}\right\} \tag{6.18}$$

For any s, $A(u)$ is given by Eq. (3.32), where $Z(w) = Z_0(w)g(w)$ and $Z_0(w)$ is given by Eq. (3.40), while $P(u)$ is given by Eq. (6.18) when s is inside γ and by the same equation with $u_0 = 0$ when s is outside γ. The function $P(u; s)$ again satisfies certain identities which make its manipulation simpler than would be expected [24, 14, 25].

7. Analytical solutions of specific problems

The theory sketched in Section 6 can be used to solve analytically problems of shear flows when the region filled by gas is the whole space or a half space. We can, for example, solve the following problem: let two half-spaces be separated by the plane $x = 0$, and assume that initially the gas has the same density ρ_0 and temperature T_0 in both regions, while the gas in the region $x > 0$ flows uniformly in the z-direction with velocity V and the gas in the region $x < 0$ flows uniformly in the same direction with velocity $-V$; we want to find the evolution of the gas comprising the smoothing out and diffusion of the velocity discontinuity. The problem can be solved [26] by using the theorem of full range completeness to construct the Laplace transform of the solution. We can even obtain an analytic inversion of the Laplace transform and write the solution for the mass velocity as follows:

$$v(x, t) = U \operatorname{sgn} x \left[1 - 2\pi^{-\frac{1}{2}} \int_{-\infty}^{\infty} H[(u/x) - (1/t)] \exp\{-x/(\theta u) + q(u)[(x/u) - t]/\theta\} \left\{ e^{-u^2} \cos\left[\frac{\pi^{\frac{1}{2}} u}{\theta}(t - x/u) e^{-u^2}\right] + \frac{1 - q(u)}{u\pi^{\frac{1}{2}}} \sin\left[\frac{\pi^{\frac{1}{2}} u}{\theta}(t - x/u)e^{-u^2}\right]\right\} du \right] \tag{7.1}$$

where H is the Heaviside step function, $\operatorname{sgn} x = H(x) - H(-x)$ and

$$q(u) = \pi^{-\frac{1}{2}} e^{-u^2} p(u) \tag{7.2}$$

The exact solution can be used to obtain asymptotic expansions for both short and long times and for the numerical tabulation of the space-time behavior of the gas. The solution shows that the velocity profile becomes more and more flattened as time increases, but a disagreement of about 10% from the Navier-Stokes equations is still present after 12 collision times.

Half-space problems are more difficult to solve, since they require using the half-range completeness theorem and, consequently, equations involving $P(u; s)$. The solution can, however, always be reduced to a double quadrature for initial value problems and a single quadrature for problems of steady oscillations, provided the boundary conditions give an explicit expression of the distribution function of the molecules entering the half-space (as in the case of complete diffusion from the wall).

As an example of a half-space problem, we consider the propagation of Rayleigh waves in a half-space: let a half-space be filled with a gas of density ρ_0 and temperature T_0 and bounded by an infinite plane wall which is oscillating in its own plane with frequency ω. We shall consider the system in a steady state when the transients have disappeared. Therefore, if the velocity of the wall is the real part of $Ue^{i\omega t}$ (U being a constant), the solution of the linearized problem will be the real part of a function h having time dependence $e^{i\omega t}$ and satisfying

$$i\omega h + \xi_1 \frac{\partial h}{\partial x} = Lh \tag{7.3}$$

The linearized boundary condition at the wall, which is assumed to diffuse the molecules according to the basic Maxwellian f_0, can be written as follows:

$$h(0, \boldsymbol{\xi}, t) = 2Ue^{i\omega t}\xi_3 \qquad (\xi_1 > 0) \tag{7.4}$$

We also require that the solution be bounded at infinity. If the BGK model is assumed to describe collisions, we can write

$$h(x, \boldsymbol{\xi}, t) = 2Ue^{i\omega t}\xi_3 \tilde{Y}(x, \xi_1) \tag{7.5}$$

where $\tilde{Y}(x, \xi)$ satisfies Eq. (6.2) (with $s = i\omega$) and the boundary condition:

$$\tilde{Y}(0, \xi) = 1 \qquad (\xi > 0) \tag{7.6}$$

We shall write s in place of $i\omega$ because the following results are valid for any complex s (Re $s \geqslant -1$) and the more general form will be useful later. Using the boundary condition, Eq. (7.6), and the half-range completeness (see Section 6) together with boundedness at infinity in space, we find [14, 25, 24]:

$$\tilde{Y}(x, \xi) = -\pi^{\frac{1}{2}}s\int_0^\infty \frac{g_u(\xi; s)[P(u; s)]^{-1}}{[p(u; s)]^2 + \pi^2u^2} \exp\left[-(s+1)\frac{x}{u} + u^2\right] du \tag{7.7}$$

when s is outside γ (then $P(u; s)$ is given by Eq. (6.18) with $u_0 = 0$). Analogously, when s is inside γ, we find:

$$\tilde{Y}(x, \xi) = -2g_+(\xi)\exp[-(s+1)x/u_0][P(u_0; s)]^{-1} -$$

$$- \pi^{\frac{1}{2}}s\int_0^\infty \frac{[P(u; s)]^{-1}g_u(\xi; s)\exp\left[-(s+1)\dfrac{x}{u} + u^2\right]}{[p(u; s)]^2 + \pi^2u^2} du \tag{7.8}$$

where $u_0 = u_0(s)$ is selected between the two possible values in such a manner that

$$Re\left[\frac{s+1}{u_0(s)}\right] > 0 \tag{7.9}$$

and $P(u; s)$ is given by Eq. (6.18).

Let us now briefly discuss the solution. First of all, we note that there is a limiting frequency ω_0 ($i\omega_0 \in \gamma$) such that for $\omega > \omega_0$ we have only the eigensolutions of the continuous spectrum. It seems, therefore, that for $\omega > \omega_0$ no plane shear wave exists. However, we are able to exhibit a discrete term for $\omega > \omega_0$ [25]; as a matter of fact, we can rotate the path of integration in Eq. (7.8) downward, provided that we add the contribution from any poles of the integrand between this half straight line and the real semi-axis. Now, it is easily seen that, at least for frequencies larger than but still close to ω_0, there is one such pole u_0 which satisfies

$$p(u_0; s) - \pi i u_0 = 0 \tag{7.10}$$

where $p(u; s)$ is given by Eq. (6.8) and the second expression in Eq. (3.41), which has a meaning for any complex u. Eq. (7.10) is the analytical continuation of Eq. (6.6) for $\omega > \omega_0$. From this point of view ω_0 loses its character of a critical frequency. Another feature of our results is that for any fixed frequency, if we go sufficiently far from the wall, the contribution from the continuous spectrum dominates the discrete term, since the former is less than exponentially damped. This feature is strictly related to the fact that the spectrum of values of v extends to ∞, and would not be present if the collision frequency increased at least linearly with molecular speed for large values of the latter. The experimental verification of this asymptotic behavior seems to be outside the available techniques. As a matter of fact, the physically relevant region (say $\frac{1}{10}$ to 10 mean free paths) appears to be characterized by the fact that the discrete term (either the genuine one or its analytical continuation) dominates, according to estimates made by Dorning and Thurber [27] for a similar problem concerning neutron waves.

Another solvable problem is the following: Let a half-space be filled with a gas of density ρ_0 and temperature T_0 and bounded by a plate; the gas is initially in absolute equilibrium and the wall is at rest; then the plate is set impulsively into motion in its own plane with constant velocity U; the propagation into the gas of the disturbance produced by the motion of the plate is to be studied. This problem is known under the name of Rayleigh's problem; we want to solve it analytically by using the linearized BGK model.

The perturbed distribution function satisfies the linearized Boltzmann equation and the following initial and boundary conditions:

$$h(x, 0, \boldsymbol{\xi}) = 0 \tag{7.11}$$

$$h(0, t, \boldsymbol{\xi}) = 2U\xi_3 \qquad (\xi_1 > 0) \tag{7.12}$$

Also, when $x \to \infty$, $h(x, t, \boldsymbol{\xi})$ must be bounded for any fixed t and ξ. If we use the BGK model, we have

$$h(x, t, \boldsymbol{\xi}) = 2U\xi_3 Y(x, t, \xi_1) \tag{7.13}$$

where $Y(x, t, \xi)$ satisfies Eq. (6.1) and the following initial and boundary conditions:

$$Y(x, 0, \xi) = 0 \tag{7.14}$$

$$Y(0, t, \xi) = 1 \qquad (\xi > 0) \tag{7.15}$$

By introducing the Laplace transform of Y, $\tilde{Y}(x, s, \xi)$, Eq. (6.1) reduces to Eq. (6.2) while the boundary condition at the wall becomes

$$\tilde{Y}(0, s, \xi) = \frac{1}{s} \qquad (\xi > 0) \tag{7.16}$$

Accordingly, $\tilde{Y}$ is obtained from the equations for the oscillating wall by multiplying the right hand side of the equations by $1/s$; the same is also true for the mass velocity and the stress. It is important to note that $\tilde{Y}$ defined by Eqs. (7.7) and (7.8) for s outside and inside γ, is an analytic function of s, not only outside and inside γ but also through this curve [24]; in other words, Eq. (7.8) is the analytic continuation of Eq. (7.7) inside γ. This follows from the fact that the expression appearing in Eq. (7.7) undergoes a discontinuity when s crosses γ and this discontinuity is equal to the limiting value of the discrete term in Eq. (7.8). Hence, when inverting the Laplace transform, the path of integration can be moved through γ, provided that in every region the appropriate expression is used. On the other hand, the segment $(-1, 0)$ of the real axis is easily seen to be a discontinuity line because of the choice of u_0 in Eq. (7.9). According to well-known theorems on the Laplace transform, $Y(x, t, \xi)$ is given by

$$Y(x, t, \xi) = \frac{1}{2\pi i}\int_{c-i\infty}^{c+i\infty} \frac{e^{st}}{s}\,\tilde{Y}(x, s, \xi)\,ds \qquad (7.17)$$

where $\tilde{Y}$ is given by Eq. (7.7) and the path of integration is a vertical straight line to the right of γ. Owing to the analyticity properties of $\tilde{Y}$, this integration path in the s-plane can be deformed to a path indented on the segment $(-1, 0)$ of the real axis, and along the vertical line $\mathrm{Re}(s + 1) = 0$. The resulting integrals can be put into a completely real form.

Thus the problem is solved in terms of quadratures, which can, in principle, be performed with any desired accuracy. But we can use our results also for obtaining interesting information by analytical manipulations. We can, for example, expand our results for short and long times [14].

For long times, the mass velocity is given by:

$$v_3(x, t)/U \simeq 1 - (\pi\nu t)^{-\frac{1}{2}} v_3(x) \qquad (t \to \infty) \qquad (7.18)$$

where ν is the kinematic viscosity ($\theta = 2\nu$ if $2RT_0 = 1$) and $v_3(x)$ the mass velocity corresponding to a unit gradient at infinity in the Kramers' problem (Eq. (4.15) with $a = 1$). Therefore the flow shows the same structure of the kinetic boundary layer both in the steady and in the time-dependent flows (for large values of the time variable); this is not surprising, because we know that the equations which describe the kinetic layer (Chapter V) do not depend upon the particular problem. In particular, the slip coefficient retains, in this time-dependent flow and for sufficiently large value of t, the same values as in the steady flows.

The analytical solution also leads to simple expressions for the velocity and the stress at the plate [14]. As an example, we quote

$$v_3(0, t)/U = 1 - \frac{1}{\pi}\int_0^1 (u^{-1} - 1)^{\frac{1}{2}} e^{-ut}\,du = \tfrac{1}{2} + \tfrac{1}{2}\int_0^{t/2} e^{-v} v^{-1} I_1(v)\,dv \quad (7.19)$$

where I_1 denotes, as usual, the modified Bessel function of the first kind of order one [28]. These expressions show that $v_3(0, t)$ slowly increases from an initial value $U/2$ to the final value U.

8. More general models

After the detailed treatment of both steady and time-dependent shear flows, it seems natural to consider Eq. (2.5) which describes steady heat transfer processes. This equation, however, contains three moments rather than one. We can copy the method of Section 3 by defining w by Eq. (3.4) and setting

$$\begin{aligned} Z^{(k)} &= \int \nu \psi_k f_0 h \frac{\partial \xi_1}{\partial w}\, d\xi_2\, d\xi_3 \\ Z_0^{(hk)}(w) &= \int \nu \psi_k \psi_h f_0 \frac{\partial \xi_1}{\partial w}\, d\xi_2\, d\xi_3 \end{aligned} \tag{8.1}$$

where, of course, ξ_1 is expressed in terms of w, ξ_2, ξ_3 before integrating. Then, if $\mathbf{Z}$ and Z_0 denote the vector with components $Z^{(k)}$ and the matrix with elements $Z_0^{(hk)}$, we have

$$w \frac{\partial \mathbf{Z}}{\partial x} + \mathbf{Z} = \mathrm{Z}_0 \int_{-k}^{k} \mathbf{Z}(x, w_1)\, dw_1 \tag{8.2}$$

and so we obtain a system of three coupled equations. By operating with vectors and matrices instead of scalars, we can repeat the procedures of previous sections as far as the construction of the elementary solutions of Eq. (8.2) and the proof of their full-range completeness are concerned. The half-range completeness is a considerably different matter as will be discussed below. Since the same methods can be applied to more general models, the only difference being the use of $n \times n$ instead of 3×3 or 2×2 matrices, there is no point in considering the collision model given by Eq. (2.2) separately from higher order models when heat transfer problems or unsteady problems with normal stresses are considered. Accordingly, in this section we shall consider the more general models, described in Section 9 of Chapter IV; in order to simplify notation and to include the time-dependent case, we shall limit ourselves to models with constant collision frequency. The most general model equation of this kind can be written as follows (one-dimensional problems):

$$\frac{\partial h}{\partial t} + \xi_1 \frac{\partial h}{\partial x} = \sum_{k,j=1}^{M} \alpha_{jk}(h, \psi_j)\psi_k - h \tag{8.3}$$

where the collision time ν^{-1} has been taken as time unit and the ψ_j are the eigenfunctions of the Maxwell collision operator.

First of all, we eliminate the transverse components of the mass velocity, by decomposing $h(x, t, \boldsymbol{\xi})$ according to

$$h = \sum_{j=1}^{n} h_j(x, t, \xi_1) g_j(\xi_2, \xi_3) + h_R(x, t, \boldsymbol{\xi}) \tag{8.4}$$

where $\{g_j\}$ denotes the set of polynomials orthogonal according to the inner product

$$(h, g)_2 = \pi^{-1} \iint_{-\infty}^{\infty} h(\xi_2, \xi_3) g(\xi_2, \xi_3) \exp(-\xi_2{}^2 - \xi_3{}^2)\, d\xi_2\, d\xi_3 \qquad (2RT_0 = 1) \tag{8.5}$$

and h_R is unambiguously defined as being orthogonal to the g_j $(j = 1, 2, \ldots, n)$; n is an integer $(\leqslant M)$ such that every ψ_k $(k = 1, 2, \ldots, M)$ can be expressed in terms of the g_j $(j = 1, \ldots, n)$ with coefficients which are polynomials in ξ_1. Then Eq. (8.3) can be transformed into the following system:

$$\frac{\partial h_j}{\partial t} + \xi \frac{\partial h_j}{\partial x} + h_j(x, t, \xi) = \sum_{r=1}^{n} \sum_{k=0}^{m} (\varphi_k, h_r)_1 X_{jkr}(\xi) \qquad (j = 1, \ldots, n) \tag{8.6}$$

$$\frac{\partial h_R}{\partial t} + \xi_1 \frac{\partial h_R}{\partial x} + h_R(x, t, \boldsymbol{\xi}) = 0 \tag{8.7}$$

where $\{\varphi_k(\xi)\}$ is the set of polynomials which are orthogonal with respect to the scalar product:

$$(h, g)_1 = \pi^{-\frac{1}{2}} \int_{-\infty}^{\infty} h(\xi) g(\xi) e^{-\xi^2}\, d\xi \tag{8.8}$$

m is an integer $(\leqslant M)$ which is given by the maximum degree of the coefficients of the expansion of the ψ_k $(k = 1, \ldots, M)$ in terms of the g_j $(j = 1, \ldots, n)$. The X_{jkr} are suitable polynomials (having degree not higher than m).

Eq. (8.7) may be immediately solved and, since it is not coupled with the system given by Eq. (8.6), will not be considered further. Separating the variables in the usual way in Eq. (8.6), we obtain

$$(s + 1)\left(1 - \frac{\xi}{u}\right) h_j(\xi; u, s) = \sum_{r=1}^{n} \sum_{k=0}^{m} (\varphi_k, h_r)_1 X_{jkr}(\xi) \tag{8.9}$$

From this equation we obtain the following moment equations:

$$u(s + 1)(\varphi_r, h_j)_1 - (\xi \varphi_r, h_j)_1 = u \sum_{k=0}^{m} \sum_{j=1}^{n} (\varphi_k, h_s)_1 (\varphi_r, X_{jks})_1$$

$$(r = 0, 1, \ldots, m - 1; j = 1, \ldots, n) \tag{8.10}$$

Since $\xi \varphi_r$ can be expressed as a combination of φ_{r-1} and φ_{r+1}, the latter system of $m \times n$ equations can be solved immediately with respect to the

$m \times n$ quantities (φ_r, h_j) $(r = 1, 2, \ldots, m;\ j = 1, \ldots, n)$ which are accordingly expressed in terms of $(\varphi_0, h_j)_1$. The coefficients of the linear combination are rational functions of u and j (polynomials for Maxwell molecules). Then Eq. (8.9) can be written as follows:

$$\left(1 - \frac{\xi}{u}\right) h_j(\xi; u, s) = \sum_{k=0}^{m} T_{jk}(\xi; u, s) A_k(u, s) \tag{8.11}$$

where the T_{jk} are polynomials in ξ, u, s and

$$A_k(u, s) = (1, h_k)_1 [D(u, s)(s + 1)]^{-1} \tag{8.12}$$

$D(u, s)$ being the determinant of Eq. (8.10) (Essentially, $D(u, s) = 1$ for Maxwell's molecules, but is a polynomial in u and s for more general models). Eq. (8.1) can be transcribed in matrix notation

$$\left(1 - \frac{\xi}{u}\right) \mathbf{h}(\xi, u, s) = \mathsf{T}(\xi; u, s) \mathbf{A}(u, s) \tag{8.13}$$

where $\mathbf{h}$ and $\mathbf{A}$ are vectors and T is a matrix. In correspondence to the usual continuous spectrum, we have now the following eigenfunctions:

$$h(\xi; u, s) = \mathsf{T}(\xi; u, s) \mathbf{A}(u, s) P \frac{u}{u - \xi} + \mathsf{p}(u; s) \mathbf{A}(u, s)\, \delta(u - \xi) \tag{8.14}$$

where because of Eq. (8.9):

$$\mathsf{p}(u; s) = e^{v^2} \left[\pi^{\frac{1}{2}} D(u; s)(s + 1) \mathsf{I} - P \int_{-\infty}^{\infty} \frac{\mathsf{T}(\xi; u, s) u e^{-\xi^2}}{u - \xi}\, d\xi \right] \tag{8.15}$$

where I is the identity matrix. The integrals appearing in Eq. (8.15) can be easily expressed in terms of the function $p(u)$ introduced in Section 3 and, therefore, in terms of tabulated functions. Let us introduce the following matrix, whose elements are functions of the complex variable z:

$$\mathsf{M}(z; s) = \pi^{\frac{1}{2}} D(u; s) \mathsf{I} + \frac{1}{s + 1} \int_{-\infty}^{\infty} \frac{\mathsf{T}(\xi; z, s) z e^{-\xi^2}}{\xi - z}\, d\xi \tag{8.16}$$

then it is easily seen that the discrete spectrum (if any) is given by the complex values of u solving

$$\operatorname{Det} \mathsf{M}(u; s) = 0 \tag{8.17}$$

The curves of the s-plane where the discrete eigenvalues ceases to exist have been investigated by R. Mason [29] in the case of the BGK model.

It is relatively easy to prove the completeness of eigensolutions in the full range $(-\infty < \xi < \infty)$ and to construct the coefficients of the expansion of a given function in terms of quadratures. The expansion for a general function

$\mathbf{g}(\xi)$ takes on the form

$$\mathbf{g}(\xi) = \sum_{i=1}^{N} \mathbf{g}_i(\xi, s)\mathbf{A}_i(s) + \int_{-\infty}^{\infty} \mathbf{g}(\xi; u, s)\mathbf{A}(u; s)\, du \tag{8.15}$$

where

$$\mathsf{g}(\xi; u, s) = \mathsf{T}(\xi; u, s)P\frac{u}{u-\xi} + \mathsf{p}(u; s)\,\delta(u-\xi) \tag{8.16}$$

and

$$\mathbf{g}_i(\xi; s) = \mathsf{T}(\xi; \mathbf{u}_i, s)P\frac{u_i}{u_i-\xi} \tag{8.17}$$

u_i $(i = 1, \ldots, N)$ being the possible solutions of Eq. (8.14). The coefficients $\mathbf{A}_i(s)$ and $\mathbf{A}(u, s)$ can be expressed explicitly in terms of quadratures involving the given function $\mathbf{g}(\xi)$. A key role is played by $\mathsf{M}(z; s)$ and its obvious relations to $\mathsf{p}(u; s)$, which allow us to solve the necessary integral equations by algebraic operations on matrices [25]. We note also that in the steady case $(s = 0)$ we have to add to the right hand side of Eq. (8.15) the solutions arising from the collision invariants as well as from the particular solutions (linear in x) which appear in Eq. (IV.7.53). In the same case there are no complex solutions of Eq. (8.14), and therefore no isolated points of the spectrum for constant collision frequency models; the latter points can occur for more general models, such as Eq. (8.2) with $k < 0$ in the ranges $-\infty < u < -k$, $k < u < \infty$ (see Refs. [30, 31] and Section 9).

It is an easy guess to anticipate that the elementary solutions have partial range completeness. This circumstance can be easily proved by indirect methods [25], but the straightforward demonstration procedure which proves useful for the case of one equation cannot be extended to the case of a system. The main problem is to find a closed form for a certain matrix (the analog of the function $P(u)$ in the scalar case), which plays a fundamental role in the analytical process of solution. A brilliant but unsuccessful attempt to overcome this difficulty was made by Darrozès [32] in the case of the BGK model; in this case conservation of mass allows us to get rid of one of the three equations and we are led to problems involving two-by-two matrices. Darrozès suggested that the Hilbert problem related to the half-range completeness proof should be diagonalized; this is possible but the diagonalization introduces additional singularities in the complex plane, leading to a difficult problem, which Darrozès was not able to master. A solution to this difficulty has been given in a very recent paper by the author [33], where it is shown that, in order to solve a certain class of systems of singular integral equations, it is useful to make use of concepts from the theory of the integrals of algebraic functions. We shall not enter into the details of the method used in this paper, because they would lead us a little too far. We only remark that these methods would also apply to multigroup theory, which arises in neutron transport when the neutrons are divided in groups of different energy (rather than using a continuous variable for the speed) [2, 3, 16].

9. Some special cases

The results of Section 8 imply that shear flow problems with the models employed in Sections 2–7 (in particular, the BGK model, if ν = const.) are very special in the sense that certain simple analytical tools can be used to a larger extent than for more general problems involving models. We can, however, consider some other cases which allow the analytical machinery to be developed to the same extent without having recourse to the more complicated ideas mentioned at the end of Section 8.

A first class of cases arises from the simplifications of the BGK model: we can drop conservation of energy for the purpose of studying the so-called isothermal waves [34, 35], or we can conserve energy but allow only one-dimensional collisions [36] or, finally, we can decouple one of the three degrees of freedom of the molecules from the remaining two [25]. All these modifications allow a simplification of the equations in such a way that we can reduce the problem to solving a single equation instead of a system.

Another interesting case is offered by the ES model (Eq. (IV.9.8) with $N = 9$). If we consider shear flows depending on a single space variable, the only difference from the BGK model is the appearance of an integral proportional to the shearing stress, but by using conservation of momentum we can eliminate it in favour of the integral proportional to mass velocity and thereby obtain an equation very similar to the BGK model. The situation is very simple in steady problems [37] and we can express the solution of shear problem with the ES model in terms of the analogous solution with the BGK model. In particular, we can show that the slip coefficient according to the ES model has exactly the same value as the BGK one.

Another special case is offered by the model used in Section 2, Eq. (2.2) when the collision frequency is proportional to speed, $\nu(\xi) = \sigma\xi$ (σ = const.). In this case the heat transfer equation, Eq. (2.5), can be reduced to equations involving a single moment, as was shown by Cassel and Williams [31] by solving a specific half-space problem by the Wiener-Hopf technique. To show this, let us take σ^{-1} as the length unit, $(2RT_0)^{\frac{1}{2}}$ as the velocity unit and consider Eq. (2.5) for this case:

$$\mu\frac{\partial h_1}{\partial x} + h_1 = \frac{\pi^{\frac{1}{2}}}{2}\int f_0'\xi' h_1'\, d\boldsymbol{\xi}' + \tfrac{3}{4}\pi^{\frac{1}{2}}\xi_1\int \xi_1'\xi' f_0' h_1'\, d\boldsymbol{\xi}' + \\ + \frac{\pi^{\frac{1}{2}}}{4}(\xi^2 - 2)\int f_0'\xi'(\xi'^2 - 2)h_1'\, d\boldsymbol{\xi}' \quad (9.1)$$

where $\mu = \xi_1/\xi$ and $f_0 = \pi^{-\frac{3}{2}}\exp(-\xi^2)$. It is convenient to use ξ, μ and φ

(the polar angle in the (ξ_2, ξ_3) plane) as variables in velocity space. Let $\mathscr{I}$ be the Hilbert space of the functions of ξ and φ, where the scalar product is defined as follows:

$$(g, h)_{\mathscr{I}} = \int f_0 g h \xi^3 \, d\xi \, d\varphi \tag{9.2}$$

and let us split h_1 as follows

$$h_1 = Y + Y_\perp \tag{9.3}$$

where Y is in the subspace of $\mathscr{I}$ spanned by 1, ξ, $\xi^2 - 2$, with $Y_\perp$ in the orthogonal complement to such a subspace.

Then:

$$\mu \frac{\partial Y}{\partial x} + Y = \frac{\pi^{\frac{1}{2}}}{2} \int f_0' Y' \xi'^3 \, d\xi' \, d\varphi' \, d\mu' + \tfrac{3}{4}\pi^{\frac{1}{2}} \xi \mu \int f_0' Y' \mu' \xi'^4 \, d\xi' \, d\varphi + \\ + \frac{\pi^{\frac{1}{2}}}{4} (\xi^2 - 2) \int f_0' Y' (\xi'^2 - 2) \xi_1'^3 \, d\xi' \, d\varphi' \, d\mu' \tag{9.4}$$

$$\mu \frac{\partial Y_\perp}{\partial x} + Y_\perp = 0 \tag{9.5}$$

The last equation is trivially solvable. In order to solve Eq. (9.4) we remark that, by definition, Y can be written as follows:

$$Y = Y_0(x, \mu) + \xi Y_1(\xi, \mu) + (\xi^2 - 2) Y_2(x, \mu) \tag{9.6}$$

Substituting Eq. (9.6) into Eq. (9.4) and equating the coefficients of 1, ξ, $\xi^2 - 2$, we obtain:

$$\mu \frac{\partial Y_0}{\partial x} + Y_0 = \tfrac{1}{2} \int_{-1}^{1} Y_0(x, \mu') \, d\mu' + \tfrac{3}{8}\sqrt{\pi} \int_{-1}^{1} Y_1(x, \mu') \, d\mu' \tag{9.7}$$

$$\mu \frac{\partial Y_1}{\partial x} + Y_1 = \frac{9}{16} \sqrt{\pi}\, \mu \int_{-1}^{1} \mu' Y_0(x, \mu') \, d\mu' + \\ + \mu \int_{-1}^{1} \mu' Y_1(x, \mu') \, d\mu' + \frac{9}{32} \sqrt{\pi}\, \mu \int_{-1}^{1} \mu' Y_2' \, d\mu' \tag{9.8}$$

$$\mu \frac{\partial Y_2}{\partial x} + Y_2 = \tfrac{3}{32}\sqrt{\pi} \int_{-1}^{1} Y_1(x, \mu') \, d\mu' + \tfrac{1}{2} \int_{-1}^{1} Y_2(x, \mu') \, d\mu' \tag{9.9}$$

Eqs. (9.7)–(9.9) give, by integration with respect to μ

$$\frac{d}{dx} \int_{-1}^{1} \mu Y_0 \, d\mu = \tfrac{3}{4}\sqrt{\pi} \int_{-1}^{1} Y_1 \, d\mu$$

$$\frac{d}{dx} \int_{-1}^{1} \mu Y_1 \, d\mu + \int_{-1}^{1} Y_1 \, d\mu = 0 \tag{9.10}$$

$$\frac{d}{dx} \int_{-1}^{1} \mu Y_2 \, d\mu = \tfrac{3}{16}\sqrt{\pi} \int_{-1}^{1} Y_1 \, d\mu$$

Using the second of these equations to eliminate $\int_{-1}^{1} Y_1\,d\mu$ from the remaining two equations, we obtain:

$$\frac{d}{dx}\left[\int_{-1}^{1} Y_0\,d\mu + \tfrac{3}{4}\sqrt{\pi}\int_{-1}^{1}\mu Y_1\,d\mu\right] = 0$$
$$\frac{d}{dx}\left[\int_{-1}^{1}\mu Y_2\,d\mu + \tfrac{3}{16}\sqrt{\pi}\int_{-1}^{1}\mu Y_1\,d\mu\right] = 0 \tag{9.11}$$

or

$$\int_{-1}^{1}\mu Y_0\,d\mu + \tfrac{3}{4}\sqrt{\pi}\int_{-1}^{1}\mu Y_1\,d\mu = A$$
$$\int_{-1}^{1}\mu Y_2\,d\mu + \tfrac{3}{16}\sqrt{\pi}\int_{-1}^{1}\mu Y_1\,d\mu = B \tag{9.12}$$

where A and B are constants. Eqs. (9.12) can be used to eliminate Y_0 and Y_2 from Eqs. (9.8) with the following result

$$\mu\frac{\partial Y_1}{\partial x} + Y_1 = \tfrac{3}{2}(1 - \tfrac{81}{56}\pi)\mu\int_{-1}^{1}\mu' Y_1(x,\mu')\,d\mu' + \tfrac{9}{16}\sqrt{\pi}\left(A + \frac{B}{2}\right)\mu \tag{9.13}$$

If we let

$$c = (1 - \tfrac{81}{56}\pi) \simeq 0.006 \tag{9.14}$$

$$Y_1(x,\mu) = \tfrac{9}{16}\frac{\sqrt{\pi}}{1-c}\left(A + \frac{B}{2}\right)\mu + \mu Z_1(x,\mu) \tag{9.15}$$

we obtain

$$\mu\frac{\partial Z_1}{\partial x} + Z_1 = \frac{3c}{2}\int_{-1}^{1}\mu'^2 Z_1(x,\mu')\,d\mu' \tag{9.16}$$

Once this equation is solved, Y_1 is obtained from Eq. (9.15) and can be inserted into Eqs. (9.7) and (9.9). Eqs. (9.16), (9.7) and (9.9) can be solved by the methods used in the previous sections. We remark that Eq. (9.16) does not possess a conservation equation and can be compared with Eq. (3.11); since, according to Eq. (9.14), $c < 1$, Eq. (9.16) is expected to possess two real discrete eigenvalues outside the interval $(-1, 1)$ (see remarks in Section 3). In fact, if $u \notin (-1, 1)$ there is a solution $e^{-x/u} g_u(\mu)$ provided

$$\frac{3c}{2}\int_{-1}^{1}\frac{\mu^2 u}{u-\mu}\,d\mu = 1 \tag{9.17}$$

or

$$3c\mu^2\left[\frac{u}{2}\log\left(\frac{u+1}{u-1}\right) - 1\right] = 1 \tag{9.18}$$

Since c is rather small, the roots can be expected to be very close to ± 1. As a matter of fact, if we let $u = \pm(1 + \varepsilon)$ and neglect terms of higher order in ε, we obtain

$$-\frac{3c}{2}\log\frac{\varepsilon}{2} \simeq 1 \tag{9.19}$$

or

$$\varepsilon \simeq 2e^{-\frac{2}{3c}} \simeq 2e^{-111.1} \simeq 3.10^{-40} \tag{9.20}$$

which is a very small number indeed.

Since Y_1, Y_0, Y_2 can be found by means of uncoupled equations, the problems involving a half-space can be solved analytically. In particular, the problem of determining the temperature jump coefficient (see Chapter V, Section 5) has been solved by Cassel and Williams [31] by means of the equivalent method of Wiener-Hopf. The result for the temperature jump coefficient is

$$\tau = \left(\nu + \frac{9\sqrt{3\pi}}{128}\,U_0\right)\bar{l} \tag{9.21}$$

where ν is the ratio of extrapolation length to the mean free path $\bar{l}$ for one-speed transport in the presence of isotropic scattering, Eq. (4.22), while U_0 is given by a similar, though more complicated expression [31]. If we take into account that c in Eq. (9.16) is small, we can simplify the latter expression to yield $U_0 = \nu/\sqrt{3}$ and

$$\tau = \frac{15}{8}\left(1 + \frac{9\sqrt{\pi}}{128}\right)\nu l/Pr \simeq 1.498(l/Pr) \tag{9.22}$$

where l is the mean free path defined by Eq. (V.1.3) and Pr is the Prandtl number ($\Pr \simeq \frac{2}{3}$ for a correct model of a monatomic gas). If the above-mentioned simplification is not introduced, a correction of order 0.15% results [31].

10. Unsteady solutions of kinetic models with velocity dependent collision frequency

In this section we consider the possibility of extending the treatment in Section 6 to the case of variable collision frequency. Eq. (6.1) is now replaced by

$$\frac{\partial Y}{\partial t} + \xi_1\frac{\partial Y}{\partial x} + \nu(\xi)Y = \nu(\xi)\int g_0(\boldsymbol{\xi}')Y(x, \boldsymbol{\xi}')\,d\boldsymbol{\xi}' \tag{10.1}$$

where $h = \psi_3 Y$ is the perturbation of the distribution function (see Eq. (2.7)) and $g_0(\boldsymbol{\xi})$ is defined by Eq. (3.2). The collision frequency is assumed to depend on the molecular speed ξ.

If we take the Laplace transform of Eq. (10.1) and disregard, as usual, a possible inhomogeneous term related to the initial datum, we have

$$[\nu(\xi) + s]\tilde{Y} + \xi_1 \frac{\partial \tilde{Y}}{\partial x} = \nu(\xi)\int g_0(\boldsymbol{\xi}')\tilde{Y}(x, \boldsymbol{\xi}')\, d\boldsymbol{\xi}' \tag{10.2}$$

where $\tilde{Y}$ is the Laplace transform of Y. The same equation (with $s = i\omega$) governs the state of a gas forced to undergo steady transverse oscillations with frequency ω. If s is real, then the treatment is very similar to the steady case; it is sufficient to introduce a variable w related to ξ_1 and ξ by Eq. (3.4) with $\nu(\xi) + s$ in place of $\nu(\xi)$. If s is complex, however, w turns out to be complex:

$$w = \alpha + i\beta = \frac{\xi_1}{\nu(\xi) + s} \tag{10.3}$$

For any given complex s, when ξ_1 and $\rho = \sqrt{\xi_2^2 + \xi_3^2}$ range from $-\infty$ to ∞ and from 0 to ∞ respectively, w covers a region $G(s)$ of the plane (α, β). This region reduces to a segment of the real axis for real values of s, but is a two-dimensional region when $\text{Im } s \neq 0$ and $\partial\nu/\partial\xi \neq 0$. It is useful to introduce

$$Z(x, w) = \frac{\nu(\xi) + s}{2\pi\nu(\xi)} \int_0^{2\pi} \tilde{Y}\, d\varphi \tag{10.4}$$

where φ is the polar angle in the (ξ_2, ξ_3) plane and the dependence of Z on s is not explicitly indicated in order to simplify the notation. It is also useful to remark that the notation $Z(x, w)$ does not mean that Z is an analytic function of the complex variable w: it is only a shorthand notation for $Z(x, \alpha, \beta)$. In terms of the new variables, Eq. (10.2) can be written as follows:

$$Z + w\frac{\partial Z}{\partial x} = \iint_{G(s)} d\alpha_1\, d\beta_1 \Phi(w_1) Z(x, w_1) \tag{10.5}$$

Here

$$\Phi(w) = \frac{|\nu(\xi) + s|^4 \rho^2 \xi}{|\text{Im } s|\, |\xi_1|\, |\nu'(\xi)|\, \bar{\nu}} e^{-\xi^2} \tag{10.6}$$

where

$$\bar{\nu} = \tfrac{4}{3}\int_0^\infty \nu(\xi) e^{-\xi^2} \xi^4\, d\xi \tag{10.7}$$

and ξ, ξ_1 are to be replaced by their expressions in terms of α and β as obtained from Eq. (10.3). Separating the variables in Eq. (10.5) gives

$$Z_u(x, w) = g_u(w)\exp(-x/u) \tag{10.8}$$

where $u = \alpha_0 + i\beta_0$ is the separation parameter and $g_u(w)$ satisfies:

$$(1 - w/u)g_u(w) = \iint_{G(s)} d\alpha_1\, d\beta_1 \Phi(w_1) g_u(w_1) \tag{10.9}$$

The right hand side does not depend on w and can be normalized to unity. Accordingly we are led to a typical division problem in complete analogy with the previous cases. If the factor $(u - w)/u$ cannot be zero, $g_u(w)$ is an ordinary function given by

$$g_u(w) = \frac{u}{u - w} \tag{10.10}$$

with the normalization condition

$$\iint_{G(s)} d\alpha\, d\beta \Phi(w) \frac{u}{u - w} = 1 \tag{10.11}$$

If, on the contrary, the factor $(u - w)/u$ is zero for some w, $g_u(w)$ must be allowed to be a generalized function. This occurs if $u \in G(s)$. We must then allow for a delta-like term which becomes zero when multiplied by $(u - w)$. If we write $\delta(u - w)$ for $\delta(\alpha - \alpha_0)\, \delta(\beta - \beta_0)$ we have

$$g_u(w) = \frac{u}{u - w} + p(u; s)\, \delta(u - w) \tag{10.12}$$

where $p(u; s)$ is given by

$$p(u; s) = \left[1 - \iint_{G(s)} d\alpha\, d\beta \Phi(w) \frac{u}{u - w}\right] \Big/ \Phi(u) \tag{10.13}$$

Eq. (10.12) gives the eigensolutions corresponding to the continuous spectrum. We note that in this case $u/(u - w)$ does not need to be interpreted through the "principal part" concept, since double integrals with a first order pole at a point of the integration domain exist in the ordinary sense.

Concerning the discrete spectrum, a discussion of Eq. (10.11) [38] shows that the solution is qualitatively the same as for the BGK model; that is, a curve γ exists such that there are two points $\pm u_0$ of the discrete spectrum when s is inside and none when γ is outside. We can obtain an implicit parameter representation of γ and show that γ is symmetric with respect to the real axis. γ has two points in common with the real axis: the abscissa of one of them is $-\nu(0)$, while the other one has a positive abscissa (0 in the limiting case $k = \infty$).

The general solution of Eq. (10.5) can now be written as follows:

$$Z(x, w) = \iint_{G(s)} \frac{A(u; s)u}{u - w}\, d\alpha_0\, d\beta_0 + p(w; s)A(w; s)e^{-x/w} + \\ + A_+ \frac{u_0}{u_0 - w} e^{-x/u_0} + A_- \frac{u_0}{u_0 + w} e^{x/u_0} \quad (10.14)$$

where $\pm u_0(s)$ are the possible eigenvalues of the discrete spectrum. $A_\pm$ are arbitrary coefficients ($= 0$ if no discrete spectrum exists) and $A(u; s)$ is an "arbitrary function".

The main difference between Eq. (10.12) and the general solutions of the models considered in previous sections, is that here we have a double integral (which exists in the ordinary sense) in place of a simple integral (of the Cauchy type). A disadvantage of the present situation is that no standard theory exists for equations having the complex Cauchy kernel $(u - w)^{-1}$ and involving two-dimensional integrations; such a theory is needed to prove the theorems of completeness and orthogonality in a constructive fashion. It is possible, however, to construct such a theory [38] by using some results from the theory of generalized analytic functions [39]. In general, if $w = \alpha + i\beta$ is a complex variable, a generalized analytic function $f = \varphi + i\psi$ is a complex function of α and β which satisfies

$$\frac{\partial f}{\partial \bar{w}} + g_1(w)f(w) + g_2(w)\bar{f}(w) = h(w) \qquad (w \in G) \qquad (10.15)$$

where the bar denotes complex conjugation, g_1, g_2, h are given functions of α and β and

$$\frac{\partial f}{\partial \bar{w}} = \frac{1}{2}\left(\frac{\partial f}{\partial \alpha} + i\frac{\partial f}{\partial \beta}\right) \qquad (10.16)$$

The latter definition is meaningful if f is differentiable with respect to α and β; otherwise, $\partial/\partial\bar{w}$ is to be understood as the Sobolev generalized derivative or the Pompeju areolar derivative [39]. It is obvious that any differentiable function of α and β can be made to satisfy Eq. (10.15) by a suitable choice of g_1, g_2, h, but this approach is useless; generalized analytic functions are useful when we have a whole class of functions which satisfy Eq. (10.15) with fixed g_1, g_2 (h can vary). The name "generalized analytic function" obviously comes from the fact that when $g_1 = g_2 = h = 0$ we obtain the Cauchy-Riemann equations for the analytic function $f(w) = \varphi(\alpha, \beta) + i\psi(\alpha, \beta)$. From any integrable function of α and β ($\alpha, \beta \in \bar{G}$, $\bar{G}$ being the closure of G), $h(w)$, we can immediately construct a generalized analytic function satisfying Eq. (10.15) with $g_1 = g_2 = 0$. This function is given by

$$f = T_G h = -\frac{1}{\pi} \iint_G \frac{h(u)}{u - w}\, d\alpha_0\, d\beta_0 \qquad (w \in G) \qquad (10.17)$$

where $u = \alpha_0 + i\beta_0$ is an integration variable. In order to prove the statement we observe that

$$\frac{\partial}{\partial \bar{w}} \frac{1}{(u - w)} = -\pi\delta(u - w) \tag{10.18}$$

Eq. (10.18) is immediately obtained by remarking that if $v_1 - iv_2 = (u - w)^{-1}$, the irrotational vector field $\mathbf{v} = (v_1, v_2)$ corresponds to a unit point source in the (α, β) plane after which we recall Eq. (10.16).

Eqs. (10.17) and (10.18) imply (for a less formal derivation see Vekua's book [39]):

$$\frac{\partial f}{\partial \bar{w}} = \iint h(u)\, \delta(u - w)\, d\alpha_0\, d\beta_0 = h(w) \qquad (w \in G) \tag{10.19}$$

which is Eq. (10.15) with $g_1 = g_2 = 0$. We note that when $w \notin G$ then f is an analytic function of w which tends to zero when $w \to \infty$. This proves that for any integrable h, $T_G h$ is analytic outside G, tends to zero at infinity and:

$$\frac{\partial}{\partial \bar{w}} T_G h = h \tag{10.20}$$

If, *vice versa*, $\partial f/\partial \bar{w} = h$ in G, f is analytic outside G and tends to zero when $w \to \infty$, then $f = T_G h$. This follows because $f - T_G h$ is analytic everywhere and is zero at infinity, which implies its vanishing according to Liouville's theorem. These results allow us to solve Eq. (10.15) by quadratures in the case that $g_2(w) = 0$ [38, 39]. On the other hand, if we want to prove full range or partial range completeness, we have to solve equations of the following kind:

$$p(w)A(w) - \pi T_H[wA(w)] = Z(w) \tag{10.21}$$

where T_H is the operator defined by Eq. (10.17) (with H in place of G) and $Z(w)$ is given. The domain H can be either the whole region $G(s)$ or a subset (typically one half of G, corresponding to $\mathrm{Re}\, w \geqslant 0$). It is clear that if we put

$$f(w) = T_H[wA(w)] \tag{10.22}$$

Eq. (10.21) can be written as follows:

$$p(w)\frac{\partial f}{\partial \bar{w}} - \pi w f = wZ(w) \qquad (w \in H) \tag{10.23}$$

which is an equation of the type (10.15) with $g_2(w) = 0$ and H in place of G. It is obvious now that this equation can be solved analytically. In fact the general solution of the homogeneous equation is given by $[X(w)]^{-1}$ where

$$X(w) = \psi(w)\exp\left\{-T_H\left[\frac{\pi w}{p(w)}\right]\right\} \tag{10.24}$$

where $\psi(w)$ is an *analytic* function in H. The general solution of Eq. (10.23) is then given by

$$f = [X(w)]^{-1}T_H\left[X(w)\frac{wZ(w)}{p(w)}\right] + \varphi(w)[X(w)]^{-1} \tag{10.25}$$

where $\varphi(w)$ is analytic in H. Eq. (10.25) follows from the results on Eq. (10.20) and the fact that Eq. (10.23) can be written as follows

$$\frac{\partial}{\partial \bar{w}}\left[X(w)f\right] = X(w)\frac{wZ(w)}{p(w)} \tag{10.26}$$

The analytic function $\psi(w)$ in Eq. (10.24) can be fixed once and for all, while $\varphi(u)$ is to be determined in such a way as to have f analytic outside $\bar{H}$ and vanishing for $w \to \infty$ as implied by Eq. (10.22). Analyticity can be obtained, as Eq. (10.25) shows, by taking $\varphi(w) = 0$ provided $X(w)$ is analytic, nonzero and bounded outside H, and no other solution can exist according to the general results about Eq. (10.20) (applied to Eq. (10.26)). At the boundary ∂H, however, $X(w)$ must be such that $A(w)$ is not too singular when $f(w)$ is regular ($wA(w)$ must be integrable). In order to investigate this condition, we note that the exponent in Eq. (10.24) can be singular only at the points of H where $p(w)$ is zero, so that according to Eq. (10.13)

$$L(w) \equiv 1 - \pi w T_G \Phi = 0 \qquad (w \in H) \tag{10.27}$$

If there is a discrete spectrum, then $L(w)$ is zero at $w = \pm u_0$; as a consequence $\log L(w)$ changes by $-4\pi i$ when w encircles the boundary ∂G in the positive direction. Hence if there is a discrete spectrum, $L(w)$ must vanish at some point in G as well; otherwise $\log L(w)$ would be single-valued and a contradiction would arise. If u_1 is such a zero of $L(w)$, then

$$L(w) = \left(\frac{\partial L}{\partial w}\right)_1 (w - u_1) + \left(\frac{\partial L}{\partial \bar{w}}\right)_1 (\bar{w} - \bar{u}_1) + O(|w - u_1|^2) \tag{10.28}$$

where $\partial L/\partial \bar{w} = -\pi w \Phi \neq 0$ in H. When w encircles a small contour enclosing u_1 then the variation of $\log L(w)$ is given by

$$\Delta \log L(w) = \Delta \log\left[\left|\left(\frac{\partial L}{\partial w}\right)_1\right| e^{i(\varphi+\alpha)} + \left|\left(\frac{\partial L}{\partial \bar{w}}\right)_1\right| e^{-i(\varphi-\beta)}\right]$$
$$\left(\alpha = \arg\left(\frac{\partial L}{\partial w}\right)_1; \beta = \arg\left(\frac{\partial L}{\partial \bar{w}}\right)_1\right) \tag{10.29}$$

and hence equals $2\pi i$, 0, $-2\pi i$ according to whether $|(\partial L/\partial w)_1| > |(\partial L/\partial \bar{w})_1|$, $|(\partial L/\partial w)_1| = |(\partial L/\partial \bar{w})_1|$, $|(\partial L/\partial w)_1| < |(\partial L/\partial \bar{w})_1|$, respectively. Following Klinç and Kuščer [40] who first pointed out the existence of zeroes of $L(w)$ in H in an explicit fashion, we shall call u_1 a normal zero, a neutral zero or an antizero, respectively. Multiple zeroes are excluded because $\partial L/\partial \bar{w} \neq 0$; vanishing upon a line can occur only if $|(\partial L/\partial w)| = |(\partial L/\partial \bar{w})|$ upon such a line.

If only isolated zeroes exists then the existence of a discrete spectrum implies that each half of G (by symmetry) contains one more antizero than normal zeroes. One can conjecture [40] that there is, in each half of G, just one antizero, and no normal zeroes, neutral zeroes and zero lines. It is to be expected that when s varies in such a way that $\pm w_0$ approach ∂H from outside, simultaneously the antizeroes $\pm u_1$ will approach the boundary from inside. Eventually they will merge there with the zeroes $\pm u_0$ and "annihilate" thereafter (see Fig. 39). This conjecture has so far been verified only by numerical calculation in particular cases [40] and will be adopted in the following. For simplicity, we shall restrict ourselves to the case when H is exactly one half of G (Re $w \geqslant 0$).

We can use now the complex form of the Greens' formula [39] to obtain

$$\begin{aligned} T_H\left[\frac{\pi w}{p(w)}\right] &= -\log L(w) + \frac{1}{2\pi i}\int_{\partial H+\Gamma}\frac{\log L(z)}{z-w}\,dz \qquad (w \in H) \\ T_H\left[\frac{\pi w}{p(w)}\right] &= \frac{1}{2\pi i}\int_{\partial H+\Gamma}\frac{\log L(z)}{z-w}\,dz \qquad (w \notin H) \end{aligned} \tag{10.30}$$

where Γ is a loop surrounding a cut C connecting u_1 with a point of ∂H, whenever u_1 exists ($\log[L(w)]$ is single valued in $H - C$) and the contour is described counterclockwise. As a consequence, when $w \to u_1$, taking into

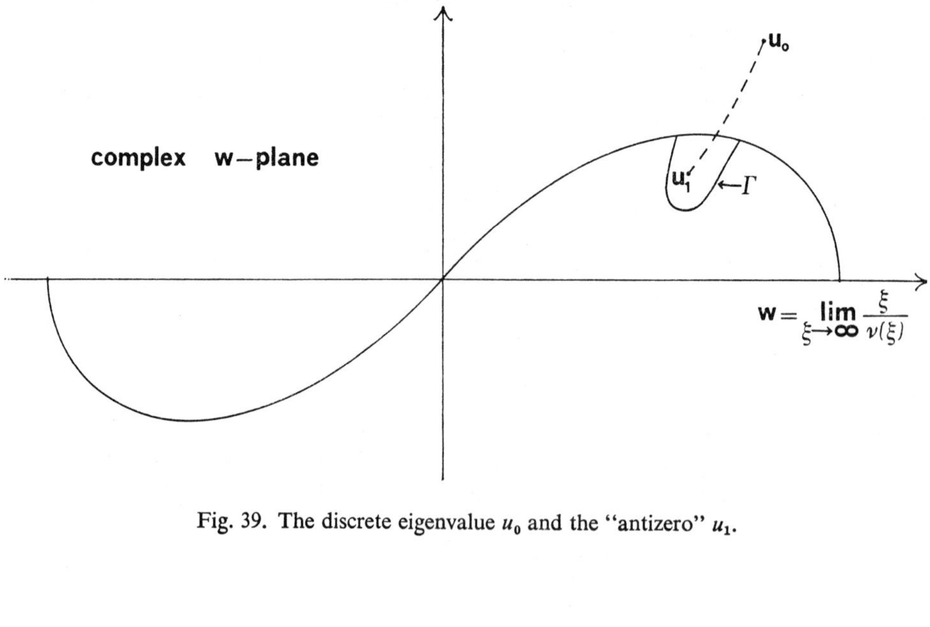

Fig. 39. The discrete eigenvalue u_0 and the "antizero" u_1.

account the fact that u_1 is an antizero and applying well-known results on singular Cauchy integrals [8; see Appendix, Eq. (A.5)] gives:

$$T_H\left[\frac{\pi w}{p(w)}\right] = \begin{cases} O(1) & \text{if no discrete spectrum exists} \\ -2\log|u_1 - w| + O(1) & \text{if a discrete spectrum exists} \end{cases} \tag{10.31}$$

Accordingly we let $\psi(w) = (u_1 - w)^{-h}$ in Eq. (10.24), where $h = 1$ or 0 depending upon whether a discrete spectrum exists or not. In this way $X(w)$ is bounded everywhere; although $X(v_1) = 0$ when the discrete spectrum exists, $[X(w)]^{-1}$ is integrable. The solution $f(w)$ of Eq. (10.23) is then given by

$$f(w) = [X(w)]^{-1}T_H\left[\frac{wX(w)Z(w)}{p(w)}\right] \tag{10.32}$$

According to Eq. (10.22), however, $f(w) \to 0$ when $w \to \infty$ thus Eq. (10.32) gives

$$\int_H \frac{wX(w)Z(w)}{p(w)}\,d\alpha\,d\beta = 0 \tag{10.33}$$

(if the discrete spectrum exists)

since $X(w) = (-w)^{-1} + O(1)$ for $w \to \infty$ when there is a discrete spectrum. Eq. (10.33) is a condition upon $Z(w)$; but this is very satisfactory because $Z(w)$ in Eq. (10.21) is the function to be expanded *minus* the term corresponding to the discrete spectrum and Eq. (10.33) fixes the coefficient A_0 of the discrete eigenfunction corresponding to u_0 ($\operatorname{Re} u_0 \geqslant 0$).

We recover $A(w)$ by differentiation according to Eqs. (10.22) and (10.20):

$$\begin{aligned} A(w) &= \frac{\partial}{\partial \bar{w}}\left\{[X(w)]^{-1}T_H\left[\frac{wX(w)Z(w)}{p(w)}\right]\right\} \\ &= \frac{wZ(w)}{p(w)} + \frac{\pi w}{p(w)X(w)}T_H\left[\frac{wX(w)Z(w)}{p(w)}\right] \end{aligned} \tag{10.34}$$

A difficulty arises from the zero u_1 of $p(w)$ and $X(w)$, which, in general, produces a second order pole in $A(w)$. Nevertheless the integral in the eigenfunction expansion exists in the principal value sense [40].

The above treatment shows that all the problems which can be solved by means of the BGK model can be solved with Eq. (10.1) as well. In this connection we remark that the function $X(u)$ satisfies several important identities, analogous to those holding in the case of the BGK model [38, 41]. We note also that it can be convenient, once we have established the basic formulas, to return to the original variables ξ_1 and ξ rather than $w = \alpha + i\beta$.

It goes without saying that the method described above can be applied to models describing neutron transport. Several papers have been devoted to the treatment of neutron waves with this technique [40, 42–44].

11. Analytic continuation

The use of generalized analytic functions sketched in Section 10 leads to an interesting explicit treatment of a continuous spectrum filling a two-dimensional region. Continuous spectra, however, are not usually apt to provide well defined information about the results to expect from experiments. It may happen, in fact, that a well defined eigenvalue comes out from experimental data even when the theory predicts a continuous spectrum. A similar situation arose already in Section 7, where plane shear waves were treated with the BGK model. There, it was shown that discrete eigenvalues could be obtained by analytically continuing the relation which determines the discrete spectrum (the so called "dispersion relation"). For the model discussed in Section 10, the dispersion relation is given by Eq. (10.9) or:

$$L(u; s) = 0 \tag{11.1}$$

where

$$L(u; s) = 1 - \int_0^\infty d\xi \int_{-1}^1 d\mu \, \frac{\xi^2 u \rho(\xi)(1 - \mu^2)}{u[\nu(\xi) + s] - \xi\mu} \tag{11.2}$$

$$\rho(\xi) = \frac{1}{\bar{\nu}} [\nu(\xi)]^2 \xi e^{-\xi^2} \tag{11.3}$$

and $\bar{\nu}$ is given by Eq. (10.7). If we perform the integration with respect to μ in Eq. (11.2), we obtain

$$\int_{-1}^1 \frac{\xi^3 u(1 - \mu^2)\, d\mu}{u[\nu(\xi) + s] - \xi\mu}$$

$$= u\{\xi^2 - u^2[\nu(\xi) + s]^2\}\log\left\{\frac{[\nu(\xi) + s]u + \xi}{[\nu(\xi) + s]u - \xi}\right\} - 2u^2\xi[\nu(\xi) + s] \tag{11.4}$$

provided s is such that u does not belong to the continuous spectrum. Consequently Eq. (11.1) can be rewritten as follows

$$L(u; s) = 1 - \int_0^\infty u\rho(\xi)\{\xi^2 - u^2[\nu(\xi) + s]^2\}\log\left\{\frac{[\nu(\xi) + s] + \xi}{[\nu(\xi) + s] - \xi}\right\} d\xi \, +$$

$$+ \, 2u^2 \int_0^\infty \xi[\nu(\xi) + s]\rho(\xi)\, d\xi \tag{11.5}$$

The first integral can now be partially integrated by letting

$$R(\xi; u; s) = \int_\xi^\infty u\rho(\xi')\{\xi'^2 - u^2[\nu(\xi') + s]^2\}\, d\xi' \tag{11.6}$$

Eq. (11.5) can then be written as follows:

$$L(u;s) = 1 + \int_0^\infty R(\xi;u;s)\left\{\frac{u\nu'(\xi)+1}{u[\nu(\xi)+s]+\xi} - \frac{u\nu'(\xi)-1}{u[\nu(\xi)+s]-\xi}\right\} + \\ + 2u^2\int_0^\infty [\nu(\xi)+s]\rho(\xi)\,d\xi \quad (11.7)$$

The first integral in this equation is such that the method of analytic continuation with respect to u can be applied, provided the collision frequency is an analytic function of ξ (more generally, it can be piecewise analytic). In fact, if we let

$$w = \pm\frac{\xi}{\nu(\xi)+s} \quad (11.8)$$

the integrals to be evaluated can be written in the form:

$$I(u,s) = \int_\Delta \frac{Q(w,u,s)}{w-u}\,dw \quad (11.9)$$

where Δ is part of the boundary of the region G in the complex w-plane (see Fig. 40) and $Q(w, u, s)$ is an analytic function of w which is locally obtained by inverting Eq. (11.8) and substituting $\xi = \xi(w, s)$ in Eq. (11.7).

Eqs. (11.7) and (11.9) do not show any presence of a two-dimensional continuous spectrum. The singularities are distributed along Δ; that is to say a *part* of the boundary of the continuous spectrum. Hence Eq. (11.7) gives an analytic continuation of L into the continuous spectrum, so that $L(u; s)$

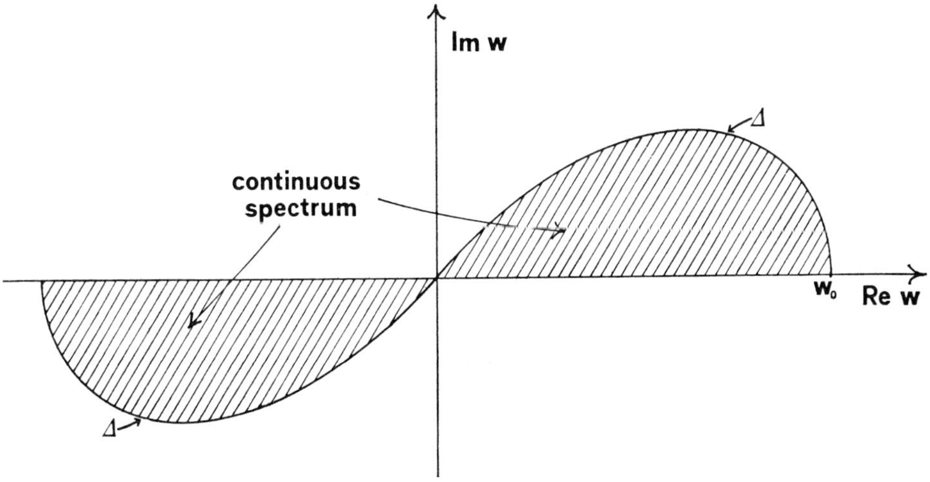

Fig. 40. The continuous spectrum and the curve Δ.

is now defined in the complex w plane with a cut along Δ. It is possible, however, to continue $L(u; s)$ across the cut by moving the path and subtracting $2\pi i Q(u, u, s)$ in agreement with Plemelj's formulas (see the Appendix).

In this fashion, we consider $L(u; s)$ as a many-valued function of w whose values are represented on a many-sheeted Riemann surface; the analytic continuation leads from the physical to another sheet (exactly as in the case of forced shear waves, treated with the BGK model in Section 7). Thus if $L_C(u; s)$ denotes the analytic continuation of $L(u; s)$ into the continuous spectrum (i.e. the branch of the many-valued $L(u; s)$ which is reached through Δ from the region outside the continuous spectrum), the equation $L_C(u; s) = 0$ may have a root u_0 even when there is no root of $L(u; s) = 0$ (where $L(u; s)$ is given by Eq. (11.2) even for $u \in G(s)$). In particular for s close to a critical value at which $u_0 = u_0(s)$ merges into the continuous spectrum, $L_C(u; s)$ will have a zero which is the analytic continuation of $u_0(s)$. If the solutions of a given boundary value problem are sufficiently smooth, we can try to use this circumstance to perform an analytic continuation of the integrand of the integral over the continuous spectrum and then use the result to shrink the contour of integration about points and lines of singularity of the analytically continued integrand.

Let us examine the process of analytic continuation for those half-space problems for which the distribution function of the molecules entering the half-space is given.

The function $X(w)$ defined by Eq. (10.24) (with H equal to one half of G) can be analytically continued by first transforming the integral in the exponent by means of the second Greens' formula, Eq. (10.30). Accordingly, since $\psi(w) = (u_1 - w)^{-h}$ ($h = 0, 1$), we obtain

$$X(w) = (u_1 - w)^{-h} \exp\left\{-\frac{1}{2\pi i}\int_{\partial H+\Gamma} \frac{\log[L(z)]}{z - w}\,dz\right\} \qquad w \notin H \quad (11.10)$$

where Γ is a loop about the cut C (see Fig. 39), which exists if and only if $h = 1$, and we omit showing the dependence upon s.

Let us transform the integral in the exponent. If w_0 denotes the intersection of the cut C with ∂H ($h = 1$), it is convenient to take w_0 to coincide with an extreme of Δ (see Fig. 40), as is always possible. Then:

$$\frac{1}{2\pi i}\int_{\partial H+\Gamma} \frac{\log L(z)}{z - w}\,dz = \frac{1}{2\pi i}\int_{\partial H} \frac{\log L(z)}{z - w}\,dz + h \log\left(\frac{w_0 - w}{u_1 - w_1}\right) \quad (11.11)$$

and Eq. (11.10) becomes

$$X(w) = (w_0 - w)^{-h} \exp\left\{-\frac{1}{2\pi i}\int_{\partial H} \frac{\log[L(z)]}{z - w}\,dz\right\} \quad (11.12)$$

If $L(z)$ is replaced by L_C we can squeeze the contour ∂H into Δ to obtain

$$X(w) = (w_0 - w)^{-h} \exp\left\{-\frac{1}{2\pi i}\int_\Delta \frac{\log[L_c^+/L_c^-]}{z - w}\,dz\right\} \qquad w \notin H \qquad (11.13)$$

If convenient, of course, Δ can now be transformed into another path Δ'. An analytic continuation of $X(w)$ is thus obtained. Note that Eq. (11.13) gives already an analytic continuation of $X(w)$ into $\bar{H} - \Delta$.

Eq. (10.33) shows that $wA(w)$ and hence also $wA(w)e^{-x/w}$ is the derivative with respect to $\bar{w}$ of a function computable in terms of $X(w)$ and the boundary data. Hence the contribution from the continuous spectrum can be transformed into a boundary integral by means of the complex Greens' formulas. The integrand is analytic outside H and can be analytically continued into H with the same technique as above if the boundary datum $Z(w)$ can be analytically continued into H. Thus analytic continuation can be performed for suitable boundary data. When moving the path of integration, however, we pick up residue contributions from the zeroes of the analytically continued dispersion relation. Hence even when Eq. (11.1) has no solution, the main contribution to the solution can arise from a "discrete eigenfunction".

It seems that the procedure of analytic continuation removes the two-dimensional continuous spectrum. This is true in the sense that the integral over G can be transformed into a boundary integral; the subsequent transformation into a cut integral is possible only if $Z(w)$ can be analytically continued into H. A typical boundary datum for the perturbation h (problem of shear waves, see Eq. (7.4)) leads to $Z(w) = $ const. and hence to the possibility of analytically continuing into H; in this case the two-dimensional continuous spectrum disappears and only poles and cut integrals contribute to the solution.

On the basis of these facts, it is frequently claimed [45–47] that the method of solution based on the use of generalized analytic functions is unnecessarily cumbersome and yields results in unwieldy form. This criticism is accompanied by a recommendation of the Wiener-Hopf technique [1–3] as a substitute for solving the half-space problems. While there is no doubt that any problem which can be solved by the Wiener-Hopf technique can also be solved by the method of elementary solutions and *vice versa*, it seems that choosing one method rather than another is a matter of personal taste. The author of this book, for example, sees an advantage in the method of elementary solutions because it leads immediately to the general solution of Eq. (10.5), valid even for problems which cannot be solved exactly by either method, while the Wiener-Hopf technique can lead to the same result only after complicated contour integration. Some, on the other hand, can see an advantage in the Wiener-Hopf method because it defines $L(u; s)$ for imaginary u only to begin with ($u = (ik)^{-1}$, where k is the variable corresponding to x

in the Fourier transform), while methods of generalized analytic functions use a $L(u; s)$ defined by Eq. (11.2) for all complex u (and hence nonanalytic in a finite area G of the complex plane). In either case, however, we have to resort to analytic continuation to put the results into a manageable form. In this connection we remark that while there is no doubt that the dispersion function $L(u; s)$ can be continued even for more complicated models and the linearized Boltzmann equation itself [48–50], the same statement is not obvious for the integrand of the integral over the continuous spectrum resulting from a given boundary value problem, because the possibility of analytic continuation depends on the smoothness of the boundary data.

To summarize, discarding the area spectrum is correct, provided the area is replaced by its own boundary: a further replacement of the latter by discrete terms plus an integral extended to a branch line different from ∂G is possible only if the boundary data are "smooth". We can claim that the latter condition is always satisfied for "physical" problems and this may be conceded; what cannot be conceded is that the area spectrum G (or, better, the boundary ∂G) is not needed in the *general* solution of Eq. (10.2). In other words, a denial of "reality" to the area spectrum, though mathematically unrigorous, might be reasonable for physical purposes.

12. Sound propagation in monatomic gases

One of the problems for which the theory expounded in Section 8 is useful, is the problem of sound propagation. A plate oscillates in the direction of its normal with frequency ω; a periodic disturbance propagates through the gas which fills the region at one side of the plate. If the frequency ω is very high, the Navier-Stokes equations are not good even at ordinary densities, because ω^{-1} can be of the order of the mean free time. It is possible to measure experimentally the phase speed and attenuation of the disturbance by assuming that the latter is locally a plane wave (it is not, in general, because there is the continuum contribution). One of the difficulties is that there is the receiver which, in principle, does not allow us to treat the problem as a half-space problem, especially because the receiver is usually kept very close to the plate (at a distance of a mean free path or less [51, 52]). If we disregard the disturbance produced by the receiver, then we can treat the problem as a half-space problem [53] by means of the method of elementary solutions. If we make the simple, but realistic, assumption that the molecules are completely diffused by the plate, the problem of computing the coefficients of the eigenfunctions discussed in Section 8 reduces to using the half-range completeness (the condition of boundedness at infinity is also to be used, of course). Now proofs of the latter do not yield useful expressions for the expansion coefficients unless use is made of the more sophisticated and recent approach described at the end of Section 8. Consequently, Buckner and Ferziger [53] solved the problem in an approximate fashion. They started from the remark that a boundary value problem may always be replaced by a suitable infinite space problem, by specifying a source distribution at the boundary. Buckner and Ferziger proposed to assign a source distribution containing adjustable parameters and thereafter choose the parameters in such a way as to minimize the solution in the half space $x < 0$ (where the exact solution is zero). Their results (see Figs. 41 and 42) are in good agreement with experiments for low and very high frequencies (continuum and free molecular limits); in the transition regime, the agreement is good as far as the phase speed is concerned but the attenuation coefficient is in error by about 30% with respect to the experimental data. This means that either the approximation to the source term chosen by Buckner and Ferziger is not sufficiently accurate, or the collision models considered by these authors (Gross and Jackson's models with 3 and 5 moments) are not appropriate, or both. This point will be discussed below.

A method which appears to have successfully fitted the full range of experimental data is the method of analytic continuation of dispersion laws used by Sirovich and Thurber [48]. In principle, the method was described in Sections 7 and 11 for the case of shear waves. In the latter case, however, the complete solution was available and the analytic continuation of the dispersion relation was only made with a view to obtaining a different representation of the solution. In particular (see, e.g., Eq. (7.10)), a discrete spectrum term can be exhibited even when the dispersion relation does not have a solution by a simple analytic continuation of the dispersion relation itself. The problem arises as to whether the contribution from the discrete spectrum (or its analytic continuation) gives an accurate representation of the

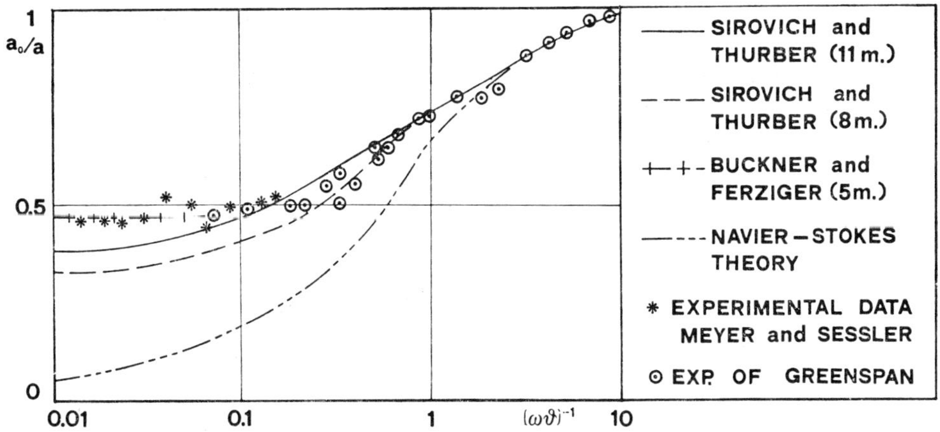

Fig. 41. Comparison of different theories of sound propagation with experiment. The inverse of the nondimensional phase speed is plotted versus the ratio of the collision frequency to the sound frequency.

whole solution. According to the estimates of Dorning and Thurber [27], there is a region where the contribution from the discrete spectrum is enough to represent the solution; we have to exclude the regions very close to and very far from the boundary. In the first region free molecular collisions prevail, in the second high speed molecules dominate and the distribution function is influenced by the high speed tail of the wall Maxwellian (see Section 7 and Refs. [36, 34, 25, 54]). The latter phenomenon occurs several mean free paths from the plate and is irrelevant to a region of the order of a mean free path; also, it would not occur for collision frequencies growing linearly for high speeds, as is the case for rigid spheres and potentials with radial cutoff. The free molecule flow region is much smaller than a mean free path; in fact, according to an estimate of Sirovich and Thurber [55], even if the emitter and the receiver are a distance apart of 1/10 of a mean free path, 25% of the molecules leaving the emitter would experience a collision before reaching the receiver. This high percentage of collisions is due to the fact that molecules have components of velocity parallel to the planes of the emitter and receiver.

The method of Sirovich and Thurber consists in bypassing the boundary value problem and relying upon the dispersion relation (or its analytic continuation). This means that we assume that the discrete spectrum terms are

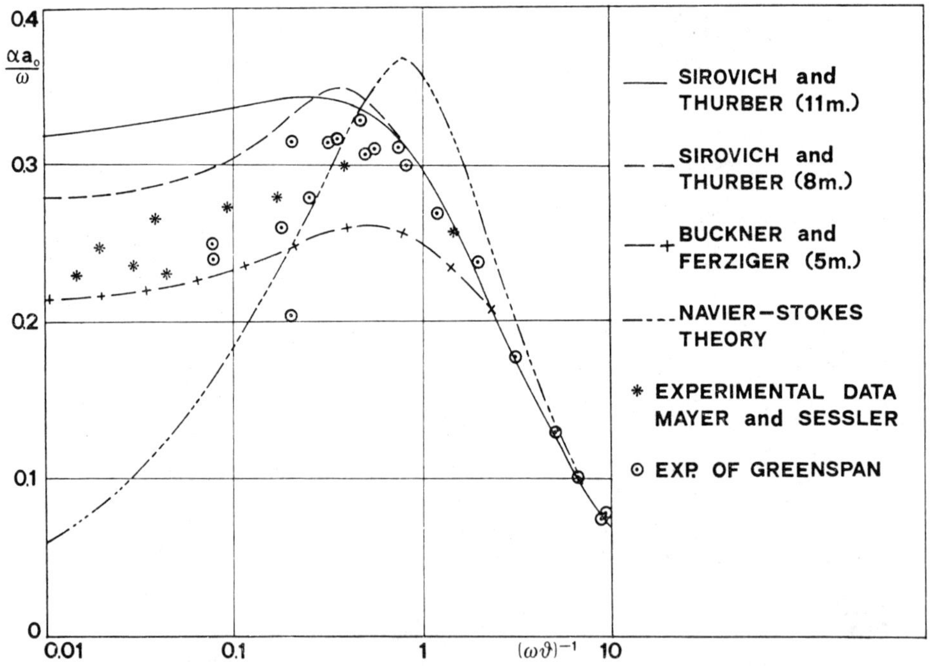

Fig. 42. Comparison of different theories of sound propagation with experiment. Attenuation rates are plotted versus the ratio of the collision frequency to the sound frequency.

largely dominant in the region relevant to experiments. According to the abovementioned estimates of Dorning and Thurber [27], this seems to be the case [56] for the experiments of Greenspan [51] and Meyer and Sessler [52].

The advantage of the method used by Sirovich and Thurber is that it requires only a study of the dispersion relation (Eq. (8.17) with $s = i\omega$) rather than the solution of a boundary value problem. Hence calculations can be carried out with fairly sophisticated models. Sirovich and Thurber [48] computed the sound speed and the attenuation rate for the Gross and Jackson models, Eq. (IV.9.8) with 3, 5, 8 and 11 moments (Maxwell's molecules) and the generalized models suggested by Sirovich, Eq. (IV.9.12) with 3, 5, 8 and 11 moments in the finite rank term of the linearized collision operator for rigid spheres. The results for the Maxwell gas (with 11 moments) are in qualitative agreement with experimental data (see Figs. 41 and 42); in particular, the behavior of the phase speed for $\omega \to \infty$ (free molecular flow) is affected by an error of order 15%, while the errors in the attenuation rate are of order 25% for $\omega\theta > 5$ (here θ is a mean free time defined by $\theta = \mu/p$; μ viscosity coefficient, p pressure). The results for the rigid sphere gas (11 moments) are in good agreement with experiments in the high frequency region ($\omega\theta > 5$) but the attenuation is in error by 20% in the transition regime.

Oddly enough, the results of Buckner and Ferziger for a model of the Maxwell gas (5 moments) are closer to the results of Sirovich and Thurber for rigid spheres than to those for the Maxwell gas. To make a precise assessment of the merits even more difficult, the results of Sirovich and Thurber for the 8-moment model of rigid spheres are closer to experiments than those for the corresponding 11 moment model.

An undisputable statement is that the only methods so far which have given a good fit with experiments are the method of elementary solutions as used by Buckner and Ferziger [53] and the method of analytic continuation as used by Sirovich and Thurber [48]. A more delicate question is to decide between the two methods. The agreement with experimental data is slightly in favor of the method of analytic continuation, according to the above discussion. It is to be remarked, however, that the method of elementary solutions should give the exact solution if carried out with a sufficiently sophisticated model (Buckner and Ferziger considered only low order models because of the cost of explicit calculations for a higher order model) and a sufficiently flexible source term (incidentally, due to an oversight, most of the calculations of Buckner and Ferziger were not performed with the approximate source term which would appear appropriate on physical grounds). The problem of the source term could be eliminated altogether by means of the method described at the end of Section 8, but very cumbersome analytic expression would certainly appear. We mention also an interesting paper by R. Mason [57], who considered sound propagation in the presence of a specularly reflecting emitter; in this case only full range completeness is to be used (by exploiting symmetry), and accordingly Mason was able to solve the problem analytically (he used the Fourier transform method rather than elementary solutions, however). His results are in worse agreement with experiments than those of Buckner and Ferziger and Sirovich and Thurber, but it is not clear whether this is due to the special form of the boundary condition (which would be an interesting result) or, more likely, to some approximations introduced by Mason in the numerical evaluation of his analytic results.

The remarkable success of the methods described above is to be particularly stressed because other methods, based on expanding the solution of the linearized Boltzmann equation into a series of orthogonal polynomials, have failed. A first method, used by Wang Chang and Uhlenbeck [58] and Pekeris and his coworkers [59], was to expand the solution into the eigenfunctions of the Maxwell operator; the results for the attenuation rate are in complete disagreement with the experiments. Since Pekeris and his coworkers used 483 moments (!), we conclude that their expansion, if convergent, does not converge to the correct solution for large values of ω. Another approach is due to Kahn and Mintzer [60]; in their solution the unknown in the linearized Boltzmann equation is expanded into a series of orthogonal polynomials whose weight function is based on a free-molecule solution rather than a Maxwellian. Unexpectedly, their results turned out to tend to the correct continuum limit. Because of this circumstance, the method of Kahn and Mintzer attracted considerable attention and favorable comment [53, 54, 61–66]. Subsequent work [67, 68] showed, however, that this unexpected accuracy is due to some mistakes made by Kahn and Mintzer; in fact Toba [67] pointed out an error in the boundary conditions and Hanson and Morse [68] a mistake in the asymptotic evaluation of certain integrals. Hanson and Morse recomputed the asymptotics correctly and found that the agreement with experiments did not improve; on the contrary, the behavior for low frequencies was completely wrong and even physically nonsensical (growing rather than damped modes). The agreement for very high frequencies is reasonably good, as was to be expected.

13. Two-dimensional and three-dimensional problems. Flow past solid bodies

As already noted (Chapter IV, Section 8), the method of separating the variables can be used not only for one-dimensional problems, but in general. The main difficulty is not to separate the variables and discuss the possible eigensolutions, but to single out a complete set and prove its completeness. In view of this fact, the only problems which seem amenable to a complete solution are full space problems or problems which can be reduced to the latter. In this case it is possible to work in terms of the Greens' function introduced in Chapter IV, Section 11. According to the results proved there, the Green' function is easily constructed once the elementary solutions are known (half-range completeness is not required). In view of the cumbersome form of the results, it is usually better to work directly in terms of the Fourier transform of the solution. If boundaries are present, the boundary values of the unknown h appear in the Fourier transformed Boltzmann equation in the form of a source term. Since h is not explicitly known at the boundary (in the simplest cases, it is known for $\boldsymbol{\xi} \cdot \mathbf{n} > 0$ but not for $\boldsymbol{\xi} \cdot \mathbf{n} < 0$) the problems in the presence of boundaries are not easily solved by this method. An exception is offered by specular reflection in an external problem (internal problems yield only trivial results for this kind of boundary condition, see Chapter III, Section 10). In this case, if the boundary is a flat plate in the (x, y) plane and the problem is symmetric with respect to the reflection $y \to -y$ then we can explicitly compute the source term in the Fourier transformed equation (see below).

In solving problems of flow past a body kept at rest and at a fixed temperature, a complication arises in connection with the use of the linearized Boltzmann equation [69]. It is the exact counterpart of the so-called Stokes paradox arising in the theory of linearized viscous flow, or Stokes flows [70]. If we linearize about the Maxwellian of the body f_0 in a two-dimensional flow, no solution bounded at infinity (except $h = 0$, i.e. $f = f_0$) exists. In order to prove this, we remark that h satisfies the linearized Boltzmann equation Eq. (IV.2.6) and the homogeneous boundary conditions at the body, Eq. (IV.4.11); hence $h = 0$ is a solution. If, in addition:

$$\iint_{\Sigma} \boldsymbol{\xi} \cdot \mathbf{n} h^2 \, d\boldsymbol{\xi} \, dS \to 0 \tag{13.1}$$

when the points of the surface Σ surrounding the body tend to infinity, then the only solution, according to the uniqueness theorem proved in Chapter IV, Section 4, is $h = 0$. Accordingly, if Eq. (13.1) is satisfied, a very unpleasant consequence follows: the only situation described by the steady linearized Boltzmann equation is a state of rest for the gas surrounding the body.

This result is related to the fact that the linearization is not uniformly valid at infinity; if we go sufficiently far from the body, the space derivatives in the Boltzmann equation are no longer larger than the neglected quadratic terms. This occurs at a distance l/M where l is the mean free path and M the Mach number ($M \ll 1$, in general, for the linearization to be valid). The problem, is however, to verify whether Eq. (13.1) holds or not at infinity; this question can be answered by examining the general solution of the linearized Boltzmann equation, discussed in Chapter IV, Section 11. In order to discuss the behavior of h at infinity, the asymptotic part of the

Greens' function is required; according to the results of Chapter IV, Section 11, the solution at infinity is always given by a truncated Chapman-Enskog expansion with velocity, pressure and temperature satisfying the steady linearized Navier-Stokes equations. It is not hard, then, to discuss whether Eq. (13.1) holds or not for solutions which tend to a linear combination of the collision invariants at infinity (the linearized version of tending to a Maxwellian). The result is that Eq. (13.1) is violated for three-dimensional flows (for which nontrivial solutions exists, as a consequence) but must hold true for two-dimensional flows. These results [69] are based on the fact that, in three dimensions, the Stokes solution is known to be well behaved at infinity, in the sense that the velocity profiles approach arbitrarily close (for sufficiently small Ma/l, a being a typical dimension of the body) to the conditions of a uniform stream before the linearization breaks down, whilst no such approach is possible for plane flows [70].

In order to avoid the unpleasant situation just described, it is necessary to imitate the procedures used in connection with the analogous situation in the theory of viscous flow [69]. We can either resort to inner-outer expansions or, alternatively, linearize about the Maxwellian at infinity (this is the equivalent of the Oseen linearization in viscous flow [70]). It is to be noted that in the second case we are led to a Boltzmann equation linearized about a nonzero velocity; when the problem is assumed to be steady, the latter equation is not equivalent to the Boltzmann equation linearized about a zero velocity, which we have used so far. In fact, we use a reference frame where the body is at rest, but linearize about the situation at infinity; if we try to unify the viewpoint by taking a reference frame at rest with respect to the flow at infinity, then the problem becomes unsteady. If we let $\boldsymbol{\xi} = \mathbf{V}_\infty + \mathbf{c}$ the Boltzmann equation to be used becomes

$$(\mathbf{V}_\infty + \mathbf{c}) \cdot \frac{\partial h}{\partial \mathbf{x}} = Lh \tag{13.2}$$

where L is linearized with respect to a Maxwellian with an average $\boldsymbol{\xi}$ equal to $\mathbf{V}_\infty$, and hence an average $\mathbf{c}$ equal to zero. In terms of $\mathbf{c}$ Eq. (13.2) is more similar to the unsteady linearized Boltzmann equation than to the steady one.

Eq. (13.2) can be discussed by analogy with the particular case $V_\infty = 0$ treated in Chapter IV, Section 11: the form of the eigensolutions is more complicated however. Thus, the general form of the asymptotic part of the solution is not easily obtained. Scharf [71] has obtained the terms up to second degree of the expansion of this asymptotic part into a power series in k. This, of course, is equivalent to a Chapman-Enskog solution truncated at the Navier Stokes level; accordingly the results which can be obtained by means of these solutions can also be obtained by means of the Navier-Stokes equations directly [72].

An application of Eq. (13.2) to moderately high Mach number flow past an almost specularly reflecting airfoil (Fig. 43) has been made by the author [73]. The distribution function is assumed to satisfy Maxwell's boundary conditions, Eq. (III.5.1) and the free stream velocity $\mathbf{V}_\infty$ to be directed along the x-axis. The accommodation coefficient α and the angle between the surface of the airfoil and the x-axis, $\varepsilon(x)$, are assumed to be small. To be precise, the following inequalities are assumed to be satisfied (M = Mach number of the free stream):

$$\alpha \ll 1; \qquad \alpha M \ll 1; \qquad \varepsilon \ll 1; \qquad \varepsilon M \ll 1 \tag{13.3}$$

If $M < 1$, of course, it is sufficient to assume the first and third of these conditions and if $M > 1$, the second and fourth. Under these assumptions, the problem can be linearized; if we let

$$f = f_\infty (1 + h) \tag{13.4}$$

where f_∞ is the free stream Maxwellian, h^2 will be negligible and h will satisfy Eq. (13.2) with $\mathbf{V}_\infty = V_\infty \mathbf{i}$ ($\mathbf{i}$ and $\mathbf{j}$ are the unit vectors along the x and y-axis, respectively). The boundary conditions, Eq. (III.5.1), take on the following form (terms of higher order are neglected):

$$h(\mathbf{x}, \mathbf{c}) - h(\mathbf{x}, \mathbf{c} - 2\mathbf{j}[\mathbf{j} \cdot \mathbf{c}]) = \alpha \frac{f_0(\mathbf{c}) - f_\infty(\mathbf{c})}{f_\infty(\mathbf{c})} - \frac{2V_\infty}{RT_0} \mathbf{c} \cdot \mathbf{j} \varepsilon_\pm(x)$$
$$(y = 0\pm; 0 < x < L; \pm\mathbf{j} \cdot \mathbf{c} > 0) \tag{13.5}$$

where $\varepsilon_+(x)$ and $\varepsilon_-(x)$ are the slopes of the upper and lower surface of the airfoil and L is the chord. It is to be remarked that, according to Eq. (13.5),

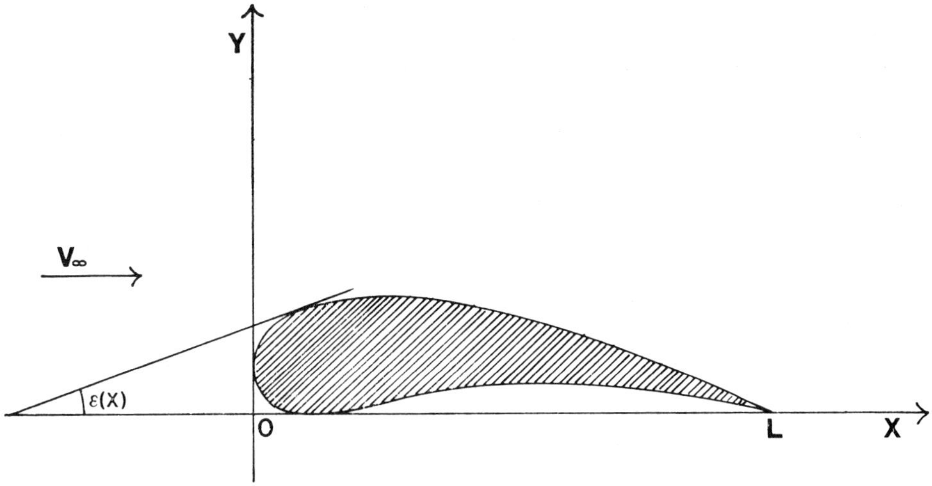

Fig. 43. Nomenclature for flow past a thin airfoil at a small angle of attack.

the boundary conditions are applied on segment $(0, L)$ of the x-axis and not on the boundary of the airfoil; this is of course connected with the smallness of $\varepsilon_{\pm}(x)$.

In order to make progress towards an analytical solution of the problem, it is expedient to assume the BGK model to describe collisions; then if $\lambda_\infty = \theta_\infty/\sqrt{2RT_\infty}$, where θ_∞ is the collision time at infinity, Eq. (11.2) becomes

$$S\frac{\partial h}{\partial x} + \mathbf{c}\cdot\frac{\partial h}{\partial \mathbf{x}} = \lambda_\infty^{-1}[r + 2\mathbf{c}\cdot\mathbf{u} + \tau(c^2 - \tfrac{3}{2}) - h] \tag{13.6}$$

where $S = V_\infty/\sqrt{2RT_\infty}$ is the speed ratio, $\mathbf{c}$ is measured in $\sqrt{2RT_\infty}$ units and

$$\begin{aligned} r &= \pi^{-\frac{3}{2}}\int e^{-c^2}h(\mathbf{x}, \mathbf{c})\, d\mathbf{c} \\ \mathbf{u} &= \pi^{-\frac{3}{2}}\int e^{-c^2}\mathbf{c}h(\mathbf{x}, \mathbf{c})\, d\mathbf{c} \\ \tau &= \tfrac{2}{3}\pi^{-\frac{3}{2}}\int e^{-c^2}(c^2 - \tfrac{3}{2})h(\mathbf{x}, \mathbf{c})\, d\mathbf{c} \end{aligned} \tag{13.7}$$

In order to solve the problem, it is necessary to consider the Fourier transform $\tilde{\tilde{h}}$ of h with respect to the space variables. As remarked above, a source term involving the boundary values of the unknown h will appear in the Fourier transformed equation. Since the boundary conditions are applied on the segment $0 < x < L$ of the x-axis, the source term depends only upon the jump of h at $y = 0$ $(0 < x < L)$. In the case of a symmetric profile with no lift $(\varepsilon_-(x) = \varepsilon_+(x) = \varepsilon(x))$, $h(x, 0_-, \mathbf{c}) = h(x, 0_+, \mathbf{c} - 2\mathbf{j}[\mathbf{j}\cdot\mathbf{c}])$ by symmetry, and hence Eq. (13.5) can be written as follows:

$$\begin{aligned} &h(x, 0_+, \mathbf{c}) - h(x, 0_-, \mathbf{c}) \\ &\qquad = \alpha\frac{f_0(\mathbf{c}) - f_\infty(\mathbf{c})}{f_\infty(\mathbf{c})}\operatorname{sgn}(\mathbf{c}\cdot\mathbf{j}) - 4S\mathbf{c}\cdot\mathbf{j}\varepsilon(x) \qquad (0 < x < L) \end{aligned} \tag{13.8}$$

so that the jump of h at $y = 0$ is explicitly known. By means of this result we can easily find that $\tilde{\tilde{h}}$ satisfies the following equation:

$$\tilde{\tilde{h}} = \frac{\tilde{\tilde{r}} + 2\mathbf{c}\cdot\tilde{\tilde{\mathbf{u}}} + \tilde{\tilde{\tau}}(c^2 - \frac{3}{2})}{1 + i\lambda_\infty(Sk_x + \mathbf{k}\cdot\mathbf{c})} + \frac{\sigma(k_x, \mathbf{c})}{1 + i\lambda_\infty(Sk_x + \mathbf{k}\cdot\mathbf{c})} \tag{13.9}$$

where

$$\begin{aligned} \sigma(k_x, \mathbf{c}) = +2S\frac{(\mathbf{c}\cdot\mathbf{j})^2}{\pi}\int_0^L \varepsilon(x)e^{-ik_x x}\, dx\, + \\ + i\alpha|\mathbf{c}\cdot\mathbf{j}|\frac{f_0(\mathbf{c}) - f_\infty(\mathbf{c})}{f_\infty(\mathbf{c})}\frac{e^{-ik_x L} - 1}{2\pi k_x} \end{aligned} \tag{13.10}$$

An equation similar to Eq. (13.9) was considered by H. Grad [74] and L. Sirovich [75] in a study of the far field in the problem of flow past a body; in their case, however, the source term was a function of **c** alone and was not known.

If we now use the expression of h given by Eq. (13.9) to compute $\tilde{\tilde{r}}$, $\tilde{\tilde{\mathbf{u}}}$, $\tilde{\tilde{\tau}}$ according to Eqs. (13.7), we obtain:

$$\sum_{\beta=0}^{4} M_{\alpha\beta} v_\beta = \sigma_\alpha \qquad (\alpha = 0, 1, 2, 3, 4) \tag{13.11}$$

where $\|M_{\alpha\beta}\|$ is a matrix, v_β and σ_α are vectors, defined by

$$M_{\alpha\beta} = \delta_{\alpha\beta} - \pi^{-\frac{3}{2}} \int \frac{e^{-c^2} \psi_\alpha \psi_\beta \, d\mathbf{c}}{1 + i\lambda_\infty (Sk_x + \mathbf{k} \cdot \mathbf{c})} \tag{13.12}$$

and

$$v_\beta = (\psi_\beta, \tilde{\tilde{h}}); \qquad \sigma_\alpha = \left(\psi_\alpha, \frac{\sigma}{1 + i\lambda_\infty (Sk_x + \mathbf{k} \cdot \mathbf{c})}\right) \tag{13.13}$$

where (,) is the scalar product in $\mathscr{H}$ and the ψ_α are normalized in the usual way ($(\psi_\alpha, \psi_\beta) = \delta_{\alpha\beta}$).

By a suitable rotation in velocity space, it can be shown that the four equations of the system (13.11) referring to the motion in the (x, y) plane ($\alpha = 0, 1, 2, 4$) can be transformed into two subsystems of one and three equations respectively (it is sufficient to take the two axes parallel and orthogonal to **k**, respectively). The determinants of these systems turn out to be the same as those of systems already studied: the vanishing of the denominator of the solution of the single equation is equivalent to the vanishing of $M(u, s)$ as defined by Eq. (6.9) (with $s = i\lambda_\infty Sk_x$, $u = -(1 + i\lambda_\infty Sk_x)/K$), the vanishing of the determinant of the system of three equations is equivalent to Eq. (8.14). The problem of studying the zeroes of these equations can be handled by means of the study of the γ-curve discussed in Section 6 and the analogous curves studied by Mason [29] and mentioned in Section 8. The discussion of these zeroes, of course, is extremely important for discussing the inversion of the Fourier transforms; they produce contributions to the structure of acoustic fronts (weak shocks), boundary layers and waves. These results are only sketched in the abovementioned paper [73] but, to date, these have not been discussed in detail. A simple observation is that the shear stress at the wall is given by:

$$P_{xz} = -\alpha \int_{\eta<0} f(\mathbf{x}, \boldsymbol{\xi}) \xi\eta \, d\boldsymbol{\xi} = -\alpha \int_{\eta<0} f_\infty \xi\eta \, d\boldsymbol{\xi} - \alpha \int_{\eta<0} f_\infty \xi\eta h \, d\boldsymbol{\xi}$$
$$(\xi = \boldsymbol{\xi} \cdot \mathbf{i}; \eta = \boldsymbol{\xi} \cdot \mathbf{j}) \tag{13.14}$$

where Eqs. (11.4) and (III.5.1) have been used. This exact formula shows that the tangential stress is made up of two parts. The first part is of order α^2 and is just the free molecular value; the second part is of order α^2 since h is of order α. Hence the correction to the free molecular value of the drag is of order α^2; thus we see that a computation of the first order term of h, which is provided by the linearized treatment sketched above, gives the expression of the drag correct to second order in α.

14. Fluctuations and light scattering

As we know (see Chapter II), the Boltzmann equation is an average equation which describes a deterministic evolution of the one-particle distribution function. This means that when we write the Boltzmann equation, we do not take into account the fluctuations of the distribution function around its average value. These fluctuations can be taken into account formally (at least at the level of the *linearized* Boltzmann equation) by adding a fluctuating term $S(\mathbf{x}, \boldsymbol{\xi}, t)$; we can however make precise statements about the average value of the latter, which must be zero, as well as about the correlation function $\langle S(\mathbf{x}, \boldsymbol{\xi}, t)S(\mathbf{x}', \boldsymbol{\xi}', t')\rangle$, where the brackets denote averaging. This problem has been investigated by Fox [76–78]. If we are only interested in fluctuations of density, velocity and temperature, however, then, due to the conservation equations, we can compute the correlation functions for the latter quantities ($\langle\rho(\mathbf{x}, t)\rho(\mathbf{x}', t')\rangle = \langle\rho(\mathbf{x}, t - t')\rho(\mathbf{x}, 0)\rangle$, etc.) by solving the linearized Boltzmann equation (without the fluctuating term) with an initial datum corresponding to an excess in the number momentum and energy of the particles.

The density fluctuations can be experimentally observed by studying the scattering of laser light from monatomic gases. The characteristics of light scattered from fluids depends on fluctuations in the dielectric constant of the material contained in a fixed volume element. In general, the dielectric constant ε depends both on the local mass density and temperature, but for gaseous systems consisting of simple non-polar molecules the dependence on temperature of ε is very small. The spectrum of the scattered light depends on the time correlation of the fluctuations in the dielectric constant and hence on the density-density correlation function $G(|\mathbf{x} - \mathbf{x}'|, t) = \langle\rho(\mathbf{x}, t)\rho(\mathbf{x}', 0)\rangle$ or, better, its Fourier transform $S(\mathbf{k}, \omega)$.

The wavelength of the light used in the experiments is usually small compared to the mean free path of the gas, but the wavenumber $|\mathbf{k}|$ appearing in $S(\mathbf{k}, \omega)$ is $2|\mathbf{k}_0|\sin(\theta/2)$ where $\mathbf{k}_0$ is the wavevector of the impinging radiation and θ is the angle between $\mathbf{k}_0$ and the wave-vector $\mathbf{k}_s$ of the scattered light. Accordingly, at each observation angle, there is a definite wave-length fluctuation and so it is possible to measure the Fourier transform of the density-density correlation function by changing the angle. For sufficiently small angles we are in the continuum regime and a hydrodynamic theory, based on the Navier-Stokes equations, can be applied, but we would expect that when the mean free path is large compared to the wavelength and θ is not so small, then the profiles predicted by the continuum theory will not agree with experiments. Thus Nelkin and Yip [79] suggested that scattering experiments could be used to test the linearized Boltzmann equation. According to the above discussion, in fact, the density correlation function $G(r, t)$ is given by

$$G(r, t) = \int f_0(\boldsymbol{\xi})h(\mathbf{x}, \boldsymbol{\xi}, t)\, d\boldsymbol{\xi} \qquad (r = |\mathbf{x}|) \tag{14.1}$$

where $f_0(\boldsymbol{\xi})$ is the equilibrium Maxwellian corresponding to the density and temperature of the gas and h satisfies the linearized Boltzmann equation in infinite space, with the initial datum

$$h(\mathbf{x}, \boldsymbol{\xi}, 0) = \delta(\mathbf{x}) \tag{14.2}$$

representing an extra particle at the origin.

This problem can be analytically solved if use is made of kinetic models, and the solution can be easily found by Fourier transforming these models and proceeding to find a solution in a way similar to that employed in the previous sections. Some of these solutions have been computed by Yip and Nelkin [79], Yip and Ranganathan [80] and Sugawara, Yip and Sirovich [81]. The validity of these results over a wide range of wavelengths and frequencies has been confirmed by the careful experiments by Greytag and Benedek [82].

Appendix

We collect here some results about singular integral equations which underlie the completeness properties of the eigensolutions studied in the main text.

We start by recalling some properties of the analytic functions regular in the complex plane with a cut along a line L; for further details and proofs, see [8].

Let $f(z)$ be an analytic function which tends to zero when $z \to \infty$ and is regular in the complex plane with a cut along an oriented open line L; let the limits of $f(z)$ when z tends to a point $t \in L$ exist and let us denote by $f^+(t)$ and $f^-(t)$ the limits when the line is approached from the left and from the right hand sides, respectively; also if c denotes either endpoint of L, let $|f(z)| < A(z - c)^{-\gamma}$ for some A and $\gamma < 1$ when $z \to c$. Let

$$\Delta(t) = f^+(t) - f^-(t) \tag{A.1}$$

denote the jump of $f(z)$ when going through the cut. Then $\Delta(t)$ is a Hölder type function on L (i.e. $|f(t_1) - f(t_2)| < A\,|t_1 - t_2|^\alpha$, $0 < \alpha \leqslant 1$ for $t_1, t_2 \in L$) and

$$f(z) = \frac{1}{2\pi i}\int_L \frac{\Delta(t)}{t - z}\, dt \tag{A.2}$$

Also

$$f^+(t) + f^-(t) = \frac{1}{\pi i} P \int_L \frac{\Delta(t')}{t' - t}\, dt' \tag{A.3}$$

where it is understood that P means that the Cauchy principal value of the integral.

If, *vice versa*, a function $\Delta(t)$ is given which is of Hölder type on the open line L, and $\Delta(t) < A\,|t - c|^{-\gamma}$ at both endpoints for some A and $\gamma < 1$, let us define $f(z)$ through Eq. (A.1.2). Then $f(z)$ is regular in the complex plane with a cut along L, tends to zero when $z \to \infty$ and its limits when approaching L from the left and from the right exists, are finite and are related to $\Delta(t)$ by Eqs. (A.1) and (A.3). These equations, which are usually referred to as Plemelj's formulas, can be written in the equivalent form:

$$f^{\pm}(t) = \frac{1}{2\pi i} P \int_L \frac{\Delta(t')}{t' - t}\, dt' \pm \tfrac{1}{2}\Delta(t) \tag{A.4}$$

The exact behavior of $f(z)$ when $z \to c$ is given by

$$f(z) = \pm \frac{\Delta(c)}{2\pi i} \log(c - z) + O(1) \tag{A.5}$$

if $\Delta(c) \neq \infty$; the sign is positive for the upper limit of the integral (A.2) and negative for the lower one.

Let is consider now a singular integral equation with a Cauchy kernel:

$$A(t)y(t) + B(t)P \int_L \frac{y(t')}{t - t'}\, dt' = C(t) \qquad (t \in L) \tag{A.6}$$

where $A(t)$, $B(t)$ and $C(t)$ are given functions and L is a given open line in the complex plane. A procedure for solving Eq. (A.6) analytically is as follows. Let us introduce the following function of the complex variable z:

$$N(z) = \frac{1}{2\pi i} \int_L \frac{y(t)}{t - z}\, dt \tag{A.7}$$

Then Eq. (A.6) can be rewritten as follows:

$$[A(t) + \pi i B(t)]N^+(t) - [A(t) - \pi i B(t)]N^-(t) = C(t) \tag{A.8}$$

where the Plemelj formulas, Eqs. (A.1) and (A.3) (with $\Delta = y, f = N$), have been used. Given the relation between $N^+(t)$ and $N^-(t)$ expressed by Eq. (A.8) and the analyticity of $N(z)$, we must determine $N(z)$; this problem is known as the Hilbert problem. We shall assume that $A \pm iB \neq 0$ on L. Once $N(z)$ is known, $y(t)$ is obtained by means of the Plemelj formulas.

In order to solve the Hilbert problem we remark that if we were to know a function $X(z)$ that was analytic and nonzero in the complex plane with a cut along L and such that

$$[A(t) + \pi i B(t)]X^-(t) = [A(t) - \pi i B(t)]X^+(t) \tag{A.9}$$

then Eq. (A.8) could be written

$$X^+(t)N^+(t) - X^-(t)N^-(t) = \frac{X^-(t)C(t)}{A(t) - \pi i B(t)} \tag{A.10}$$

and because of the Plemelj formulas applied to $X(z)N(z)$:

$$N(z) = \frac{1}{X(z)}\frac{1}{2\pi i}\int_L \frac{X^-(t)C(t)}{A(t) - \pi i B(t)}\frac{dt}{t - z} \tag{A.11}$$

provided $X(z)N(z)$ tends to zero for $z \to \infty$. Thus we can solve the inhomogeneous Hilbert problem related to Eq. (A.6) provided we know a solution of the homogeneous Hilbert problem (A.9). But Eq. (A.9) can be written as follows:

$$\log X^+(t) - \log X^-(t) = \log\frac{A(t) + \pi i B(t)}{A(t) - \pi i B(t)} \tag{A.12}$$

and the Plemelj formulas applied to $\log X(z)$ give:

$$X(z) = (z - \alpha)^\gamma (z - \beta)^\delta \exp\left\{\frac{1}{2\pi i}\int_L \log\left[\frac{A(t) + \pi i B(t)}{A(t) - \pi i B(t)}\right]\frac{dt}{t - z}\right\} \tag{A.13}$$

where α and β are the endpoints of L; γ and δ are two integers which must be chosen in such a way that $|X(z)| \sim k(z - c)^\sigma$ $(-1 < \sigma < 1)$ at both endpoints. This is required for Eq. (A.11) to be valid. Accordingly, it can happen that $N(z)$ is not completely determined if more than one pair (γ, δ) can be chosen or, on the contrary, a restriction on the function $C(t)$ must be imposed since the behavior at infinity of $X(z)$ can be such that $N(z)$, as given by Eq. (A.11), does not tend to zero when $z \to \infty$ for a general $C(t)$. The latter case occurs for the equations considered in the main text whenever the eigensolutions of the continuous spectrum must be complemented by discrete terms; the restriction on $C(t)$ determines the coefficients of these discrete terms. A particular but important case is when one of the endpoints is at infinity. We consider only the case where $[A(t) + \pi i B(t)]/[A(t) - \pi i B(t)]$ tends to 1 when $t \to \infty$; then the branch of the logarithm in Eq. (A.13) is to be chosen in such a way that the logarithm tends to zero when $t \to \infty$. Also, $\delta = 0$ if α is the endpoint different from infinity.

We finally remark that theorems I and II of Section 3 are obtained by applying the method described above to Eqs. (3.25) and (3.30). In particular $P(w)$ is simply $[X(-w)]^{-1}$, given by Eq. (A.13) with $A(t) = p(t)$, $B(t) = t$, $\alpha = 0$, $\gamma = -1$, $\delta = 0$. This fact can be exploited to prove Eqs. (3.35)–(3.37) by applying Plemelj's formulas to $X(z)$ [4–6].

References

[1] B. NOBLE, "The Wiener-Hopf Technique", Pergamon Press, Oxford (1950).
[2] B. DAVISON, "Neutron Transport Theory", Oxford University Press (1957).
[3] M. M. R. WILLIAMS, "Mathematical Methods in Particle Transport Theory", Butterworths (1971).
[4] K. M. CASE, "Ann. Phys." (NY) **9,** 1 (1960).
[5] C. CERCIGNANI, "Ann. Phys." (NY) **20,** 219 (1962).
[6] C. CERCIGNANI, "Ann. Phys." (NY), **40,** 469 (1966).
[7] I. M. GELFAND and co-authors, "Generalized Functions", Academic Press, New York (1964).
[8] N. I. MUSKHELISHVILI, "Singular Integral Equations", Noordhoff, Groningen (1953).
[9] V. N. FADDEYEVA and N. M. TERENT'EV, "Tables of the Values of the Function, etc.", Pergamon Press, London (1961).
[10] B. D. FRIED and S. D. CONTE, "The Plasma Dispersion Function: The Hilbert Transform of the Gaussian", Academic Press, New York (1961).
[11] C. CERCIGNANI, "J. Math. Anal. Appl.", **10,** 93 (1965).
[12] S. ALBERTONI, C. CERCIGNANI and L. GOTUSSO, "Phys. Fluids", **6,** 993 (1963).
[13] C. CERCIGNANI, P. FORESTI and F. SERNAGIOTTO, "Nuovo Cimento", **X,** 57B, 297 (1968).
[14] C. CERCIGNANI and F. SERNAGIOTTO, in "Rarefied Gas Dynamics", C. L. Brundin, ed., vol. I, p. 381, Academic Press, New York (1967).
[15] J. C. MAXWELL, "Phil. Trans. Royal Soc. I", Appendix (1879); reprinted in "The Scientific Papers of J. C. Maxwell", Dover, New York (1965).
[16] K. M. CASE and P. ZWEIFEL, "Linear Transport Theory", Addison-Wesley, Reading (1967).
[17] C. CERCIGNANI, "J. Math. Anal. Appl.", **11,** 93 (1965).
[18] C. CERCIGNANI, "J. Math. Ann. Appl.", **12,** 234 (1965).
[19] C. CERCIGNANI, in "Rarefied Gas Dynamics", J. A. Laurmann, ed., Vol. II, p. 92, Academic Press, New York.
[20] C. CERCIGNANI, in "Rarefied Gas Dynamics", J. A. Laurmann, ed. Vol. II, p. 92, Academic Press, New York (1963).
[21] M. KNUDSEN, "Ann. d. Physik", **28,** 75 (1909).
[22] G. J. MITSIS, "Nucl. Sci. and Eng.", **17,** 55 (1963).
[23] R. L. BOWDEN and C. D. WILLIAMS, "J. Math. Phys.", **5,** 1527 (1964).
[24] C. CERCIGNANI and S. SERNAGIOTTO, "Ann. Phys. (N.Y.)", **30,** 154 (1964).

[25] C. CERCIGNANI, "Elementary Solutions of Linearized Kinetic Models and Boundary Value Problems in the Kinetic Theory of Gases", Brown University Report (1965).
[26] C. CERCIGNANI and R. TAMBI, "Meccanica" **2,** 25 (1967).
[27] J. J. DORNING and J. K. THURBER, in "Neutron Transport Theory Conference", ORO, 3858-1, USAEC (1969).
[28] G. N. WATSON, "Theory of Bessel Functions", Cambridge University Press (1958).
[29] R. J. MASON, "Phys. Fluids", **13,** 467 (1970).
[30] M. M. R. WILLIAMS and J. SPAIN, "J. Fluid Mech.", **42,** 85 (1970).
[31] J. S. CASSEL and M. M. R. WILLIAMS, "Transport Theory and Statistical Physics", **2,** 81 (1972).
[32] J. S. DARROZES, "La Recherche Aérospatiale", **119,** 13 (1967).
[33] C. CERCIGNANI, "Transport Theory and Statistical Physics", (to appear).
[34] H. S. OSTROWSKI and D. J. KLEITMAN, Nuovo Cimento, *XLIV* B49
[35] R. J. MASON, in "Rarefied Gas Dynamics", C. L. Brundin ed., Vol. I, p. 395, Academic Press, New York (1967).
[36] H. WEITZNER, in "Rarefied Gas Dynamics", J. H. de Leeuw, ed., Vol. I, p. 1, Academic Press, New York (1965).
[37] C. CERCIGNANI and G. TIRONI, "Nuovo Cimento", **43,** 64 (1966).
[38] C. CERCIGNANI, "Ann. Phys. (NY)", **40,** 454 (1966).
[39] I. N. VEKUA, "Generalized Analytic Functions", Pergamon Press, Oxford (1962).
[40] T. KLINÇ and I. KUŠČER, "Transport Theory and Statistical Physics", **1,** 41 (1971).
[41] C. CERCIGNANI and F. SERNAGIOTTO, in "Rarefied Gas Dynamics", C. L. Brundin, ed., Vol. I, p. 381, Academic Press, New York (1967).
[42] H. J. KAPER, "J. Math. Phys.", **10,** 286 (1969).
[43] J. J. DUDERSTADT, Ph.D. Thesis, California Institute of Technology, Pasadena (1968).
[44] H. G. KAPER, J. H. FERZIGER and S. K. LOYALKA, in "Neutron Thermalization and Reactor Spectra", IAEA, Vienna (1968).
[45] K. M. CASE, in "The Boltzmann Equation", F. A. Grünbaum, ed., New York University, New York (1972).
[46] K. M. CASE and R. D. HAZELTINE, "J. Math. Phys.", **11,** 1126 (1970).
[47] K. M. CASE and R. D. HAZELTINE, "J. Math. Phys.", **12,** 1970 (1971).
[48] L. SIROVICH and J. K. THURBER, "J. Acoust. Soc. Am.", **37,** 329 (1965).
[49] B. NICOLAENKO, J. K. THURBER and J. J. DORNING, "J. Quant. Spectrosc. Radiat. Transfer", **11,** 1007 (1971).
[50] B. NICOLAENKO, in "The Boltzmann Equation", F. A. Grünbaum, ed., p. 173, New York University, New York, 1972.
[51] M. GREENSPAN, "J. Acoust. Soc. Am.", **28,** 644 (1956).
[52] E. MEYER and G. SESSLER, "Z. Physik", **149,** 15 (1957).

[53] J. K. Buckner and J. H. Ferziger, "Phys. Fluids", **9**, 2315 (1966).
[54] H. Grad, "SIAM J. Appl. Math.", **14**, 935 (1966).
[55] L. Sirovich and J. K. Thurber, "J. Acoust. Soc. Am.", **38**, 478 (1965).
[56] J. K. Thurber, in "The Boltzmann Equation", F. A. Grünbaum, ed., p. 211, New York University, New York (1972).
[57] R. J. Mason, in "Rarefied Gas Dynamics", J. H. de Leeuw, ed., vol. I, 48, Academic Press, New York (1965).
[58] C. S. Wang Chang and G. E. Uhlenbeck, Univ. of Michigan Engineering Research Institute Project M 999. Reprinted in: C. S. Wang Chang and G. E. Uhlenbeck, "The Kinetic Theory of Gases", Studies in Statistical Mechanics, vol. V, North-Holland, Amsterdam.
[59] C. L. Pekeris, Z. Alterman, L. Finkelstein and K. Frankowski, "Phys. Fluids", **5**, 1608 (1962).
[60] D. Kahn and D. Mintzer, Phys. Fluids, **8**, 1090 (1965).
[61] S. S. Abarbanel, in "Rarefied Gas Dynamics", C. L. Brundin, ed., Vol. I, p. 369, Academic Press, New York (1967).
[62] G. Sessler, "J. Acoust. Soc. Am.", **38**, 974 (1965).
[63] G. Maidanik, H. L. Fox and M. Heckl, "Phys. Fluids", 259 (1965).
[64] L. H. Holway, "Phys. Fluids", **10**, 35 (1967).
[65] L. Lees, "SIAM J. Appl. Math.", **13**, 278 (1965).
[66] C. Cercignani, "Mathematical Methods in Kinetic Theory", Plenum Press, New York (1969).
[67] K. Toba, "Phys. Fluids", **11**, 2495 (1968).
[68] F. B. Hanson and T. F. Morse, "Phys. Fluids", **12**, 1564 (1969).
[69] C. Cercignani, "Phys. Fluids", **11**, 303 (1968).
[70] M. Van Dyke, "Perturbation Methods in Fluid Mechanics", Academic Press, New York (1964).
[71] G. Scharf, "Phys. Fluids", **13**, 848 (1970).
[72] C. Cercignani, "Phys. Fluids", **15**, 957 (1972).
[73] C. Cercignani, Presented at the 8th Symposium on Rarefied Gas Dynamics, Stanford University, Stanford (1972).
[74] H. Grad, New York Univ. NYO-2543 (1959).
[75] L. Sirovich, in "Rarefied Gas Dynamics", L. Talbot, ed., Academic Press, New York (1961).
[76] R. F. Fox, Ph.D. Thesis, Rockefeller University, New York (1969).
[77] R. F. Fox and G. E. Uhlenbeck, "Phys. of Fluids", **13**, 1893 (1970).
[78] H. H. Szu, Ph.D. Thesis, Rockefeller University, New York (1971).
[79] S. Yip and M. Nelkin, "Phys. Rev." **136**, A1241 (1964).
[80] S. Yip and S. Ranganathan, "Phys. Fluids", **8**, 1956 (1965).
[81] A. Sugawara, S. Yip and L. Sirovich, "Phys. Fluids", **11**, 925 (1968).
[82] T. J. Greytag and G. B. Benedek, "Phys. Rev. Letters", **17**, 179 (1966).

VII THE TRANSITION REGIME

1. Introduction

The complexity of the results obtained in Chapter VI even for relatively simple problems and collision models suggests that for more complicated problems and accurate collision models, we should look for less sophisticated procedures, yielding approximate but essentially accurate results. For either large or small Knudsen numbers such methods have been examined in Chapter V; in this chapter the transition regime, intermediate between the nearly continuum and nearly free molecular regimes, will be examined.

Systematic, accurate and relatively simple methods of solution can be constructed for *linearized* transition flows. These methods are based on variational techniques (Section 3), and discrete ordinate methods (Section 2). In general the results obtained by these methods have produced predictions which are in spectacular agreement with experiment and have shed considerable light on the basic structure of transition-flow theory whenever nonlinear effects (in particular shock waves) can be neglected. For nonlinear problems, "moment methods" can be introduced (Section 2), which lead to the solution of a system of nonlinear partial differential equations. The latter, in general, are harder to handle than the Navier-Stokes equations and, as a consequence, it is necessary to resort to numerical procedures to solve them. It may be found convenient then to try a direct numerical method based on a discrete ordinate technique on the Boltzmann equation itself (Section 2). Finally, a great deal of research has been devoted to Monte Carlo simulation techniques (Section 4).

In connection with all these methods, we can simplify the calculations by the use of suitable collision models, but it is to be remarked that the accuracy of kinetic models in nonlinear problems is less obvious than in the linearized ones.

2. Moment and discrete ordinate methods

If we multiply both sides of the Boltzmann equation by the functions $\varphi_i(\boldsymbol{\xi})$ $(i = 1, \ldots, N, \ldots)$ forming a complete set and then integrate over the molecular velocity, we obtain infinitely many relations to be satisfied by the distribution function:

$$\frac{\partial}{\partial t}\int \varphi_i(\boldsymbol{\xi})f(\mathbf{x}, \boldsymbol{\xi}, t)\, d\boldsymbol{\xi} + \frac{\partial}{\partial \mathbf{x}} \cdot \int \varphi_i(\boldsymbol{\xi})f(\mathbf{x}, \boldsymbol{\xi}, t)\boldsymbol{\xi}\, d\boldsymbol{\xi}$$
$$= \int \varphi_i(\boldsymbol{\xi})Q(f, f)\, d\boldsymbol{\xi} \qquad (i = 1, \ldots, N, \ldots) \quad (2.1)$$

This system of infinitely many relations (Maxwell's transfer equations) is equivalent to the Boltzmann equation, because of the completeness of the set $\{\varphi_i\}$. The common idea of the so called moment methods is to satisfy only a finite number of transfer equations or moment equations. This leaves the distribution function f largely undetermined, since only the infinite set (2.1) with proper initial and boundary conditions can determine f. This means that we can choose, to a certain extent, f arbitrarily and then let the moment equations determine the details which we have not specified. The different "moment methods" differ in the choice of the set φ_i and the arbitrary input for f. Their common feature is that f is chosen to be a given function of $\boldsymbol{\xi}$ containing N undetermined parameters depending upon $\mathbf{x}$ and t, $M_i(i = 1, \ldots, N)$, this means that, if we take N moment equations, we obtain N partial differential equations for the unknowns $M_i(\mathbf{x}, t)$. In spite of the large amount of arbitrariness, it is hoped that any systematic procedure yields, for sufficiently large N, results essentially independent of the arbitrary choices. On practical grounds, another hope is that, for sufficiently small N and a judicious choice of the arbitrary elements, we can obtain accurate results.

The simplest choice [1] is to assume f to be a Maxwellian f_0 times a polynomial:

$$f = f_0 \sum_{k=0}^{N-1} Q_k(\mathbf{x}, t)H_k(\boldsymbol{\xi}) \qquad (2.2)$$

where, for convenience, the polynomial is expressed in terms of the three-dimensional Hermite polynomials $H_k(\boldsymbol{\xi})$ orthogonal with respect to the weight f_0. It is also convenient to choose f_0 to be the local Maxwellian (this implies that $Q_0 = 1$, $Q_1 = Q_2 = Q_3 = Q_4 = 0$. There are N arbitrary quantities which may be identified with the basic moments (ρ, v_i, T, p_{ij}, q_i and higher order moments) and can be determined by solving the partial differential equations obtained by taking $\varphi_i = H_i$ in Eq. (2.1) ($i = 0, \ldots, N - 1$). A reasonable choice is $N = 13$; in such a case the unknowns are ρ, v_i, T, $p_{ij} - p\,\delta_{ij}$, q_i and the equations are known under the name of Grad's thirteen moment equations.

There is an obvious disadvantage in Grad's choice, which is the fact that the distribution function is assumed to be continuous in the velocity variables and this is not true, for example, at a flat boundary. Also, even the question of convergence of polynomial approximations can be given a definite negative answer in certain nonlinear flows; in fact Holway [2] noted that for molecules with a finite interaction range (such as rigid spheres), the series of Hermite polynomials representing the distribution function in a shock wave does not converge for an upstream Mach number M larger than 1.851.

In order to avoid the first of the above mentioned consequences, we can take for f a distribution function which already takes into account the discontinuities in velocity space, for example, a function which piecewise reduces to Eq. (2.2). Then we can choose between piecewise continuous functions [3–6] or continuous functions [7] for the set $\{\varphi_i\}$. The second method seems to be more rewarding and easier to handle: in its crudest version it reduces to assuming that the distribution function is piecewise Maxwellian the discontinuities being located exactly where predicted by the free molecular solution.

In order to avoid the absence of convergence in nonlinear flows, especially those involving shock waves, we can use linear combinations of Maxwellians [8] and take suitable moments (see Section 6). Alternatively we can use for f_0 a Maxwellian which is not the local one [9]. Also, the results of Chorin [10] seem to indicate that, in the problem of the structure of a shock wave for moderate Mach numbers, the difficulty of absence of convergence can be removed by searching for the solution of a steady problem as the limit for $t \to \infty$ of a time dependent problem.

We can obtain a more accurate (but usually more complex) approximation to the distribution function by exploiting the information gained by an analytic study of the singularities of the solution; that is by choosing approximating functions which reproduce the correct singular behaviour in the appropriate limits [11].

It is obvious that the moment methods can be used to obtain reasonably good results by a judicious choice of the arbitrary elements, but it is to be remembered that the equations to be solved eventually become very complicated and can be handled only by numerical methods.

A particular feature is offered by Maxwell's molecules; in fact, if we choose for $\{\varphi_i\}$ a set of polynomials, the integrals on the right hand side of Eq. (2.1) can be evaluated explicitly in terms of the full range moments $\int \varphi_i f \, d\boldsymbol{\xi}$, and only moments of degree not larger than the degree of φ_i appear in the right hand side of the i-th equation. This means that in the space homogeneous case we can evaluate exactly the time evolution of subsequent moments by solving ordinary differential equations (essentially linear, even in the nonlinear case) [12].

All the above methods can be, and have been, applied to the linearized Boltzmann equation [13–19, 3–5]. Similar methods are well-established and reliable for dealing with neutron transport and radiative transfer. In particular, a great deal of work has been done with expansions in spherical harmonics or, in particular, Legendre polynomials [20, 22] and with the so-called double-P_1 method, based on a piecewise continuous approximation [22–23].

A problem of importance for the moment methods is the choice of the boundary conditions to be imposed upon the solution of the moment equations. No special problems arise for methods based on piecewise continuous functions of $\boldsymbol{\xi}$, provided the discontinuities are so located that they occur for $\boldsymbol{\xi} \cdot \mathbf{n} = 0$ at the boundaries. In the case of continuous approximations to the distribution function such as Eq. (2.2), we encounter the difficulty that the boundary conditions assign the emerging distribution function in terms of the impinging one; accordingly we can obtain only half-range moment relations from the boundary conditions. Apart from the little accuracy to be expected from such a procedure, the question arises as to how many and which conditions we must take for a given problem on a certain part of the boundary. It is clear that we should not take the number of conditions simply equal to the number of moments or to the order of the equations. This question has been discussed in detail in neutron transport, where several sets of boundary conditions have been given, among which the most generally known are those of Marshak and those of Mark [22].

Boundary conditions analogous to those of Marshak were adopted by Grad [1] and Wang Chang and Uhlenbeck [14–16] for gas-kinetic problems.

It is to be remarked that the moment methods based on continuous approximations to f do not behave well in the free-molecular limit, due to the critical importance of the boundary conditions in this limit.

A method closely related to the moment methods is the method of discrete ordinates (or discrete velocities). We choose a set of values of the velocity $\boldsymbol{\xi}_i$ where $i = 0, 1, \ldots, N - 1$; then, by some interpolation method, we express the distribution function in terms of its values corresponding to the velocities $\boldsymbol{\xi}_i$ and then use Eq. (2.1) with $\varphi_i = \delta(\boldsymbol{\xi} - \boldsymbol{\xi}_i)$ as a system of equations determining the values $f_i(\mathbf{x}, t)$ of $f(\mathbf{x}, \boldsymbol{\xi}, t)$ for $\boldsymbol{\xi} = \boldsymbol{\xi}_i$. Thus, the integro-differential Boltzmann equation is replaced by a system of simultaneous nonlinear differential equations for the N functions f_i. The solution of the resulting equations (by numerical methods) has been carried out for the BGK and ES models [24–31] and for the linearized Boltzmann equation and kinetic models [30, 32–38]. The method is, of course, well known also for problems of radiative transfer and neutron transport [23, 39, 40]. If we choose the velocities $\boldsymbol{\xi}_i$ to be located at the zeroes of the Hermite polynomials $H_k(\boldsymbol{\xi})$, and a related interpolation formula is used, then the method of discrete ordinates is essentially equivalent to a moment method based on an expansion of the form (2.2), but about a fixed rather than a local Maxwellian f_0. This result has been proved in detail by Gast [41] for the case of the one speed neutron transport equation.

The equations obtained by the method of discrete ordinates always contain a simple linear differential operator on the left hand side, while in the method of moments quasilinear differential expressions are obtained for the full Boltzmann equation (unless an expansion such as the one given by Eq. 2.2 is based on a fixed Maxwellian).

Several other types of discrete ordinate methods can be devised; we just mention the so called S_N method of Carlson [42] widely used for neutron transport calculations.

Moment and discrete ordinate methods can of course be applied to the integral form of the Boltzmann equation (Chapter IV, Section 12) rather than to the standard integrodifferential form [43–46]. This circumstance is to be duly considered when a model, such as the BGK model, is used to describe collisions. In this case in fact we can use the integral form of the Boltzmann equation to obtain a finite system of exact integral equations involving only a finite number of moments (five, in the case of the BGK model). These equations can be very complicated but have the essential advantage that the independent variables are only four ($\mathbf{x}$, t) instead of seven ($\mathbf{x}$, $\boldsymbol{\xi}$, t) which is a very important feature for numerical computations. These exact integral equations can now be solved by making discrete the space and time variables (the only independent variables which have been left). In one dimensional linearized problems [47–53] this reduces to a system of a few integral equations with one independent variable; we can then achieve a practically exact result with limited amounts of computing time.

3. The variational method

When the Boltzmann equation is solved by a moment method or the collision term is replaced by a simple model, we give up any intent of accurately investigating the distribution function and restrict ourselves to the study of the space variation of some moments of outstanding physical significance, such as density, mass velocity, temperature and heat flux. However, it is to be noted that even such restricted knowledge is not required for the purpose of comparison with some experimental results. As a matter of fact, the typical output of an experimental investigation of Poiseuille flow is a plot of the flow rate versus the Knudsen number. Analogously, the outstanding quantity is the stress constant in Couette flow, the heat flux constant in heat transfer problems and the drag on the body in the flow past a body. From the point of view of evaluating these overall quantities, any computation of the flow fields appears to be a waste of time. Of course, the knowledge of flow fields is always interesting and illuminating, but frequently it happens that we have such a clear qualitative insight of the space behaviour of the unknowns, that we can imagine, for them, simple approximate analytical expressions containing a small number of adjustable constants. It appears, therefore, that a method which would succeed in giving both a precise rule for determining the above mentioned constants and a highly accurate evaluation of the overall quantities of outstanding interest should turn out to be most useful. The features which we have just mentioned are typical of variational procedures; it is, therefore, in this direction that we must look for the desired method.

The variational technique is based on the existence of a variational principle that is characteristic of the equation to be solved. A function $\tilde{h}$ containing some undetermined constants c_i (trial function) is then inserted into the functional $J(\tilde{h})$ whose variation is zero, according to the variational principle, if and only if evaluated at the solution $\tilde{h} = h$ of the problem to be solved. The expression of J becomes a function of the constants c_i; setting equal to zero the partial derivatives of this function with respect to the c_i, we obtain a system which determines the "best" values of the c_i according to the variational principle. The value of the method is enhanced by the circumstance that usually a global quantity of great interest can be related to the value attained by J for $\tilde{h} = h$, and by the fact that if h is approximated with an average error of order ε, then J is approximated with an error or order ε^2.

Variational principles for the linearized Boltzmann equation have been treated in Sections 10 and 12 of Chapter IV. For the principle based on the integrodifferential equation (Chapter IV, Section 10) it is not easy to make simple but reasonable guesses about the distribution function, but whenever these are made we are led to simple expressions for the approximate solution. The use of kinetic models in connection with the integral equation (Chapter IV, Section 12) leads to lengthy calculations, and cumbersome results even for very simple trial function but the results are very rewarding even if the guess is a poor one. In fact the use of kinetic models in the integral form implies that a guess about a finite number of moments leads to a guess about the distribution function which automatically satisfies the boundary conditions; in addition, no matter how bad the guess is, the result is correct by construction in the free molecular limit.

As an example of the use of the variational technique, we consider the solution of the Kramers' problem (Chapter VI, Section 4) with the BGK model [54]; as we know, this problem can be solved exactly but we shall ignore this at present. If we follow the procedure indicated in Chapter IV, Section 12, we can express the perturbation of the distribution function in terms of a single moment, the z component of the mass velocity, $v_3(x)$; in fact, density, temperature and the remaining components of $\mathbf{v}$ remain unperturbed in a linearized treatment (as we know from the splitting procedure used in Chapter VI). Accordingly, we can construct an integral equation for v_3 by taking the corresponding moment of the expression of h in terms of v_3. If we put $\varphi(x) = [v_3(x) - ax]/(a\theta)$ so that we subtract the asymptotic behaviour and make the unknown nondimensional, we obtain the following integral equation:

$$\pi^{\frac{1}{2}}\varphi(x) = T_1(x) + \int_0^\infty T_{-1}(|x - y|)\varphi(y)\, dy \qquad (x \geqslant 0) \qquad (3.1)$$

where θ has been taken to be unity, while the velocity unit is, as usual, $(2RT)^{\frac{1}{2}}$ and the transcendental functions $T_n(x)$ are defined by Eq. (IV.12.3).

Let us put:

$$\mu = \lim_{x\to\infty} \varphi(x) = \pi^{\frac{1}{2}}\zeta/(2l) \tag{3.2}$$

where ζ is the slip coefficient and l the mean free path; then, if we define

$$\psi(x) = \varphi(x) - \mu \tag{3.3}$$

we have

$$\pi^{\frac{1}{2}}\psi(x) - \int_0^\infty T_{-1}(|x-y|)\psi(y)\,dy = T_1(x) - \mu T_0(x) \qquad (x \geqslant 0) \tag{3.4}$$

where Eq. (IV.12.24) has been used.

Since $\psi(x)$ is integrable on $(0, \infty)$, we can integrate Eq. (3.4) from x to ∞ twice and obtain:

$$\int_0^\infty T_1(|x-y|)\psi(y)\,dy = \mu T_2(x) - T_3(x) \qquad (x \geqslant 0) \tag{3.5}$$

where Eqs. (IV.12.24) and (IV.12.25) have been used.

For this problem, the basic functional is

$$J(\tilde{\varphi}) = \pi^{\frac{1}{2}}\int_0^\infty dx\left\{[\tilde{\varphi}(x)]^2 - \int_0^\infty T_{-1}(|x-y|)\tilde{\varphi}(x)\tilde{\varphi}(y)\,dx - 2\tilde{\varphi}(x)T_1(x)\right\} \tag{3.6}$$

This functional attains a minimum when $\tilde{\varphi} = \varphi$, φ being the solution of Eq. (3.1). Eq. (3.5) with $x = 0$ gives

$$\int_0^\infty T_1(y)\psi(y)\,dy = \tfrac{1}{4}\mu\pi^{\frac{1}{2}} - \tfrac{1}{2} \tag{3.7}$$

and using Eq. (3.2) we have:

$$\zeta = l\left[2/\pi + (4/\pi)\int_0^\infty \varphi(y)T_1(y)\,dy\right] \tag{3.8}$$

We note now that for $\tilde{\varphi} = \varphi$ we have

$$J(\varphi) = \min[J(\tilde{\varphi})] = -\int_0^\infty \varphi(y)T_1(y)\,dy \tag{3.9}$$

and Eq. (3.8) becomes

$$\zeta = l\{2/\pi - (4/\pi)\min[J(\tilde{\varphi})]\} \tag{3.10}$$

In such a way we have found a direct connection between the slip coefficient and the minimum value attained by the functional $J(\tilde{\varphi})$; this means that even a poor estimate for $\tilde{\varphi}$ can give an accurate value for ζ. Accordingly we make the simplest choice for the trial function: $\tilde{\varphi} = c$ (constant).

We find by straightforward computation that:

$$J(c) = \tfrac{1}{2}c^2 - \tfrac{1}{2}\pi^{\frac{1}{2}}c \tag{3.11}$$

The minimum value is attained when

$$c = \tfrac{1}{2}\pi^{\frac{1}{2}} \tag{3.12}$$

corresponding to

$$J(\tfrac{1}{2}\pi^{\frac{1}{2}}) = -\tfrac{1}{8}\pi \tag{3.13}$$

Eq. (3.10) then gives for ζ:

$$\zeta = [\tfrac{1}{2} + (2/\pi)]l = (1.1366)l \tag{3.14}$$

to be compared with the exact value $\zeta = (1.1466)l$ given by Eq. (VI.4.14). We see that even a very simple choice for $\tilde{\varphi}$, a choice which we know to be very inadequate to describe the kinetic boundary layer, yields a rather accurate estimate of the slip coefficient ζ.

The same procedure can be applied to models with velocity dependent collision frequency [55, 56].

From the above example we see that if a variational principle is available, a small amount of computation yields reasonably accurate results; this is not a coincidence as will become clear by the results indicated in subsequent sections (Sections 5 and 7). This fact makes it desirable to find a variational principle for the nonlinear Boltzmann equation. Unfortunately this is not so simple. As a matter of fact, given any equation

$$N(f) = 0 \tag{3.15}$$

we can construct a variational principle by the least square method by setting

$$J(\tilde{f}) = \int [N(\tilde{f})]^2 \, d\Omega; \qquad \delta J = 0 \tag{3.16}$$

where $d\Omega$ denotes the volume element in an appropriate space ($d\Omega = W(\mathbf{x}, \boldsymbol{\xi}) \, d\mathbf{x} \, d\boldsymbol{\xi}$ for the steady Boltzmann equation, where $W(\mathbf{x}, \boldsymbol{\xi})$ is any function, e.g. $W = 1$). In general, however, the equations determining the approximate solutions are rather complicated; in addition the value attained by J for $\tilde{f} = f$ is zero and cannot give the value of a physically interesting quantity. We shall call these variational principles trivial. A non-trivial variational principle usually reflects some basic symmetry of the equation to be solved; these symmetries are easier to find for linear equations, though systematic methods of investigation are available for the nonlinear case as well [57].

Alternatively, we can introduce an additional unknown φ and consider the following functional

$$J(f, \varphi) = \int \varphi N f \, d\Omega \tag{3.17}$$

Then, varying f and φ separately, $\delta J = 0$ if and only if:

$$Nf = 0$$
$$\tilde{N}_f \varphi = 0 \tag{3.18}$$

where N_f is the Gateaux derivative [57] of N with respect to f (N_f is a linear operator depending, nonlinearly in general, upon f) and $\tilde{N}_f$ is the adjoint of N_f. In the case of the steady Boltzmann equation:

$$\tilde{N}_f \varphi = -\boldsymbol{\xi} \cdot \frac{\partial f}{\partial \mathbf{x}} - \int f_*(\varphi' - \varphi_*' - \varphi - \varphi_*) B(\theta, |\boldsymbol{\xi}_* - \boldsymbol{\xi}|) \, d\theta \, d\varepsilon \, d\boldsymbol{\xi}_* \tag{3.19}$$

where the standard notation (Chapter II, Section 4) has been used.

The approach based on Eq. (3.17) is useful only if we have a good guess available not only for f but also for φ (solution of Eq. (3.19)). The second guess is made difficult by the fact that φ does not have a simple physical significance.

Variational methods have been widely applied to problems of neutron transport. There, of course, the equation is linear and hence the method works quite well, though the choice of the trial functions is nontrivial when capture or fission strongly dominate [22, 58–66].

We mention also the so called method of boundary conditions [67–69] which seems to have a connection with the variational method [69] and might be extended to the treatment of nonlinear problems. It consists of using the Navier-Stokes equations (for any Knudsen number!) with suitable boundary conditions which simulate the effect of Knudsen layers. By its nature the method is useful for computing overall quantities. Some unnoticed numerical errors produced results that induced the originators of the method to dismiss it. The subsequent contribution of Loyalka [69] seems to suggest that the method deserves further study.

4. Monte Carlo methods

The name Monte Carlo method is applied to a considerable variety of computational procedures whose common element is a mathematical simulation of the physical phenomenon on a high speed computer rather than the solution of the Boltzmann equation as such.

The simplest case is offered by free molecular flow past a concave body [70]. As mentioned in Section 8 of Chapter V, we are here led to solving integral equations. According to the Monte Carlo method, the trajectories of individual molecules are followed and the distribution function computed from the actual number of test particles present in each of the discrete cells into which the phase space has been subdivided. When a particle collides against the body surface, it is replaced by a particle with a velocity chosen at random according to the probability distribution dictated by the boundary conditions.

The method can be extended to hypersonic nearly free-molecule flow by taking into account not only the collisions with the solid boundaries, but also the first collisions of the re-emitted molecules with the free stream molecules [71, 72].

At lower Knudsen numbers, when it is not possible to confine attention only to first collisions, the analysis becomes essentially more complicated. Accordingly, several Monte Carlo approaches have been proposed.

A first method is an extension of the "test particle" method, and is based on an iterative process. In this case the collisions of the test particles with a "target" gas are also taken into account. The distribution function of the latter gas is assigned on the basis of some preconceived idea at the first step, but is replaced, at each step of the iterative process, by the estimate of the distribution function established by the cumulative history of the test particles. The iterative process continues until the test particle distribution function reproduces the target-particle distribution function to an acceptable accuracy. This method was originated by Haviland and Lavin [73, 74] and extended and modified by other authors [75–78].

A second method is the direct-simulation method originated by Bird [79–89]. The gas is represented by a few thousand particles, which are initially distributed uniformly and assigned velocities drawn at random from a drifting Maxwellian distribution. The space region in which their flow is to be computed is divided into a number of contiguous cells, sized so that the gas properties should be nearly uniform throughout each cell at any stage of the motion. The boundary conditions depend on the particular problem to be studied. For example, the steady flow past a solid body is obtained as the long time solution of the unsteady flow generated by the instantaneous insertion of the body into a uniform flow at zero time. The representative molecules are allowed to move with their initial velocities for a time interval Δt_m which is small compared with the mean free time in the unperturbed gas. Those particles that hit the perturbing boundary during the time interval Δt_m are immediately re-emitted with a new velocity chosen at random according to the adopted re-emission law.

At time Δt_m, all the molecules are stopped and all those within a given configuration cell, regardless of their position within the cell, are considered as possible collision partners.

The actual collision probability, for a random pair, however, depends on the molecular model. Hence a random pair is retained or rejected for a collision computation with a retention probability proportional to the product of relative speed V and collision cross section $\sigma(V, \theta)$. When a collision is to be computed, the azimuth angle ε and the impact parameter b are picked at random (with uniform and linearly increasing distribution, respectively).

An angle cutoff is introduced for power law potentials so that $b \leqslant b_{\max}(V)$. The post-collision velocities are then computed and memorized.

Each collision computation advances a time counter for that cell by an amount $\Delta t = (2/Nc)[\pi(b_{\max})^2 Vn]^{-1}$ where n is the number density and N_c is the actual number of simulated molecules in the cell. The second factor in the expression of Δt is the average time between collisions as a beam of molecules of density n moves past a single scattering center at relative speed V. The factor $2/N_c$ reduces the mean time between collisions in proportion to the number of scattering centers (the factor 2 avoids counting each representative molecule simultaneously as a scattering center and as a member of the beam).

Collisions continue to be computed until the time increments add up approximately to Δt_m. When this procedure has been carried out for every cell, the overall time is advanced through Δt_m and each molecule is moved through a distance equal to the product of its velocity and Δt_m. Then follows another collision interval and so on.

At selected time intervals, the number density n in each cell is sampled. This is done because n is required in the collision computation. Other quantities such as mass, momentum and energy fluxes at the boundaries may be sampled. For flows that eventually become steady, successive samples may be averaged after the establishment of steady conditions in order to build up the sample size and reduce the statistical fluctuations in the final results. A similar effect is achieved for unsteady flows by averaging the results from a number of separate runs.

Similar procedures have been widely used in neutron transport [90–92], where it is possible to exploit the fact that the target species has a fixed distribution. Hence the test particle method can be applied without the necessity of an iteration.

It is to be remarked that the name Monte Carlo method is also used for a statistical method for evaluating the collision integrals [93–95] in a numerical method of solution of the Boltzmann equation.

5. Problems of flow and heat transfer in regions bounded by planes or cylinders

The simplest problems of gas dynamics are those which occur in a gas between two parallel plates. Such are the plane Couette and Poiseuille flows described in Section 5 of Chapter VI and the heat transfer which occurs when a gas at rest is placed between parallel plates kept at different temperatures. Next in complexity are the corresponding problems in cylindrical geometry: Couette flow of a gas between two rotating coaxial cylinders, Poiseuille flow in tubes of cylindrical and annular cross section and heat transfer between coaxial cylinders.

The main purpose of the present section is to describe the results obtained for these problems by means of the different methods of solution described in the previous sections.

Let us start with Couette flow between two parallel plates (see Chapter VI, Section 5). The problem has been accurately solved with the linearized BGK model. In this case, if $\varphi(x)$ denotes the ratio of the mass velocity to $U/2$ ($\pm U/2$ being the velocities of the plates), we obtain the following integral equation [47]:

$$\pi^{\frac{1}{2}}\varphi(x) = T_0(\delta/2 - x) - T_0(\delta/2 + x) + \int_{-\delta/2}^{\delta/2} T_{-1}(|x - y|)\varphi(y)\, dy \quad (5.1)$$

where δ is the distance between the plates in θ units.

Analogously we can find:

$$\pi_{xz} = T_1(\delta/2 - x) + T_1(\delta/2 + x) - \int_{-\delta/2}^{\delta/2} \operatorname{sgn}(x - y) T_0(|x - y|)\varphi(y)\, dy \quad (5.2)$$

where π_{xz} is the ratio of the stress tensor p_{xz} to its free molecular value $(-\rho U\pi^{-\frac{1}{2}}/2)$. Note that π_{xz} is constant, in spite of the fact that Eq. (5.2) shows an x-dependence; the constancy of π_{xz} is obvious from conservation of momentum and can be recovered from Eq. (5.2), by differentiating and comparing with Eq. (5.1). Therefore we can take Eq. (5.2) at any point for evaluating π_{xz}; the simplest choice is to take the arithmetical mean of the expression given by Eq. (5.2) for $x = \pm\delta/2$. We obtain:

$$\pi_{xz} = \tfrac{1}{2} + T_1(\delta) + \tfrac{1}{2}\int_{-\delta/2}^{\delta/2} [T_0(\delta/2 + y) - T_0(\delta/2 - y)]\varphi(y)\, dy \quad (5.3)$$

Eq. (5.1) was solved numerically by Willis [47] for δ ranging from 0 to 20. His results for π_{xz} are given in Table I (second column). In connection with Eq. (5.1) the variational procedure has also been used [54]. Let us consider the functional:

$$J(\tilde{\varphi}) = \int_{-\delta/2}^{\delta/2} [\tilde{\varphi}(x)]^2\, dx - \pi^{\frac{1}{2}}\int_{-\delta/2}^{\delta/2}\int_{-\delta/2}^{\delta/2} T_{-1}(|x - y|)\tilde{\varphi}(x)\tilde{\varphi}(y)\, dx\, dy +$$
$$+ 2\pi^{-\frac{1}{2}}\int_{-\delta/2}^{\delta/2} [T_0(\delta/2 + x) - T_0(\delta/2 - x)]\tilde{\varphi}(x)\, dx \quad (5.4)$$

which attains its minimum value

$$J(\varphi) = \min J(\tilde{\varphi}) = \pi^{-\frac{1}{2}}\int_{-\delta/2}^{\delta/2} [T_0(\delta/2 + x) - T_0(\delta/2 - x)]\varphi(x)\, dx \quad (5.5)$$

when $\tilde{\varphi} = \varphi(x)$ solves Eq. (5.1). Comparing Eq. (5.3) and (5.5), we deduce:

$$\pi_{xz} = \tfrac{1}{2} + T_1(\delta) + \frac{\pi^{\frac{1}{2}}}{2}\min J(\tilde{\varphi}) \quad (5.6)$$

In such a way, the stress has a direct connection with the minimum value attained by $J(\tilde{\varphi})$. As a trial function, we can take the continuum solution,

$$\tilde{\varphi}(x) = Ax \tag{5.7}$$

where A is an indeterminate constant. After some easy manipulations based on the properties of the T_n functions, we find [54] an expression for $J(Ax)$ in terms of $T_n(\delta)$ $(1 \leqslant n \leqslant 3)$. The minimum condition is easily found, the corresponding $A = A_0(\delta)$ and $J(A_0 x)$ being again expressed in terms of the $T_n(\delta)$. Inserting $J(A_0 x)$ into Eq. (5.6) we obtain π_{xz}; the resulting values are tabulated versus δ in Table I (third column). It will be noted that the agreement with Willis' results is excellent. But we can see something more, namely that the results obtained by the variational procedure are more accurate than Willis'. As a matter of fact, the variational procedure gives for π_{xz} a value approximated from above, and it is noted that values in the third column are never larger than the corresponding ones in the second column. The fourth column gives the results obtained by using a cubic trial function:

$$\tilde{\varphi}(x) = Ax + Bx^3 \tag{5.8}$$

In this case the algebra is more formidable but still straightforward. It is to be noted that the resulting values for π_{xz} are only slightly different from the previous ones: this is another test of the accuracy of the method.

A variational solution of the integrodifferential form of the linearized Boltzmann equation for plane Couette flow has been given in Ref. [96]

TABLE I. Stress versus the inverse Knudsen number for Couette flow (from Ref. 54)

δ	Willis' result (Ref. 47)	Linear trial function	Cubic trial function
0.01	0.9913	0.9914	0.9914
0.10	0.9258	0.9258	0.9258
1.00	0.6008	0.6008	0.6008
1.25	0.5517	0.5512	0.5511
1.50	0.5099	0.5097	0.5096
1.75	0.4745	0.4743	0.4742
2.00	0.4440	0.4438	0.4437
2.50	0.3938	0.3935	0.3933
3.00	0.3539	0.3537	0.3535
4.00	0.2946	0.2945	0.2943
5.00	0.2526	0.2524	0.2523
7.00	0.1964	0.1964	0.1963
10.00	0.1474	0.1474	0.1473
20.00	0.0807	0.0805	0.0805

where the following approximate expression for the stress at the plate is obtained for any molecular model:

$$\pi_{xz} = \frac{a + \sqrt{\pi}\,\delta}{a + b\delta + c\delta^2} \tag{5.9}$$

where a, b, c are constants which can be computed by quadratures for any molecular model. If we make lengths nondimensional by means of $(\mu/p)(2RT)^{\frac{1}{2}} = 2l/\sqrt{\pi}$ (μ = viscosity coefficient, l mean free path defined by Eq. (V.1.3)), the following values are obtained for a, b, c [96]:

$$a = \frac{4 - \pi}{\pi - 2} \qquad b = \frac{\pi\sqrt{\pi}}{\pi - 2} \qquad c = 1 \qquad \text{(BGK model)} \tag{5.10}$$

$$a = 0.2225 \qquad b = 2.1400 \qquad c = 1 \quad \text{(Maxwell molecules)} \tag{5.11}$$

$$a = 0.3264 \qquad b = 2.1422 \qquad c = 1 \quad \text{(rigid spheres)} \tag{5.12}$$

The maximum disagreement between Eq. (5.9) (with a, b, c given by Eq. (5.10)) and Willis' results is about 0.5%. The results for the BGK models and Maxwell's molecules are in very close agreement except for high Knudsen numbers ($\delta < 2$), where a difference of order 3% arises. Surprisingly enough, the results for rigid spheres are much closer to the BGK results.

Moment methods have been widely used but their accuracy is not high in the nearly free molecular flow. This is due to the fact that the moment methods do not reproduce the correct analytic behavior for $\delta \to 0$. As a matter of fact, as we know (Chapter V, Section 9) logarithmic factors appear in the correct analytic expressions of velocity and shear stress, while moment methods exhibit only rational functions (the same criticism applies, of course, to the variational solution leading to Eq. (5.9) but not to the variational solution obtained by applying the variational method to the integral equation). The simplest version of the method leads to the following formula for the shearing stress:

$$\pi_{xz} = \frac{1}{1 + \pi^{-\frac{1}{2}}\,\delta} \tag{5.13}$$

which differs from the exact numerical solution by about 5%. An accurate but cumbersome solution which exhibits the correct free molecular behavior can be obtained by a proper account of the nonanalytic dependence of f on the mean free path [11].

The same methods have been applied to the problem of heat transfer between flat plates in the linearized approximation [15, 5, 53, 30, 97–99]. The heat flux according to the accurate numerical solution of the BGK model [53] is compared with the experimental data of Teagan and Springer

[100] in Fig. 44. The calculated results are always lower than the experimental data. The same fact occurs for variational solutions based on different collision models (rigid spheres, Maxwell's molecules) [99] and this seems to exclude that the discrepancy is due to the use of the BGK model. It is possible, according to a private comment made by G. S. Springer, that the discrepancy may be explained by a difference between the chamber pressure and the pressure between the plates, while the experimental data are deduced by the assumption that the two pressures are equal. The same kind of discrepancy is found by Huang and Hwang [30], who solved the Holway model [101] for the case of Couette flow of a diatomic gas by the discrete ordinate method.

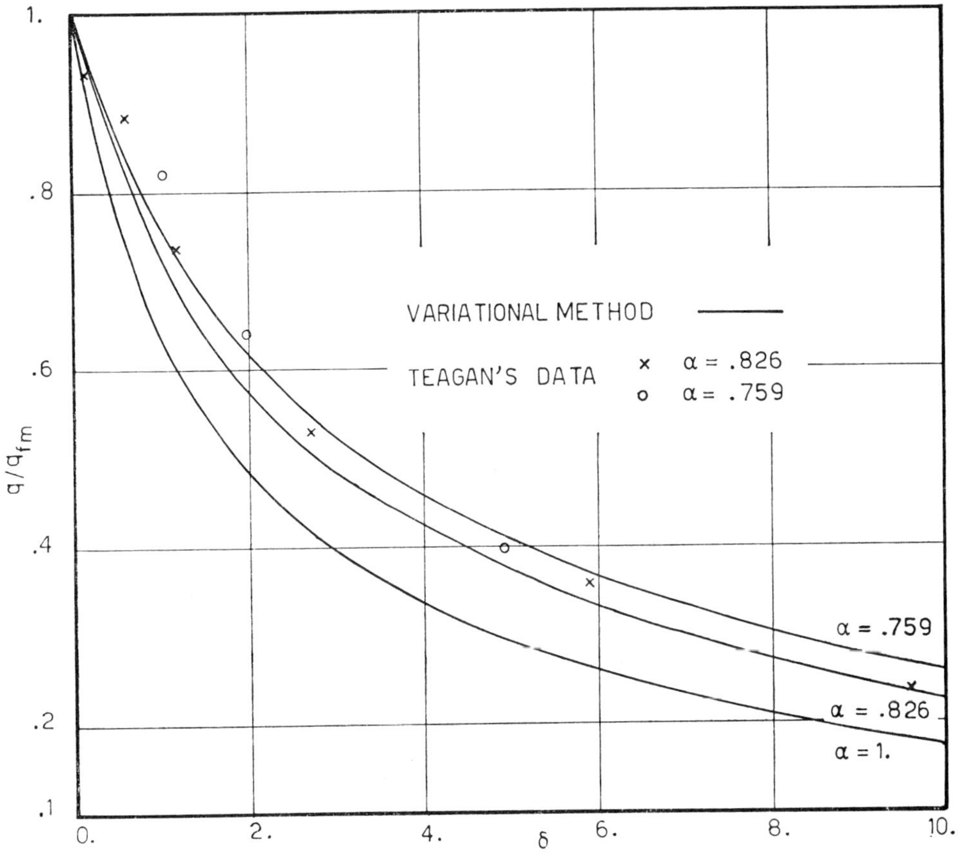

Fig. 44. Heat flux between parallel plates. Comparison of the variational theory (Ref. 53) with the experimental data of Teagan and Springer (Ref. 100); q is the normal heat flux, q_{fm} its free-molecular value.

The full nonlinear problems of Couette flow and heat transfer between parallel plates have also been treated by different authors. The methods of solution include Lees' moment method [102], numerical solution of the BGK and ES integral equations [103, 104, 46], discrete ordinate methods [25, 30] and Monte Carlo methods [74]. Significant comparisons with experiments are not available to the best of the author's knowledge.

Another example is offered by Poiseuille flow. The starting point is Eq. (VI, 5.4). If we adopt the BGK model, we can obtain an integral equation for the mass velocity $v_3(\mathbf{x})$ where $\mathbf{x}$ is the two dimensional vector describing the cross-section of the channel. If we put $v_3(\mathbf{x}) = \frac{1}{2}k\theta[1 - \varphi(\mathbf{x})]$, where $k = p^{-1}\,\partial p/\partial z$, the integral equation to be solved can be written as follows [49, 50]:

$$\varphi(\mathbf{x}) = 1 + \pi^{-1}\iint_{\Sigma(\mathbf{x})} T_0(|\mathbf{x}-\mathbf{y}|)\,|\mathbf{x}-\mathbf{y}|^{-1}\,\varphi(\mathbf{y})\,d\mathbf{y} \tag{5.14}$$

where $\mathbf{x}$ is measured in θ units and $\Sigma(\mathbf{x})$ is the part of the cross-section whose points can be reached from $\mathbf{x}$ by straight lines without intersecting boundaries ($\Sigma(\mathbf{x})$ is the whole cross section Σ when the boundary curvature has constant sign). If the variational method is used, the functional to be considered is now:

$$J(\tilde{\varphi}) = \int_\Sigma [\tilde{\varphi}(\mathbf{x})]^2\,d\mathbf{x} - \int_\Sigma \tilde{\varphi}(\mathbf{x})\int_{\Sigma(\mathbf{x})} T_0(|\mathbf{x}-\mathbf{y}|)\,|\mathbf{x}-\mathbf{y}|^{-1}\,\tilde{\varphi}(\mathbf{y})\,d\mathbf{y} - 2\int_\Sigma \tilde{\varphi}(\mathbf{x})\,d\mathbf{x} \tag{5.15}$$

The value of this functional attains a minimum when $\tilde{\varphi} = \varphi$, where φ satisfies Eq. (5.14) and the minimum value is

$$J(\varphi) = -\int_\Sigma \varphi(\mathbf{x})\,d\mathbf{x}, \tag{5.16}$$

a quantity obviously related to the flow rate. Actual calculations with the variational method have been performed for the case of a slab [54] and a cylinder [105] by assuming a parabolic profile (i.e. a continuum-like solution). In the plane case it is possible to obtain the flow rate in terms of T_n functions, while in the cylindrical case it is also necessary to perform numerical quadratures involving these functions.

In both cases we can evaluate the flow rate with great accuracy and compare the results with the numerical solutions by Cercignani and Daneri [48] and Cercignani and Sernagiotto [49]. As shown by Tables II and III the agreement is very good. The results also compare very well with experimental data of Dong [106] and Knudsen [107] (Figs. 45 and 46) and, in particular, the flow rate exhibits a minimum for $\delta \simeq 1.1$ in the plane case (δ = distance of the plates in θ units) and for $\delta \simeq 0.3$ in the cylindrical case

TABLE II. Flow rate versus the inverse Knudsen number for plane Poiseuille flow (from Ref. 54)

δ	Results from Ref. 48 from above	from below	Variational method
0.01	3.0499	—	3.0489
0.10	2.0331	2.0326	2.0314
0.50	1.6025	1.6010	1.6017
0.70	1.5599	1.5578	1.5591
0.90	1.5427	1.5367	1.5416
1.10	1.5391	1.5352	1.5379
1.30	1.5441	1.5390	1.5427
1.50	1.5546	1.5484	1.5530
2.00	1.5963	1.5862	1.5942
2.50	1.6497	1.6418	1.6480
3.00	1.7117	1.7091	1.7092
4.00	1.8468	1.8432	1.8440
5.00	1.9928	1.9863	1.9883
7.00	2.2957	2.2851	2.2914
10.00	2.7669	2.7447	2.7638

TABLE III. Flow rate versus the inverse Knudsen number for cylindrical Poiseuille flow (from Ref. 105)

δ	Numerical solution (Ref. 49)	Variational method
0.01	1.4768	1.4801
0.10	1.4043	1.4039
0.20	1.3820	1.3815
0.40	1.3796	1.3788
0.60	1.3982	1.3971
0.80	1.4261	1.4247
1.00	1.4594	1.4576
1.20	1.4959	1.4937
1.40	1.5348	1.5321
1.60	1.5753	1.5722
2.00	1.6608	1.6559
3.00	1.8850	1.8772
4.00	2.1188	2.1079
5.00	2.3578	2.3438
7.00	2.8440	2.8245
10.00	3.5821	3.5573

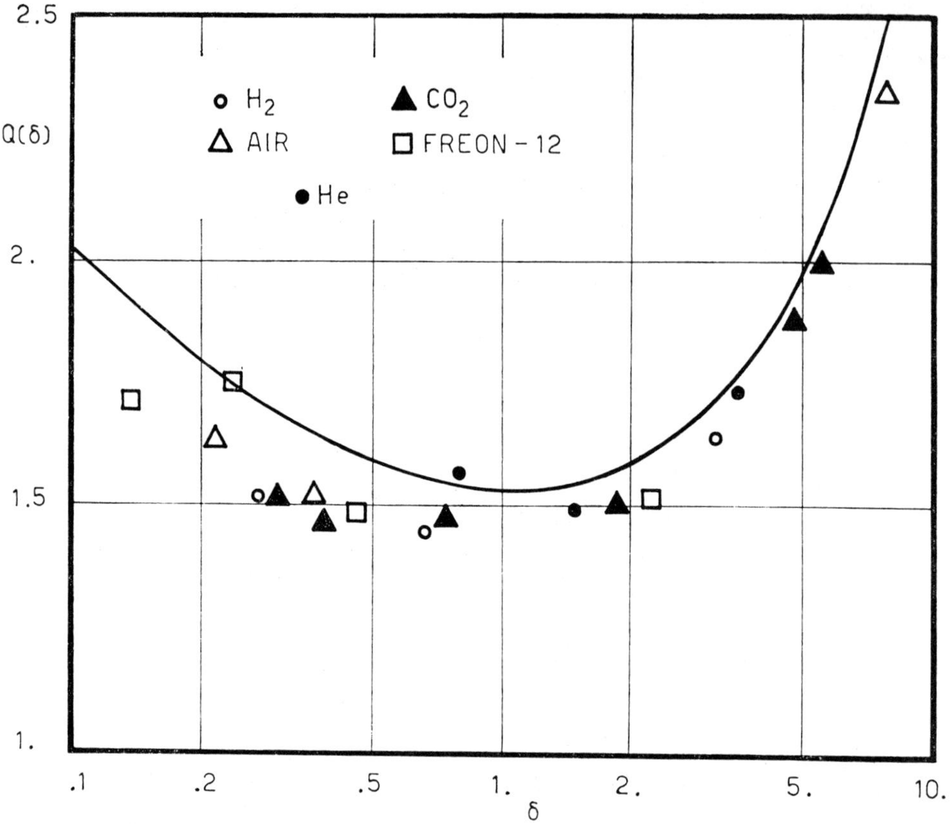

Fig. 45. Comparison of theory (Ref. 48) with experiment (Ref. 108) for the nondimensional flow rate in Plane Poiseuille flow. Here δ is the ratio of the distance d between the plates to $\theta(2RT)^{\frac{1}{2}}$ and Q the ratio of the mass flow to $-d^2(2RT)^{-\frac{1}{2}}(dp/dz)$.

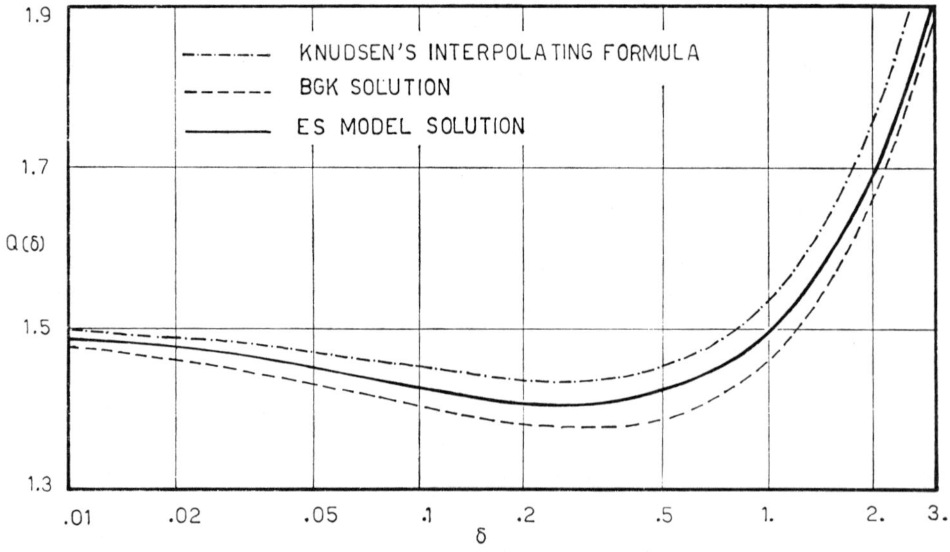

Fig. 46. Comparison between theory and experiments for cylindrical Poiseuille flow. Here δ is the ratio of the radius a of the cylinder to $\theta(2RT)^{\frac{1}{2}}$, and Q the ratio of the mass flow to $-\pi a^3(2RT)^{-\frac{1}{2}}(dp/dz)$.

(δ = radius of the cylinder in θ units). The location of the minimum is in excellent agreement with experimental data, while the values of the flow rate show deviations of order 3% from experimental data in the cylinder case. It is to be noted that the same problem can be treated with the ES model [108] to show that, once the BGK solution is known (Pr = 1) we can obtain the ES for any Prandtl number; in particular the nondimensional flow rate in the cylindrical case is given by

$$Q(\delta, \mathrm{Pr}) = Q(\delta\,\mathrm{Pr}, 1) + (1 - \mathrm{Pr})\,\delta/4 \tag{5.17}$$

If the value $\mathrm{Pr} = \frac{2}{3}$ is chosen, as appropriate for a monatomic gas, we obtain results deviating from experimental data by only 1% or 2% (see Ref. [109] and Fig. 46).

The problem of Poiseuille flow in annular tubes has also been accurately solved by means of the integral equation approach [50]. Moment methods [110] and discrete ordinate techniques [35] do not yield satisfactory results for this kind of problem. Other problems in cylindrical geometry which have been solved with the BGK model are cylindrical Couette flow and heat transfer between concentric cylinders [111, 45].

6. Shock-wave structure

As is well known from the theory of compressible inviscid fluids, shock waves may develop in supersonic flows. In the inviscid fluid approximation, the shock waves are described as discontinuity surfaces. If the Navier-Stokes equations are used, then the shock wave is a region across which physical quantities vary smoothly but rapidly and the shock has a finite thickness, generally of the order of a mean free path. This small thickness indicates that the Navier-Stokes equations, strictly speaking, should not be used. In order to obtain reliable results for the structure of shock waves, use has to be made of the Boltzmann equation.

The simplest case is offered by a steady normal plane shock wave: the gas flows parallel to the x-axis and all the quantities do not depend upon time or the remaining two space coordinates. The gas is in equilibrium at infinity upstream and downstream so that the corresponding distribution functions, f_- and f_+ are Maxwellians:

$$f_\pm = \rho_\pm(2\pi RT_\pm)\exp[-(2RT_\pm)^{-1}(\boldsymbol{\xi} - v_\pm\mathbf{i})^2] \tag{6.1}$$

Conservation of mass, momentum and energy yield:

$$\begin{aligned} \rho v &= A \\ \rho v^2 + p_{11} &= B \\ \rho v(\tfrac{1}{2}v^2 + \tfrac{3}{2}RT) + p_{11}v + q_1 &= C \end{aligned} \tag{6.2}$$

where A, B, C are constants. In particular when $x \to \pm\infty$, ρ, v, p_{11}, T, q_1 tend to $\rho_\pm$, $v_\pm$, $p_\pm = \rho_\pm RT_\pm$, $T_\pm$, 0 respectively, and Eqs. (6.2) yield the well-known Rankine-Hugoniot relations:

$$\begin{aligned} \rho_+ v_+ &= \rho_- v_- \\ \rho_+(v_+^2 + RT_+) &= \rho_-(v_-^2 + RT_-) \\ \rho_+ v_+(\tfrac{1}{2}v_+^2 + \tfrac{5}{2}RT_+) &= \rho_- v_-(\tfrac{1}{2}v_-^2 + \tfrac{5}{2}RT_-) \end{aligned} \tag{6.3}$$

The problem of shock-wave structure is the simplest problem in which the nonlinear structure of the Boltzmann equation plays an essential role.

If the shock wave is weak, so that the downstream and upstream values of density, velocity and temperature are not widely different from each other then we may expect that the derivative $\partial f/\partial x$ is small, of the order of the strength ε of the shock (defined, e.g., by $\varepsilon = (p_- - p_+)/(p_- + p_+)$), and try to expand the solution into a series of powers of ε [112, 113]. The result is similar to the Hilbert expansion described in Chapter V, Section 2, since we are led to

$$Q(f_0, f_0) = 0 \tag{6.4}$$

$$\sum_{k=0}^{n} Q(f_k, f_{n-k}) = \xi_1 \frac{\partial f_{n-1}}{\partial x} \qquad (n = 1, 2, 3) \tag{6.5}$$

Since ε is small the solution must describe a slight departure from a uniform state. Hence we satisfy Eq. (6.4) as well as the orthogonality conditions which arise at the next step by taking the density ρ_0, the velocity v_0, and the temperature T_0 in f_0 to be constant. Hence

$$f_1 = f_0\left[\frac{\rho_1}{\rho_0} + \frac{\xi_1 - v_0}{RT_0} v_1 + \left(\frac{c^2}{2RT_0} - \frac{3}{2}\right)\frac{T_1}{T_0}\right] \tag{6.6}$$

where ρ_1, v_1, T_1 are the first order contribution to ρ, v, T. The next orthogonality conditions lead to

$$\begin{aligned} & v_0 \frac{d\rho_1}{dx} + \rho_0 \frac{dv_1}{dx} = 0 \\ & v_0^2 \frac{d\rho_1}{dx} + 2\rho_0 v_0 \frac{dv_1}{dx} + RT_0 \frac{d\rho_1}{dx} + R\rho_0 \frac{dT_1}{dx} = 0 \\ & \tfrac{3}{2}RT_0 \frac{d\rho_1}{dx} + \rho_0 v_0 \frac{dv_1}{dx} + \tfrac{3}{2}\rho_0 v_0 R \frac{dT_1}{dx} + \tfrac{1}{2}v_0^2 \frac{d\rho_1}{dx} + \\ & \quad + \tfrac{3}{2}\rho_0 v_0^2 \frac{dv_1}{dx} + \rho_0 RT_0 \frac{dv_1}{dx} + v_0 RT_0 \frac{d\rho_1}{dx} + v_0 R\rho_0 \frac{dT_1}{dx} = 0 \end{aligned} \tag{6.7}$$

These equations are an algebraic system for $d\rho_1/dx$, dv_1/dx, dT_1/dx; in order to have a nonvanishing solution the determinant

$$D = \rho_0^2 v_0[\tfrac{5}{3}RT_0 - v_0^2] \tag{6.8}$$

must vanish. If we exclude $v_0 = 0$, we are led to

$$v_0 = \sqrt{\tfrac{5}{3}RT_0} = c_0 \tag{6.9}$$

where c_0 is the sound speed. This result means that a weak shock wave can be the perturbation of a uniform flow only if the latter has Mach number unity; this is quite clear from the inviscid theory of shock waves. If we proceed further, we can compute the expression of f_2 [113]. The next orthogonality condition leads to an inhomogeneous algebraic system for $d\rho_2/dx$, dv_2/dx, dT_2/dx, whose determinant is D and hence vanishes. Then the source term of the algebraic system must satisfy an orthogonality condition which leads to:

$$\frac{d}{dx}\left(\frac{\rho_1}{\rho_1^+}\right) = -L\frac{d^2}{dx^2}\left(\frac{\rho_1}{\rho_1^+}\right) \tag{6.10}$$

where [113]:

$$L = (\tfrac{3}{2}\kappa T_0 + \tfrac{4}{3}\mu\tfrac{5}{3}RT_0)\Big/\rho_1^+\left(c_0^3 + \tfrac{2}{3}\rho_0\left[\frac{c_0^3}{2\rho_0} - \frac{p_0 c_0}{\rho_0^2}\right]\right) \tag{6.11}$$

Here κ and μ denote the heat conduction and viscosity coefficients. Integration of Eq. (6.10) yields:

$$\frac{\rho_1}{\rho_1^+} = \mathrm{Th}\left(\frac{x - x_0}{L}\right) \tag{6.12}$$

Notice that the constant L is a measure of the shock thickness and is of the order of the ratio of the mean free path to the strength of the shock ($\rho_1^+ \simeq \varepsilon$). The Navier-Stokes equations also lead to Eq. (6.12) for weak shocks [114] and hence are correct to the lowest order in such a case.

The other extreme case is the infinitely strong shock profile. H. Grad [115] suggested that the limit of f for $\varepsilon \to \infty$ exists (for collision operators with finite collision frequency) and is given by a multiple of the delta function centered at the upstream velocity plus a comparatively smooth function for which it is not hard to derive an equation. The latter seems more complicated than the Boltzmann equation itself, but the presumed smoothness of the solution should allow a simple approximate solution to be obtained. The simplest choice for the smooth remainder is a Maxwellian [115] whose parameters are determined by the conservation equations.

Grad's approach is perhaps the most recent analytical method for dealing with the problem of shock wave structure. In its simplest version it is strictly related to, and perhaps inspired by, the first approach to the same problem used by H. M. Mott-Smith [8] in 1954. The latter author represented the approximating distribution function in the form

$$f(x, \boldsymbol{\xi}) = a_+(x)f_+(\boldsymbol{\xi}) + a_-(x)f_-(\boldsymbol{\xi}) \tag{6.13}$$

where f_+ and f_- are given by Eq. (6.1) and a_+, a_- are to be determined by suitable moment equations. The obvious choice for the first few moment equations are the conservation equations leading to Eq. (6.2), that is to:

$$\begin{aligned}
&a_+(x)\rho_+v_+ + a_-(x)\rho_-v_- = \rho_\pm v_\pm \\
&a_+(x)\rho_+v_+{}^2 + a_+(x)RT_+\rho_+ + a_-(x)\rho_-v_-{}^2 + a_-(x)RT_-\rho_- \\
&\qquad\qquad = \rho_\pm v_\pm + \rho_\pm RT_\pm \qquad (6.14)\\
&a_+(x)v_+\rho_+(\tfrac{1}{2}v_+{}^2 + \tfrac{5}{2}RT_+) + a_-(x)\rho_-v_-(\tfrac{1}{2}v_-{}^2 + \tfrac{5}{2}RT_-) \\
&\qquad\qquad = \rho_\pm v_\pm(\tfrac{1}{2}v_\pm{}^2 + \tfrac{5}{2}RT_\pm)
\end{aligned}$$

Dividing by $\rho_+v_+ = \rho_-v_-$ the first of these equations leads to

$$a_+(x) + a_-(x) = 1 \tag{6.15}$$

Hence the remaining two equations are automatically satisfied provided, of course, the upstream and downstream values satisfy Eqs. (6.3). An additional moment equation is thus needed. A possible choice is to take $\varphi_i = \xi_i{}^2$ in Eq. (2.1) (with $\partial/\partial t = \partial/\partial y = \partial/\partial z = 0$ and f given by Eq. (6.13)). The result is:

$$\rho_-v_-(3RT_- + v_-{}^2)\frac{da_-}{dx} + \rho_+v_+(3RT_+ + v_+{}^2)\frac{da_+}{dx} = -\alpha a_+a_- \tag{6.16}$$

where

$$\alpha = \int(\xi_1{}^2 - \xi_1'^2)f_+(\boldsymbol{\xi})f_-(\boldsymbol{\xi}_*)B(\theta, V)\,d\theta\,d\varepsilon\,d\boldsymbol{\xi}_*\,d\boldsymbol{\xi} \tag{6.17}$$

If we eliminate a_+ by means of Eq. (6.15) we obtain an equation of the form

$$\frac{da_-}{dx} = -\beta a_-(1 - a_-) \tag{6.18}$$

where, if use is made of the third Rankine-Hugoniot relation:

$$\beta = \frac{\alpha}{2\rho_-v_-R(T_- - T_+)} \tag{6.19}$$

Eq. (6.18) can be immediately integrated to yield

$$a_-(x) = \frac{1}{1 + e^{\beta x}} \tag{6.20}$$

where an arbitrary constant has been set equal to zero since this is equivalent to fixing the position of the center of the shock (which is arbitrary). Hence

$$a_+(x) = \frac{e^{\beta x}}{1 + e^{\beta x}} \tag{6.21}$$

and

$$\rho(x) = a_-\rho_- + a_+\rho_+ = \frac{\rho_- + \rho_+ e^{\beta x}}{1 + e^{\beta x}} = \rho_- \frac{1 + \left(\dfrac{4M_-^2}{3 + M_-^2}\right) e^{\beta x}}{1 + e^{\beta x}} \tag{6.22}$$

$$v(x) = \frac{\rho_- v_-}{\rho(x)} = v_- \frac{1 + e^{\beta x}}{1 + \left(\dfrac{4M_-}{3 + M_-^2}\right) e^{\beta x}} \tag{6.22a}$$

$$T(x) = T_-(1 + \tfrac{5}{9} M_-^2)a_- + T_+(1 - \tfrac{5}{9} M_+^2)a_+ - \frac{5}{9}\left(\frac{v}{v_-}\right)^2 M_-^2 \tag{6.23}$$

where

$$M_- = v_-\sqrt{\frac{3}{5RT_-}} \qquad M_+ = v_+\sqrt{\frac{3}{5RT_+}} \tag{6.24}$$

are the upstream and downstream Mach numbers.

The calculation of the constant β is complicated except for Maxwell's molecules; in the latter case

$$\beta = \left[\frac{90M_-}{3 + M_-^2} \frac{(M_-^2 - 1)^2}{16M_-^4 - (3 + M_-^2)^2} \sqrt{\frac{2\pi}{15}}\right] l_-^{-1}$$

where l_- is the upstream mean free path defined by Eq. (V.1.3) for $p = p_-$, $\mu = \mu_-$, $T = T_-$. For $M_- \to \infty$, $\beta l^- \simeq (6/\sqrt{(2\pi/15)})/M_-$ tends to zero; that is, the ratio of the upstream mean free path l_- to the shock thickness ($\simeq \beta^{-1}$) tends to zero. This occurs for any power-law molecular interaction; for rigid spheres, however, the same ratio tends to a finite limit.

If we select another moment equation in the place of that corresponding to $\varphi_i = \xi_1^2$, we are led to the same results except for a change in the expression and value of the constant β. The latter is, however, very important because it is related to the thickness of the shock wave. Choosing $\varphi_i = \xi_1^3$ rather than $\varphi_i = \xi_1^2$, for example, leads to a change of about 25% in the shock thickness. This fact indicates that the Mott-Smith method although qualitatively correct is not quantitatively adequate. In addition the results for weak shocks ($M \to 1$) are not in agreement with the weak shock theory or, equivalently, the Navier-Stokes results. Because of this fact several authors have presented modifications of Mott-Smith's method. The most interesting ones appear to be those due to Salwen et al. [116] and Holway [117]. According to the first method, Eq. (6.13) is replaced by

$$f(x, \boldsymbol{\xi}) = a_+(x)f_+(\boldsymbol{\xi}) + a_-(x)f_-(\boldsymbol{\xi}) + a_0(x)f_0(\boldsymbol{\xi}) \tag{6.25}$$

where f_+ and f_- are as above and

$$f_0 = (\xi_1 - u_0)(h/\pi)^3\exp\{-h[(\boldsymbol{\xi} - v_0\mathbf{i})^2]\} \tag{6.26}$$

where u_0 and v_0 are constants. These constants turn out to be determined by the conservation equations, which also give a relation among the three functions $a_0(x)$, $a_+(x)$, $a_-(x)$.

Accordingly, two further moment equations are necessary. Salwen et al. [116] considered two choices: $\varphi_i = (\xi_1^2, \xi_1^3)$ and $\varphi_i = (\xi_1^2, \xi_1\xi^2)$. The resulting differential equations were then solved numerically. The discrepancy between the solutions corresponding to the two choices is considerably less than for the "bimodal distribution" of Mott-Smith. The most remarkable improvement, however, is the removal of the disagreement with weak shock theory for $M_- \to 1$.

Holway's method [117] amounts to using Eq. (6.25), where f_- is given by Eq. (6.1) but f_+ is as follows:

$$f_+ = (h_1/\pi)^{\frac{1}{2}}(h_2/\pi)\exp[-h_1(\xi_1 - Q)^2 - h_2(\xi_2^2 + \xi_3^2)] \tag{6.27}$$

where Q, h_1, h_2 are three functions to be determined together with $a_\pm(x)$. The results of all these methods give a temperature maximum larger than T_+. The maximum is not very marked, and it is not clear whether it is due to an inaccuracy in the approximation. It is notable however that the sign of the heat flux does not change when the temperature gradient does. The different results for the shock thickness defined by

$$L = \frac{\rho_+ - \rho_-}{\left(\dfrac{d\rho}{dx}\right)_{\max}} \tag{6.28}$$

are compared in Fig. 47; a comparison of the "bimodal distribution" for a Lennard-Jones potential between the molecules [118] with the experimental data [119] is shown in Fig. 48.

It is clear that the Navier-Stokes results are not in agreement with experiments for $M_- > 2$. The situation is not improved by considering the Burnett equations (cf. Chapter V, Section 3) or Grad's thirteen moment equations. These sets of equations do not have a solution, in fact, for $M_- > 2.1$ and $M_- > 1.65$, respectively. In general, as mentioned in Section 2, moment methods present difficulties for $M_- > 1.851$.

A general method, extending Mott-Smith's approach, would seem to use a linear combination of a number of Maxwellians; in this connection the analytic results of Deshpande and Narasimha [120] should be very useful.

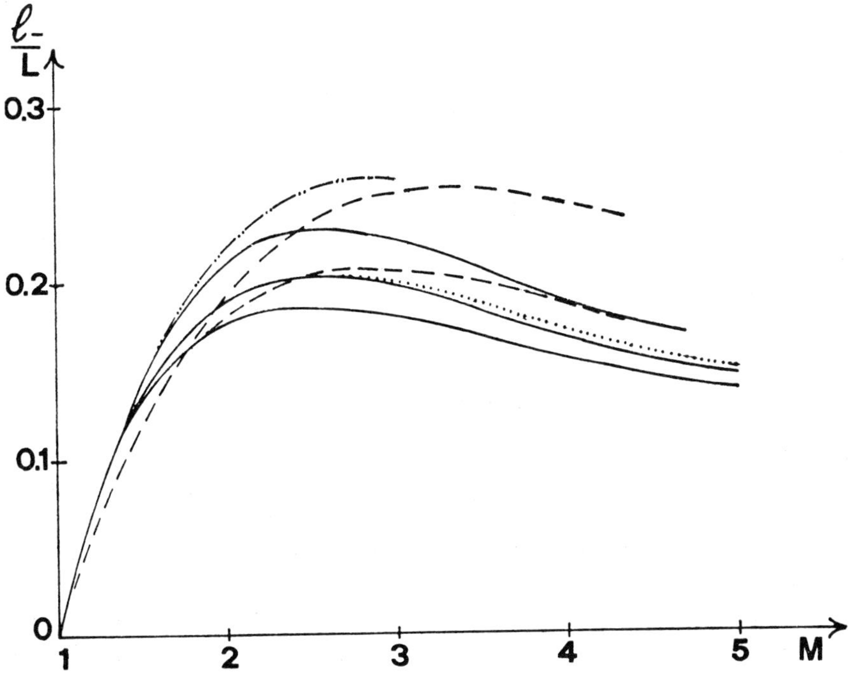

Fig. 47. Different results for the shock thickness. The full line curves correspond to the results of Salwen, Grosch, Ziering for different choices of the moment equations (from above; $(\xi_1^3, \xi_1\xi^2)$, $(\xi_1^2, \xi_1\xi^2)$, (ξ_1^2, ξ_1^3)). The dashed line corresponds to Mott-Smith's results for two choices of moments (from above: ξ_1^2 and ξ_1^3). The dotted line corresponds to the results of Holway. The dot and dash line is the inverse shock thickness according to Navier-Stokes theory.

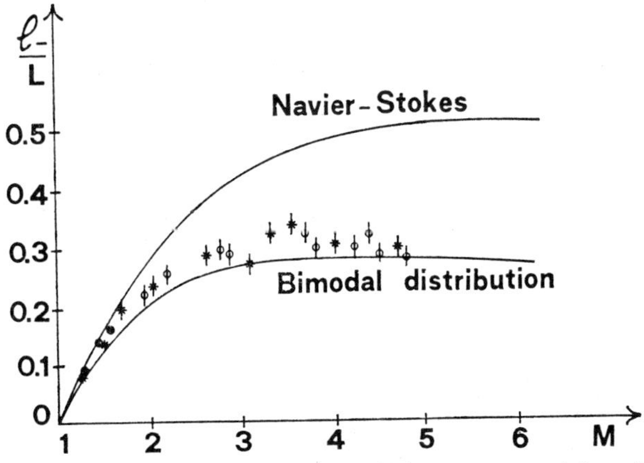

Fig. 48. Comparison between the inverse shock thickness computed for a bimodal distribution with a Lennard-Jones potential (Ref. 118) and experimental data from different authors (see Ref. 119).

The main problem for the methods of the type due to Mott-Smith remains the choice of moment equations; a way out of the difficulty would be to use a variational method, but, as discussed in Section 3, the situation in this connection is not very encouraging for nonlinear problems, though Oberai [121] and Narasimha et al. [122] obtained interesting results for the shock structure by using the least square method (Eq. (3.16)).

An exact numerical solution of the problem for the BGK model was obtained by Liepmann et al. [43], who used the integral form of the BGK model equation; thus they obtained three integral equations for the three macroscopic quantities ρ, v, T. Those equations were solved by the method of successive approximations with the Navier-Stokes solution as zero-th iterate. The BGK solution does not give the maximum in the temperature curve noted above and the density and velocity profiles are much less antisymmetrical about the appropriate shock center than are those obtained by the bimodal and trimodal approximations to the solution of the Boltzmann equation.

The BGK solution was recalculated with greater accuracy by Anderson [104]. More recently, Segal and Ferziger [123] studied the problem using several models of the Boltzmann equation: the BGK and ES models described in Chapter II, Section 10, a so-called polynomial model of the class of Sirovich's nonlinear models [124] and a new model, the so called trimodal gain function model. The latter was developed supposedly for dealing with the problem of shock structure; it is essentially based upon replacing Φ in Eq. (II.10.3) by an expression of the form (6.25), where the coefficients, and parameters are suitably expressed in terms of ρ, v, T and the upstream and downstream values. Thus the qualitative features of the Mott-Smith approach are embodied in the model, but no committment is made to a detailed validity of the multimodal approximation for the solution.

The problem of shock wave structure has been treated by several authors using Monte Carlo methods [73, 76, 87, 95]. The Monte Carlo results as well as the numerical results for the trimodal model of Segal and Ferziger [123] seem to agree in showing no evidence of the temperature overshoot mentioned above.

Concerning comparison with experiments, it is not clear at present whether the experimental data (including those on the properties of the gas) are known to a sufficient accuracy to discriminate between different results. In fact, as remarked by Bird [87], a comparison based on either the shock wave thickness or even the complete density profile does not provide a verdict on which of the methods provides the best description of the shock structure, and the prospects for a sufficiently accurate experimental determination of the higher moments are not hopeful.

A great deal of research has also been devoted to the problem of the shock structure for mixtures and polyatomic gases. In particular, for a binary mixture a shock wave may lead to a diffusive separation of the species. Sherman [125] worked out the continuum theory of this effect. According to his results, through the shock, the heavier gas goes faster than the mixture, which, accordingly, becomes richer in the light species. For a fairly small concentration of the heavier component (2% for an Argon-Helium mixture, with mass ratio 10), Sherman found that, even in fairly weak shocks, the heavy species is first accelerated, then decelerated, reaching a maximum in velocity of 18% in excess of the upstream value. These results suggested that a continuum theory was not adequate and a kinetic theory was required. Thus Oberai [126], by means of the Mott-Smith method, obtained the expected result that the so-called Sherman anomaly was physically not significant and was in fact due to the use of the continuum equations for a situation in which they do not apply. The problem was reconsidered by several authors [83, 127, 128] and Oberai's conclusions were confirmed.

The study of shock waves in polyatomic gases leads to most interesting problems concerning the processes of adjustment of rotational and translational energy. We restrict ourselves to quoting some of the basic contributions [129–132].

7. External flows

In this section we shall discuss the results available for flows and heat transfer processes occurring in an infinite expanse of gas limited by solid surfaces. Mathematically the simplest problems are pure heat transfer (for small temperature differences) and flow around slowly rotating bodies. The problem of heat transfer from a sphere was treated by the variational method with results in good agreement with experiment [133]. The problem of a slowly rotating cylinder (limiting case of Couette flow) has been solved numerically as well as by an approximate analytical method [51].

The problem of torque upon a slowly rotating sphere was solved [68] by the method of boundary conditions mentioned at the end of Section 3.

When we pass to the more interesting problems of flow past a solid body, it is natural to study the low Mach number case first. Here we meet the analogue of Stokes' paradox, as discussed in Chapter VI, Section 13, which forbids using the simplest linearization for two-dimensional flows. We can, however, treat flows past three-dimensional bodies. The low Mach number flow past on axisymmetric body can be treated [134] by the variational method applied to the integral form of the BGK equation. Explicit variational calculations have been made for the case of a sphere [135]. Fig. 49 shows a comparison between the variational results for the drag coefficient C_D of a sphere and the semiempirical formula proposed by Millikan [136] to interpolate his experimental data. In the same plot Stokes' classical result and Sherman's interpolating formula for the drag are also reported. The latter formula reads as follows:

$$C_D(R)/C_{DFM} = [1 + 0.685R]^{-1} \tag{7.1}$$

where R is the radius of the sphere in $\theta(2RT)^{\frac{1}{2}}$ units and $C_{DFM} = C_D(0)$ is the drag coefficient in free molecular flow. Eq. (7.1) is a particular case of a universal formula [137] relating a quantity as a function of the Knudsen

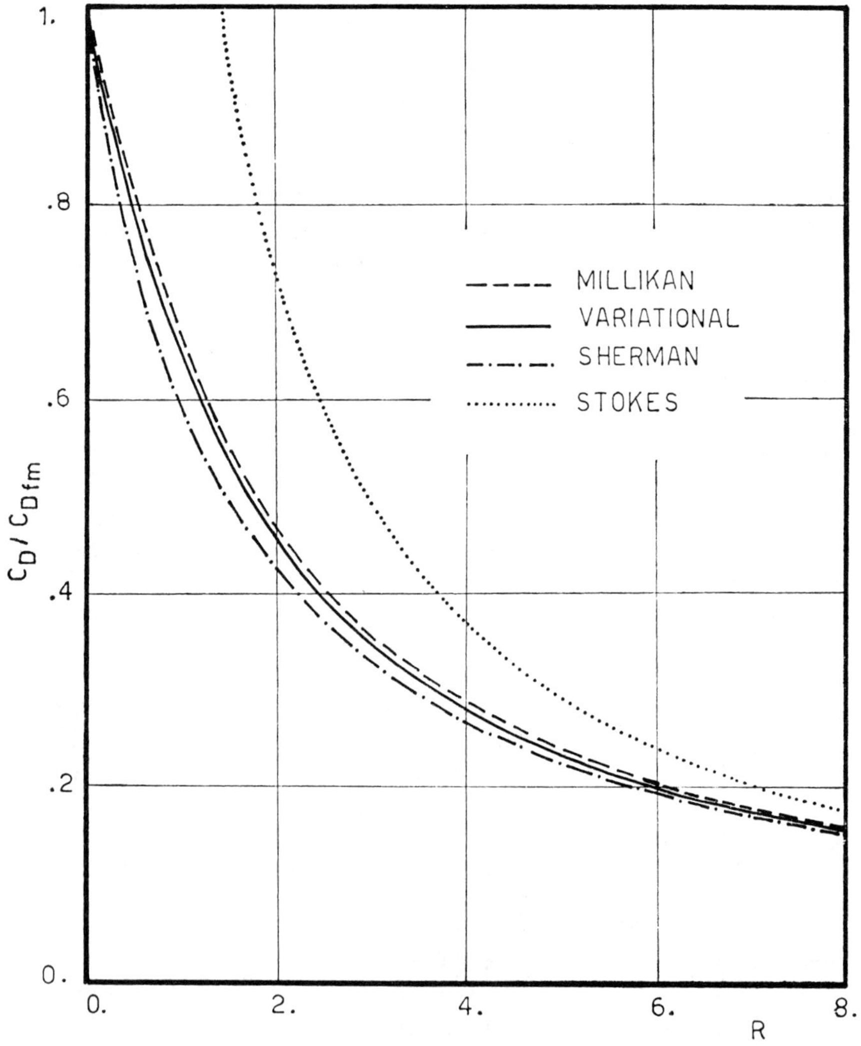

Fig. 49. Low speed drag coefficient of a sphere versus the nondimensional radius R. Millikan's curve interpolates his experimental data.

number Kn (here Kn $= 1/R$) to its free-molecular value F_{FM} and its value according to continuum theory, $F_c(\mathrm{Kn})$, provided $F_c(\mathrm{Kn})/F_{FM} \to 0$ as $\mathrm{Kn} \to 0$. Sherman's general formula reads as follows:

$$F/F_{FM} = (1 + F_{FM}/F_c)^{-1} \tag{7.2}$$

and is in reasonably good agreement with experimental data for most subsonic experiments. A notable exception is offered by Poiseuille flows in long tubes, for which Sherman's formula (applied to $1/Q(\delta)$) does not predict a minimum in the flow rate. This exception may be due to the fact that $1/Q(\delta)$ is at variance with other overall quantities such as drag or heat transfer coefficients and is not simply related to the flow of a conserved quantity.

A treatment of flow past thin airfoils was sketched in Section 13 of Chapter VI. The treatment there applied to almost specularly reflecting surfaces and moderate Mach numbers.

When the Mach number is not small and the gas surface interaction is far from specular reflection the problem of flow past a solid body becomes extremely difficult even for the simplest shapes. Apart from analytical studies of the far field [138, 139], the approaches which have been used are discrete ordinate methods and interpolation between nearly-free molecular and slip regimes. Particular attention has been paid to hypersonic flow ($M_\infty \gg 1$) since the results are of the utmost importance for upper atmosphere flight. We cannot deal here with the details of the calculations and results obtained by the different authors in the slip regime, since it would be impossible to do justice to all the authors without exceeding the limit of the present treatment. We refer the reader to the books of Shidolovski [140] and of Hayes and Probstein [141] as well as to the surveys of Schaaf [142], Sherman [137] and Potter [143]. In connection with nearly free molecular calculations, the work of Willis [144–146], as well as the asymptotic results for slender bodies obtained by Grad and Hu [147], Hamel and Cooper [148] and Pan [149] are particularly worth mentioning.

A basic problem arises in connection with the flow of a gas over a very sharp flat plate, parallel to the oncoming stream. When the Reynolds number $\mathrm{Re} = \rho_\infty V_\infty L/\mu_\infty$ based on the plate length is very great, the usual picture of a potential flow plus a boundary layer is valid everywhere except near the leading and trailing edges. Estimates based on the recent work of Stewartson [150] and Messiter [151] show that the Knudsen number at the trailing edge is of order $M_\infty \mathrm{Re}^{-\frac{3}{4}}$ where M_∞ is the upstream Mach number. As a result kinetic theory is not needed to investigate the main behavior near the trailing edge. For the leading edge, the Knudsen number is of order M_∞; hence in supersonic or, even more, hypersonic flow, the flow in the region about the leading edge should be considered as a typical problem in kinetic theory.

In particular, the boundary layer and the outer flow are no longer distinct from each other although a shock-like structure may still be identified [152–154]. The name of merged layer regime is used in such a case. There are several methods based on simplified continuum models, represented by the papers of Oguchi [155], Shorenstein and Probstein [156], Chow [157, 158], Rudman and Rubin [159], Cheng et al. [160] and Kot and Turcotte [161], which usefully predict surface and other gross properties in the merged layer regime. The good agreement between this kind of theory and experiment gives new evidence of the importance of the Navier-Stokes equations. Nevertheless, if we go sufficiently close to the leading edge, the Navier-Stokes equations must be given up in favour of the Boltzmann equation. A full kinetic theory treatment, however, is very difficult and very few papers deal with this problem. Kogan and Degtyarev [162] studied the flow over a very short flat plate, so that the whole flow is in the nearly free-molecular regime; this makes it difficult to draw significant conclusions about the leading edge effects. Charwat [163] assumed that the leading edge region is in free-molecular conditions, which is very doubtful. Huang and coworkers [28, 29, 164] carried out extensive computations based on the discrete ordinate technique for the BGK model and were able to show the process of building the flow picture assumed in the simplified continuum models mentioned above. The Monte Carlo computations of Vogenitz et al. [165] give a solution of the leading edge problem for a sufficiently long plate. All of these papers show evidence of a shear flow region near the plate, shielded from the upstream flow by a compression region. The results for the surface pressure (or, rather, the normal stress p_{22}) show that the value at the leading edge is at least twice the free-molecule result (for $5.5 < M_\infty < 30$), which is evidence of a significant upstream influence (the experiments of Joss and Bogdonoff [166] indicate an upstream influence for at least five mean free paths at $M_\infty = 26$).

8. Expansion of a gas into a vacuum

A fundamental theoretical and experimental problem of rarefied gas dynamics is the free expansion of a gas into a vacuum. This expansion occurs, for example, in the discharge from an orifice into a low pressure chamber.

This problem embodies, within a simple framework, a transition from continuum almost to free-molecule flow without the usual complication of the effects of gas-surface interaction. As remarked by Ashkenas and Sherman [167], the flow along the axis of the expanding jet can be to some extent simulated by the spherically symmetric expansion of a monatomic gas from a near continuum source. Accordingly, several authors have studied the spherical source flow. According to inviscid gas theory such an expansion should accelerate the gas to any Mach number and to an arbitrarily low density. However, at large distances from the sources the density will be so low, that collisions will be unable to support the continuum expansion. In fact the average energy due to random motion perpendicular to the streamlines is continually decreasing and is being fed into the mean motion of the gas as well as the random motion parallel to the streamlines. Since random motion is connected with temperature, this circumstance is often referred to as the "freezing" of the parallel temperature.

The first careful attempts at quantifying such a picture for a spherical expansion by means of a kinetic theoretic formulation were by Hamel and Willis [168] and Edwards and Cheng [169], who both used moment equations and the so-called hypersonic approximation. The latter is based on an expansion in powers of S^{-1} of all moments of the distribution function where S is the speed ratio. Thus a rational truncation of the infinite set of moment equations is achieved. A solution of these equations can be found, which asymptotically approaches the isentropic flow solution for small values of a suitably scaled radial distance. The mathematical theory was formalized by Freeman [170] and applied to unsteady problems by Freeman and Grundy [171].

The technique introduced by Freeman amounts to an asymptotic expansion in powers of the source Knudsen number. At large distances from the source, the standard (Chapman-Enskog) expansion breaks down, but it is possible to re-scale the Boltzmann equation in this outer region, and the resulting moment equations form a closed set.

The set of moment equations to be solved is:

$$\begin{aligned} \frac{d}{dr}(r^2 p_{\|}) &= \frac{\bar{\nu}}{V_\infty} r^2(\rho RT - p_{\|}) \\ r\frac{d}{dr}(RT) &= \frac{2}{3}\left(\frac{p_{\|}}{\rho} - 3RT\right) \end{aligned} \tag{8.1}$$

where $p_{\|}$ is the normal stress in radial direction and $\bar{\nu}$ is an average collision frequency. The radial velocity is constant to order $O(1/r^2)$ and hence $\rho = \rho_1(r_1/r)^2$, where the subscript 1 designates a reference upstream point where the flow is hypersonic, but still isentropic. Since $\bar{\nu}$ is proportional to ρ Eq. (8.1) can be written as follows:

$$\frac{dT_{\|}}{dr} = (\beta/r^2)(T - T_{\|}) \tag{8.2}$$

$$r\frac{dT}{dr} = \tfrac{2}{3}(T_{\|} - 3T) \tag{8.3}$$

where

$$T_{\|} = \frac{p_{\|}}{\rho R} \tag{8.4}$$

and

$$\beta = \frac{\rho_1 r_1^{\,2}}{V_\infty}(\bar{\nu}/\rho) \tag{8.5}$$

depends on T alone, at least for Maxwell's molecules (more generally, a dependence on both T and $T_{\|}$ could be allowed [168]). $T_{\|}$ can be eliminated between Eqs. (8.2) and (8.3) to yield:

$$r^2\frac{d^2T}{dr^2} + (3r + \beta)\frac{dT}{dr} + \frac{4}{3r}\beta T = 0 \tag{8.6}$$

For inverse power potentials β can be taken to vary according to

$$\beta = \beta_\infty\left(\frac{T}{T_\infty}\right)^\alpha \tag{8.7}$$

where α is an exponent related to the force exponent. Hence if we let

$$\theta = T/T_\infty \qquad s = r/\beta_\infty \tag{8.8}$$

Eq. (8.6) can be rewritten as follows:

$$s^2\frac{d^2\theta}{ds^2} + (3s + \theta^\alpha)\frac{d\theta}{ds} + \frac{4}{3s}\theta^{1-\alpha} = 0 \tag{8.9}$$

If $\alpha = 0$ (Maxwell's molecules), Eq. (8.9) can be solved in terms of confluent hypergeometric functions and a unique solution can be obtained which satisfies the conditions $\theta(s) \to s$ for $s \to \infty$, $\theta = O(s^{-\frac{4}{3}})$ for $s \to 0$ (approach to the isentropic solution). Otherwise, Eq. (8.9) can be integrated numerically [168, 169]. Some results for $\alpha = 0$ (Maxwell's molecules) and $\alpha = \frac{1}{2}$ (rigid spheres) are given in Figs. 50 and 51. The asymptotic behavior of θ can be obtained by letting

$$\theta = 1 + \frac{c}{s} + \cdots \tag{8.10}$$

Inserting Eq. (8.10) and comparing the coefficients of the terms of order $1/s$ yields $c = \frac{4}{3}$. We may now compute the asymptotic behavior of the transversal temperature:

$$T_\perp = \frac{3T - T_{\|}}{2} \tag{8.11}$$

Eq. (8.3) shows that

$$\theta_\perp = \frac{T_\perp}{T_\infty} = -\frac{3}{4}s\frac{d\theta}{ds} = \frac{3}{4}\frac{c}{s} + \cdots = \frac{1}{s} + \cdots \tag{8.12}$$

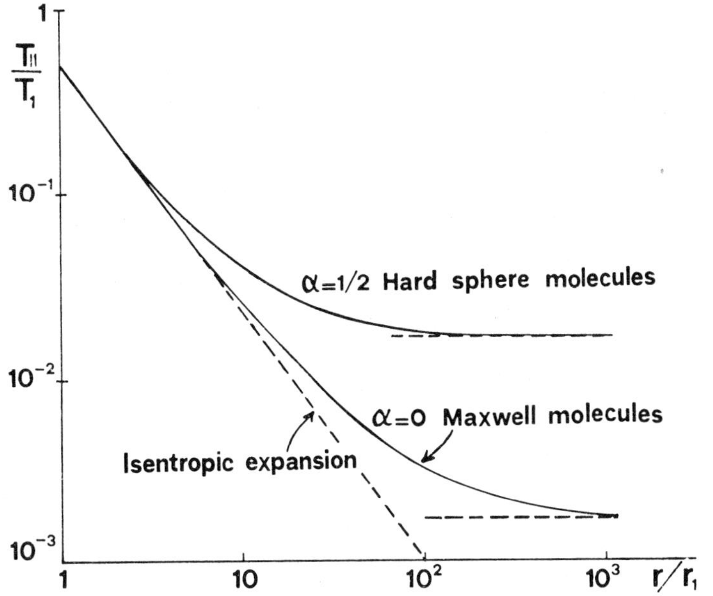

Fig. 50. The results of Hamel and Willis for the parallel temperature in the expansion from a spherical source.

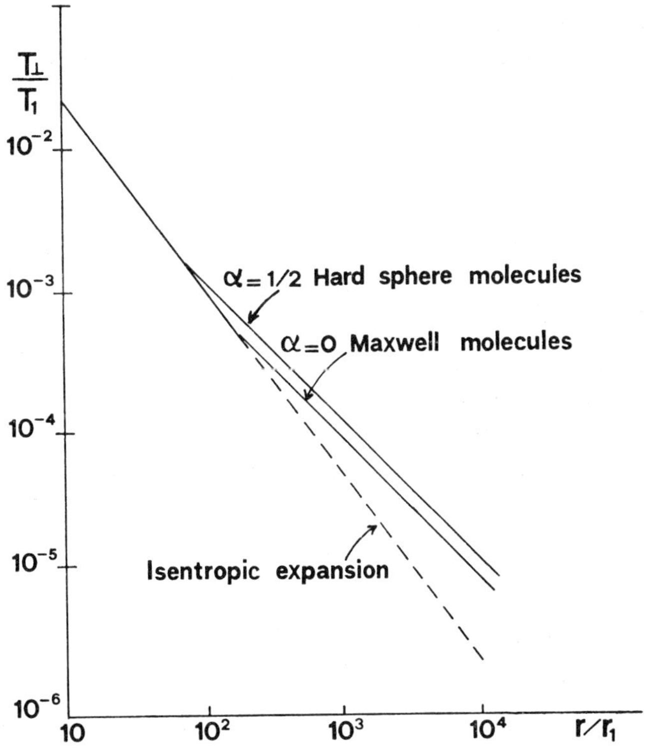

Fig. 51. The results of Hamel and Willis for the transverse temperature in the expansion from a spherical source.

Hence for a spherical source, a freezing of the parallel temperature is found; no such freezing is found, however, for a cylindrical source, in contrast with a previous, extremely simplified analysis of Brook and Oman [172]. No accurate prediction of the observed freezing phenomena for the spherical case is obtained by the Navier-Stokes equations in the hypersonic approximation, since the solution starting out along the isentrope at small distances from the source eventually violates the hypersonic assumption on which it is based.

When the kinetic theory descriptions of the translational freezing process were first published, they appeared to contradict molecular-beam findings concerning the distribution of random velocities perpendicular to the streamlines. Specifically, according to Eq. (8.12), it was predicted that the energy in these degrees of freedom eventually diminished as r^{-1}, while experiments seemed clearly to indicate an r^{-2} dependence [173]. This apparent contradiction was elucidated by Edwards and Cheng [174]. It was found that, as $r \to \infty$ the distribution of random velocities perpendicular to the streamlines has a fairly narrow central spike with rather thick wings. The spike width decreases as $1/r$ and, since the perpendicular temperature measurements are based on the spike width, it explained why they indicate a temperature decay proportional to $1/r^2$. However, experiments measuring the energy content of the wings, if possible, should indicate the r^{-1} behaviour. A study of the fourth order moments of the distribution function for Maxwell's molecules was performed by Freeman and Thomas [175]. Their results were in qualitative agreement with the previous results of Edwards and Cheng [174].

The same conclusions are indirectly confirmed by the results of Miller and Andres [176] who, in order to treat the intermolecular potential in a realistic fashion, assumed a distribution function of the form (6.27) which, according to Willis, Hamel and Lin [177], is valid only for $T_{\parallel}/T_{\perp} \lessdot 2$. The resulting behaviour of $T_{\perp}$ does not follow the r^{-1} law predicted by Eq. (8.12). The validity of the assumption that the free jet can be approximated by a spherical source was supported by Grundy [178] in an investigation of the solution of the Boltzmann equation for the complete axisymmetric free jet. Grundy's analysis used matched asymptotic expansions to deduce a valid solution of the Boltzmann equation for Maxwell's molecules. The most recent study of the problem of a spherical expansion into vacuum was that of Bird, who evaluated the behavior of both Maxwell's molecules and rigid spheres by his Monte Carlo method [89]. The behavior of $T_{\parallel}$ again displays a gradual freezing in substantial agreement with the results of previous authors. The results for $T_{\perp}$, however, do not attain any asymptotic behavior, at variance with the r^{-1} dependence predicted by Eq. (8.12). The failure of the Monte Carlo method has been attributed by Bird to computational limitations on the simulation of far field collisions. In the simulation very weak collisions were neglected for economic reasons; also, the random sampling may not have properly represented the wings of the distribution function, which, according to the analytical results quoted above, hold most of the transversal kinetic energy in the far field.

References

[1] H. Grad, "Comm. Pure and Appl. Math.", **2,** 33 (1949).
[2] L. W. Holway, Jr., "Phys. Fluids", **7,** 911 (1964).
[3] E. P. Gross, E. A. Jackson and S. Ziering, "Ann. Phys.", **1,** 141 (1957).
[4] E. P. Gross and S. Ziering, "Phys. Fluids", **1,** 215 (1958).
[5] E. P. Gross and S. Ziering, "Phys. Fluids", **2,** 701 (1959).
[6] C. Cercignani, "Nuovo Cim.", *X*, **27,** 1240 (1963).
[7] L. Lees, "GALCIT Hypersonic Research Project Memo. No. 51" (1959).
[8] H. M. Mott Smith, "Phys. Rev.", 82, 885 (1951).
[9] D. S. Butler and W. M. Anderson, in "Rarefied Gas Dynamics", C. L. Brundin, Ed., vol. I, p. 731, Academic Press, New York (1967).
[10] A. J. Chorin, in "The Boltzmann Equation", F. A. Grünbaum, Ed., Courant Institute, New York University (1972).
[11] S. F. Shen, in "Rarefied Gas Dynamics", J. A. Laurman, Ed., Vol. II, p. 112, Academic Press, New York (1963).
[12] E. Ikenberry and C. Truesdell, "J. Rat. Mech. Anal.", **5,** 1 (1956).,
[13] C. S. Wang Chang and G. E. Uhlenbeck, "Engineering Res. Institute, University of Michigan. Project M 999" (1952). Reprinted in "The Kinetic Theory of Gases", vol. V, p. 43 of "Studies in Statistical Mechanics", J. de Boer and G. E. Uhlenbeck, Eds. North Holland, Amsterdam (1970).
[14] C. S. Wang Chang and G. E. Uhlenbeck, "Engineering Res. Inst., University of Michigan, Rep. 1999-I-T" (1954).
[15] C. S. Wang Chang and G. E. Uhlenbeck, "Engineering Res. Inst., University of Michigan, Project M 999" (1953).
[16] C. S. Wang Chang and G. E. Uhlenbeck, "Engineering Res. Inst. University of Michigan, Project 2457" (1956).
[17] H. M. Mott Smith, "M.I.T. Lincoln Lab. Rep. V, 2" (1954).
[18] C. L. Pekeris, Z. Alterman, L. Finkelstein and K. Frankowski, "Phys. Fluids", **5,** 1608 (1962).
[19] K. Frankowski, Z. Alterman and C. L. Pekeris, "Phys. Fluids", **8,** 245 (1965).
[20] M. M. R. Williams, "The Slowing Down and Thermalisation of Neutrons", North Holland, Amsterdam (1966).
[21] W. R. Conkie, "Nucl. Sci. Eng.", **7,** 295 (1960).
[22] B. Davison, "Neutron Transport", Oxford University Press (1957).

[23] J. J. THOMPSON, "Australian Atomic Energy Comm. Rep. AAEC/E 107" (1963).
[24] A. B. HUANG, "Georgia Tech. Aerospace Engineering Rarefied Gas Dynamics Rep. No. 4" (1967).
[25] A. B. HUANG and D. L. HARTLEY, "Phys. Fluids", **11,** 1321 (1868).
[26] A. B. HUANG and D. L. HARTLEY, "AIAA Journal", **6,** 2023 (1968).
[27] A. B. HUANG and D. L. HARTLEY, "Phys. Fluids", **12,** 96 (1969).
[28] A. B. HUANG and P. F. HWANG, "Phys. Fluids", **13,** 309 (1970).
[29] A. B. HUANG, in "Rarefied Gas Dynamics", L. Trilling and H. Wachman, Eds., vol. I, 529, Academic Press, New York (1969).
[30] A. B. HUANG and P. F. HWANG, "Phys. Fluids", **16,** 466 (1973).
[31] M. WACHMAN and B. B. HAMEL, in "Rarefied Gas Dynamics", C. L. Brundin, Ed., Vol. I, 675, Academic Press, New York (1967).
[32] B. B. HAMEL and M. WACHMAN, in "Rarefied Gas Dynamics", J. H. de Leeuw, Ed., vol. I, 370, Academic Press, New York (1965).
[33] A. B. HUANG and D. P. GIDDENS, in "Rarefied Gas Dynamics", C. L. Brundin, Ed., vol. I, 481, Academic Press, New York (1967).
[34] A. B. HUANG, "Phys. Fluids", **11,** 61 (1968).
[35] A. B. HUANG, D. P. GIDDENS and C. W. BAGNAL, "Phys. Fluids", **10,** 498 (1967).
[36] A. B. HUANG and D. P. GIDDENS, "A.I.A.A. Journal", **5,** 1354 (1967).
[37] A. B. HUANG and D. P. GIDDENS, "Phys. Fluids", **10,** 232 (1967).
[38] A. B. HUANG and D. P. GIDDENS, "Phys. Fluids", **11,** 446 (1968).
[39] G. C. WICK, "Z. Phys.", **121,** 702 (1963).
[40] S. CHANDRASEKHAR, "Radiative Transfer", Dover, New York (1960).
[41] R. C. GAST, Bettis Report, WAPD-TM, 118 (1958).
[42] B. G. CARLSON and K. D. LATHROP, in "Computing Methods in Reactor Physics", H. Greenspan, C. W. Kalber, and D. Okrent, Eds., p. 171, Gordon and Breach, New York (1968).
[43] H. W. LIEPMAN, R. NARASIMHA and M. T. CHAHINE, "Phys. Fluids," **5,** 1313 (1962).
[44] D. R. WILLIS, in "Rarefied Gas Dynamics", J. A. Laurman, Ed., Vol. I, p. 205, Academic Press, New York (1963).
[45] D. ANDERSON, "J. Plasma Phys.", **1,** 255 (1968).
[46] C. CERCIGNANI and G. TIRONI, in "Rarefied Gas Dynamics", C. L. Brundin, Ed., Vol. I, p. 441, Academic Press, New York (1967).
[47] D. R. WILLIS, "Phys. Fluids", **5,** 127 (1962).
[48] C. CERCIGNANI and A. DANERI, "J. Appl. Phys.", **34,** 3509 (1963).
[49] C. CERCIGNANI and F. SERNAGIOTTO, "Phys. Fluids", **9,** 40 (1966).
[50] P. BASSANINI, C. CERCIGNANI and F. SERNAGIOTTO, "Phys. Fluids", **9,** 1174 (1966).
[51] P. BASSANINI, C. CERCIGNANI and P. SCHWENDIMANN, in "Rarefied Gas

Dynamics", C. L. Brundin, Ed., vol. I, p. 505, Academic Press, New York (1967).
[52] C. CERCIGNANI and F. SERNAGIOTTO, "Phys. Fluids", **10,** 1200 (1967).
[53] P. BASSANINI, C. CERCIGNANI and C. D. PAGANI, "Int. Jour. of Heat and Mass Transfer", **10,** 447 (1967).
[54] C. CERCIGNANI and C. D. PAGANI, "Phys. Fluids", **9,** 11 (1966).
[55] C. CERCIGNANI, P. FORESTI and F. SERNAGIOTTO, "Nuovo Cimento", **X,** 57B, 297 (1968).
[56] S. K. LOYALKA and J. FERZIGER, "Phys. Fluids", **10,** 1833 (1967).
[57] M. M. VAINBERG, "Variational Methods for the Study of Nonlinear Operators", Holden Day, San Francisco (1954).
[58] G. ROWLANDS, "J. Nucl. Energy", **A13,** 176 (1961).
[59] R. KLADNIK and I. KUŠČER, "Nucl. Sci. Eng.", **11,** 116 (1961).
[60] R. KLADNIK and I. KUŠČER, "Nucl. Sci. Eng.", **13,** 149 (1963).
[61] T. LE CAINE, "Can. J. Res., **A28,** 242 (1950).
[62] G. C. POMRANING, "J. Math. Phys.", **8,** 2096 (1967).
[63] G. C. POMRANING, "J. Math. Phys.", **8,** 155 (1968).
[64] G. C. POMRANING and M. CLARK, JR., "Nucl. Sci. Eng.", **16,** 147 (1963).
[65] G. C. POMRANING and M. CLARK, Jr., "Nucl. Sci. Eng.", **16,** 155 (1963).
[66] C. CERCIGNANI, in "Fisica del Reattore", p. 633, Consiglio Nazionale delle Ricerche, Roma (1966).
[67] C. CERCIGNANI and G. TIRONI, "J. Plasma Phys.", **2,** part 3, 293 (1968).
[68] C. CERCIGNANI and G. TIRONI, in "Rarefied Gas Dynamics", L. Trilling and H. Wachman, Eds., vol. I, p. 281, Academic Press, New York (1969).
[69] S. K. LOYALKA, "Z. Naturforsch.", **26a,** 1708 (1971).
[70] J. O. STEWART, in "Rarefied Gas Dynamics", L. Trilling and H. Wachman, Eds., vol. I, p. 545, Academic Press, New York (1969).
[71] M. N. KOGAN, "Rarefied Gas Dynamics", Plenum Press, New York (1969).
[72] S. J. ROBERTSON, in "Rarefied Gas Dynamics", L. Trilling and H. Wachman, Eds., vol. I, p. 767, Academic Press, New York (1969).
[73] J. K. HAVILAND and M. L. LAVIN, "Phys. Fluids", **5,** 1399 (1962).
[74] J. K. HAVILAND, in "Rarefied Gas Dynamics", J. A. Laurmann, Ed., Vol. I, p. 274, Academic Press, New York (1963).
[75] M. PERLMUTTER, in "Rarefied Gas Dynamics", C. L. Brundin, Ed., vol. I, p. 455, Academic Press, New York (1967).
[76] M. PERLMUTTER, in "Rarefied Gas Dynamics", L. Trilling and H. Wachman, Eds., vol. I, p. 327, Academic Press, New York (1969).
[77] Y. YOSHIZAWA, in "Rarefied Gas Dynamics", L. Trilling and H. Wachman, Eds., vol. I, p. 177, Academic Press, New York (1969).

[78] J. KONDO and K. KOURA, in "Rarefied Gas Dynamics", L. Trilling and H. Wachman, Eds., vol. I, p. 181, Academic Press, New York (1969).
[79] G. A. BIRD, "Phys. Fluids", **6,** 1518 (1963).
[80] G. A. BIRD, in "Rarefied Gas Dynamics", J. H. de Leeuw, Ed., Vol. I, 216, Academic Press, New York (1965).
[81] G. A. BIRD, "AIAA Journal", **4,** 55 (1966).
[82] G. A. BIRD, "J. Fluid Mech.", **30,** 479 (1967).
[83] G. A. BIRD, "J. Fluid Mech.", **31,** 657 (1968).
[84] F. W. VOGENITZ, G. A. BIRD, J. E. BROADWELL and H. RUNGALDIER, "AIAA Journal", **6,** 2388 (1968).
[85] G. A. BIRD, in "Rarefied Gas Dynamics", L. Trilling and H. Wachman, Eds., Vol. I, p. 301 and p. 85, Academic Press, New York (1969).
[86] G. A. BIRD, "J. Fluid Mech.", **36,** 571 (1969).
[87] G. A. BIRD, "Phys. Fluids", **13,** 1172 (1970).
[88] G. A. BIRD, "Phys. Fluids", **13,** 2676 (1970).
[89] G. A. BIRD, "AIAA Journal", **8,** 1998 (1970).
[90] G. GOERTZEL and M. H. KALOS, "Monte Carlo in Transport Problems" in "Progress in Nuclear Energy", Sect. I: Physics and Mathematics, Pergamon Press, London (1958).
[91] E. D. CASHWELL and C. J. EVERETT, "The Monte Carlo Method for Random Walk problems", Pergamon Press, New York (1959).
[92] J. SPANIER and E. M. GELBARD, "Monte Carlo Principles and Neutron Transport Problems", Addison-Wesley, Reading, Mass. (1969).
[93] A. NORDSIECK and B. L. HICKS, in "Rarefied Gas Dynamics", C. L. Brundin, Edt., vol. I, p. 695, Academic Press, New York (1967).
[94] B. L. HICKS and S. M. YEN, in "Rarefied Gas Dynamics", L. Trilling and H. Wachman, Eds., vol. I, p. 313, Academic Press, New York (1969).
[95] B. L. HICKS, S. M. YEN and B. J. RAILLY, "J. Fluid Mech.", **53,** 85 (1972).
[96] C. CERCIGNANI, "J. Stat. Phys.", **1,** 297 (1969).
[97] S. ZIERING, "Phys. Fluids", **3,** 503 (1960).
[98] M. L. LAVIN and J. K. HAVILAND, "Phys. Fluids", **5,** 274 (1962).
[99] C. CERCIGNANI and J. CIPOLLA, Presented at the VII Symposium on Rarefied Gas Dynamics (Pisa, 1970).
[100] W. P. TEAGAN and G. S. SPRINGER, "Phys. Fluids", **11,** 497 (1968).
[101] L. H. HOLWAY, JR., "Phys. Fluids", **9,** 1658 (1966).
[102] C. LIN and L. LEES, in "Rarefied Gas Dynamics", L. Talbot, Ed., p. 391, Academic Press, New York (1961).
[103] D. R. WILLIS, in "Rarefied Gas Dynamics", J. A. Laurman, Ed., Vol. I, p. 209, Academic Press, New York (1963).
[104] D. ANDERSON, "J. Fluid Mech.", **25,** 271 (1966).

[105] C. CERCIGNANI and C. D. PAGANI, in "Rarefied Gas Dynamics", C. L. Brundin, Ed., vol. I, p. 555, Academic Press, New York (1967).
[106] W. DONG, "Univ. of Calif. Rept. UC RL, 3353" (1956).
[107] M. KNUDSEN, "Ann. der Physik", **28,** 75 (1909).
[108] C. CERCIGNANI and G. TIRONI, "Proceedings of the AIDA-AIR Meeting, p. 174, AIDA-AIR", Roma (1967).
[109] C. CERCIGNANI, "Phys. Fluids," **10,** 1859 (1967).
[110] A. B. HUANG and R. L. STOY, JR., "Phys. Fluids", **9,** 2327 (1966).
[111] P. BASSANINI, C. CERCIGNANI and C. D. PAGANI, "Int. J. Heat and Mass Transfer", **11,** 1359 (1968).
[112] P. N. HU, Unpublished work quoted in Ref. **115**.
[113] C. CERCIGNANI, "Meccanica", **5,** 7 (1970).
[114] G. I. TAYLOR, "Proc. Roy. Soc. London", **A84,** 371 (1950).
[115] H. GRAD, in "Transport Theory", R. Bellman et al. Eds., p. 209, American Mathematical Society Providence, R.I. (1969).
[116] H. SALWEN, C. GROSCH and S. ZIERING, "Phys. Fluids", **7,** 180 (1969).
[117] L. H. HOLWAY, JR., in "Rarefied Gas Dynamics", J. H. de Leeuw, Ed., Vol. I. p. 193, Academic Press, New York (1965).
[118] C. MUCKENFUSS, "Phys. Fluids", **5,** 1325 (1962).
[119] M. LINZER and D. F. HORNIG, "Phys. Fluids", **6,** 166 (1963).
[120] S. M. DESHPANDE and R. NARASIMHA, "J. Fluid Mech.", **36,** 545 (1969).
[121] M. M. OBERAI, "J. de Mec.", **6,** 317 (1967).
[122] R. NARASIMHA, S. M. DESHPANDE and M. R. ANANTHASAYANAM, in "Rarefied Gas Dynamics", L. Trilling and H. Wachman, Eds., Vol. I, p. 417, Academic Press, New York (1969).
[123] B. M. SEGAL and J. H. FERZIGER, "Phys. Fluids", **15,** 1233 (1972).
[124] L. SIROVICH, "Phys. Fluids", **5,** 908 (1962).
[125] F. S. SHERMAN, "J. Fluid Mech.", **8,** 465 (1960).
[126] M. M. OBERAI, "Phys. Fluids", **9,** 1634 (1966).
[127] L. SIROVICH and E. GOLDMAN, in "Rarefied Gas Dynamics", L. Trilling and H. Wachman, Eds., vol. I, p. 407, Academic Press, New York (1969).
[128] K. ABE and H. OGUCHI, in "Rarefied Gas Dynamics", L. Trilling and H. Wachman, Eds., vol. I, p. 425, Academic Press, New York (1969).
[129] L. TALBOT and M. SCALA, in "Rarefied Gas Dynamics", L. Talbot, Ed., p. 603, Academic Press, New York (1961).
[130] D. L. TURCOTTE, in "Rarefied Gas Dynamics", L. Trilling and H. Wachman, Eds. Vol. I, p. 331, Academic Press, New York (1969).
[131] C. A. BRAU, G. A. SIMONS and H. K. MACOMBER, in "Rarefied Gas Dynamics", L. Trilling and H. Wachman, Eds., vol. I, p. 343, Academic Press, New York (1969).
[132] R. VENKATARAMAN and T. F. MORSE, in "Rarefied Gas Dynamics",

L. Trilling and H. Wachman, Eds., vol. I, 353, Academic Press, New York (1969).
[133] C. CERCIGNANI and C. D. PAGANI, in "Rarefied Gas Dynamics", C. L. Brundin, Ed., vol. I, p. 555, Academic Press, New York (1967).
[134] C. CERCIGNANI and C. D. PAGANI, "Phys. Fluids", **11,** 1395 (1968).
[135] C. CERCIGNANI, C. D. PAGANI and P. BASSANINI, "Phys. Fluids", **11,** 1399 (1968).
[136] R. A. MILLIKAN, "Phys. Rev.", **22,** 1 (1923).
[137] F. S. SHERMAN, in "Rarefied Gas Dynamics", J. A. Laurman, Ed., vol. II, p. 228, Academic Press, New York (1963).
[138] H. GRAD, in "Proceedings of the Conference on Aerodynamics and the Upper Atmosphere", Rand. Corp. (1959).
[139] L. SIROVICH, in "Rarefied Gas Dynamics", L. Talbot, Ed., Academic Press, New York (1961).
[140] V. P. SHIDLOVSKI, "Introduction to the Dynamics of Rarefied Gases", Elsevier, New York (1967).
[141] W. D. HAYES and R. F. PROBSTEIN, "Hypersonic Flow Theory", Academic Press, New York (1957).
[142] S. A. SCHAAF, in "Handbuch der Physik", S. Flugge, Ed., vol. VIII, 591, Springer Verlag Berlin (1963).
[143] J. L. POTTER, in "Rarefied Gas Dynamics", C. L. Brundin, Ed., vol. II, 881, Academic Press, New York (1967).
[144] D. R. WILLIS, "Princeton Univ. Aeronaut. Eng. Lab. Rept. No. 440" (1958).
[145] D. R. WILLIS, "J. Fluid Mech.", **21,** 21 (1965).
[146] D. R. WILLIS, "Phys. Fluids", **8,** 1908 (1965).
[147] H. GRAD and P. N. HU, in "Rarefied Gas Dynamics", L. Trilling and H. Wachman, Eds., vol. I, p. 561, Academic Press, New York (1969).
[148] B. B. HAMEL and A. L. COOPER, in "Rarefied Gas Dynamics", L. Trilling and H. Wachman. Eds., vol. I, p. 433, Academic Press, New York (1969).
[149] Y. S. PAN, in "Rarefied Gas Dynamics", L. Trilling and H. Wachman, Eds. vol. I, p. 779, Academic Press, New York (1969).
[150] K. O. STEWARTSON, "Mathematika", **16,** part 1, 106 (1969).
[151] A. F. MESSITER, "SIAM J. Appl. Math.", **18,** 241 (1970).
[152] W. J. MCCROSKEY, S. M. BOGDONOFF and J. G. MCDOUGALL, "AIAA J.", **4,** 1580 (1966).
[153] P. J. HARBOUR and J. H. LEWIS, in "Rarefied Gas Dynamics", C. L. Brundin, Ed., vol. II, p. 1031, Academic Press, New York (1967).
[154] W. W. JOSS, I. E. VAS and S. M. BOGDONOFF, "AIAA Paper 68-5" (January 1968).
[155] H. OGUCHI, in "Rarefied Gas Dynamics", L. Talbot, Ed., p. 501, Academic Press, New York (1961).

[156] M. Shorenstein and R. F. Probstein, "AIAA J.", **6,** 1898 (1968).
[157] W. L. Chow, "AIAA J.", **5,** 1549 (1967).
[158] W. L. Chow, "AIAA J.", **6,** 1 (1968).
[159] S. Rudman and S. G. Rubin, "AIAA J.", **6,** 1883 (1968).
[160] H. K. Cheng, S. Y. Chen, R. Mobly and C. Huber, in "Rarefied Gas Dynamics", L. Trilling and H. Wachman, Eds., vol. I, p. 451, Academic Press, New York (1969).
[161] S. S. Kot and D. L. Turcotte, "AIAA J.", **10,** 291 (1972).
[162] M. N. Kogan and L. M. Degtyariev, "Astron. Acta", **11,** 36 (1965).
[163] A. F. Charwat, in "Rarefied Gas Dynamics", L. Talbot, Ed., p. 553, Academic Press, New York (1961).
[164] A. B. Huang and P. F. Hwang, "IAF Paper RE 63" (October 1968).
[165] F. W. Vogenitz, J. E. Broadwell and G. A. Bird, "AIAA J.", **8,** 504 (1970).
[166] W. W. Joss and S. M. Bogdonoff, in "Rarefied Gas Dynamics", L. Trilling and H. Wachman, Eds., vol. I, p. 483, Academic Press, New York (1969).
[167] H. Ashkenas and F. Sherman, in "Rarefied Gas Dynamics", J. H. De Leeuw, Ed., Vol. II, 84, Academic Press, New York (1965).
[168] B. B. Hamel and D. R. Willis, "Phys. Fluids", **9,** 829 (1966).
[169] R. H. Edwards and H. K. Cheng, "AIAA J.", **4,** 558 (1966).
[170] N. C. Freeman, "AIAA J.", **5,** 1696 (1967).
[171] N. C. Freeman and R. E. Grundy, "J. Fluid Mech.", **31,** 723 (1968).
[172] J. W. Brook and R. A. Oman, in "Rarefied Gas Dynamics", J. H. de Leeuw, Ed., vol. I, 125, Academic Press, New York (1965).
[173] N. Abuaf, J. B. Anderson, R. D. Andres, J. B. Fenn and D. R. Miller, in "Rarefied Gas Dynamics", C. L. Brundin, Ed., vol. II, 1317, Academic Press, New York (1967).
[174] R. H. Edwards and H. K. Cheng, in "Rarefied Gas Dynamics", C. L. Brundin, Ed. Vol. I, 819, Academic Press, New York (1967).
[175] N. C. Freeman and D. R. Thomas, in "Rarefied Gas Dynamics", L. Trilling and H. Wachman, Eds., vol. I, 163, Academic Press, New York (1969).
[176] D. R. Miller and R. P. Andres, in "Rarefied Gas Dynamics", L. Trilling and H. Wachman, Eds., vol. II, p. 1385, Academic Press, New York (1969).
[177] D. R. Willis, B. B. Hamel and J. T. Lin, "Univ. of Calif. Berkeley Aero. Sci. Rept. AS-70-8" (1970).
[178] R. E. Grundy, "Phys. Fluids", **12,** 2011 (1969).

VIII THEOREMS ON THE SOLUTIONS OF THE BOLTZMANN EQUATION

1. Introduction

In this chapter we shall describe the mathematically more advanced part of the theory of the Boltzmann equation, based on the use of techniques of functional analysis. The main results refer to existence and uniqueness theorems and the qualitative behaviour of the solutions of the Boltzmann equation.

The development of such theory serves several purposes. The first is to ascertain whether the solution of a specific problem or the general solution of a model equation are representative of more general cases or whether they may be exceptional. Such a failure is always conceptually possible, no matter how convincing are the physical arguments which are used to derive the equations, and no matter how plausible are the results derived by heuristic approximate methods.

Although the first proof of existence and uniqueness for the Boltzmann equation goes back to Carleman [1], the most important progress has taken place in the last twenty years, thanks to the development of a detailed theory of the linearized Boltzmann equation and a systematic exploitation of monotonicity arguments, rigorous perturbation theory and other tools of modern functional analysis. This development has led to a rather complete theory of existence and uniqueness in the linearized case and to several interesting results for nonlinear problems.

An interesting relation between the existence and uniqueness theory and the rigorous derivation of the Boltzmann equation (see Chapter II), came into focus with the important paper by Lanford [2] to be discussed in Section 5.

2. The space homogeneous case

The simplest kind of problem which can be considered in connection with the Boltzmann equation is the initial value problem for an infinite expanse of gas with initial data independent of the space variables (the spatially homogeneous problem). The first existence and uniqueness theorem for this problem was obtained by Carleman [1, 3] in the case of hard sphere molecules. He obtained a strong nonlinear result in the large, under the rather weak assumptions that the initial value φ of the distribution function be integrable, nonnegative and bounded from above by $A_1(1+|\xi|)^{-\alpha}$ where $A_1>0$ and $\alpha \geqslant 6$ are constants. The solution f is shown to exist as a positive function bounded by $A(1+|\xi|)^{-\alpha}$ for all nonnegative values of the time variable.

The H-functional defined by Eq. (III.9.1) is shown to exist as a nonincreasing function of t and the distribution function to tend to a Maxwellian for $t\to\infty$. The result of Carleman has been extended to inverse power potentials by Maslova and Tchubenko [4, 5], who showed the existence of a unique solution $f\in C^1(\mathbb{R}_+\times\mathbb{R}^3)$ with

$$\sup_{t,\xi}(1+|\xi|^2)^{\alpha/2}f(t,\xi)<\infty \tag{2.1}$$

provided

$$0<C_1\leqslant B(\theta,V)/V^{\lambda}\cos\theta\leqslant C_2, \qquad \lambda>0 \tag{2.2}$$

$$\alpha>\max\{5, 2+4C_1/C_2\}, \tag{2.3}$$

and $\varphi\in C(\mathbb{R}^3)$ with

$$\sup_{\xi}(1+|\xi|^2)^{\alpha/2}\varphi(\xi)<\infty \tag{2.4}$$

The case of a negative value of λ has been considered by Gluck [6].

In 1951, Wild [7] devised a monotone iterative scheme to prove existence and uniqueness of spatially homogeneous solutions for Maxwellian molecules. He considered initial data bounded by a Maxwellian distribution, and was able to prove convergence on a finite time interval only. In 1954, Morgenstern [8] proved that Wild's scheme could be used to prove the existence and uniqueness of spatially homogeneous solutions in $L^1(\mathbb{R}^3)$ for all t. His method was later extended by Arkeryd [9, 10] to cover all the collision operators whose kernel $B(\theta, V)$ satisfies

$$B(\theta,|\xi-\xi_*|)\leqslant A(1+|\xi|^{\lambda}+|\xi_*|^{\lambda}), \qquad 0\leqslant\lambda<2 \tag{2.5}$$

Arkeryd studied the problem in $L^1(\mathbb{R}^3)$ and provided two different proofs of existence. Both proofs start from the existence theorem for the Boltzmann equation with a bounded collision kernel ($\lambda=0$ in Eq. (2.5)). This theorem is based on a slight generalization of Morgenstern's argument. In fact, in this case

$$Q(f,f)=G(f,f)-fR(f) \tag{2.6}$$

where

$$G(f,f)=\frac{1}{m}\int f(\xi')f(\xi_*')B(\theta,V)\,d\xi_*\,d\theta\,d\varepsilon \tag{2.7}$$

$$R(f)=\frac{1}{m}\int f(\xi_*)B(\theta,V)\,d\xi_*\,d\theta\,d\varepsilon \tag{2.8}$$

and

$$R(f) \leqslant C_1 \|f\| \tag{2.9}$$

The norms, here and later, refer to $L^1(\mathbb{R}^3)$. In Morgenstern's paper [8], where Maxwellian molecules are considered, the equality sign holds in Eq. (2.9).

The iteration scheme used by Wild, Morgenstern and Arkeryd is based on the fact that, since conservation of mass holds, at least formally, we can expect $\|f\|$ to equal $\|\varphi\|$, where $\varphi \in L^1(\mathbb{R}^3)$ is the initial datum. We may then consider the following iteration method:

$$\frac{\partial f^{n+1}}{\partial t} + C\|\varphi\| f^{n+1} = \hat{Q}(f^n, f^n), \qquad f^{n+1}(0, \xi) = f_0(\xi)$$
$$f^1 = 0 \tag{2.10}$$

where the time derivative represents differentiation of a Banach space-valued function and

$$\hat{Q}(f, f) = Q(f, f) + C\|f\| f = G(f, f) + [C\|f\| - R(f)]f \tag{2.11}$$

We remark that C can be always chosen in such a way as to make $\hat{Q}(f, f)$ a positive monotone operator. The approximating sequence $\{f^n\}$ is nondecreasing in n and formed by nonnegative functions, with uniformly bounded norms; in fact, if $\rho^n = \|f^n\|$, we have

$$\frac{\partial \rho^{n+1}}{\partial t} + C\rho_\varphi \rho^{n+1} = C(\rho^n)^2, \qquad \rho^{n+1}(0) = \rho_\varphi$$
$$\rho^1 = 0 \tag{2.12}$$

and $\rho^n \leqslant \rho_\varphi$ implies $\rho^{n+1} \leqslant \rho_\varphi$ so that, by induction, all the norms ρ^n are bounded by ρ_φ. Hence by Levi's theorem [11], $\{f^n\}$ converges to some f in L^1, strongly for any t. By virtue of the continuity of $\hat{Q}(f, f)$, f satisfies

$$\frac{\partial f}{\partial t} + C\|\varphi\| f = Q(f, f) + C\|f\| f \tag{2.13}$$

Then if we let $\rho = \|f\|$ we have

$$\frac{\partial \rho}{\partial t} + C\rho_\varphi \rho = C\rho^2 \tag{2.14}$$

and by the uniqueness theorem for ordinary differential equations we have $\rho(t) = \rho_\varphi$ or $\|f\| = \|\varphi\|$, so that f satisfies the Boltzmann equation

$$\frac{\partial f}{\partial t} = Q(f, f) \tag{2.15}$$

It is then easy to prove the boundedness of moments of the solution (for finite

t) provided they are bounded at $t=0$, by utilizing an inequality proved by Povzner [12]

$$|\xi'|^s + |\xi_*'|^s - |\xi|^s - |\xi_*|^s \leqslant \kappa_s(|\xi_*|^2|\xi|^{s-2} + |\xi|^{s-2}|\xi_*|^2) \tag{2.16}$$

which implies

$$\int (1+|\xi|^2)^{\alpha/2} Q(f,f)\, d\xi \leqslant C_\alpha(\|f\|_{\alpha+\lambda-\beta}\|f\|_\beta + \|f\|_{\alpha-\beta}\|f\|_{\lambda+\beta}) \tag{2.17}$$

where we let

$$\|f\|_\alpha = \int_{\mathbb{R}^3} (1+|\xi|^2)^{\alpha/2}|f|\, d\xi \tag{2.18}$$

If $\lambda=0$, as appropriate for a bounded collision kernel, then one can take $\beta=0$ in Eq. (2.17) and the proof of the boundedness of $\|f\|_\alpha$ is trivial. For Maxwellian molecules, Truesdell [13] proved that the moments converge exponentially to their equilibrium values.

Arkeryd also gave a rigorous proof of the H-theorem, which implies the boundedness of the H-functional, the bound being simply the initial value of H. The proof differs from the formal one in that one has to show that values of f close to zero or very large do not cause problems with the factor $\log f$.

As mentioned before, Arkeryd gave two distinct proofs of existence in the case of unbounded kernels ($\lambda > 0$). In the first proof $B(\theta, V)$ is approximated by a sequence of bounded kernels

$$B_n(\theta, V) = \min\{B(\theta, V), n\} \qquad (n=1,2,\ldots) \tag{2.19}$$

The corresponding collision operators $Q_n(f,f)$ are associated with L^1 solutions satisfying

$$\frac{\partial f_n}{\partial t} = Q_n(f_n, f_n); \qquad f_n(0)=\varphi \qquad (n=1,2,\ldots) \tag{2.20}$$

Further, $\|f_n\|_0 = \|\varphi\|_0$, $\|f_n\|_2 = \|\varphi\|_2$, $H_n(t) \leqslant H_0$, where H_n is the H-functional associated with f_n. The Dunford-Pettis theorem [14] can then be used to show [9] that $\{f_n\}$ is relatively weakly compact in $L^1(\mathbb{R}^3)$ for any t, and accordingly contains a subsequence converging weakly to some $f \in L^1$, Since the sequence is also equicontinuous in t, we can take the subsequence to be the same for all t, i.e. there is a subsequence of functions $f_n(t)$ converging weakly to $f(t)$ in L^1. This function is also a weak solution of the Boltzmann equation, Eq. (2.15), because $Q_n(f_n, f_n)$ converges weakly to $Q(f,f)$ when f_n converges weakly to f.

In his second proof Arkeryd introduced

$$M(f) = A(1+|\xi|^2)\int (1+|\xi_*|^2) f(\xi_*)\, d\xi_* \tag{2.21}$$

and the equation

$$\frac{\partial \hat{f}_n}{\partial t} + M(f_0)\hat{f}_n = Q_n(\hat{f}_n, \hat{f}_n) + M(\hat{f}_n)\hat{f}_n; \qquad \hat{f}_n(0) = \varphi \tag{2.22}$$

Since Eq. (2.5) with $\lambda = 2$ holds, the monotonicity argument used before indicates that Eq. (2.22) has a solution $\hat{f}_n$ such that

$$\| \hat{f}_n(t) \|_2 \leqslant \| f_n(t) \|_2 = \| \varphi \|_2 \tag{2.23}$$

where f_n solves Eq. (2.20). The fact that Levi's theorem holds in L^1 (with weight) implies the existence of $f(t) = \lim_{n\to\infty} \hat{f}_n(t)$ in this space. Further, because of Eq. (2.23)

$$\| f(t) \| \leqslant \| \varphi \|_2 \tag{2.24}$$

and f satisfies

$$\frac{\partial f}{\partial t} + M(\varphi)f = Q(f, f) + M(f)f, \qquad f(0) = \varphi \tag{2.25}$$

Assuming now the existence of the fourth moment of φ, i.e. $\| \varphi \|_4 < \infty$, one can show, thanks to Povzner's inequality (2.17) with $\lambda = \beta = 2$, that $\| f \|_4$ is also bounded (for $0 \leqslant t \leqslant T$, T arbitrary) by a constant which only depends on $\| \varphi \|_2$, $\| \varphi \|_4$ and T. Then one can show that the equality sign must hold in Eq. (2.24). Hence $M(f) = M(f_0)$ and f satisfies the Boltzmann equation. Under the same assumptions one can prove uniqueness, which is lacking in the first proof (not requiring $\| \varphi \|_4 < \infty$).

More recently, Arkeryd [15] attacked the problem of existence for noncutoff collision operators. He defined a weak solution to mean a function $f = f(\xi, t)$ such that

$$\int f(\xi, t)\psi(\xi, t)\, d\xi = \int \varphi(\xi)\psi(\xi, 0)\, d\xi + \int_0^t \int f(\xi, s) \frac{\partial \psi}{\partial s}(\xi, s)\, d\xi\, ds + $$
$$+ \int_0^t \iiint [\psi(\xi', s) - \psi(\xi, s)] f(\xi, s) f(\xi_*, s) B(\theta, V)\, d\xi\, d\xi_*\, ds \tag{2.26}$$

for any $\varphi \in C^{1,\infty}$, where

$$C^{1,\infty} = \left\{ \psi \in C^1(\mathbb{R}^3 \times \mathbb{R}_+) \colon \sup_{\xi,t} |\psi(\xi, t)| + \sup_{\xi,t} \left| \frac{\partial \psi}{\partial t}(\xi, t) \right| + \sup_{\xi,t} \left| \frac{\partial \psi}{\partial \xi}(\xi, t) \right| < \infty \right\} \tag{2.27}$$

$B(\theta, V)$ is chosen to be the collision kernel appropriate for inverse sth power forces (with $s > 3$).

The idea of the proof of existence of solutions to Eq. (2.26) is again to approximate $B(\theta, V)$ by a sequence of kernels $B_n(\theta, V)$ with cutoff. Then a sequence of approximate solutions f_n is obtained with the properties

$$\|f_n\|_2 = \|\varphi\|_2 \qquad H_n(t) \leqslant H_\varphi \tag{2.28}$$

Then Arkeryd used the same weak compactness argument used in his first proof [9] for unbounded kernels with angular cutoff to show that there is a subsequence $\{f_n\}$ converging weakly to an f satisfying Eq. (2.26). The solution is also shown to conserve mass and momentum and to have bounded energy.

Starting from Arkeryd's result, Elmroth [16, 17] proved that if $s \geqslant 5$ the moments of higher order are bounded for any t if they are at $t = 0$. The case $3 < s \leqslant 5$ had been already proved by Arkeryd [15]. The proof by Elmroth rests on the following inequality:

$$|\xi'|^\alpha + |\xi_*'|^\alpha - |\xi|^\alpha - |\xi_*|^\alpha \leqslant K_\alpha(|\xi|^{\alpha-1}|\xi_*| + |\xi_*|^{\alpha-1}|\xi|) \cos\theta \sin\theta - \\ - C_\alpha(|\xi|^\alpha + |\xi_*|^\alpha) \cos^2\theta \sin\theta \tag{2.29}$$

In another paper, Elmroth [18] also proved that the H-theorem holds.

Further results on the space homogeneous case are discussed by Gustafsson [19] who has proved existence in any L^p $(1 \leqslant p \leqslant \infty)$ for cutoff potentials. Another existence theorem can be obtained as a by-product of a result of Povzner [8] in the space inhomogeneous case (see Section 3). Truesdell and Muncaster [20] improved upon Arkeryd's result for the cutoff case by showing that, in fact, $t \to f(t, \xi)$ is differentiable in t for almost all $\xi \in \mathbb{R}^3$. Another approach through nonlinear semigroups has been given by Di Blasio [21]. None of these papers considers any external force acting on the molecules. The case of a constant external force (such as gravity) lends itself to a possible generalization of these results; a particular, but significant, case in which the extension has been carried out can be found in a paper by Boffi and Spiga [22].

A final question connected with the space homogeneous problem is the trend to equilibrium. The fact that H decreases and is bounded from below by H_M (where H_M is the equilibrium value associated with the Maxwellian f_M having the density, bulk velocity and temperature dictated by the initial datum) is not sufficient to conclude that $H \to H_M$ and $f \to f_M$ [20, 23]. Both Carleman and Arkeryd were able to show, however, that the trend to equilibrium holds for their solutions. This is proved by considering a converging subsequence $f(t_n)$ and finding a contradiction by assuming that the subsequence (no matter how chosen) does not converge to f_M. Arkeryd uses weak convergence in L^1 and so proves that $f(t)$ converges weakly to f_M. Elmroth [18] extended this result to noncutoff collision operators.

Another result by Elmroth [18] is that *if* $H \to H_M$ then f converges *strongly* to f_M. His proof is rather long, but a simpler proof can be based on the elementary inequality

$$f \log f - f \log f_M + f_M - f \geqslant c\, g\left(\frac{|f - f_M|}{f_M}\right) |f - f_M| \tag{2.30}$$

where c is a purely numerical constant (independent of f) and

$$g(x) = \begin{cases} x & \text{if } 0 \leqslant x \leqslant 1 \\ 1 & \text{if } x > 1 \end{cases} \tag{2.31}$$

Integrating both sides of Eq. (2.30) gives

$$H - H_M \geqslant c\left[\int_{L_t} |f - f_M|\, d\xi + \int_{S_t} \frac{|f - f_M|^2}{f_M}\, d\xi\right] \tag{2.32}$$

where L_t and S_t denote the sets (depending on t) where f is larger (resp. smaller) than f_M. Since H is assumed to tend to H_M, it follows that both integrals tend to zero when $t \to \infty$. The fact that the second integral tends to zero implies, by Schwarz's inequality, that

$$\int_{S_t} |f - f_M|\, d\xi \to 0 \tag{2.33}$$

Then

$$\int |f - f_M|\, d\xi = \int_{L_t} |f - f_M|\, d\xi + \int_{S_t} |f - f_M|\, d\xi \tag{2.34}$$

also tends to zero and the strong convergence of f to f_M is proved.

3. Mollified and other modified versions of the Boltzmann equation

Although the mathematics employed in the space homogeneous case is far from elementary, the results described in the previous section are physically trivial in the sense that they include no fluid dynamics; density, bulk velocity and temperature retain their initial values.

The situation becomes more interesting and complicated if we allow space variation of the initial datum. Here all the arguments to obtain a global existence theorem for sufficiently general initial data have failed (see, however, Section 4). The reason is that the *a priori* inequalities available from conservation laws and the H-theorem are not sufficient to carry on the procedures used in the space homogeneous case. In fact, these inequalities are less powerful now because the nonlinear collision term does not contain an integration with respect to a space variable and the operators appearing in the Boltzmann equation lack sufficient "compactness" to be amenable to a rigorous treatment as far as global existence is concerned. This was noticed long ago by Morgenstern [24] and Povzner [12], who solved the Cauchy problem for certain altered versions of the Boltzmann equation, containing a "mollifier", i.e. an operator smoothing out the space dependence. These equations were chosen for mathematical convenience; they should not be confused with the model equations which have been repeatedly used in this book and do not appreciably simplify nonlinear existence theory. In particular, the mollified collision operators may no longer possess five collision invariants, except in the spatially homogeneous case.

Morgenstern [24] proposed modifying the collision integral so as to transform it to

$$Q_M(f, f) = \iiiint B(\theta, V)[f'f_*' - ff_*]K(\mathbf{x}, \mathbf{x}_*)\, d\mathbf{x}_*\, d\boldsymbol{\xi}_*\, d\theta\, d\varepsilon \quad (3.1)$$

where the notation is as usual with the exception that the space arguments in f_*' and f_* are $\mathbf{x}_*$ rather than $\mathbf{x}$. Then for a bounded $B(\theta, V)K(\mathbf{x}, \mathbf{x}_*)$ a global existence theorem is obtained in $L^1(W \times \mathbb{R}^3)$ where W may be either $\mathbb{R}^3$ or a bounded domain with specular or periodic boundary conditions. Morgenstern used arguments that are similar to those that he had employed in the space homogeneous case [8] and, in fact, the methods of Arkeryd described in Section 2 might be used to cover this case as well.

Povzner [12] considered the mollified operator

$$Q_P(f, f) = \int K(\mathbf{x} - \mathbf{x}_*, \boldsymbol{\xi} - \boldsymbol{\xi}_*)[f'f_*' - ff_*]\, d\boldsymbol{\xi}_*\, d\mathbf{x}_* \quad (3.2)$$

where $\boldsymbol{\xi}'$ and $\boldsymbol{\xi}_*'$ are related to $\boldsymbol{\xi}$, $\boldsymbol{\xi}_*$ and $\mathbf{n}$ in the usual way, but $\mathbf{n}$ is the unit vector directed as $\mathbf{x} - \mathbf{x}_*$. Here, as above, the space argument in f_*' and f_* is $\mathbf{x}_*$ rather than $\mathbf{x}$. Then for a kernel K such that $|K| < M(1 + V)$, an existence theorem is given. If the fourth moment exists, a uniqueness theorem is also proved. The case of an external force field with potential energy bounded from below is included in Povzner's treatment. Although the solutions are sought in the space of Borel measures (the dual of the space of bounded continuous functions), an L^1 treatment of the kind described in the previous section for the space homogeneous case could be given.

It is notable that Povzner's mollification is close to physical reality, where actual collisions do not occur at one point in space. As a matter of fact, the Enskog equation, used to describe dense gases (see Section 4 of Chapter II, at the end), has a form similar to Povzner's equation, except for the fact that the integral is fivefold, the integration over the distance between $\mathbf{x}$ and $\mathbf{x}_*$ being absent. This destroys the possibility of using Povzner's argument to produce a global existence theorem. It is possible, however, to show such a theorem for the case of solutions depending on just one space variable [25]. The case of solutions of the Enskog equation depending on two space variables is more difficult; in this case, a modified Enskog equation has been treated by Arkeryd [26].

A different approach to the matter of global existence for the Boltzmann equation in the space inhomogeneous case was proposed by Cercignani, Greenberg and Zweifel [27] in 1979 and was based on introducing a spatially discrete Boltzmann equation. The space variable, represented by an index i is assumed to vary over a discrete set $\mathbb{Z}_n^3$ and the distribution function $f_i(\boldsymbol{\xi}, t) \in L^1(\mathbb{R}^3 \times \mathbb{Z}^3)$ satisfies the equation

$$\frac{\partial f_i}{\partial t} + (Af)_i = G(f_i, f_i) - R(f_i) f_i \tag{3.3}$$

where A, the streaming operator, is a finite difference approximation to $\boldsymbol{\xi} \cdot \partial/\partial \mathbf{x}$ based on forward differences in the direction of $\boldsymbol{\xi}$. Existence and uniqueness of global solutions for Maxwellian molecules and a periodic lattice were proved [27] by essentially the same argument used by Morgenstern [24] for his mollified equation, but any other argument good for mollified equations or space homogeneous problems would work. The basic point is that the L^1 and L^∞ norms for functions defined on a finite lattice are equivalent, so that the noncompactness difficulties are overcome; in addition, the *a priori* inequality on total mass automatically produces an estimate for the density. A different proof of the existence theorem was obtained by Spohn [28].

The approach was extended to non-Maxwellian molecules by Greenberg, Voigt and Zweifel [29] who considered kernels bounded by

$$B(\theta, V) \leqslant A(1 + V^\lambda), \qquad 0 \leqslant \lambda < 2 \tag{3.4}$$

Following Arkeryd's proof, the collision kernel is first approximated by a sequence of bounded functions, and a corresponding sequence of approximate solutions $f_1{}^m$ is thus produced. A weakly converging subsequence exists thanks to the compactness argument of Arkeryd's first proof, once the H-theorem is proved.

Arkeryd's second proof can also be extended to this case if the initial datum φ_0 has a finite k-th moment for $k > 2$, $k > 2\lambda$. Also, the same weak compactness argument of his first proof can be used to show that there is a weakly converging subsequence when the number of points in the lattice, n^3, tends to infinity [29]. It has not been possible, however, to prove that the weak limit satisfies the Boltzmann equation.

There is also a large literature devoted to studying models of the Boltzmann equation with discrete velocity variables. These were introduced by Gatignol [30] in 1975 following some earlier examples by Carleman [3] and Broadwell [31]. The theory of these discrete velocity models proves to be easier than that of the continuous velocity equation only for special models or situations.

There is a complete theory for the Carleman model [32] which is one-dimensional and has only two discrete velocities (± 1)

$$\begin{cases} \dfrac{\partial f_1}{\partial t} + \dfrac{\partial f_1}{\partial x} = f_2{}^2 - f_1{}^2 \\[2ex] \dfrac{\partial f_2}{\partial t} - \dfrac{\partial f_2}{\partial x} = f_1{}^2 - f_2{}^2 \end{cases} \tag{3.5}$$

Nishida and Mimura have introduced a method to prove the global existence of solutions for the Broadwell model [33]. The method has been further extended. Tartar [34] has shown existence of global solutions in L^1 for sufficiently general models when the L^1 norm of the initial datum φ is sufficiently small and depends on just one space variable. If the datum is in L^∞, as well as in L^1, then it is possible to show that there is a solution in L^∞, provided the datum is small. Then, using the fact that the speed of propagation is finite and the H-theorem holds, one can deduce that if the initial data are in L^∞ then, no matter how large they are, there exists a (unique) L^∞ solution. Thus the matter of existence is essentially solved for models when the data depend on just one space variable. Various applications of the method have been given by Cabannes [35–37].

A general existence theorem for data sufficiently small has been given by Illner [38]. Recently, Shizuta and Kawashima [39] gave a general criterion for global existence when the data are close to equilibrium and Cercignani [40] gave a simplified version of this criterion when the discrete velocity has exactly the same collision invariants as the continuous velocity equation.

Cercignani and Shinbrot [41], for the particular case when the discrete velocity model

$$\frac{\partial f_i}{\partial t} + \xi_i \cdot \frac{\partial f_i}{\partial x} = \sum_{j,k} c_{ijk} f_j f_k - \sum_j \kappa_{ij'} f_j f_i \tag{3.6}$$

has the property that c_{ijk} is symmetric in the three indices, obtained a global existence result in L^∞ for arbitrarily large initial data. The condition on the coefficients seems to exclude the presence of more than one collision invariant. Carleman's model (3.5) is included in this case with the choice $c_{122} = c_{212} = c_{221} = c_{211} = c_{112} = c_{121} = 1$, $c_{111} = c_{222} = 0$, $k_{11} = k_{12} = k_{21} = k_{22} = 1$.

Further existence results on discrete velocity models resemble, in one way or another, those found for the continuous equation and their discussion will be omitted. For a survey, we refer to a paper by Platkowski and Illner [42]. Recently, discrete velocity models have been mentioned by Frisch, Hasslacher and Pomeau [43] as providing motivation and a method of analysis for cellular automata simulations of the Navier–Stokes equations.

4. Nonstandard analysis approach to the Boltzmann equation

A different approach to the problem of finding a global existence theorem for the space inhomogeneous Boltzmann equation, with no restriction on the size of the initial data, has been given by Arkeryd [44], who proposed to use a nonstandard analysis interpretation of the Boltzmann equation. While there is no room here to give the reader a serious introduction to nonstandard analysis, it seems appropriate to give a few details on the basic ideas so that he may be able to appreciate at least the flavour of Arkeryd's results; for a more complete introduction, the reader is urged to consult one of the references at the end of the chapter [45–47].

Nonstandard analysis considers, as the basic set of numbers, not $\mathbb{R}$ but $^*\mathbb{R}$, a proper ordered extension of $\mathbb{R}$. We can form an intuitive image of $^*\mathbb{R}$ by considering, along with the real numbers of $\mathbb{R}$, infinitesimals, i.e. numbers a such that $|a| < \varepsilon$ for any positive $\varepsilon \in \mathbb{R}$, and infinitely large numbers $1/a$ where a is infinitesimal. Each number a in $^*\mathbb{R}$ either differs from a real number c by an infinitesimal (in which case one writes $a - c \simeq 0$), or is an infinitely large number. One can define a mapping st: $^*\mathbb{R} \to \mathbb{R}$, called the standard part, by letting

$$st(a) = {}^0a = \begin{cases} c & \text{if } a \in {}^*\mathbb{R}, c \in \mathbb{R}, a - c \simeq 0 \\ +\infty & a > n, n \in \mathbb{N} \\ -\infty & a < -n, n \in \mathbb{N} \end{cases} \tag{4.1}$$

The sets considered in usual analysis are typically sets consisting of all sets obtained from $\mathbb{R}$ in a finite number of steps by the usual operations of set theory; those considered in nonstandard analysis are built from $^*\mathbb{R}$ in a completely analogous way. Among the two set-theoretic structures, $V(\mathbb{R})$ and $V(^*\mathbb{R})$, there is, as was shown by Robinson [48], an injective map $*$: $V(\mathbb{R}) \to V(^*\mathbb{R})$ such that

$$^*r = r \qquad \text{if} \quad r \in \mathbb{R} \tag{4.2}$$

and satisfying an important transfer principle that will be described presently. Let E be an elementary statement about the sets $S_1, \ldots, S_n \in V(\mathbb{R})$, i.e. a statement on these sets built only from $\in$, $=$, propositional connectives and bounded quantifiers ($\forall x \in S$, $\exists x \in S$). The transfer principle states that if E is true for $S_1, \ldots, S_n$ in $V(\mathbb{R})$, then E is also true for $S_1{}^*, S_2{}^*, \ldots, S_n{}^*$ in $V(^*\mathbb{R})$. A set in $V(^*\mathbb{R})$ is called standard if it is the $*$-image of a set in $V(\mathbb{R})$. It is called internal if it is an element of a standard set. Typical objects in $V(\mathbb{R})$ used in the theory of the Boltzmann equation are $\mathbb{R}^n$, subsets of $\mathbb{R}^n$, the Lebesgue measure on $\mathbb{R}^n$, $L^1(\mathbb{R}^n)$, $L^\infty(\mathbb{R})$. In $V(^*\mathbb{R})$ there are corresponding "star-objects". It is a rather simple consequence of the transfer principle that $^*(\mathbb{R}^n) = (^*\mathbb{R}^n)$ $(n \in \mathbb{N})$ and, if (X, A, m) is the triplet of a set X, a σ-algebra A and a measure m, defining a measure space, then $^*(X, A, m) = (^*X, {}^*A, {}^*m)$. In particular, $^*A$ is the $^*\sigma$-algebra of $^*(X, A, m)$.

Such a measure space is not σ-additive in an "ordinary" sense, because $^*A$ is not a σ-algebra but a $^*\sigma$-algebra. Loeb [49], however, has indicated how to obtain an "ordinary" σ-additive measure space over a set $X \in V(^*\mathbb{R})$ from an internal measure algebra over the same set X. For this, one needs the property that $V(^*\mathbb{R})$ is a denumerably comprehensive superstructure; i.e. if $S \in V(\mathbb{R})$, $A_n \in {}^*S$, then the sequence $\{A_n; n \in \mathbb{N}\}$ is the restriction to $\mathbb{N}$ of an internal function from $^*\mathbb{N}$ to $^*S$.

Now let A be an internal algebra of subsets on the internal set X, and let m be an internal measure on A. Define the Loeb measure $Lm(Y)$ on $Y \in A$ by $Lm(Y) = {}^0m(Y)$ and take for σA the smallest σ-algebra of subsets of X with $\sigma A \supset A$. Letting

$$Lm(Z) = \inf_{Y \in A,\, Z \subset Y} Lm(Y) \qquad (Z \in \sigma A) \tag{4.3}$$

we obtain a σ-additive measure space $(X, \sigma A, Lm)$. The Loeb space (X, LA, Lm) is, by definition, the completion of $(X, \sigma A, Lm)$.

Now let b be the Lebesgue measure on $\mathbb{R}^n$, then $*b$ will be an internal measure over $X \in {}^*\mathbb{R}^n$ and L^*b will be a σ-additive measure on ${}^*\mathbb{R}^n$. If $Y \subset \mathbb{R}^n$, then $L^*b(st^{-1}Y) = b(Y)$. In other words, the Loeb-Lebesgue measure gives the Lebesgue measure to standard sets.

Having introduced a measure, we can now introduce an integral. If $f: X \to {}^*\mathbb{R}$ is

(a) A-measurable,

(b) $\displaystyle\int_X (|f| - \min(|f|, w))\, dm \simeq 0,\ w \in {}^*\mathbb{N}\backslash\mathbb{N}$,

(c) $\displaystyle\int_X \min(f, w^{-1})\, dm \simeq 0,\ w \in {}^*\mathbb{N}\backslash\mathbb{N}$,

then f is called S-integrable and several properties may be proved similar to those of the usual integral. Thus, e.g. an A-measurable function whose absolute value is never larger than the corresponding value of an S-integrable function is S-integrable, the theorem on the change of variables and the analogue of Fubini's theorem hold. Further, if f is S-integrable then its standard part is Lm-integrable and the standard part of the S-integral equals the Lm-integral.

Before describing the application of these concepts to the Boltzmann equation, we need to introduce the near-standard vectors in ${}^*\mathbb{R}^n$; these are the elements of the set

$$ns{}^*\mathbb{R}^n = \{\xi \in {}^*\mathbb{R}^n;\ |\xi| < k, \text{ for some } k \in \mathbb{N}\} \tag{4.4}$$

It is to be remarked that $ns{}^*\mathbb{R}^n$ is a Loeb-measurable set, since

$$ns{}^*\mathbb{R}^n = \bigcup_{k \in \mathbb{N}} (\xi \in {}^*\mathbb{R}^n,\ |\xi| < k) \tag{4.5}$$

When applying nonstandard analysis to the Boltzmann equation, one considers the Loeb-Lebesgue measures over ${}^*D \times (ns{}^*\mathbb{R}^3)$, ${}^*S \times (ns{}^*\mathbb{R}^3)$ and $ns{}^*R_+$, where $D \subseteq \mathbb{R}^3$ is a bounded open set and S is the parameter set of the unit vector $\mathbf{n}$ appearing in the collision integral; the above sets are the natural ranges, in the nonstandard analysis approach, of $(\mathbf{x}, \xi)$, $(\mathbf{n}, \xi)$ and t. The measures on these sets are denoted by $L(\mathbf{x}, \xi)$, $L(\mathbf{n}, \xi)$ and $L(t)$.

In this Loeb space treatment, the collision operator $Q_L(f, f)$ is defined by

$$Q_L(f, f) = \int_{*S} \int_{ns^*\mathbb{R}} (f'f_*' - ff_*)\ {}^{0*}B(\theta, V)\, dL(n, \xi_*) \tag{4.6}$$

where $f \in \text{Loeb-}L^1({}^*D \times ns^*\mathbb{R}^3)$. The Boltzmann equation is written as follows (mild form):

$$f(x, \xi, t) = {}^*U(t) f_0(x, \xi) + \int_0^t {}^*U(t-s) Q_L(f, f)(s)\, dL(s) \qquad (t \in ns^*\mathbb{R}_+) \tag{4.7}$$

where $U(t)$ is the semigroup generated by collisionless flow in ${}^*D \times (ns^*\mathbb{R}^3)$.

If the initial datum $\varphi \in \text{Loeb-}L^{1,2}({}^*D \times (ns^*\mathbb{R}^3))$ (where $L^{1,s}$ means that $(1+|\xi|^2)^s f \in L^{1,s}$, $\varphi \geqslant 0$, with $\varphi \log \varphi \in \text{Loeb-}L^1({}^*d \times (ns^*\mathbb{R}^3))$, then Arkeryd shows that there exists a solution $f(t) \in \text{Loeb-}L^{1,2}({}^*D \times (ns^*\mathbb{R}^3))$ of Eq. (4.7). The solution conserves mass; energy and the H-functional are bounded. Elmroth [50] has generalized the treatment of Arkeryd, valid for periodic boundary conditions and no external forces, to the case when an external force acts on the molecules and the motion occurs in a bounded set D with a C^1 boundary; the only difference arises in the semigroup $U(t)$.

In order to prove the existence theorem, one truncates the collision integral in the following way:

$$Q_n(f, f) = \int_{\mathbb{R}^3} \int_S (\langle f'f_*' \rangle - \langle ff_* \rangle) B_n\, d\mathbf{n}\, d\boldsymbol{\xi}_* \tag{4.8}$$

where

$$\langle g \rangle = \begin{cases} g & \text{if } |g| \leqslant n \\ n \operatorname{sgn} g & \text{if } |g| \geqslant n \end{cases} \tag{4.9}$$

and

$$B_n = \begin{cases} B(V, \mathbf{n}) & \text{if } |\boldsymbol{\xi}|^2 + |\boldsymbol{\xi}|^2 \leqslant n^2 \\ 0 & \text{otherwise} \end{cases} \tag{4.10}$$

It is an elementary result that the solution of the mild Boltzmann equation associated with the truncated collision operator (4.8) exists in $L^{1,2} \cap L^\infty(D \times \mathbb{R}^3)$ for all nonnegative f_0 in the same space; it is nonnegative, conserves mass and has a globally bounded energy and H-functional.

At this point one transfers this result to the nonstandard context by means of the transfer principle. Let $\hat{\varphi}$ be an S-integrable lifting of φ, the initial condition in the existence theorem for Eq. (4.7). If χ_{V_n} is the characteristic function of $V_n = (\xi \in {}^*\mathbb{R}^3,\ |\xi| \leqslant n)$, then there is $n \in {}^*\mathbb{N} \backslash \mathbb{N}$ such that φ_n

$= \min(n, \hat{\varphi} \cdot \chi_{V_n}) + e^{-|\xi|^2}/n$ is *Lebesgue measurable and

$$ {}^{0}\!\!\int_{*D \times *\mathbb{R}^3} (1 + |\xi|^2)\varphi_n{}^* \, d\mathbf{x} \, d\xi = \int_{*D \times ns\mathbb{R}^3} (1 + |\xi|^2)\varphi \, dL(\mathbf{x}, \xi) $$

$$ {}^{0}\!\!\int_{*D \times \mathbb{R}^3} \varphi_n \log \varphi_n{}^* \, d\mathbf{x} \, d\xi = \int_{*D \times ns*\mathbb{R}^3} \varphi \log \varphi \, dL(\mathbf{x}, \xi) \tag{4.11} $$

If we consider the "star" version of the existence theorem with the truncated collision term (4.8), we find that there is a solution f of the "star" version of the truncated Boltzmann equation. This f together with the integrand in the collision operator are S-integrable. That leads to the fact that f satisfies Eq. (4.7), as was required to show.

5. Local existence and validity of the Boltzmann equation

The discussion in the previous two sections indicates that the problem of global existence and uniqueness for the unmodified Boltzmann equation in the space inhomogeneous case is a hard problem, still to be solved; even if one uses the tools of nonstandard analysis a uniqueness proof is lacking and the solution is not easily interpreted in a standard way. The problem becomes tractable, however, if the requirement of global existence is dropped and existence is looked for in a finite interval of time (determined by the size of the initial data).

The first existence theorem for the spatially dependent equation was presented by Grad in 1958 [51]. He considered Maxwellian molecules with cutoff and established the existence of local solutions for initial data bounded by a Maxwellian f_0; in other words, he looked for solutions in a weighted L^∞ with norm

$$ \|f\| = \sup_{\mathbf{x}, \xi, t} (|f|/f_0) \tag{5.1} $$

In this space a straightforward proof based on the Lipschitz continuity of the collision term is available. In fact, taking $T_0 = \frac{1}{2}\rho_0$ where ρ_0 is the density of f_0, it is easy to show (by contraction) that a unique solution exists for $0 \leqslant t \leqslant T_0$ and satisfies $f \leqslant 2f_0$. His result was extended to all cutoff potentials by Glickson [52], who considered continuous initial data bounded by f_0 and included an external force term.

Kaniel and Shinbrot [53] presented a constructive local existence proof for a general interaction in a bounded domain W with, for example, specular reflection law on the boundary ∂W. They worked in a space of functions bounded uniformly by a Maxwellian, and proved existence of a mild solution by constructing upper and lower sequences ($\{u_n\}$ and $\{l_n\}$), which bound the solution and converge monotonically to it.

The idea is to look for solutions of the following set of equations:

$$\frac{\partial u_{n+1}{}^{\#}}{\partial t} + [R(l_n)]^{\#} u_{n+1}{}^{\#} = [G(u_n, u_n)]^{\#}$$

$$\frac{\partial l_{n+1}{}^{\#}}{\partial t} + [R(u_n)]^{\#} l_{n+1}{}^{\#} = [G(l_n, l_n)]^{\#}$$

$$u_{n+1}(0) = \varphi$$

$$l_{n+1}(0) = \varphi \tag{5.2}$$

where φ is the initial datum and $u^{\#}$ denotes $U(-t)u$ where $U(t)$ is the semigroup associated with free molecular flow.

The zeroth iterates u_0 and l_0 remain to be specified. They must be chosen in such a way that $l_0 \leqslant l_1$ and $u_1 \leqslant u_0$; then it is easy to show, by induction, that the above scheme provides two sequences, one of them, $\{l_n\}$, increasing monotonically, the other, $\{u_n\}$, decreasing monotonically and hence having limits l and u. In addition, all the l_n's are less than the u_n's and l and u are shown to coincide to give the (unique) solution of the Boltzmann equation with initial datum φ.

Now, while the choice of l_0 is immediately clear, i.e. $l_0 = 0$, that of u_0 poses serious problems because u_0 must satisfy the inequality $u_1 \leqslant u_0$, i.e.

$$\varphi + \int_0^t [G(u_0, u_0)]^{\#}(s)\, ds \leqslant u_0{}^{\#} \tag{5.3}$$

This is a quadratic integral inequality that, in general, will not possess a global solution, especially if u_0 is assumed to depend on time only. It is certainly possible, as was indicated by Kaniel and Shinbrot [53], to find such a u_0 if one restricts oneself to a finite time. A case when Eq. (5.3) has a global solution will be discussed in Section 7.

If one renounces considering global proofs, it is possible to carry out the program of deriving the Boltzmann equation from the Liouville equation in a completely rigorous way. In fact, Lanford [2] showed, locally in time, that the formal steps described in Chapter II, Section 2 and 3, are rigorous. In fact, he proved that in the Boltzmann-Grad limit the solution of the finite hierarchy (II.2.22) converges, in a suitable sense, to the solution of the infinite hierarchy, Eq. (III.3.5), usually called the Boltzmann hierarchy. He further proved that the assumption of molecular chaos, Eq. (II.3.3), at $t = 0$, implies the same property at later times; the one-particle distribution function being the solution of the Boltzmann equation, (II.3.6), with the prescribed initial datum $P^{(1)}(\mathbf{x}, \boldsymbol{\xi}, 0)$. Lanford worked in a space of continuous functions bounded by a (time-dependent) Maxwellian and defined in a domain W with periodicity boundary conditions. Thus one may expect some similarity between his result and Kaniel and Shinbrot's analysis of the Boltzmann equation. As a matter of fact, Shinbrot [54] showed that one may use arguments having the flavour of those of his previous work with Kaniel [53] to reobtain Lanford's results. He first proved that the Boltzmann hierarchy has a unique local solution. Then the sequence of finite hierarchies possesses, in the Boltzmann-Grad limit, a converging subsequence of solutions, convergence being weak* in L^∞ weighted with a (time-dependent) Maxwellian. The limiting functions can then be redefined on sets of measure zero so as to become continuous and, once this is done, they satisfy the Boltzmann hierarchy.

The work of Lanford [2] has been extended by his student, King [55], to cover the case of molecules interacting with (cutoff) repulsive potentials. A global validity proof by means of the techniques of nonstandard analysis, briefly discussed in the previous section, has been announced by Hurd [56].

Spohn [57, 58] explicitly considered the question of the meaning of the Boltzmann hierarchy and indicated that the admissible initial data for the hierarchy have the form

$$P_s(\mathbf{x}_j, \xi_j, 0) = \int_Q \nu(dP) \prod_{j=1}^{s} P(\mathbf{x}_j, \xi_j, 0) \tag{5.4}$$

where $\nu(dP)$ is a probability measure on the set of functions $Q = \{P | P\colon W \times \mathbb{R}^3 \to \mathbb{R}, 0 \leqslant P \leqslant f_0, P \text{ continuous}\}$ and the solution becomes

$$P_s(\mathbf{x}_j, \xi_j, t) = \int_Q \nu(dP) \prod_{j=1}^{s} P(\mathbf{x}_j, \xi_j, t) \tag{5.5}$$

where $P(\mathbf{x}, \xi, t)$ is the solution of the Boltzmann equation with datum $P(\mathbf{x}, \xi, 0)$. This shows, in particular, that in a perfect gas of hard spheres, correlations, if present, do not decay in time.

Cercignani [59] showed that the problem of the derivation of the Boltzmann equation can be carried out globally in time, if one first smooths the hard sphere collision by assuming that the spheres do not have a given radius but only a given probability of having a continuous set of radii. The resulting "Boltzmann" equation is mollified in the way considered by Povzner and discussed in Section 3.

6. Global existence near equilibrium

There are classes of initial data for which it is possible to prove global existence theorems in the space inhomogeneous case. The first of these classes is given by data in the neighbourhood of an (absolute) Maxwellian f_0. In this case, of course, one expects the linearized Boltzmann equation to provide a good first approximation to the solution. What is needed is then a good existence theory for the linearized equation and good estimates for the nonlinear term $Q(f_0 h, f_0 h)$ when we let $f = f_0(l + h)$ in the full Boltzmann equation.

The study of existence and uniqueness theorems for the linearized equation begins with Carleman [3], who established existence and uniqueness for the hard sphere equation, including spatial variation, but with a growth estimate $e^{t/\theta}$ (θ being the mean free time). Later Grad [60, 61] was able to construct a rather general theory of the purely initial value problem for the linearized equation in the case of hard spheres and potentials with angular cutoff. In particular, he showed existence and global boundedness of the solution uniformly in the mean free path and even an ultimate decay to equilibrium [61].

An outline of Grad's proof is as follows. First the linearized equation is transformed into an integral equation:

$$h(\mathbf{x}, \boldsymbol{\xi}, t) = h_0(\mathbf{x} - \boldsymbol{\xi}t, \boldsymbol{\xi}) \exp(-\nu t) + \\ + \int_0^t \exp[-\nu(t-s)]k(\mathbf{x} - \boldsymbol{\xi}[t-s], s)\, ds \tag{6.1}$$

where the linearized collision operator L has been assumed to be decomposable according to Eq. (IV.5.20) and

$$k = Kh \tag{6.2}$$

while $h_0(\mathbf{x}, \boldsymbol{\xi})$ is the given initial value of h. Eq. (6.1) can be solved by iteration. The crude estimate obtained by setting the exponentials in Eq. (6.1) equal to unity yields convergent iterations with a growth e^{kt} in any norm of the following classes:

$$N[h] = \left[\int f_0 |h|^2 \, d\boldsymbol{\xi}\, d\mathbf{x}\right]^{\frac{1}{2}} = |||h||| \tag{6.3}$$

$$N_r[h] = \max_{\mathbf{x}, \boldsymbol{\xi}} [f_0(1+\xi^2)^r] |h(\mathbf{x}, \boldsymbol{\xi})| \tag{6.4}$$

$$N_r[h] = \max_{\boldsymbol{\xi}} (1+\xi^2)^{r/2} \left[\int f_0 |h|^2 \, d\mathbf{x}\right]^{\frac{1}{2}} \tag{6.5}$$

or even more complicated norms based on the space derivatives. The constant k in the exponential e^{kt} is the bound of K in the chosen norm. For any smooth solution h, however, we can obtain the *a priori* estimate

$$N(h) \leqslant N(h_0) \tag{6.6}$$

which is nothing other than the linearized version of the H-theorem (see Section 4 of Chapter IV). Use of Eq. (6.6) leads to a uniform L^2 estimate which can be extended to the other norms listed above. Analogous results for the linearized equation were obtained by Arsen'ev [62] by means of Fourier transform techniques and a preliminary study of the spectral problem of Eq. (IV.8.4) with real $\mathbf{k}$.

Scharf [63] again used the Fourier transform to prove a global existence theorem and a weak form of tendency toward equilibrium, under weak conditions on the initial value h_0. He used the theory of linear semigroups [64] and the fact that the operator

$$A = L - i\mathbf{k} \cdot \boldsymbol{\xi} \qquad (\mathbf{k} \text{ real}) \tag{6.7}$$

is dissipative together with its adjoint in the Hilbert space $\tilde{\mathscr{H}}$ of the functions of $\mathbf{k}$ and $\boldsymbol{\xi}$ where the norms are weighted with f_0. This follows easily from the definition of a dissipative operator [64–65] since A is densely defined, and for any $\tilde{h} \in \tilde{\mathscr{H}}$ in the domain of A,

$$\mathrm{Re}((\tilde{h}, A\tilde{h})) = ((\tilde{h}, L\tilde{h})) \leqslant 0 \tag{6.8}$$

Then, since the range of $\lambda - A$ coincides with $\tilde{\mathscr{H}}$ for any $\lambda > 0$, a family of linear operators T^t exists with the property that the solution of the equation

$$\frac{d\tilde{h}}{dt} = A\tilde{h} + \tilde{g} \tag{6.9}$$

(which exists and is unique) is given by

$$\tilde{h} = T^t \tilde{h}_0 + \int_0^t T^{t-s} \tilde{g}(s)\, ds \tag{6.10}$$

The same semigroup approach has been used by Fetz and Shen [66], who avoided using Fourier transforms. The advantage of the semigroup approach is that it can be applied even in the presence of boundaries, provided the boundary conditions are homogeneous [63, 66]. Inhomogeneous boundary conditions will be dealt with in Section 9. Semigroup techniques have also been used by Grünbaum [67] for the weakly nonlinear spatially homogeneous case.

Similar results have been proved, of course, for the initial value problem with homogeneous boundary conditions which arise in neutron transport. The works of Marti [68] and Mika [69, 70] seem particularly worth mentioning, together with the papers by Albertoni and Montagnini [71, 72] and Bednarz [73]. Though the latter authors did not prove existence theorems, they found important spectral properties of the transport operator. For a more detailed discussion of these results, as well as other applications of functional analysis to neutron transport, we refer to the treatises of Ribariç [74] and Kaper, Lekkerkerker and Hejtmanek [75].

The first attempt to extend Grad's result to the nonlinear case was performed by Grad himself [76], who provided a local existence theorem; the inequalities that he obtained were, however, the starting point for all subsequent work on global existence for data close to a Maxwellian f_0. In terms of the perturbation h, the nonlinear Boltzmann equation reads as follows:

$$\frac{\partial h}{\partial t} + \boldsymbol{\xi} \cdot \frac{\partial h}{\partial x} = Lh + f_0^{-1} Q(f_0 h, f_0 h) \tag{6.11}$$

with the initial datum $h_0 = h_0(\mathbf{x}, \boldsymbol{\xi})$.

Let us consider a cutoff hard potential and let h be in the Banach space B_r of continuous functions defined on $W \times \mathbb{R}^3$ (W parallelepiped with periodicity conditions) with the norm $N_r(h)$ defined above. The basic inequality proved by Grad [76] is

$$N_{r-1}[f_0^{-1} Q(f_0 h, f_0 g)] \leqslant c_r N_r(h) N_r(g) \tag{6.12}$$

where c_r is a constant depending on r. Accordingly, the operator $N(h) = f_0^{-1} Q(f_0 h, f_0 h)$ carries B_r into B_{r-1} and is locally Lipschitzian. Concerning the linear operator K appearing in the usual decomposition, Eq. (IV.5.20), of the linearized collision operator L, we have the following smoothing property:

$$\| Kh \|_r \leqslant c_r' \| h \|_{r-1} \tag{6.13}$$

The first complete global existence theorem for the full nonlinear equation was proved by Ukai [77], and was generalized by Nishida and Imai [78], Shizuta and Asano [79] and Caflisch [80]. The new property used by these authors with respect to Grad is that the semigroup U^t generated by the linearized operator $-\boldsymbol{\xi} \cdot \partial/\partial \mathbf{x} + L$ has a smoothing property when acting on the nonlinear term. In fact, if we consider the Banach space $B_{r,a}$ with norm

$$\| h \|_{r,a} = \sup_{t \geqslant 0} (e^{at} N_r(h)) \qquad (h \in B^r) \tag{6.14}$$

and consider functions such that

$$\sup_{\mathbf{x}} (1 + |\boldsymbol{\xi}|^2)^{r/2} f_0^{\frac{1}{2}} |h(\mathbf{x}, \boldsymbol{\xi})| \to 0 \qquad \text{as } |\boldsymbol{\xi}| \to \infty \tag{6.15}$$

then we have a bound

$$\left\| \int_0^t U^{t-s} N(h)(s) \, ds \right\|_{r,a} \leqslant C_1 \| h \|_{r,a}^2 \tag{6.16}$$

provided $h \in B_{r,a}$ is orthogonal to the collision invariants. In Eq. (6.16), of course, the integral is a Riemann integral in $B_{r-1,a}$ (see Eq. (6.12)). Actually, the condition in Eq. (6.15) was omitted in Ukai's first paper [77], and the resulting linear semigroup was incorrectly taken to be strongly continuous.

The solution may now be obtained by successive approximations if the initial datum φ is in B_r, satisfies Eq. (6.15) and is orthogonal to the five-dimensional space $\mathscr{F}$ of the collision invariants. In fact, if we let P denote the projection into $\mathscr{F}$, then

$$\| U^t (I - P) \|_r \leqslant M_r e^{-at} \qquad (t \geqslant 0) \tag{6.17}$$

for some positive constants M_r and a and the iterative scheme

$$h_{n+1} = U^t\varphi + \int_0^t U^{t-s}N(h^n)\,ds \qquad (n \geqslant 1) \tag{6.18}$$

converges for

$$\|\varphi\|_r < (4M_rC_1)^{-1} \tag{6.19}$$

In fact, we have

$$\|h_{n+1} - h_n\|_{r,a} \leqslant \alpha\|h_n - h_{n-1}\|_{r,a} \tag{6.20}$$

where

$$\alpha = 1 - (1 - (1 - 4M_rC_1\|\varphi\|_r)^{\frac{1}{2}} \tag{6.21}$$

and the convergence of the iteration to a (unique) solution is insured by the contraction mapping theorem.

Similar inequalities were proved by Nishida and Imai [78], who avoided the restriction that the initial density, bulk velocity and temperature must be carried by f_0 (and hence be constant), but assumed $W = \mathbb{R}^3$ and considered the Banach space $B_{r,l}$ with norm

$$N_{r,l}(h) = \sup(1 + |\xi|^2)^{r/2} f_0^{\frac{1}{2}} \|h(\cdot, \zeta)\|_{H_l} \qquad (r, l \geqslant 0) \tag{6.22}$$

where H_l is the Sobolev space with norm

$$\|h\|_{H_l} = \left[\int (1 + |\mathbf{k}|^2)^l |\hat{h}|^2\,d\mathbf{k}\right]^{\frac{1}{2}} \tag{6.23}$$

where $\hat{h}$ is the Fourier transform of h. The condition in Eq. (6.15) is correspondingly replaced by

$$(1 + |\xi|^2)^{r/2} f_0^{\frac{1}{2}} \|h(\cdot, \xi)\|_{H_l} \to 0 \qquad \text{as} \quad |\xi| \to \infty \tag{6.24}$$

For $r \geqslant 3/2$, $l \geqslant 3/2$, each $h \in B_{r,l}$ is, thanks to the Sobolev lemma, continuous in both $\mathbf{x}$ and ξ. An inequality similar to (6.12) was proved by Grad [76] for this norm as well; it has the form

$$N_{r,l}(f_0^{-1}Q(f_0h, f_0g)) \leqslant C_{r,l}\|h\|_{r,l}\|g\|_{r,l} \tag{6.25}$$

and holds for $r > 5/2$, $l > 3/2$. The semigroup U^t is strongly continuous in $B_{r,l}$ with the specification (6.24) for $m \geqslant 0$, $l > 0$; and an inequality of the form (6.16) insures that the iteration works in $C([0, \infty], B_{r,l})$ for $m \geqslant 3$, $l \geqslant 2$.

Shizuta and Asano [79] proved that h converges exponentially to zero as $t \to \infty$, but this is, of course, not true if the perturbation is not orthogonal to the collision invariants. However, Nishida and Imai were able to prove algebraic decay estimates under additional assumptions on φ. Shizuta [81] considered the problem of showing that the solutions discussed above were indeed classical solutions and proved some regularity results. Caflisch [81] considered the weakly nonlinear problem for soft potentials with periodicity conditions, and was able to prove existence of global solutions in the norm

$$\|f\|_{b,l} = \sup e^{b|\xi|^2} \|h(\cdot, \xi)\|_{H_l} \tag{6.26}$$

7. Perturbations of vacuum

A recent development in the theory of existence for the Boltzmann equation was introduced by Illner and Shinbrot [82], who proved global existence under conditions that include the case of a finite volume of gas in an infinite vacuum when the gas is sufficiently rarefied. Their starting point is the method of Kaniel and Shinbrot described in Section 5. We recall that the basic point there (see Eq. (5.3)) was to find a function u_0 satisfying the inequality

$$\varphi + \int_0^t [G(u_0, u_0)]^{\#}(s)\, ds \leqslant u_0 \tag{7.1}$$

where in this case

$$u_0^{\#} = u_0(\mathbf{x} + \boldsymbol{\xi} t, \boldsymbol{\xi}, t) \tag{7.2}$$

They considered hard sphere molecules and initial data satisfying

$$\varphi(\mathbf{x}, \boldsymbol{\xi}) \leqslant \psi(\boldsymbol{\xi}) e^{-\beta|\mathbf{x}|^2} \tag{7.3}$$

Under these assumptions they looked for a solution of Eq. (7.1) in the form

$$u_0(\mathbf{x}, \boldsymbol{\xi}, t) = e^{-\beta|\mathbf{x} - \boldsymbol{\xi} t|^2} w(\boldsymbol{\xi}) \tag{7.4}$$

The motivation for this assumption comes from the results of Tartar [34] in the discrete velocity case.

Then Eq. (7.1) is certainly satisfied if

$$\psi(\boldsymbol{\xi}) + \frac{\sigma^2}{m} \int_0^t \int_{\mathbb{R}^3} \int_0^{\pi} \int_0^{2\pi} e^{-\beta|\mathbf{x} + \tau(\boldsymbol{\xi} - \boldsymbol{\eta})|^2} \times$$
$$\times\, w(\boldsymbol{\xi}') w(\boldsymbol{\eta}') \mathbf{n} \cdot (\boldsymbol{\xi} - \boldsymbol{\eta})\, d\boldsymbol{\eta} \sin\theta\, d\theta\, d\varepsilon \leqslant w(\boldsymbol{\xi}) \tag{7.5}$$

where σ, as usual, is the sphere diameter and m the molecular mass. Since $w \geqslant 0$, this inequality will be satisfied for any $t \geqslant 0$ if it is satisfied for $t = \infty$. Also,

$$\int_0^{\infty} e^{-\beta|\mathbf{x} + \tau(\boldsymbol{\xi} - \boldsymbol{\eta})|^2} d\tau \leqslant \int_{-\infty}^{\infty} e^{-\beta|\mathbf{x} + \tau(\boldsymbol{\xi} - \boldsymbol{\eta})|^2} \leqslant \sqrt{\frac{\pi}{\beta}} \frac{e^{-\beta|\mathbf{x}_{\perp}|^2}}{|\boldsymbol{\xi} - \boldsymbol{\eta}|}$$
$$\leqslant \sqrt{\frac{\pi}{\beta}} \frac{1}{|\boldsymbol{\xi} - \boldsymbol{\eta}|} \tag{7.6}$$

where $\mathbf{x}_\perp$ is the part of $\mathbf{x}$ perpendicular to $\boldsymbol{\xi}-\boldsymbol{\eta}$. Then Eq. (7.5) is satisfied if

$$\psi(\boldsymbol{\xi}) + \frac{\sigma^2}{m}\int_{\mathbb{R}^3}\int_0^{\pi}\int_0^{2\pi}\sqrt{\frac{\pi}{\beta}}\,w(\boldsymbol{\xi}')w(\boldsymbol{\eta}')\,d\boldsymbol{\eta}\sin\theta\,d\theta\,d\varepsilon = w(\boldsymbol{\xi}) \tag{7.7}$$

and we are thus led to solving an integral equation and we can try to use the contraction mapping theorem to obtain the existence of a solution. A simple calculation shows that if $\psi, w \in L^1$, the nonlinear operator on the left-hand side, i.e.

$$W(w) = \psi(\boldsymbol{\xi}) + \frac{\sigma^2}{m}\sqrt{\frac{\pi}{\beta}}\iiint w(\boldsymbol{\xi}')w(\boldsymbol{\eta}')\sin\theta\,d\theta\,d\varepsilon\,d\boldsymbol{\eta} \tag{7.8}$$

satisfies

$$\|W(w)\| = \|\psi\| + 2\pi\frac{\sigma^2}{m}\sqrt{\frac{\pi}{\beta}}\|w\|^2 \tag{7.9}$$

when $\psi, w \geqslant 0$ and the norms are, of course, in L^1. This shows that if we take a ball of radius R in L^1 such that

$$\|\psi\| + \frac{2\pi\sigma^2}{m}\sqrt{\frac{\pi}{\beta}}R^2 \leqslant R \tag{7.10}$$

W will map the set of nonnegative functions in the ball into itself. Further,

$$\|W(w_1) - W(w_2)\| \leqslant \frac{2\pi\sigma}{m}\sqrt{\frac{\pi}{\beta}}(\|w_1\| + \|w\|_2)\|w_1 - w_2\| \tag{7.11}$$

and W is a contraction on the said set if

$$\frac{4\pi\sigma^2}{m}\sqrt{\frac{\pi}{\beta}}R < 1 \tag{7.12}$$

If

$$\delta \equiv \frac{8\pi\sigma^2}{m}\sqrt{\frac{\pi}{\beta}}\|\psi\| < 1 \tag{7.13}$$

then the inequalities on R can both be satisfied by choosing R so as to have

$$1 - \sqrt{1-\delta} \leqslant R < 1 \tag{7.14}$$

The number δ in Eq. (7.13) is an inverse Knudsen number for the problem. In fact, $\|\psi\|$ is a measure of the density at $t=0$, and $\beta^{-\frac{1}{2}}$ a measure of the radius of a sphere in R^3 where the density is not too close to zero. Eq. (7.13) and the contraction mapping theorem tell us that there is a nonnegative function w in L^1 satisfying Eq. (7.8). However, one needs a little more because, for u_0 to be a good zeroth iterate in the Kaniel-Shinbrot method, one needs to show that $G(u_0, u_0)$ and $u_0 L(u_0)$ are in the space S_β with norm

$$\|f\|_\beta = \int_{\mathbb{R}^3} \max_x |e^{\beta|x|^2} f(x, \xi)| \, d\xi \tag{7.15}$$

and $|\xi|w$ is thus required to be in L^1; this is true if $\psi \in L^1$, $|\xi|\psi \in L^1$, since one can prove the following estimate:

$$\| |\xi| w \| \leqslant \| |\xi| \psi \| \frac{\sqrt{2}}{\sqrt{2} - 1} \tag{7.16}$$

Then it is possible to show that the iterates l_n and u_n converge to limits l and u in $C^1(0, T; S^+)$ and these limits can be shown to be equal if $\delta < \frac{1}{2}$; then the function $l = u$ is a solution of the Boltzmann equation with initial datum φ.

Extensions of this result to molecules, other than hard spheres, have been made by several authors [83–85], and Shinbrot [86] applied the same method to the case when the cloud flows past a fixed obstacle reflecting the molecules in a specular way.

8. Homoenergetic solutions

At about the same time, 1956, Truesdell [13] and Galkin [87] independently investigated the steady homoenergetic flows of a gas of Maxwellian molecules according to the infinite system of moments associated with the Boltzmann equation. Later Galkin [88–90] extended his analysis to some typical unsteady homoenergetic affine flows. These analyses are discussed and summarized by Truesdell and Muncaster [20].

It is convenient, at this point, to recall the basic ideas about homoenergetic affine flows. The defining properties are the following ones:

(a) The body force (per unit mass) $\mathbf{X}$ acting on the molecules is constant

$$\mathbf{X} = \text{constant} \tag{8.1}$$

(b) the density ρ, the internal energy per unit mass e, the stress tensor p and the heat flux $\mathbf{q}$ may be functions of time but not of the space coordinates.

(c) The bulk velocity $\mathbf{v}$ is an affine function of position $\mathbf{x}$:

$$\mathbf{v} = \mathsf{K}(t)\mathbf{x} + \mathbf{v}_0(t) \tag{8.2}$$

This definition holds for a general material; for a gas described by kinetic theory, a natural extension of property (b) is spontaneous:

(b′) The moments formed with the peculiar velocity:

$$\mathbf{c} = \boldsymbol{\xi} - \mathbf{v} \tag{8.3}$$

may be functions of time but not of space coordinates.

We remark that (b′) holds for the solutions obtained by Truesdell [13] and Galkin [87–90]. When one works with the distribution function f, condition (b′) transforms itself into

(b″) The variable $\mathbf{x}$ appears in f only through $\mathbf{c}$, given by Eq. (8.2), i.e.

$$f = f(\mathbf{c}, t) \tag{8.4}$$

An analysis of the balance equations based on (a), (b), (c) immediately leads to the time dependence of K and $\mathbf{v}_0$ [20]

$$\begin{aligned} \mathsf{K}(t) &= [\mathsf{I} + t\mathsf{K}(0)]^{-1}\mathsf{K}(0) \\ \mathbf{v}_0(t) &= [\mathsf{I} + t\mathsf{K}(0)]^{-1}[\mathbf{v}_0(0) + t\mathbf{X} + \tfrac{1}{2}t^2\mathsf{K}(0)\mathbf{X}] \end{aligned} \tag{8.5}$$

where I is the 3×3 identify matrix. This solution exists globally for $t > 0$ if the eigenvalues of $\mathsf{K}(0)$ are nonnegative; otherwise, the solution ceases to exist for $t = t_0$, where $-t_0{}^{-1}$ is the largest, in absolute value, among the negative eigenvalues of $\mathsf{K}(0)$. In particular, if

$$[\mathsf{K}(0)]^2 = 0 \tag{8.6}$$

then $\mathsf{K}(t)$ is independent of time, as is obvious from Eq. (8.5) (see also [91, 20]). $\mathbf{v}$ is then a steady flow if and only if, in addition to Eq. (8.6), the following conditions are satisfied:

$$\mathsf{K}(0)\mathbf{X} = 0 \qquad \mathsf{K}(0)\mathbf{v}_0(0) = \mathbf{X} \tag{8.7}$$

These conditions can be satisfied, in particular, by taking $\mathbf{X} = 0$ and $\mathbf{v}_0(0) = 0$. We remark the elementary result that Eq. (8.6) is satisfied if and only if a coordinate system exists for which the matrix $\mathsf{K}(0)$ has only one nonzero entry in nondiagonal position (for a simple proof, see the Appendix of Ref. [91]).

The form (8.5) for $\mathsf{K}(t)$ and $\mathbf{v}_0$ is certainly necessary for a homoenergetic flow to exist. In order to show that this is also sufficient, we first transform the Boltzmann equation to the t, $\mathbf{c}$ variables, according to condition (b″) (Eq. (8.4)). An elementary calculation [91] gives

$$\frac{\partial f}{\partial t} - \frac{\partial f}{\partial \mathbf{c}} \cdot \mathsf{K}\mathbf{c} = Q(f, f) \tag{8.8}$$

and the space variable no longer appears explicitly. Thus the existence of homoenergetic flows in kinetic theory is reduced to proving an existence theorem for Eq. (8.8).

It is to be emphasized that the flows under-study are conceptually important but rather impractical. In particular, there is no possibility of the energy being removed by heat conduction, so that in some of the flows it grows exponentially in time. This is possible because work is done on the gas at infinity at an infinite rate, in the sense that the integral of $\mathsf{K}\mathsf{p}$ on a large sphere diverges with the radius of the sphere itself.

The analyses of Truesdell and Galkin have the great advantage of leading to explicit solutions, which lend themselves to a detailed discussion of their properties. They suffer, however, two drawbacks:

(1) they are restricted to Maxwellian molecules, and
(2) they provide solutions to the system of moment equations, but no proof is given of the existence of a corresponding solution of the Boltzmann equation itself.

An existence theorem for the Boltzmann equation and general molecular models (rigid spheres and angle cutoff potential) for initial data compatible with a homoenergetic flow is given in Ref. [91]. Some of the data lead to an implosion and infinite density in a finite time, in agreement with the physical picture of the associated flows and the remark after Eq. (8.5). It is to be remarked that the theorem to be proved delivers solutions for the space inhomogeneous Boltzmann equation without restrictions on the size of the initial data.

The problem is now reduced to proving an existence theorem for Eq. (8.8). We notice that the latter can be cast into an integral form, provided we determine the semigroup corresponding to collisionless flow. This is easily done [91], and if we let

$$f^{\#}(\mathbf{c}, t) = f([\mathsf{I} + t\mathsf{K}]^{-1}\mathbf{c}, t) \tag{8.9}$$

we obtain the following integral form of Eq. (8.8):

$$f^{\#}(\mathbf{c}, t) = f^{\#}(\mathbf{c}, 0) + \int_0^t [Q(f, f)]^{\#}(\mathbf{c}, s)\, ds \tag{8.10}$$

Other integral forms are possible when $Q(f, f)$ can be split into two separate contributions (gain and loss terms), as is the case for hard sphere molecules and cutoff interactions. It is clear that any solution of these equations is also a weak solution of the Boltzmann equation.

It is now possible to obtain estimates for the density and internal energy, using the conservation equations. This is crucial for global existence or to estimate the time of existence of the solution. We have

$$\rho(t) = \rho(0) \exp\left[-\int_0^t \mathrm{Tr}\,\mathsf{K}(s)\, ds\right] \tag{8.11}$$

$$e(t) \leqslant e(0) \exp\left[6\int_0^t k(s)\, ds\right] \tag{8.12}$$

where k is the largest element of K. We remark that $|\mathrm{Tr}\,\mathsf{K}|$ and k are bounded for $0 \leqslant t \leqslant \bar{t} < t_0$, where $-t_0{}^{-1}$ is, as above, the largest of the negative eigenvalues of $\mathsf{K}(0)$. If there are no negative eigenvalues, then k and $|\mathrm{Tr}\,\mathsf{K}|$ are bounded for any positive t ($t_0 = \infty$).

In order to prove the existence theorem, we proceed exactly as in Arkeryd [9, 10] and as described in Section 2, the only difference being that the density is not a constant (in general) but a known function of time. A solution is first proved to exist in L^1 for a bounded collision kernel. The solution has energy and H-function bounded by explicit functions of time depending only on the initial data. Using these facts, it is easy to prove an existence theorem which is global if the eigenvalues of $\mathsf{K}(0)$ are nonnegative; otherwise, the solution ceases to exist for $t = t_0$.

The case of an unbounded kernel is treated by means of cutoff expressions and the weak compactness criterion used in Arkeryd's first proof [9; see Section 2]. The following existence theorem holds:

THEOREM [91] *There exists a solution f of Eq.* (8.10) *where the kernel $B(\theta, |\mathbf{c} - \mathbf{c}_*|)$ of the collision term $Q(f, f)$ is bounded by a constant times $(1 + |\mathbf{c}|^2 + |\mathbf{c}_*|^2)$, and the initial density, energy density and H-functional are finite. These functionals are bounded for $0 \leqslant t \leqslant t_0$.*

If we add the assumption that the fourth moment of φ exists, then we can prove a uniqueness theorem. To this end, one has to prove that the fourth moment remains finite; this can be easily done by using a special case of Povzner inequality (Eq. (2.17)) to prove that the fourth moment for any finite time interval is bounded by a constant depending only on the initial data and the length of the interval. (In the case of a finite time of existence, this interval must, of course, be less than t_0.) It is now easy to prove the following.

UNIQUENESS THEOREM *Let f be the solution whose existence is guaranteed by the previous theorem. If the fourth moment exists initially, then it exists for any t for which that solution exists, and the solution f is unique among those having this property.*

9. Boundary value problems. The linearized and weakly nonlinear cases

As shown by the discussion in the previous sections, the existence and uniqueness theory of the initial value problem for the full Boltzmann equation is far from trivial, but the corresponding linearized problem is close to being trivial, in the sense that a considerable effort is required only if precise estimates of the ultimate decay to equilibrium are being investigated. A subtler penetration into the properties of the linearized Boltzmann equation is required when the steady boundary value problem (or the unsteady problem with inhomogeneous boundary conditions) is considered.

In fact, a naive use of the iterative method used for the initial value problem leads to existence if and only if restrictions are placed on the size of the domain [92, 93]. The first proofs of existence and uniqueness for a domain of arbitrary size [94, 95] are based on a detailed study of the free streaming operator $D = \boldsymbol{\xi} \cdot \partial/\partial \mathbf{x}$ and, in particular, on an inequality of the form

$$N(Dh) \geqslant cN(h) \tag{9.1}$$

where N denotes a suitable norm, c is a constant depending on the size of the domain and h satisfies homogeneous boundary conditions of the form shown in Eq. (IV.4.1), i.e. (IV.2.8) with $h_0 = 0$. It would be nice to prove Eq. (9.1) for, say, any domain R, whose boundary ∂R has finite curvature at any point and any boundary conditions of the form (IV.2.8), where the kernel $B(\boldsymbol{\xi}' \to \boldsymbol{\xi})$ is restricted only by Eqs. (IV.2.10), (III.3.9), (III.1.8), (III.1.10).

A proof for bounded domains in one, two or three dimensions was given by the author [94–96], using the properties of the operators D and L that were already used in Section 4 of Chapter IV to prove the uniqueness theorem. In order to prove existence, one has to use the explicit form of the linearized collision operator $L = K - \nu I$, valid for cutoff potential or hard spheres, in order to devise a converging approximation scheme [94–96]. To this end, one first transforms the linearized Boltzmann equation into the following integral form:

$$h = h_\lambda + U(K + \lambda \nu) h \tag{9.2}$$

where h_λ is obtained by solving

$$\boldsymbol{\xi} \cdot \frac{\partial h_\lambda}{d\mathbf{x}} + (\lambda + 1)\, \nu h_\lambda = 0 \tag{9.3}$$

with, generally speaking, inhomogeneous boundary conditions, and where U is the inverse of the differential operator acting on h in the left-hand side of Eq. (9.2) (subject to the corresponding homogeneous boundary conditions). The constant λ is, of course, arbitrary but we choose it in such a way that $K + \lambda \nu$ is a positive operator; this is always possible since we can take λ to be the constant in Eq. (IV.6.21).

The boundary conditions have the form (IV.2.8) and we assume that Eq. (IV.4.7) is strengthened to

$$|||(\xi \beta)^{\frac{1}{2}} A h|||_B \leqslant \lambda_0 |||(\xi \beta)^{\frac{1}{2}}|||_B \qquad (0 \leqslant \lambda_0 < 1) \tag{9.4}$$

where β is a weight function and h is a function defined on ∂R for $\xi \cdot \mathbf{n} > 0$ satisfying

$$((\psi_0, h))_B = 0 \qquad (\psi_0 = \text{const.}) \tag{9.5}$$

The notations $((\ ,\ ))_B$ and $|||\quad |||_B$ were defined in Eqs. (IV.4.5) and (IV.4.6). The physical meaning of Eq. (9.4) is not obvious but such a relation is satisfied, e.g. by Maxwell's boundary conditions, Eq. (III.5.1). Then it is possible to prove [94–96] the following:

THEOREM *There is a positive constant η (depending on the size of the domain where the solution is sought for) such that Eq. (9.2) has a unique solution in the Hilbert space $\mathscr{K}$ of the functions which are square integrable with respect to the weight $f_0\{(\lambda+1)^2[\nu(\xi)]^2+\eta^2\xi^2\}^{\frac{1}{2}}$ (with respect to both space and velocity variables), provided the source term h_λ also belongs to $\mathscr{K}$. This unique solution can be obtained in principle by a converging iteration method.*

For solutions depending on just one or two space variables whose range is bounded, the above theorem holds provided $\eta^2\xi$ is replaced by $\eta^2\xi_p{}^2$, where ξ_p is the magnitude of the projection of $\boldsymbol{\xi}$ on the relevant axis or plane.

The case of one space variable x and the special boundary conditions for which the incoming distribution is completely assigned was considered in Ref. [94], and the result was used by Pao [97] to prove existence in other Banach spaces $\mathscr{V}$ and $\mathscr{E}$ with norms

$$|||h|||_{\mathscr{V}}{}^2 = \max_{\mathbf{x}} \int d\boldsymbol{\xi}\{(\lambda+1)^2[\nu(\xi)]^2+\eta^2\xi_1{}^2\}^{\frac{1}{2}} f_0\xi)h^2(\mathbf{x},\boldsymbol{\xi}) \tag{9.6}$$

$$|||h|||_{\mathscr{E}}{}^2 = \max_{\mathbf{x},\boldsymbol{\xi}} (1+|\boldsymbol{\xi}|^2) f_0|h(\mathbf{x},\boldsymbol{\xi})| \tag{9.7}$$

He also proved that the moments are continuous in x if $h_0 \in \mathscr{E}$ or $h_0 \in \mathscr{V}$ (where h_0 is just h_λ with $\lambda = 0$). He further proved a result on the weakly nonlinear equation which provides a rigorous justification for the use of the linearized Boltzmann equation. In fact, he proved the following:

THEOREM *There exists a $c > 0$ such that the boundary value problem for the nonlinear Boltzmann equation has a unique solution $f = f_0(1+\bar{h})$ with $\bar{h} \in \mathscr{E}$, for all h_0 satisfying $\|h_0\|_{\mathscr{E}} < c$. Moreover, if h is the solution of the linearized Boltzmann with the same h_0, then as $\|h_0\|_{\mathscr{E}} \to 0$, we have*

$$\|h-\bar{h}\|_{\mathscr{E}}/\|h\|_{\mathscr{E}} \to 0 \tag{9.8}$$

By means of techniques strictly related to those used above, it is possible to prove the convergence of the solutions of suitable sequences of model equations to the solution of the corresponding (linearized) Boltzmann equation [95, 96].

Further results on the boundary value problem have been obtained by Guiraud [98–102], who studied both steady and unsteady problems. Guiraud avoided using Eq. (9.4), which was replaced by other assumptions on the operator A appearing in the boundary conditions. In particular, Guiraud [102] proved exponential tendency toward equilibrium in the case of the gas of rigid spheres and homogeneous boundary conditions.

The external problems lead to some difficulties as shown by the discussion of Chapter VI, Section 13. These difficulties were circumvented, for one space dimension, by Rigolot-Turbat who proved existence and uniqueness theorems for several steady and unsteady, linear and weakly linear, problems [103–105]. Some cases of existence and uniqueness in the steady case for neutron transport are discussed in the book by Case and Zweifel [106]. More recently, Ukai and Asano, using perturbation techniques of the same kind as for the initial value problem (Section 6), proved existence for the weakly nonlinear flow past a body in the three-dimensional case [107], but not in two dimensions.

Recently, the solution of half-space problems has produced much interest in connection with the study of evaporation and condensation problems. In this case, one assigns the distribution function for molecules entering the half-space at $x = 0$ (usually a Maxwellian) and looks for a bounded solution for $x > 0$. Such a solution is certainly not unique, unless one specifies more about the behaviour of the solution at infinity. One can expect (on the basis of the H-theorem) that f tends to a Maxwellian f_∞ when $x \to \infty$. Assigning the Mach number

$$M_\infty = \frac{|v_\infty|}{\sqrt{5RT_\infty/3}} \tag{9.9}$$

can make the solution unique. On the basis of a heuristic argument [108]], one can conjecture that if $v_\infty > 0$, then there is a unique solution for $M_\infty \leqslant 1$ and none for $M_\infty > 1$. This circumstance entirely depends on the behaviour at infinity (in complete analogy with the Stokes paradox), and thus one is led to conjecture [108, 109] that linearization about the Maxwellian at infinity can clarify the matter. Results on models and abstract transport equations [110–113] have proved the conjecture as far as the linearized equation is concerned. In particular, Greenberg and Van der Mee [113] have considered the following general equation:

$$(\xi_1 + v_\infty)\frac{\partial h}{\partial x} = Lh \tag{9.10}$$

where L is the usual linearized operator about a Maxwellian with zero bulk velocity and density and temperature appropriate to f_∞; Eq. (9.10) can be obtained by linearizing about f_∞ and then changing ξ_1 into $\xi_1 + v_\infty$.

Their general result is the following: if we denote by $\hat{\psi}$ the basic collision invariants chosen in such a way that

$$(\hat{\psi}_\alpha(\xi_1 + v_\infty), \hat{\psi}_\beta) = 0 \qquad (\alpha, \beta = 0, 1, 2, 3, 4; \alpha \neq \beta) \tag{9.11}$$

then the numbers

$$N_\alpha = (\hat{\psi}_\alpha(\xi_1 + v_\infty), \hat{\psi}_\alpha) \qquad (\alpha = 0, 1, 2, 3, 4) \tag{9.12}$$

determine the possibility of solving the problem. In fact, the number of negative values among the N_α gives the number of additional conditions which can be imposed on a solution h bounded at infinity. A simple calculation indicates that one can take

$$\psi_0 = 1, \qquad \psi_1 = \xi^2 - 3v_\infty \xi_1, \qquad \psi_2 = \xi_2, \qquad \psi_3 = \xi_3, \qquad \psi_4 = \xi^2 - 3RT_\infty \tag{9.13}$$

and

$$\begin{aligned} N_0 &= v_\infty, \qquad N_1 = 9v_\infty RT_\infty(v_\infty{}^2 - \tfrac{5}{3}RT_\infty), \\ N_2 &= N_3 = v_\infty RT_\infty, \qquad N_4 = 6v_\infty(RT_\infty)^2 \end{aligned} \tag{9.14}$$

Obviously, if $v_\infty > 0$, then there is one negative value for $v_\infty < (5RT_\infty/3)^{\frac{1}{2}}$, i.e. $M_\infty < 1$ and none if v_∞ $(5RT_\infty/3)^{\frac{1}{2}}$, or $M_\infty > 1$. Thus in the subsonic case one can obtain solutions with one free parameter (which can be the Mach number itself) by imposing that h does not modify the bulk velocity v_∞.

The case $v_\infty < 0$ was not mentioned in the quoted papers [108–113] but is briefly discussed in a survey paper [114]. It is clear that if $v_\infty < 0$, the number of additional conditions is four for $M_\infty < 1$ and five for $M_\infty > 1$. (Two of these can be disposed of by letting the motion in the y- and z-direction vanish at infinity.)

These results have a bearing on the problem of evaporation from or condensation on a flat plate bounding a half-space. (Evaporation can be replaced by blowing and condensation by suction through a porous boundary.) They indicate that evaporation is governed by just one parameter (v_∞) and can exist only for a subsonic flow of the vapor in the Knudsen layer, while condensation is governed by four parameters in the subsonic case, by five in the supersonic case. (These parameters reduce to two and three, respectively, when bulk motions parallel to the plate are absent.) The additional parameter in the supersonic case seems to indicate that when a vapour flows supersonically toward a cold plate, it must first slow down to subsonic speeds through a shock layer. Approximate solutions for evaporation and condensation are discussed in the Appendix of this book.

A detailed treatment of the half-space problem with linearization about a zero bulk velocity Maxwellian has been performed by Bardos, Caflisch and Nicolaenko [115] in the case of rigid spheres. Their results have been extended to general hard potentials by Cercignani [114]. Another proof had been previously presented by Maslova [116]. Essentially, one can prove the following

THEOREM *Eq.* (9.10), *with* $v_\infty = 0$ *and the boundary condition*

$$h = g \qquad at \quad x = 0, \xi_1 > 0 \tag{9.15}$$

where g *is given with* $|\xi_1|^{\frac{1}{2}} g \in H^+$ *(the Hilbert space of square summable functions with weight* f_0*, restricted to* $\xi_1 \geqslant 0$*), has a family of solutions* *parametrized by the value of the constant*

$$j = (\xi_1, h) \tag{9.16}$$

and such that $|\xi_1|^{\frac{1}{2}}h$ *and* $v^{\frac{1}{2}}n$ *are in* $L^\infty(R_x^+, L^2(R_\xi^3))$, *while* $\xi_1\, \partial h/\partial x \in L^2(R_x^+, L^2(R_\xi^3))$.

The theorem can be proved by first obtaining uniform estimates on the asymptotic behaviour of the solution of the problem in $L^\infty(R_x^+, L^2(R_\xi^3))$ (they must approach a linear combination of collision invariants when $x \to \infty$) and then passing to the limit from a suitable two-plate problem. The inclusion of the effect of weak nonlinearities was announced by Bardos, Caflisch and Golse [117].

10. Nonlinear boundary value problems

Very little is known on boundary value problems with data arbitrarily large and removed from equilibrium. There are several difficulties even in the simplest case (two-plate problems). At first, one is struck by the analogy between the problem

$$\xi_1 \frac{\partial f}{\partial x} = Q(f, f) \tag{10.1}$$

(plus suitable boundary conditions at, say, $x = -d$ and $x = d$) and the initial value problem in the space homogeneous case; in both problems, in fact, the unknown f depends on the velocity $\boldsymbol{\xi}$ and another variable (x or t); in both cases, first-order differentiation with respect to the last variable occurs in the linear part of the equation, while the dependence on $\boldsymbol{\xi}$ is paramount in the nonlinear collision operator.

There are, however, two basic differences

(1) The factor ξ_1 in front of the derivative (absent in the initial value problem) can be positive, negative or zero. This implies that, in the boundary value problem, only one of the three constant quantities

$$j = \int \xi_1 f\, d\boldsymbol{\xi} \qquad p = \int \xi_1{}^2 f\, d\boldsymbol{\xi} \qquad q = \int \xi_1 \xi^2 f\, d\boldsymbol{\xi} \tag{10.2}$$

is positive, and the fact that they are bounded does not imply that the density ρ is finite. An analogous remark applies to the inequality given by the H-theorem.

(2) In the simplest case the data are assigned at one point ($x = -d$) for positive values at ξ_1, at another ($x = d$) for negative values of ξ_1. This means that the three constants j, p, q are not known *a priori*.

The main difficulty is, presumably, related to the presence of the value zero among the values taken by ξ_1. Having this in mind, Cercignani, Illner and Shinbrot have recently solved [118] the problem in the case of a discrete velocity model, when none of the velocities has a zero component along the x-axis.

Eq. (10.1) is then replaced by

$$\xi_1{}^i \frac{df^i}{dx} = Q^i(f, f) \tag{10.3}$$

where $\xi_k{}^i$ is the k-th component of the i-th discrete velocity and f^i the corresponding value of the distribution function. The density is obtained by simply summing the values of the n functions f_i, while the constant quantities in Eq. (10.2) are replaced by

$$j = \sum_i \xi_1{}^i f^i, \qquad p = \sum_i \xi_1{}^{i2} f^i, \qquad q = \sum_i \xi_1 \xi^2 f^i \tag{10.4}$$

For any $K > 0$ we define

$$Q_K{}^i(f, f) = Q^i(f, f) + K\rho(f) f^i \qquad (\rho(f) = \sum_i f^i) \tag{10.5}$$

If we denote the nonnegative continuous functions by $C_+{}^0$ and denote the Cartesian product of n copies of $C_+{}^0$ by $C_+{}^{0n}$, then for K large enough we can assume that $Q_K{}^i$ maps $C_+{}^{0n}$ continuously into itself (with the usual topology of uniform convergence). Here the functions are assumed to be defined for $0 \leqslant x \leqslant d$, to satisfy Eq. (10.3) in $0 < x < d$ and the boundary conditions

$$\begin{aligned} f^i(0) &= a^i \quad \text{if } \xi_1{}^i > 0 \\ f^i(d) &= b^i \quad \text{if } \xi_1{}^i < 0 \end{aligned} \tag{10.6}$$

In order to prove that a solution exists, we replace Eq. (10.3) by

$$\xi_1{}^i \frac{df^i}{dx} + K\rho(f) f^i = Q_K{}^i(f, f) \tag{10.7}$$

In order to prove that this equation (and hence Eq. (10.3)) has a positive solution (at least for sufficiently large K) we first consider the linear equation

$$\xi_1{}^i \frac{df^i}{dx} + K\rho(|g|) f^i = Q_K{}^i(|g|, |g|) \tag{10.8}$$

with the boundary conditions (10.6); here g is a given function of C^{0n}, the space of n-tuples of continuous functions.

In fact, each f^i is the solution of an initial value problem for an ordinary differential equation. In this way, for given a^i and b^i, for any g there is an f and thus a nonlinear operator S such that $f = S(g)$ is defined from C^{0n} to $C_+{}^{0n}$. A solution of Eq. (10.3) with boundary conditions (10.6) is a fixed point of S. S is continuous and its restriction to bounded sets is compact. The first part of this statement is obvious, the second follows from the fact that, if g is bounded, f is bounded and, by Eq. (10.8) and the fact that $\xi_1{}^i$ is nonzero, all the derivatives df^i/dx are bounded; as a consequence, the image of a bounded set is bounded and equicontinuous and is compact.

Now, let f be any solution of $f=\lambda Sf$, with $0<\lambda<1$. Then f is nonnegative, differentiable and satisfies

$$\xi_1{}^i\frac{df^i}{dx}+K\rho(f)f^i=\lambda Q_K{}^i(f,f) \tag{10.9}$$

and

$$\begin{aligned} f^i(0)&=\lambda a^i \quad \text{if } \xi_1{}^i>0\\ f^i(d)&=\lambda b^i \quad \text{if } \xi_1{}^i<0 \end{aligned} \tag{10.10}$$

Using the fact that 1 and $\xi_1{}^i$ are collision invariants, Eq. (10.9) gives

$$\frac{dj}{dx}=(\lambda-1)K\rho^2 \tag{10.11}$$

$$\frac{dp}{dx}=(\lambda-1)Kj\rho \tag{10.12}$$

Since $\lambda<1$, Eq. (10.11) shows that j is nonincreasing. Thus

$$j(d)\leqslant j(x)\leqslant j(0) \tag{10.13}$$

On the other hand, writing, in an obvious notation, $j=j^+-j^-$ gives

$$j(d)\geqslant -j^-(d)=-\sum{}^-|\xi_1{}^i|b^i \tag{10.14}$$

and

$$j(0)\leqslant j^+(0)=\sum_i{}^+\xi_1{}^i a^i \tag{10.15}$$

Thus

$$-\sum{}^-|\xi_1{}^i|b^i\leqslant j(x)\leqslant \sum{}^+\xi_1{}^i a^i \tag{10.16}$$

Next, using the fact that $j=j^+-j^-$ along with (10.15) it is easy to see that $j^-(0)$ and $j^+(d)$ are both bounded in terms of the boundary data in a way independent of λ.

To bound $p=p(x)$, we first note that

$$p(x)\leqslant c[j^+(x)+j^-(x)] \tag{10.17}$$

where c is the maximum value of $|\xi_1{}^i|$. One can distinguish three cases:

(a) $j(x)\geqslant 0$ for all x between 0 and d. Then Eq. (10.12) gives that p is nonincreasing. Then $p(x)\leqslant p(0)$ and the latter is bounded, according to

Eq. (10.17), and the previous bound on $j^+(0)$ (as well as the obvious one on $j^-(0)$) gives a bound on $p(x)$ in terms of the boundary data and in a way independent of λ.

(b) $j(x) \leqslant 0$ for all x between 0 and d. Then we argue as in case (a) using the bounds at $x = d$.

(c) $j(x)$ changes sign. This can occur at most once, since j is nonincreasing. Then Eq. (10.12) easily gives that $p(x)$ is not larger than the largest between $p(0)$ and $p(d)$ and we can argue as in the first two cases.

Summarizing, we have shown that $p(x)$ is uniformly bounded. So then are all the f^i. In order to show that a solution of $f = Sf$ exists, we use the following version of the Leray-Schauder theorem [19, 120]:

THEOREM (*Schaefer*) *Let* $\mathbb{B}$ *be a normed space and* S *a continuous mapping from* $\mathbb{B}$ *into* $\mathbb{B}$ *which is compact on each bounded set of* $\mathbb{B}$. *Then, either*

(i) *the equation* $f = \lambda Sf$ *has a solution for* $\lambda = 1$; *or*
(ii) *the set of all solutions of* $f = \lambda Sf$ *for* $0 < \lambda < 1$ *is unbounded.*

Since (ii) has been shown not to occur, (i) must be true and we have proved the following.

THEOREM Eq. (10.3) *with the boundary conditions defined by* Eq. (10.6) *has a solution in* $C_+{}^{0n}$.

In the same paper [118], Cercignani, Illner and Shinbrot prove, in an analogous way, an existence theorem for a case when the current j vanishes at the wall (and hence everywhere); the boundary conditions are analogous to those of perfect diffusion at the wall with conservation of mass, while those indicated in Eq. (10.6) are appropriate for an evaporation–condensation problem.

So far, no uniqueness result has been proved for the same kind of problem, in general; it is easy, however, to prove uniqueness (by contraction) if the ratio of d to the mean free path is sufficiently small.

11. Concluding remarks

The theory of validity, existence and uniqueness of the Boltzmann equation is under intense study and new results appear every month. If the goal of establishing far-reaching theorems of general validity appears to be still far away, it is clear that new results will be obtained soon by exploiting the existing tools and completing some missing details.

One obvious area of research is the study of the perturbations of space homogeneous solutions, different from Maxwellians. Results on this problem for discrete velocity models have been published by Kawashima [121] and a result for the continuous velocity Boltzmann equation has been announced [122]. Another area is the perturbation of exact solutions of a particular nature, such as space inhomogeneous Maxwellians satisfying the Boltzmann equation (see Chapter III, Section 10).

Any result on the existence of solutions should stimulate a search for a related validity result. This has already happened for the perturbation of a vacuum [123–125]. The case of perturbation of equilibrium appears to be harder and has not been considered so far.

References

[1] T. Carleman, "Acta Math.", **60,** 91 (1933).

[2] O. Lanford, III, in "Proceedings of the 1974 Battelle Rencontre on Dynamical Systems", J. Moser, Ed., Lecture Notes in Physics, **35,** 1, Springer, Berlin (1975).

[3] T. Carleman, "Problèmes Mathématiques dans la Théorie Cinétique ses Gaz", Almqvist and Wiksells, Uppsala (1957).

[4] H. B. Maslova and R. P. Tchubenko, "Dokl. Akad. Nauk SSSR", **202,** no. 4, 800 (1972).

[5] H. B. Maslova and R. P. Tchubenko, "Vestnik Leningrad Univ., no. 7, 109 (1976).

[6] P. Gluck, "Transport Theory Statist Phys.", **9,** 43 (1980).

[7] E. Wild, "Proc. Camb. Philos. Soc.", **47,** 602 (1951).

[8] D. Morgenstern, "Proc. Nat. Acad. Sci. U.S.A.", **40,** 719 (1954).

[9] L. Arkeryd, "Arch. Rational Mech. Anal.", **45,** 1 (1972).

[10] L. Arkeryd, "Arch. Rational Mech. Anal.", **45,** 17 (1972).

[11] B. Levi, "Rend. Istit. Lombardo Sci. Lett. (2)", **39,** 775 (1906).

[12] A. Ya. Povzner, "Mat. Sbornik", **58,** 65 (1962).

[13] C. Truesdell, "J. Rational Mech. Anal.", **5,** 55 (1956).

[14] R. E. Edwards, "Functional Analysis", Holt, Rinehart and Winston, New York (1965).

[15] L. Arkeryd, "Arch. Rational Mech. Anal.", **77,** 11 (1981).

[16] T. Elmroth, "Arch. Rational Mech. Anal.", **2,** 1 (1983).

[17] T. Elmroth, in "Kinetic Theories and the Boltzmann Equation", C. Cercignani, Ed., Lecture Notes in Mathematics, **1048,** 192, Springer, Berlin (1984).

[18] T. Elmroth, "SIAM J. Appl. Math.", **44,** 150 (1984).

[19] T. Gustafsson, "Arch. Rational Mech. Anal.", **92,** 23 (1986).

[20] C. Truesdell and R. G. Muncaster, "Fundamentals of Maxwell's Kinetic Theory of a Simple Monatomic Gas", Academic Press, New York (1980).

[21] G. Di Blasio, "Comm. Math. Phys.", **38,** 331 (1974).

[22] V. Boffi and G. Spiga, "J. Math. Phys.", **23,** 1859 (1982).
[23] C. Cercignani, "Arch. Mech.", **34,** 231 (1982).
[24] D. Morgenstern, "J. Rational Mech. Anal.", **4,** 533 (1955).
[25] C. Cercignani, "Transport Theory Statist. Phys." (to appear, 1987).
[26] L. Arkeryd, "Transport Theory Statist. Phys.", **15,** 673 (1986).
[27] C. Cercignani, W. Greenberg and P. Zweifel, "J. Stat. Phys.", **20,** 449 (1979).
[28] H. Spohn, "J. Stat. Phys.", **20,** 463 (1979).
[29] W. Greenberg, J. Voigt and P. Zweifel, "J. Stat. Phys.", **21,** 649 (1979).
[30] R. Gatignol, "Théorie Cinétique des Gaz à Répartition Discrète de Vitesses", Lectures Notes in Physics, **36,** Springer, Berlin (1975).
[31] J. E. Broadwell, "Phys. Fluids", **7,** 1243 (1964).
[32] M. Shinbrot, in "Mathematical Problems in the Kinetic Theory of Gases", D. C. Pack and H. Neunzert, Eds., Lang, Frankfurt (1980).
[33] T. Nishida and M. Mimura, "Proc. Japan. Acad.", **50,** 812 (1974).
[34] L. Tartar, "Séminaire Goulaouic–Schwartz", Ecole Polytechnique, Paris (1975).
[35] H. Cabannes, "J. Méc. Théor. Appl.", **17,** 1 (1978).
[36] H. Cabannes, "C.R. Acad. Sci, Paris", sèrie A, **284,** 269 (1977).
[37] H. Cabannes, "Mech. Res. Comm.", **10,** 317 (1983).
[38] R. Illner, "Habilitationsschrift," Universität Kaiserslautern, (1981).
[39] Y. Shizuta and S. Kawashima, "Hokkaido Math. J.", **14,** 249 (1985).
[40] C. Cercignani, "C.R. Acad. Sci. Paris", sèrie I, **301,** 89 (1985).
[41] C. Cercignani, and M Shinbrot, "Comm. Partial Differential Equations" (to appear, 1987).
[42] T. Platkowski and R. Illner, "SIAM Rev." (to appear, 1987).
[43] U. Frisch, B. Hasslacher and Y. Pomeau, "Phys. Rev. Lett.", **56,** 1505 (1986).
[44] L. Arkeryd, "Arch. Rational Mech. Anal.", **86,** 85 (1984).
[45] S. Albeverio, J. E. Fenstad, R. Høoegh-Krohn and T. Lindstrøm, "Nonstandard Methods in Stochastic Analysis and Mathematical Physics", Academic Press, New York (1986).
[46] K. D. Stroyan and W. A. J. Luxenburg, "Introduction to the Theory of Infinitesimals", Academic Press, New York (1976).
[47] A. E. Hurd and P. Loeb, "An Introduction to Nonstandard Real Analysis", Academic Press, New York (1985).
[48] A. Robinson, "Non-standard Analysis", North-Holland, Amsterdam (1966).
[49] P. Loeb, "Trans. Math. Soc., **211,** 113 (1975).
[50] T. Elmroth, Chalmers University of Technology, Preprint 1984–23 (1984).
[51] H. Grad, in "Handbuch der Physik", S. Flugge, Ed., Vol. XII, Sect. 26, Springer, Berlin (1958).

[52] A. GLICKSON, "Arch. Rational Mech. Anal.", **45,** 5 and **47,** 389 (1972).

[53] S. KANIEL and M. SHINBROT, "Comm. Math. Phys.", **58,** 65 (1978).

[54] M. SHINBROT, "Math. Methods Appl. Sci.", **6,** 539 (1984).

[55] F. G. KING, Ph.D. Dissertation, Dept. of Mathematics, Univ. of California at Berkeley (1975).

[56] A. HURD, Preprint (1986).

[57] H. SPOHN, in "Kinetic Theories and the Boltzmann Equation", C. Cercignani, Ed., Lecture Notes in Mathematics, **1048,** 207 Springer, Berlin (1984).

[58] H. SPOHN, in "Nonequilibrium Phenomena I. The Boltzmann Equation", J. L. Lebowitz and E. W. Montroll, Eds., North–Holland, Amsterdam (1983).

[59] C. CERCIGNANI, "Comm. Pure Appl. Math.", **36,** 479 (1983).

[60] H. GRAD, in "Rarefied Gas Dynamics", J. A. Laurmann, Ed., Vol. I, p. 26, Academic Press, New York (1963).

[61] H. GRAD, "SIAM J.", **13,** 259 (1965).

[62] A. A. ARSEN'EV, "USSR Comput. Math. and Math. Phys.", **5,** 110 (1965).

[63] G. SCHARF, "Helv. Phys. Acta", **40,** 929 (1967).

[64] E. HILLE and R. S. PHILLIPS, "Functional Analysis and Semigroups", American Mathematical Society Colloquium Publications, Providence, RI (1957).

[65] T. KATO, "Perturbation Theory for Linear Operators", Springer, Berlin (1966).

[66] B. FETZ and S. F. SHEN, in "Rarefied Gas Dynamics", D. Dini, Ed., Vol. II, p. 729, Editrice Tecnico Scientifica, Pisa (1971).

[67] F. A. GRÜNBAUM, in "The Boltzmann Equation", F. A. Grünbaum, Ed., p. 103, Courant Institute, New York University, New York (1972).

[68] J. T. MARTI, "Nukleonik", **8,** 159 (1966).

[69] J. MIKA, "Nukleonik", **9,** 200 (1967).

[70] J. MIKA, "Nukleonik", **9,** 303 (1967).

[71] S. ALBERTONI and B. MONTAGNINI, in "Pulsed Neutron Research", Vol. I, p. 239, IAEA, Vienna (1966).

[72] S. ALBERTONI and B. MONTAGNINI, "J. Math. Anal. Appl.", **13,** 19 (1966).

[73] R. BEDNARZ, in "Pulsed Neutron Research", Vol. I, p. 259, IAEA, Vienna (1966).

[74] M. RIBARIC, "Functional Analytic Concepts and Structures in Neutron Transport Theory", Slovenska Akademya Znanosti in Unnetnosti, Ljubljana (1973).

[75] H. G. KAPER, C. G. LEKKERKERKER and J. HEITMANEK, "Spectral Methods in Linear Transport Theory", Birkhäuser, Basel (1982).

[76] H. GRAD, "Proc. Appl. Math.", **17,** 154 (1965).

[77] S. Ukai, "Proc. Japan. Acad. Ser. A, Math. Sci.", **50,** 179 (1974).
[78] T. Nishida and K. Imai, "Publ. Res. Inst. Math. Sci, Kyoto Univ.", **12,** 229 (1977).
[79] Y. Shizuta and K. Asano, "Proc. Japan Acad, Ser. A, Math. Sci.", **53,** 3 (1977).
[80] R. E. Calflisch, "Comm. Math. Phys.", **74,** 71 and 97 (1980).
[81] Y. Shizuta, Preprint (1979).
[82] R. Illner and M,. Shinbrot, "Comm. Math. Phys., **95,** 217 (1984).
[83] K. Hamdache, "Comptes Rendus", Sèrie I, **299,** 431 (1984).
[84] N. Bellomo and G. Toscani, "J. Math. Phys,", **26,** 334 (1985).
[85] G. Toscani, "Arch. Rational Mech. Anal.", **95,** 37 (1986).
[86] M. Shinbrot, "Transport Theory Statist. Phys,.", **15,** 317 (1986).
[87] V. S. Galkin, "Prikl. Mat. Mekh.", **20,** 445 (1956).
[88] V. S. Galkin, "Prikl. Mat. Mekh.", **22,** 532 (1958).
[89] V. S. Galkin, "Prikl, Mat. Mekh.", **28,** 226 (1964).
[90] V. S. Galkin, "Fluid Dynamics", **1,** 29 (1966).
[91] C. Cercignani, submitted to "Arch. Rational Mech. Anal." (1986).
[92] D. R. Willis, in "Rarefied Gas Dynamics", L. Talbot, Ed., p. 429, Academic Press, New York (1961).
[93] H. Grad, "SIAM J.", **14,** 935 (1966).
[94] C. Cercignani, "J. Math. Phys.", **8,** 1653 (1967).
[95] C. Cercignani, "J. Math. Phys.", **9,** 633 (1968).
[96] C. Cercignani, "Mathematical Methods in Kinetic Theory", Plenum Press, New York (1969).
[97] Y. P. Pao, "J. Math. Phys.", **9,** 1893 (1967).
[98] J. P. Guiraud, "J. Méc. Théor. Appl.", **7,** 171 (1968).
[99] J. P. Guiraud, "J. Méc. Théor. Appl.", **9,** 443 (1970).
[100] J. P. Guiraud, "J. Méc. Théor. Appl.", **11,** 2 (1972).
[101] J. P. Guiraud, "C.R. Acad. Sci. Paris", **274,** 417 (1972).
[102] J. P. Guiraud, "C.R. Acad. Sci. Paris", **275,** 1259 (1972).
[103] C. Rigolot-Turbat, "C.R. Acad. Sci. Paris", **272,** 617 (1971).
[104] C. Rigolot-Turbat, "C.R. Acad. Sci. Paris", **272,** 763 (1971).
[105] C. Rigolot-Turbat, "C.R. Acad. Sci. Paris", **273,** 58 (1971).
[106] K. M. Case and P. F. Zweifel, "Linear Transport", Addison–Wesley, Reading, MA. (1967).
[107] S. Ukai and K. Asano, "Arch. Rational Mech. Anal.", **84,** 249 (1983).
[108] C. Cercignani, in "Mathematical Problems on the Kinetic Theory of Gases", D. C. Pack and H. Neunzert Eds., p. 129, Lang, Frankfurt (1980).
[109] M. D. Arthur and C. Cercignani, "Z. Angew. Math. Phys.", **31,** 634 (1980).
[110] C. E. Siewert and J. R. Thomas, "Z. Angew. Math. Phys.", **32,** 421 (1981).

[111] C. E. SIEWERT and J. R. THOMAS, "Z. Angew. Math. Phys.", **33,** 202 (1982).

[112] C. E. SIEWERT and J. R. THOMAS, "Z. Angew. Math. Phys.", **33,** 626 (1982).

[113] W. GREENBERG and C. V. M. VAN DER MEE, "Z. Angew. Math, Phys.", **35,** 156 (1984).

[114] C. CERCIGNANI, in "Trends in Applications of Pure Mathematics to Mechanics", E. Kröner and K. Kirchgässner, Eds., Lecture Notes in Physics, **249,** p. 35 Springer, Berlin (1986).

[115] C. BARDOS, R. E. CAFLISCH and B. NICOLAENKO, "Comm. Pure. Appl. Math.", **39,** 323 (1986).

[116] N. B. MASLOVA, "USSR Comput. Math. Math. Phys.", **22,** 208 (1982).

[117] C. BARDOS, R. E. CAFLISCH and F. GOLSE (to appear, 1986).

[118] C. CERCIGNANI, R. ILLNER and M. SHINBROT "Duxe Math. J." (Sept. 1987).

[119] H. SCHAEFER, "Math. Ann.", **129** (1955).

[120] S. R. SMART, "Fixed Point Theorems", Cambridge University Press, New York (1974).

[121] S. KAWASHIMA, in "Recent Topics in Nonlinear PDE (Hiroshima, 1983)", p. 59, North-Holland Mathematical Studies, **98,** North-Holland, Amsterdam (1984).

[122] L. ARKERYD, R. ESPOSITO and M. PULVIRENTI (to appear in "Comm. Math. Phys.", 1987).

[123] R. ILLNER and M, PULVIRENTI, "Comm. Math. Phys.", **105,** 189 (1986).

[124] R. ILLNER and M. PULVIRENTI (to appear in "Transport Theory Statist. Phys.", 1987).

[125] M. PULVIRENTI (to appear in "Comm. Math. Phys.", 1987).

APPENDIX

The aim of this Appendix is to update and complete the material of the first seven chapters of the book, which were written for the first edition of the book [1] in 1974.

There is not much to say concerning the material of the first chapter, except correcting the statement of Section 6 that an ergodic theorem for a system of rigid spheres had been proved by Sinai. In fact, the proof of Ref. [8] of Chapter I was not available in English in 1974 and, on closer scrutiny, turned out to refer to systems of just two rigid spheres and not arbitrarily many. Accordingly, the result retains its importance for the theory of dynamical systems, but requires further developments in order to be useful for the foundations of equilibrium kinetic theory.

Concerning the material of Chapter II, the main results after 1974 were the rigorous proofs of the derivation of the Boltzmann equation, discussed in Chapter VIII. In addition, classical (i.e. nonquantum) equations for polyatomic gases have been justified [2, 3].

The subject of gas-surface interaction, treated in Chapter III, has also undergone progress in the direction of polyatomic gases [4, 5]. In addition, a proof of the H-theorem for a classical polyatomic gas has been obtained without invoking particular symmetries in the interaction [6]. This was a result that Boltzmann obtained in an incorrect way [7], and which was criticized by Lorentz [8]. Later a quantum-mechanical proof was obtained [9–11], but a proof for classical models was thought to be missing [12], in spite of Boltzmann's efforts in this direction [13, 14]. Concerning monatomic gases, we remark that a detailed discussion of the mathematical aspects of the H-theorem is contained in the book by Truesdell and Muncaster [15] (see also [16]).

The theory of the linearized collision operator has witnessed a few changes. Drange [17] has introduced a new cutoff, called radial integral cutoff, simply consisting of cutting all the interactions with impact parameters larger than an assigned value, and has developed a spectral theory for the linearized collision operator with either the radial cutoff discussed in the main text or the new one. Concerning noncutoff operators, the proof by Pao [18], mentioned in Chapter IV and presented in more detail in two subsequent papers [19, 20] was found to be wrong by Klaus [21]. In fact, Pao's discussion of the essential self-adjointness of L appears to be erroneous, since it is based on the symmetry and negativeness of the operator without proving another condition which is necessary for the conclusion. In order to avoid this difficulty, as well as the tedious estimates of the so-called symbols of the pseudodifferential operator used by Pao, Klaus [21] constructs the collision operator for an infinite range potential as the limit of operators with a cutoff, thus transforming the intuitive argument of the main text into a rigorous proof. In his paper, the approximation is in the sense of strong resolvent convergence. Klaus is also able to describe the spectrum in the case of a radial integral cutoff.

It is also to be mentioned that Jenssen [22] generalized the method of Kuščer and Williams [23] to an arbitrary value of the index l of spherical harmonics and was able to show that L has infinitely many discrete eigenvalues for $l = 0, 1, 2$, but at most a finite number for each value of $l \geq 3$. Jenssen also gave a numerical evaluation of the largest eigenvalues and the corresponding eigenfunctions for $l = 0, 1, 2$. His numerical computations for $l = 3$ led him to conjecture that for $l \geq 3$, L has no eigenvalues at all. The integral version of the variational principle was developed by Cole and Pack [24], who introduced modified functionals leading to upper and lower bounds for the computed quantity.

The main development concerning the Hilbert and Chapman-Enskog expansions was in the direction of proving their validity, i.e. the agreement between the Boltzmann equation and fluid dynamic equations; in fact, rigorous proofs have been that the first few terms of the expansions agree asymptotically (for a vanishing mean free path ε) with the corresponding solutions of the Boltzmann equation in certain cases. The main results are the following:

(1) For the linearized Boltzmann equation the Hilbert or Chapman-Enskog expansions (with modification) are asymptotic to the Boltzmann equation [25, 26]. These papers completed the work of Grad [27], mentioned in Chapter V.

(2) For initial data close to a global Maxwellian, the nonlinear Boltzmann and Navier-Stokes equations agree to leading order as $t \to \infty$, and ε is held constant [28, 29].

(3) For initial data close to a global Maxwellian, the nonlinear Boltzmann and Euler equations agree for at least a short time as $\varepsilon \to 0$ [30, 31].

(4) If the nonlinear Euler equations have a smooth solution in some time interval, then there is a solution of the Boltzmann which agrees with the Euler solution as $\varepsilon \to 0$ [32].

The treatment of the Knudsen layers for kinetic models has been considered by many authors. Part of this work was stimulated by the experimental procedures developed [33, 34] to measure velocity profiles in the Knudsen layer on a flat wall. In particular, Reynolds *et al.* [33] reported that in the Knudsen layer the deviation of the velocity defect (i.e. the deviation of the actual velocity from the continuum velocity profile) shows a behaviour quantitatively different from the results obtained by the BGK model and described in Section 4 of Chapter V.

Loyalka [35] pointed out that such discrepancy could be due to a basic deficiency of the BGK model, in that it does not allow for the velocity dependence of the collision frequency. With this in view, he carried out a detailed numerical study of a model with velocity-dependent collision frequency, used to find an analytical solution for the structure of the kinetic layer as early as 1966 [36]. The latter solution (given in Chapter VI, Section 4) was considered by Loyalka not to be useful for the purpose of numerical evaluation, for which he preferred to use a direct numerical technique. His results show that the velocity dependence of the collision frequency appropriate to rigid spheres does indeed have an important effect on the velocity defect, and in fact the numerical solution practically coincides with the upper boundary of the region containing 80% of the experimental data reported by Reynolds *et al.* [33]. Another well-known deficiency of the BGK model is that it yields a Prandtl number Pr not appropriate to a monatomic gas (Pr $= 1$ instead of Pr $= 2/3$). This unsatisfactory aspect motivated the use of the ES model to investigate the structure of the Knudsen layer as early as 1966 [37]. Unfortunately this solution contained a trivial mistake, in that a numerator and a denominator were erroneously interchanged. When the mistake is corrected the results are in reasonably good agreement with the experiments; in fact, almost indistinguishable from the results presented by Loyalka [35]. This fact was pointed out in 1976 [38]. At about the same time, Abe and Oguchi [39] examined the Knudsen layers with a model equation of the Gross and Jackson hierarchy, involving thirteen moments in the collision term. The model contains two nondimensional parameters, the Prandtl number and a further parameter taking the value unity when the model reduces to the linearized ES model. They conclude that the solution indicates a "weak dependence" on this parameter (which should take the value 4/9, according to Gross and Jackson's prescription).

Gorelov and Kogan [40] reported the results of Monte Carlo calculations which appear to fall on the lower boundary of the aforementioned region containing 80% of the data. These Monte Carlo calculations were confirmed by Bird [41]. Then it was pointed out [42] that one can concoct a new model having the desirable properties of both a correct Prandtl number and variable collision frequency, while amenable to a simple solution. The velocity profile computed by means of this model turns out to be in exceptionally good agreement [42] with the experiments of Reynolds *et al.* [33].

So far we have dealt with the problem of the velocity profile in the presence of a velocity gradient outside the kinetic layer. Analogous problems arise in connection with temperature gradients either normal or tangential to the wall. The first problem, the so-called temperature jump problem, leads to a mathematically more complex situation than the velocity slip problem; there is no difficulty in obtaining numerical solutions but an analytical treatment is far from easy. As mentioned in Chapter VI, Section 9, Cassel and Williams [43] discovered that the problem can be reduced to a standard one if the collision frequency is taken to be proportional to the molecular speed, while the solution with the BGK model requires considerable effort. A method of dealing with this problem was mentioned in Chapter VI, Section 8, but was only published in 1977 [44]. The problem was worked out in more detail by Siewert and Kelley in 1980 [45]. In both papers, use is made of the diagonalization idea of Darrozès [46] in order to solve the relevant matrix Riemann-Hilbert problem, followed by a careful study of the new singularities (branch cuts) arising from the diagonalization procedure. The cancellation of these singularities, a problem that Darrozés was not able to master, is the essentially new feature of the method, which was brought to completion in 1982 [47], when a procedure to compute analytically the partial indices of the Riemann-Hilbert problem was indicated. The method was later extended to time-dependent problems [48, 49] and used to treat the half-range completeness problem for sound propagation in a closed form. This problem was discussed in Section 12 of Chapter VI.

An accurate numerical method for the half-range analysis of the BGK model had been previously developed by Siewert and Burniston [50], and applied to the problem of sound propagation by Thomas and Siewert [51] in order to improve upon the results of Buckner and Ferziger [52] discussed in Chapter VI. There is a problem when comparing theoretical results with experimental data, in that for high frequencies the solution remains wavelike, but is no longer a classical plane wave, with the consequence that the phase speed and the attenuation parameter are no longer clearly defined. The solutions [51, 52] produce nearly linear plots for the real and imaginary parts of the logarithm of the pressure, but the average slopes are determined by the range of the space coordinate within which the calculations are performed. Buckner and Ferziger correctly chose the range chosen by Meyer and Sessler [53] to produce the results which were mentioned above. This work seems not to have been done, so far, for the results of Thomas and Siewert [51].

An important feature of all these problems is that they are amenable to a linear analysis. This is not true for the study of a completely different kind of kinetic layer, which shares the features of both a shock wave and a Knudsen layer, and is met in the study of the evaporation from a surface into a low-pressure ambient. The problem of investigating the layer of evaporating gas close to the wall has been the subject of several researches because of its importance in various fields of physics, chemistry and engineering. The aim of these researches is to improve on the approximate results obtained by Hertz [54] and Knudsen [55].

Anisimov [56], in 1968, suggested the use of a trimodal *ansatz* for the molecular distribution function:

$$f(x, \xi) = a_L^{+}(x) f_L^{+} + a_M^{+}(x) f_M^{+} + a_M^{-}(x) f_M^{-} \tag{A.1}$$

where x is the space coordinate normal to the evaporating surface, f_L^{+}, f_M^{+}, f_M^{-} are half-range Maxwellians. He solved the conservation equations for one set of flow conditions (sonic) only, and estimated the thickness of the corresponding Knudsen layer from the BGK model. Ytrehus [57, 58] carried this method to completion by solving appropriate moment equations of the Boltzmann equation (Maxwell molecules) with a distribution function given by Eq. (A.1). The choice of Eq. (A.1) is such that the boundary conditions at infinity, and at the evaporating boundary, can be satisfied exactly (in the case of an absorption coefficient equal to unity). In addition to the three conservation equations, another moment equation is required in order to determine the three functions $a_L^{+}(x)$, $a_M^{+}(x)$, $a_M^{-}(x)$. Ytrehus was able to show that the problem is solvable if and only if M_M, the Mach number associated with the drift velocity of the downstream Maxwellian, is not larger than a critical value M_c, which turns out to be approximately unity ($M_c \simeq 0.992$). This result is in agreement with the results obtained by Murakami and Oshima [59] by a Monte Carlo method. These authors computed the solution of the one-dimensional evaporation problem in an unsteady situation and found that no steady state is reached when the downstream Mach number is larger than unity. It seems reasonable to conjecture that $M_c = 1$ exactly for the exact solution of the problem [60]. As a consequence of the limitation of the downstream Mach number, no steady one-dimensional solution to the evaporation problem exists for a pressure ratio larger than (approximately) 4.8. In particular, no steady one-dimensional solution is possible for a gas evaporating into a vacuum.

This conjecture has been confirmed by an analysis based on exact solutions of both a one-dimensional and three-dimensional BGK model [61–63] linearized about a drifting Maxwellian. More general results have been obtained by an abstract approach [64], as mentioned in Chapter VIII. In the papers on the three-dimensional model [62, 63], the matrix technique introduced in [44] was used. The same technique was used in a paper dealing with Rayleigh scattering in radiative transfer [65].

One of the basic new results coming from the studies on evaporation and condensation by means of kinetic theory is the circumstance that the classical Hertz-Knudsen formula [54, 55] for the mass flow-rate at an evaporating wall disagrees, by as much as 70%, with the more accurate treatments in the limit of weak evaporation (small pressure difference) [57, 58, 66, 67]. A more spectacular result is the fact that in a two-plate experiment the temperature of the vapour at the hot wall can be below that at the cold wall, if the ratio between the latent heat and RT_w is larger than a critical value of the order of 4.75, in the case of a monatomic gas. To appreciate the issue, it is to be remarked that this ratio is 13.1 for water at 100°C. The critical value of the ratio was later computed for polyatomic molecules [68] and grows with the number of internal degrees of fredom. However, one should assign more than eight *internal* degrees of freedom to a water molecule in order to avoid the paradox. For finite Reynolds numbers and small Mach numbers, the entire phenomenon takes place in the Knudsen layer [69]. The effect of changing the boundary conditions has also been investigated [68]. No experimental results appear to be available.

Analytical, numerical and Monte Carlo solutions of transition flow problems continued after the appearance of the first edition of this book, and it would not be appropriate to mention all of them here. The Proceedings of the Biannual Symposia on Rarefied Gas Dynamics [70–75] together with the book by Bird [76] should give a fairly good idea of the developments and trends in this area. Here attention will be restricted to some new results concerning the typical problems discussed in Chapter VII. Applications of variational techniques have been considered by several authors [77–79]. Among the new problems treated with this technique we mention a model of the gas centrifuge [79].

Nicolaenko [80] investigated the existence and uniqueness of shock wave solutions for the nonlinear Boltzmann equation in the case of Grad's angular cutoff. He was able to show that the shock wave solution arises by bifurcation from the constant Maxwellian distribution. The critical value of the bifurcation parameter corresponds to the sonic regime. These results had been previously demonstrated in a simpler but nonrigorous fashion [81]. The Chapman-Enskog solutions as well as the formal expansion using "stretched variables" are shown to fail to yield uniform asymptotics (in the x-variable) beyond the Navier-Stokes level. This is related to the circumstance that the Navier-Stokes and Fourier constitutive relations are not valid at the upstream and downstream point, even though $|f - f_\infty|$ becomes vanishingly small there, a fact discussed in detail by Elliot *et al.* [82, 83]. These authors proposed new closure relations for the thirteen-moment theory, which models the upstream flow quite well. The picture downstream remains incomplete although superior to anything previously available. In addition, since the upstream singular point is the more critical, the theory of Elliot *et al.* removes the difficulties previously encountered in computing shock wave profiles at the thirteen-moment level.

As discussed in Section 6 of Chapter VII, one of the main difficulties of the Mott-Smith method for the problem of shock wave structure is the choice of a moment equation. Lampis [84] suggested using entropy balance in place of a moment equation. The same idea occured later [85] in a paper using the Navier-Stokes equations to deal with the problem of shock structure. In this paper, the profile is assumed to be given by a hyperbolic tangent and the constant related to the shock thickness is determined by a global entropy balance. The agreement with the experiments of Alsmeyer [86] is rather good. Hosokawa and Inage [87] used a similar idea, i.e. they assumed the hyperbolic tangent profile, and used the equation proposed by Lampis [84] at just one point to determine the thickness. The agreement with experiments is reasonably good.

As shown in Section 8 of Chapter VII, the hypersonic approximation yields a solution of the problem of spherical expansions valid for small values of the source Knudsen number Kn. As Kn increases the method loses validity. This remark led Soga and Oguchi to analyze the source expansion flow by means of the Krook model [88]. According to their results, the hypersonic approximation is very accurate for $\mathrm{Kn} \leq 10^{-3}$, reasonably accurate for $\mathrm{Kn} \simeq 10^{-2}$, while it is decidedly unacceptable for $\mathrm{Kn} \leq 10^{-1}$. Later, Abe and Oguchi [89] analyzed the problem by means of a kinetic model with a correct Prandtl number, but found no appreciable deviation from the results based on the BGK model.

The subject of jet expansions lends itself to a remark on the evaporation problem, discussed above. Since a steady one-dimensional solution is not possible for a gas evaporating into a vacuum, the question arises whether a steady solution for this problem is possible when the limitation to a one-dimensional geometry is suppressed. The answer is positive, as can be shown by the method of matched asymptotic expansions [90, 91]. The flow from, say, an evaporating disk takes the form of an initially one-dimensional expansion (if border effects are neglected) which reaches sonic conditions, then develops as an inviscid gas jet expansion from a sonic disk, and finally behaves as the frozen jet occurring in a source expansion. Extensions to polyatomic gases and mixtures have also been considered [91, 92].

A subject which developed in the last ten years deals with exact solutions of the space homogeneous Boltzmann equation. Although Boltzmann's H-theorem guarantees that any assigned initial space-independent distribution function will decay to a Maxwellian and this result can even be proved in all rigor, the details of this decay are not known explicitly. An exception is offered by Maxwell molecules, for which Maxwell [93] derived a set of exact equations satisfied by the moments. He did not write out all the terms which were first published by Grad [94], and scholarly discussed by Ikenberry and Truesdell [95] and Truesdell [96]. In the latter paper, the general solution for the moments of the distribution function is given by the following expression:

$$M_\theta = \sum_{i=1}^{N_\theta} A_{\theta i}(t)\, e^{-k_{\theta i} t} + M_\theta^{(\circ)} \tag{A.2}$$

where M_θ is any moment and θ a suffix to identify it, $M_\theta^{(\circ)}$ is the equilibrium value of $M_\theta^{(\circ)}$, $A_{\theta i}(t)$ is a polynomial in t, $k_{\theta i}$ is a positive constant, and N_θ is a positive integer; $M_\theta^{(\circ)}$, $k_{\theta i}$ and the coefficients in $A_{\theta i}(t)$ depend on the constant values of density, velocity and temperature. It is not proved by Truesdell's theorem, but only conjectured that the $A_{\theta j}(t)$ are in fact constants, i.e. zero degree polynomials. The explicit expressions for $A_{\theta i}$ can be found by recursion; easy, albeit tedious, computations for the first few moments seem to confirm the conjecture.

It is to be remarked that, although a knowledge of all the moments is theoretically equivalent to a knowledge of the distribution function, no simple expression of the latter is obtained from Eq. (A.2). This explains the interest produced by the publication [97] of an exact solution of the nonlinear Boltzmann equation for Maxwell molecules. This solution corresponds to a special initial datum, i.e.:

$$f = e^{-a\xi^2}(b + c\xi^2) \tag{A.3}$$

where ξ is, as usual, the molecular speed, and a, b, c positive constants. The basic property of a distribution function of the form (A.3) is that it evolves preserving its shape, the only change being in the coefficients a, b, c, which vary in time till, ultimately, c tends to zero, a and b to positive constants and f to a Maxwellian distribution. It is to be noted that the same solution was independently published by Bobylev [98], and had appeared in an unpublished master thesis as early as 1967 [99]. Also, if one is aware of the fact that the Boltzmann equation for Maxwell molecules has a solution of the form indicated in Eq. (A.2), it is not difficult to check this and compute explicitly a, b and c as functions of time t [100].

The solution found by Bobylev [98] and Krook and Wu [97] (frequently called the BKW mode in the literature) approaches an equilibrium distribution when $t \to \infty$ in a nonuniform fashion; this is due to the high speed tail of the distribution and indicates that linearization does not hold for high speeds even when we are close to a Maxwellian in some sense. This was already known from other facts (loss of positivity of certain linearized solutions), but it is immediately obvious here. Physically, as remarked by Krook and Wu, this is due to the fact that, at most, the total kinetic energy of two molecules after a collision can be concentrated in one of them. If the tail is initially absent above a certain energy (cutoff energy), then this value can at most double after each favourable collision. Thus the time to reach a higher cutoff can be expected to grow logarithmically with the latter.

The importance of the BKW solution was overestimated initially, because of the following conjecture formulated by Krook and Wu: An arbitrary initial state tends towards a BKW mode; then a relaxation according to the latter takes place. This conjecture can be rephrased in more mathematical terms, but this is not necessary, since analytical and numerical evidence against this conjecture was found by many authors. An enormous literature on this subject and related solutions of other molecules than Maxwell is available. For a survey, we refer to the papers by Ernst [101, 102].

Before closing this Appendix it is appropriate to mention the developments which took place in the kinetic theory of polyatomic gases, with particular concern for polarization phenomena. The early theory of these phenomena was in the continuum regime, and accordingly mainly dealt with general properties and the computation of transport coefficients (see [103–105]). More recently, problems in the slip and transition regime have been considered [106–108].

References

[1] C. CERCIGNANI, "Theory and Application of the Boltzmann Equation", Scottish Academic Press, Edinburgh and Elsevier, New York (1975).

[2] V. D. BORMAN, A. S. BRUEV and L. A. MAKSIMOV, "Zh. Èksper. Teoret. Fiz." **67,** 951 (1974); translated in "Soviet Phys. JETP", **40,** 472 (1975).

[3] I. KUŠČER, H. F. P. KNAAP and J. J. M. BEENAKKER, "Physica", **108A,** 265 (1981).

[4] C. CERCIGNANI, "Physica", **97A,** 440 (1979).

[5] H. F. P. KNAAP and I. KUŠČER, "Physica", **104A,** 95 (1980).

[6] C. CERCIGNANI and M. LAMPIS, "J. Statist. Phys.", **26** 795 (1981).

[7] L. BOLTZMANN, "Wien. Ber.", **66**, 275 (1872).

[8] H. A. LORENTZ, "Wien. Ber.", **95**, 115 (1887).

[9] E. C. G. STUECKELBERG, "Helv. Phys. Acta.", **25**, 577 (1952).

[10] L. WALDMANN, in "Handbuch der Physik", S. Flügge, Ed., Vol. 12, p. 484, Springer, Berlin (1958).

[11] L. WALDMANN, in "The Boltzmann Equation: Theory and Applications", E. G. D. Cohen and W. Thirring, Eds., "Acta Phys. Austriaca Suppl. X", **107** (1973).

[12] G. E. UHLENBECK, *ibid.* p. 107.

[13] L. BOLTZMANN, "Wien. Ber.", **95**, 153 (1887).

[14] L. BOLTZMANN, "Vorlesungen über Gastheorie", Barth, Leipzig (1898).

[15] C. TRUESDELL and R. G. MUNCASTER, "Fundamentals of Maxwell's Kinetic Theory of a Simple Monatomic Gas", Academic Press, New York (1980).

[16] C. CERCIGNANI, "Arch. Mech.", **34,** 3 (1982).

[17] H. DRANGE, "SIAM J. Appl. Math.", **29,** 4 (1975).

[18] Y. P. PAO, in "Rarefied Gas Dynamics", M. Becker and M. Fiebig, Eds., Vol. I, p. A.6-1, DFVLR Press, Porz-Wahn (1974).

[19] Y. P. PAO, "Comm. Pure Appl. Math.", **27,** 407 (1974).

[20] Y. P. PAO, "Comm. Pure Appl. Math.", **27,** 559 (1974).

[21] M. KLAUS, "Helv. Phys. Acta", **50,** 893 (1977).

[22] O. O. JENSSEN, "Phys. Norvegica", **6,** 179 (1972).

[23] I. KUŠČER and M. M. R. WILLIAMS, "Phys. Fluids", **10,** 1922 (1967).

[24] R. J. Cole and D. C. Pack, in "Rarefied Gas Dynamics", R. Campargue, Ed., Vol. I, p. 187, CEA, Paris (1979).
[25] R. Ellis and M. Pinsky, "J. Math. Pure Appl.", **54,** 125 (1975).
[26] R. Ellis and M. Pinsky, "J. Math. Pure Appl.", **54,** 157 (1975).
[27] H. Grad, "Phys. Fluids", **6**, 145 (1963).
[28] S. Kawashima, A. Matsumura and T. Nishida, "Comm. Math. Phys." **70,** 97 (1979).
[29] J. Nishida and K. Imai, "Publ. Res. Inst. Math. Sci. Kyoto", **12**, 229 (1976).
[30] H. Grad, "Proc. Sympos. Appl. Math." **17,** 154, (1965).
[31] T. Nishida, "Comm. Math. Phys." **61,** 119 (1978).
[32] R. Caflisch, "Comm. Pure Appl. Math." **33,** 651 (1980).
[33] M. A. Reynolds, J. J. Smolderen and J. F. Wendt, in "Rarefied Gas Dynamics", M. Becker and M. Fiebig, Eds., Vol. I, p. A21, DFVLR Press, Porz-Wahn (1974).
[34] W. Rixen and F. Adomeit, in "Rarefied Gas Dynamics", M. Becker and M. Fiebig, Eds., Vol. I, p. B18, DFVLR Press, Porz-Wahn (1974).
[35] S. K. Loyalka, "Phys. Fluids", **18,** 1666 (1975).
[36] C. Cercignani, "Ann. Phys.", **40,** 469 (1966).
[37] C. Cercignani and G. Tironi, "Nuovo Cimento", **43,** 64 (1966).
[38] C. Cercignani, in "Rarefied Gas Dynamics", J. L. Potter, Ed., Vol. II, p. 795, AIAA, New York (1977).
[39] T. Abe and H. Oguchi, "ISAS Rept. No. 553", Vol. 42, No. 8, Tokyo (1977).
[40] S. L. Gorelov and M. N. Kogan, "Izv. Akad Nauk SSSR, Mekh. Zhidk. Gaza", **3,** 136 (1968), translated in "Fluid Dynamics", **3,** 96 (1968).
[41] G. A. Bird, in "Rarefied Gas Dynamics", L. Potter, Ed., Vol. I, p. 323, AIAA, New York (1977).
[42] C. Cercignani, in "Recent Developments in Theoretical and Experimental Fluid Mechanics", U. Müller, K. G. Roesner and B. Schmidt, Eds., Springer, Berlin (1979).
[43] J. S. Cassel and M. M. R. Williams, "Transport Theory Statist. Phys.", **2,** 81 (1972).
[44] C. Cercignani, "Transport Theory Statist. Phys.", **6,** 29 (1977).
[45] C. E. Siewert and C. T. Kelley, "Z. Angew. Math. Phys.", **31,** 344 (1980).
[46] J. S. Darrozès, "Rech. Aérospat.", **119,** 13 (1967).
[47] C. Cercignani and C. E. Siewert, "Z. Angew. Math. Phys.", **33,** 297 (1982).
[48] K. Aoki and C. Cercignani, "Z. Angew. Math. Phys.", **35,** 127 (1984).
[49] K. Aoki and C. Cercignani, "Z. Angew. Math. Phys.", **35,** 345 (1984).
[50] C. E. Siewert and E. E. Burniston, "J. Math. Phys.", **18,** 376 (1973).

[51] J. R. THOMAS, Jr. and C. E. SIEWERT, "Transport Theory Statist. Phys.", **8,** 219 (1979).
[52] J. K. BUCKNER and J. H. FERZIGER, "Phys. Fluids", **9,** 2315 (1966).
[53] E. MEYER and G. SESSLER, "Z. Physik", **149,** 15 (1957).
[54] H. HERTZ, "Ann. Physik", **17,** 177 (1882).
[55] M. KNUDSEN, "Ann. Physik", **47,** 697 (1915).
[56] S. I. ANISIMOV, "Soviet Phys. JETP", **27,** 182 (1968).
[57] T. YTREHUS, "Von Karman Institute Technical Note 112" (1975).
[58] T. YTREHUS, in "Rarefied Gas Dynamics", L. Potter, Ed. Vol. II, p. 1197, AIAA, New York (1977).
[59] M. MURAKAMI and K. OSHIMA, in "Rarefied Gas Dynamics", M. Becker and M. Fiebig, Eds., Vol. II, p. F6 DFVLR Press, Pouz-Wahn (1974).
[60] C. CERCIGNANI, in "Mathematical Problems in the Kinetic Theory of Gases", D. C. Pack and H. Neunzert, Eds., p. 129, Lang, Frankfurt (1980).
[61] M. D. ARTHUR and C. CERCIGNANI, "Z. Angew. Math. Phys.", **31,** 634 (1980).
[62] C. E. SIEWERT and J. R. THOMAS, "Z. Angew. Math. Phys.", **32,** 421 (1981).
[63] C. E. SIEWERT and J. R. THOMAS, "Z. Angew. Math. Phys.", **33,** 202 (1982).
[64] W. GREENBERG and C. V. M. VAN DER MEE, "Z. Angew. Math. Phys.", **35,** 156 (1984).
[65] K. AOKI and C. CERCIGNANI, "Z. Angew. Math. Phys.", **36,** 61 (1985).
[66] Y. P. PAO, "Phys. Fluids", **14,** 1340 (1971).
[67] J. W. CIPOLLA, Jr., H. LANG and S. K. LOYALKA, "J. Chem. Phys.", **61,** 69 (1974).
[68] C. CERCIGNANI, W. FISZDON and A. FREZZOTTI, "Phys. Fluids", **28,** 3237 (1985).
[69] K. AOKI and C. CERCIGNANI, "Phys. Fluids", **26,** 1163 (1983).
[70] L. POTTER, Ed., "Rarefied Gas Dynamics", Vols. I and II, AIAA, New York (1977).
[71] R. CAMPARGUE, Ed., "Rarefied Gas Dynamics", Vols. I and II, CEA, Paris (1981).
[72] S. S. FISHER, Ed., "Rarefied Gas Dynamics", Vols. I and II, AIAA, New York (1981).
[73] O. M. BELOTSERKOVSKII, M. N. KOGAN, S. S. KUTATELADZE, A. K. REBROV, Eds., "Rarefied Gas Dynamics", Vols. I and II, Plenum Press, New York (1985).
[74] H. OGUCHI, Ed., "Rarefied Gas Dynamics", Vols. I and II, University of Tokyo Press, Tokyo (1984).
[75] V. BOFFI and C. CERCIGNANI, Eds., "Rarefied Gas Dynamics", Vols. I and II, Teubner, Stuttgart (1986).

[76] G. A. BIRD, "Molecular Gas Dynamics", Clarendon Press, Oxford (1976).
[77] R. J. COLE, in Ref. 72, Vol. II, p. 1007 (1981).
[78] R. J. COLE, in Ref. 73, Vol. I, p. 439 (1985).
[79] C. CERCIGNANI and M. LAMPIS, in Ref. 74, Vol. I, p. 89 (1984).
[80] B. NICOLAENKO, in "Théories Cinétiques Classiques et Relativistes", G. Pichon, Ed., p. 127, CNRS, Paris (1975).
[81] C. CERCIGNANI, "Meccanica", **5,** 7(1970).
[82] J. P. ELLIOT, D. BAGANOFF and R. D. MCGREGOR, in Ref. 70, Vol. II, p. 703 (1977).
[83] J. P. ELLIOT and D. BAGANOFF, "J. Fluid Mech.", **65,** 603 (1974).
[84] M. LAMPIS, "Meccanica", **12,** 171 (1977).
[85] P. A. THOMPSON, T. W. STROCK and D. S. LIM, "Phys. Fluids", **26,** 48 (1983).
[86] H. ALSMEYER, "J. Fluid Mech.", **74,** 497 (1976).
[87] I. HOSOKAWA and S. INAGE, "J. Phys. Soc. Japan", **55,** 3402 (1986).
[88] T. SOGA and H. OGUCHI, "Phys. Fluids", **12,** 2011 (1969).
[89] T. ABE and H. OGUCHI, "ISAS Report No. 554", Vol. 42, No. 9 (1977).
[90] C. CERCIGNANI, Report EUR 6843 EN, Commission of the European Communities, Brussels and Luxembourg (1980).
[91] C. CERCIGNANI, in Ref. 72, Vol. I, p. 305 (1981).
[92] A. FREZZOTTI, in Ref. 75, Vol. II, p. 313 (1986).
[93] J. C. MAXWELL, "Scientific Papers", Part 2, pp. 26 and 680, Dover, New York (1965).
[94] H. GRAD, "Comm. Pure Appl. Math.", **3,** 331 (1949).
[95] E. IKENBERRY and C. TRUESDELL, "J. Rational. Mech. Anal.", **5,** 1 (1956).
[96] C. TRUESDELL, "J. Rational Mech. Anal.", **5,** 55 (1958).
[97] M. KROOK and T. T. WU, "Phys. Rev. Lett.", **36,** 1107 (1976) and "Phys. Fluids", **20,** 1589 (1977).
[98] A. V. BOBYLEV, "Soviet Phys. Dokl.", **20,** 820 and 822 (1976) and **21,** 632 (1977).
[99] R. KRUPP, Master Thesis, MIT, Cambridge, MA. (1967).
[100] C. CERCIGNANI, in Ref. 71, Vol. I, p. 141 (1979).
[101] M. H. ERNST, "Phys. Rep.", **78,** 1 (1981).
[102] M. H. ERNST, in "Nonequilibrium Phenomena I: The Boltzmann Equation", J. L. Lebowitz and E. W. Montroll, Eds., North-Holland, Amsterdam (1983).
[103] J. J. M. BEENAKKER, "Festkörperprobleme", **8,** 275 (1968).
[104] J. J. M. BEENAKKER and F. R. MCCOURT, "Ann. Rev. Phys. Chem.", **21,** 47 (1970).
[105] J. H. FERZIGER and H. G. KAPER, "Mathematical Theory of Transport Processes in Gases", North-Holland, Amsterdam (1972).

[106] C. Cercignani, "Physica", **115A,** 143 (1982).
[107] C. Cercignani, L. Geppert and M. Lampis, "Physica", **121A,** 531 (1983).
[108] C. Cercignani and M. Lampis, in Ref. 75, Vol. I. p. 336 (1986).

AUTHOR INDEX

SUBJECT INDEX

Applied Mathematical Sciences

cont. from page ii